Maintenance-Free Batteries

SECOND EDITION

ELECTRONIC & ELECTRICAL ENGINEERING RESEARCH STUDIES

POWER SOURCES TECHNOLOGY SERIES

Series Editor: **N. E. Bagshaw, MA CEng FIM FBIM**

2. Batteries for Automotive Use
   P. Reasbeck *and* **J. G. Smith**

3. Maintenance-Free Batteries
   Lead-Acid, Nickel/Cadmium, Nickel/Metal Hydride
   A HANDBOOK OF BATTERY TECHNOLOGY, SECOND EDITION
   D. Berndt

4. Batteries for Electric Vehicles*
   D. A. J. Rand, R. Woods *and* **R. M. Dell**

* Forthcoming

Maintenance-Free Batteries

Lead-Acid, Nickel/Cadmium, Nickel/Metal Hydride

A HANDBOOK OF BATTERY TECHNOLOGY

SECOND EDITION

D. Berndt

RESEARCH STUDIES PRESS LTD.
Taunton, Somerset, England

JOHN WILEY & SONS INC.
New York · Chichester · Toronto · Brisbane · Singapore

RESEARCH STUDIES PRESS LTD.
24 Belvedere Road, Taunton, Somerset, England TA1 1HD

Marketing and Distribution:

Australia and New Zealand:
JACARANDA WILEY LTD.
Sydney Office, Suite 4A, 113 Wicks Road, North Ryde, NSW 2113, Australia

Canada:
JOHN WILEY & SONS CANADA LIMITED
22 Worcester Road, Rexdale, Ontario, Canada

Europe, Africa, Middle East and Japan:
JOHN WILEY & SONS LIMITED
Baffins Lane, Chichester, West Sussex, UK, PO19 1UD

North and South America:
JOHN WILEY & SONS INC.
605 Third Avenue, New York, NY 10158, USA

South East Asia:
JOHN WILEY & SONS (ASIA) PTE. LIMITED
2 Clementi Loop #02-01
Jin Xing Distripark, Singapore 129809

Library of Congress Cataloging-in-Publication Data

Berndt, D. (Dietrich), 1926-
Maintenance-free batteries : lead-acid, nickel/cadmium, nickel/metal hydride : a handbook of battery technology / D. Berndt. --2nd ed.
p. cm. -- (Electronic & electrical engineering research studies. Power sources technology ; 3)
Includes bibliographical references and index.
ISBN 0-471-97018-2 (alk. paper)
1. Lead-acid batteries--Design and construction. 2. Nickel -cadmium batteries--Design and construction. 3. Nickel-metal hydride batteries--Design and construction. I. Title. II. Series.
TK2941.B47 1997
621.31'2424--dc21 97-5956
CIP

British Library Cataloguing in Publication Data
A catalogue record for this book is available from the British Library.

ISBN 0 86380 198 6 (Research Studies Press Ltd.) *[Identifies the book for orders except in America.]*
ISBN 0 471 97018 2 (John Wiley & Sons Inc.) *[Identifies the book for orders in USA.]*

Printed in Great Britain by SRP Ltd., Exeter

Editorial Foreword

The aim of the present series of books, with the general title Power Sources Technology, is to present an up-to-date picture in rapidly developing areas of electrochemical power sources and their applications.

The subject of the present book is Maintenance-Free Batteries and the author is Dr Dietrich Berndt, a colleague and friend over many years in the battery industry. Dr Berndt has an international reputation in the field of batteries and for some years has paid particular attention to the subject of maintenance-free batteries and their applications, giving numerous papers at conferences and symposia throughout the world.

The first edition of this book was published in 1993 and was immediately given a good reception from both industrial battery technologists and academics. It was particularly welcomed because it started from basic principles of electrochemistry and physics, and demonstrated their effects on the performance and characteristics of batteries, following their implications through to actual applications of batteries in the field.

Maintenance-free batteries is a rapidly changing subject and, since 1993, there have been advances and developments, particularly in the new valve-regulated lead-acid and nickel-metal hydride types. It was therefore decided that it was time for a second edition of the book.

In this edition, there is a wealth of new information. Dr Berndt has also taken the opportunity of reorganising the format to some extent, to give as logical an approach as possible to the subject and including more detailed explanations for greater clarity.

I am sure that the new book will be welcomed by all those working in battery research, by the industry itself, and by battery users in the electrical and electronic industries.

N.E. Bagshaw
Editor
April 1997

Foreword to the Second Edition

The First Edition of this book was well accepted and there was a predominantly positive response. Nevertheless, four years after its publication, I had the wish to prepare a completely revised edition for several reasons. Some of the items should be presented more clearly; a number of items and details should be added. The theoretical background of the valve-regulated lead-acid battery as well as the practical experience with these batteries have largely been extended during these years. The nickel/metal hydride battery is now well established on the market, while it was still open whether it would be a successful system, when the First Edition was written.

As a consequence, the central chapters have been more or less re-written. The 'Fundamental Laws' are now concentrated in Chapter 2, while 'Battery Parameters and Definitions' form a separate Chapter 3. The chapters on 'Battery Design' of the First Edition are now included in the chapters on 'Battery Technology' since they only represent examples of how the battery system concerned is realised in practice.

A new maintenance-free and rechargeable battery system, the lithium-ion battery, appeared on the market during recent years and is fast gaining importance. It is not treated in this book, since it is not based on aqueous electrolyte. So the internal oxygen cycle that characterises the maintenance-free version of lead-acid and nickel/cadmium batteries and the nickel/metal hydride battery, is not present in lithium-ion batteries. Thus it would require a widely different treatment that would go beyond the limits of this book.

As with the First Edition, I have to thank quite a number of people who supported me. After the unexpected death of Dr. R. Gereth, Dr. F. J. Kruger allowed me the continued use of the excellent library and documentation system of the VARTA Research and Development Centre at Kelkheim. My former colleagues, Dr. W. Gann and Dr. D. Dennstedt, read most of the manuscript and gave me much valuable advice. Chris Ferguson as a 'native English speaker' helped to improve the text. As with the First Edition, Norman Bagshaw scrutinised each chapter. Last but not least I have to thank the Research Studies Press Limited, where Mrs Wallace made it possible for me to write this Second Edition and the company made a meticulous proof read and prepared the printed layout.

D. Berndt
Kronberg, April 1997

Foreword to the First Edition

When N.Bagshaw asked me to write this book, it appeared to me as a great challenge to summarize thirty years of professional experience with batteries. My idea was, to impart the understanding of the principles of valve-regulated lead-acid and sealed nickel/cadmium batteries. So, mainly general aspects are treated: the rules that determine battery behaviour, and the influence of operational conditions on this behaviour. Questions of battery design are only touched, and the batteries, shown in the book, are more or less examples that demonstrate the various ways of technical realization.

Batteries are quite complex systems, because charge and discharge are based on chemical reactions which means continual and far-reaching changes of chemical composition and structure of the main components. The course of such reactions is changed even at temperatures only slightly above or below the normal room temperature. For this reason, batteries are more sensitive to temperature deviations than usual electronic components. Furthermore, the reaction rate, i.e. the charging or discharging current itself, influences the reaction parameters and thereby the battery performance data. Additional decisive parameters are caused in the maintenance-free design by secondary reactions, like the internal oxygen cycle.

The intention of my book is, to enable the reader to understand the specific behaviour of batteries and to interpret observed deviations from expected data. So it may help to improve the operational conditions of batteries, especially of the maintenance-free design, and it may furthermore help to avoid (often unnoticed) abuse and its consequence of disappointing results.

My idea of this book could only be realized by the help of many people. First of all I have to thank the VARTA Batterie AG, especially Dr. R.Gereth, for the approval to write this book and to use the facilities of the VARTA Research and Development Centre at Kelkheim. Furthermore, I am indebted to many colleagues at the VARTA Research Centre and the VARTA plants, who revised parts of the book or the whole manuscript and contributed much helpful advice. I am also obliged to the librarian, Mr. B.Gnida, for his support and to Mr. D.Metzeltin who prepared most of the photographs.

Norman Bagshaw, the Editor of this book series, scrutinised each chapter and gave a great amount of advice, mainly to improve my English text.

Last not least I have to thank the Research Studies Press Limited where Mrs. V. Wallace accepted most of my wishes and arranged for a further thorough edition of the text that was a great help to me.

D. Berndt
Kronberg, December 1992

CONTENTS

Major Symbols

Symbol	Meaning	Dimension	Section
a	Constant in the TAFEL equation		2.3.3.1
a_j	Activity of the reacting component j	mole/dm^3	2.2.1.1
b	Slope of the TAFEL line	V	2.3.3.1
c_j	Concentration of species j	mole/dm^3	2.3
C	Capacity of a battery (usually in Ah)	Ah, Wh	3.2
C_{Ah}	Capacity under constant-current load	Ah	3.2
$C_{Batt.}$	Heat capacity of a battery	kJ/K	2.5.1
C_r	Rated capacity (r usually means 20, 10, 5 or 3 hours)	Ah	3.3.1
C_{Wh}	Capacity under constant-power load	Wh	3.2
D_j	Diffusion coefficient of species j	cm^2·s^{-1}	2.3
E	Cell voltage or electrode potential	V	2.2.1
E^o	Equilibrium voltage or potential	V	2.2.1
E_A	Activation energy in the ARRHENIUS relation	J	2.3.3
E_o	Potential referred to the standard hydrogen electrode (SHE)	V	2.2.1
$E^{o,s}$	Standard value of E^o	V	2.2.1
E_{cal}	Calorific voltage	V	2.2.4
E_h	Potential ref. to the hydrogen electrode in the same solution (HESS)	V	2.2.3.1

Symbol	Meaning	Dimension	Section
F	Faraday constant, 96485 As/equiv. = 26.802 Ah/equiv	As/equiv.	2.2.1
G	Free enthalpy equals Gibbs free energy	$J \cdot mole^{-1}$	2.2
ΔG	Free enthalpy of reaction, change of Gibbs free energy (per multiple of the reaction equation)	J	2.2
$\Delta G^{o,s}$	Standard value of ΔG	J	2.2.1.1
H	Enthalpy	$J \cdot mole^{-1}$	2.2
ΔH	Enthalpy of reaction (per reaction equation)	J	2.2
i	Current density (*often related to capacity 'A/100 Ah'*)	$A \cdot cm^{-2}$	
i_o	Exchange-current density	$A \cdot cm^{-2}$	2.3.3
$i(H_2$ evol.)	Hydrogen-evolution rate at the neg. electrode	current	8.1
i(float)	Overcharging or float current	current	8.1
$i(O_2$ escape)	Rate of oxygen excape	current	8.1
$i(O_2$ evol.)	Oxygen-evolution rate at the pos. electrode	current	8
$i(O_2$ reduct.)	Oxygen-reduction rate at the neg. electrode	current	8
I_{10}	Current that causes 10 hour discharge duration. Correspondingly I_5; $I_{0.5}$	A	3.3.1
k	Reaction constant		2.3.3
m_i	Weight of component i	kg	2.5.1
n	Number of exchanged electronic charges in electrochemical reactions	equiv.	2.2.1
n	Number of moles in a chemical reaction		2.6.2
n	Number of cells in a blockbattery (Monobloc)		13.1.1.1
N.C. or n.c.	Nominal capacity, see C_r (rated capacity)	Ah	3.3.1
p	Gas pressure	Pa	2.6.3
Δp	Opening pressure of a valve	Pa	10.6
pH	Acidity, negative logarithm of H^+ concentration		2.2.3.1
p_{H_2}	Hydrogen pressure	Pa	2.2.3.1

Symbol	Meaning	Dimension	Section
Q_{rev}	Reversible heat effect (per reaction equation)	J or Wh	2.2.4
Q_{joule}	Joule effect	J or Wh	2.4.1
Q_{total}	Heat generation in total $Q_{total} = Q_{joule} + Q_{rev}$	J or Wh	2.4.2
Q_{gen}	Heat generation $Q_{gen} = -Q_{total}$	J or Wh	2.5.1
R	Molar gas constant 8.3145 $J \cdot mole^{-1} \cdot K^{-1}$		2.2.1.1
R_i	Internal resistance of a battery	Ω	3.5.1
S	Entropy	$J \cdot K^{-1} \cdot mole^{-1}$	2.2
ΔS	Entropy of reaction (per reaction equation)	$J \cdot K^{-1}$	2.2
T	Temperature	K or °C	2.2.1.1
ΔT	Temperature difference	K	2.5.2
t	Time	s, hours	2.6.2
Δt	Time period	s, hours	2.6.2
t_+, t_-	Transference numbers in binary electrolytes		2.3.2
η	Overvoltage = $E - E^o$	V	2.3.3
η	Efficiency of the internal oxygen cycle	%	8.3.5

1. Introduction

In this book, the term 'maintenance-free batteries' means batteries that are based on the well known systems lead-acid and nickel/cadmium battery or related to them, but are distinguished by special features. Such maintenance-free designs of the nickel/cadmium battery and the lead-acid battery are in widespread use today as sealed nickel/cadmium batteries or valve-regulated lead-acid batteries. The nickel/metal hydride battery can only be made in the sealed version. All three designs are based on aqueous electrolytes, dilute sulphuric acid or dilute potassium or sodium hydroxide. A general disadvantage of aqueous electrolytes is the decomposition of their component 'water' into hydrogen and oxygen if the cell voltage exceeds 1.23 V. For this reason, conventional lead-acid and nickel/cadmium batteries require regular refilling of lost water. Although for modern stationary batteries the refill intervals cover years, refilling requires continual maintenance, distilled water has to be provided, and the cells opened from time to time. Furthermore, a mixture of hydrogen and oxygen escapes from the battery and a certain ventilation is required to avoid the hydrogen concentration in the surroundings of the battery reaching 4% per volume. Higher concentrations form an explosive mixture with air. As a third disadvantage, fumes of the acid or hydroxide solution are formed that escape with the gas from the cell and may cause corrosion problems in the neighbourhood of the battery.

These drawbacks of the aqueous electrolyte are completely evaded with sealed nickel/cadmium and nickel/metal hydride batteries that (practically) do neither lose water nor generate gas that escapes from the battery. Maintenance-free lead-acid batteries cannot be made completely sealed; small amounts of hydrogen escape from time to time. But water loss is so far reduced that they need no refill during their service life, and they also can be installed close to electrical equipment, due to the low hydrogen-generation rate and the fact that acid fumes do not escape.

All three maintenance-free battery designs are based on the same principle, namely the internal oxygen cycle. It means that during overcharging oxygen is generated at the positive electrode and subsequently reduced at the negative. A closed cycle of oxygen thus replaces hydrogen and oxygen evolution and escape.

Note: *The generally applied term 'maintenance-free' should not mislead the user of such batteries to abandon all kinds of supervision. To ensure its proper function, the maintenance-free battery has to be observed and checked regularly, if it is not monitored automatically (cf. Chapter 14).*

The book begins with a summary of electrochemical and physical laws that are required to describe battery systems (Chapter 2). Specific parameters and definitions that concern batteries are compiled in Chapter 3. In the subsequent Chapters 4 to 6, the fundamental features of the three battery systems are described.

The central problem of electrolytic water decomposition is studied in Chapter 7, but also measures to overcome the problem of water loss are described. In Chapter 8, the principle of the internal oxygen cycle is treated to some detail.

Chapters 9 - 12 concern the technical realisation of the principle and show typical battery designs.

Charging methods are treated in Chapter 13. Ways to supervise or monitor the battery and its charging system are considered in Chapter 14.

Chapter 15 covers a number of more general problems connected with the use of batteries, like storage problems, life expectancy and the question of what happens to the battery after it is run down.

A brief survey of laws and regulations concerning maintenance-free batteries is given at the end in Chapter 16.

1.1.1 Sealed Nickel/Cadmium and Nickel/Hydride Batteries

The sealed nickel/cadmium, as well as the (necessarily sealed) nickel/metal hydride battery, represent the absolutely maintenance-free battery. No water loss occurs in normal operating conditions; neither oxygen nor hydrogen is released from the battery. The electrochemical system remains practically unaltered. If the battery is equipped with a valve, this valve never opens except in an emergency, e.g. when the battery is charged with too high a current and the internal pressure exceeds tolerable limits, because of this abuse.

Note: *Today in the field of nickel/cadmium batteries only button or coin cells are really 'sealed'. Cylindrical and prismatic nickel/cadmium cells were also virtually sealed during the first decades of their manufacture. They were closed by a membrane that was destroyed by inadmissible internal overpressure (cf. Fig. 11.9, p.379). For practical reasons, nowadays in most sealed nickel/cadmium batteries this emergency vent is replaced by a resealing valve similar to that used with lead-acid batteries (cf. Section 11.4). Nevertheless, the fundamental difference remains that in sealed nickel/cadmium and nickel/hydride batteries the valve opens only in an emergency, while this happens with lead-acid batteries from time to time also in normal operational conditions.*

1.1.2 Valve-Regulated Lead-Acid Batteries

In the lead-acid battery, complete sealing can never be achieved because the generation of hydrogen can never be avoided completely, according to fundamental electrochemical parameters. The valve has to open periodically to let small amounts of gas, mainly hydrogen, escape from the cell. Otherwise the internal pressure would exceed technically tolerable limits. Gradual water loss is connected with this gas evolution, and this water loss causes slight changes in the cell parameters during service life. But the rate of water loss can be kept so low that the initial amount of electrolyte is sufficient for a service life of ten years or more.

The '**maintenance-free starter battery**' is beyond the scope of this book, when it is a flooded lead-acid battery, with special constructional features designed to outlast the usual service life in a motor car without refilling. The principles of this battery type are described briefly in Section 7.5.1.1.

1.1.2.1 The Name 'Valve-Regulated Lead-Acid Battery'

Different names are in use for this kind of lead-acid battery. An earlier proposal has been 'sealed valve-regulated lead-acid battery' ([1]), but the term 'sealed' is redundant in this connection, because a valve can operate only when the battery itself is sealed. The term now mostly accepted is '**valve-regulated lead-acid battery**' ([2]) or its abbreviation **VRLA battery**. This name describes quite well the special feature of this design, namely:

- The interior of the valve-regulated lead-acid battery does not have a continuous connection with the surrounding atmosphere to exchange gases, as is characteristic of the vented lead-acid battery.
- The valve-regulated lead-acid battery cannot be sealed completely but has to be equipped with a valve that opens even in normal operation conditions from time to time for gas escape. This is basically different from 'sealed' batteries.

In the literature, however, a number of different names are used instead of 'valve-regulated lead-acid battery'. One reason may be that marketing people want to emphasise the advantages of this battery design by the name. The term 'sealed lead-acid battery' or 'sealed maintenance-free lead-acid battery' is often used, although neither name is correct, as already mentioned. 'Recombinant lead-acid battery ' is rather common, but this name does not point out that this battery cannot be sealed, as is possible with the 'recombinant nickel/cadmium battery'. Sometimes the name 'maintenance-free lead-acid battery" is applied. However, this name is already used for the above mentioned vented starter batteries that do not require water refilling during their service life in the special operating conditions in a motor car. A further version of the name is 'starved-electrolyte lead-acid battery', which indicates that the electrolyte has to be immobilised to achieve the internal oxygen cycle. But 'starved electrolyte' could as well mean a flooded battery with extremely reduced acid content.

On the whole, 'valve-regulated lead-acid battery' seems to be the most appropriate name to express the special features of the lead-acid battery under consideration. This term will be used in this book.

1.2 History

The history of the lead-acid battery goes back to 1854 when Sinsteden published performance data of this battery system for the first time (cf. ([3])). However, the practical importance of the lead-acid battery system was detected independently of Sinsteden's work in 1859 by Planté who produced a rechargeable battery by alternately charging and discharging lead sheets, immersed in sulphuric acid ([4]). Planté plates in principle produced by this method are still in use ([5]). Separate production of the active material that is the basis of the pasted-plate design, mainly used today, was introduced by Fauré in 1881 (cf. ([6]). p.4). The introduction of electric lighting favoured the production of batteries, which started at that time in the industrialised countries (cf., e.g., ([7])).

In the beginning, lead-acid batteries were operated in open jars. Plates could easily be cleaned or exchanged, but much water was lost by evaporation.

The maintenance expenditure on refilling the water was dramatically reduced when the cells were provided with covers. Only small holes in the vent plug were left for gas outlet, and evaporation could then be neglected.

Experiments to design a spill-proof lead-acid battery with the aid of gelled electrolyte were carried out at the beginning of this century under the influence of primary batteries, often called 'dry batteries' ([8]). But the market introduction of this lead-acid battery design started virtually in the late fifties of this century when small lead-acid batteries were made spill-proof by using silica gel (SiO_2). Initially, the methods to immobilise the electrolyte were improved to achieve simply spill-proof lead-acid batteries with a certain cycle stability ([9]). Later, the combination with a valve yielded the characteristic features of the maintenance-free lead-acid battery (cf. ([10])). A further very important step was the development of glass felts based on fibres with diameters in the µm range ([11]). The good wettability of this material for sulphuric resulted in the absorbed type of immobilised electrolyte ([12]). The GATES round cell (Fig. 10.24, p.344) has been the first successfully operating battery with this absorbent glass-mat separator (AGM separator). This new separator material initiated an intensive development in the field of starter batteries ([13]), which, however, did not prove as successful as expected; while the parallel development for stationary batteries ([14]) initiated the widespread introduction of this type of battery.

The fundamental inventions in the field of nickel/iron and nickel/cadmium batteries were made around 1900 ([15]). The batteries with alkaline electrolyte had to be provided with cell covers from the very beginning, which contributed to the reputation of superior mechanical sturdiness of the nickel/cadmium battery. The reason was that in open cells the electrolyte (potassium hydroxide (KOH)) would absorb carbon dioxide from the air, which would reduce the conductivity of the electrolyte (cf. Section 11.1.1.1).

A first step in the development of a sealed nickel/cadmium battery was the discovery that oxygen was reduced at the negative electrode ([16]). The principles of the sealed nickel/cadmium battery were described in remarkable detail by G.Neumann ([17]), including the internal oxygen cycle and the avoidance of hydrogen evolution by oversizing the negative-electrode capacity (charge reserve). This principle was patented world wide (cf., e.g., ([18])) giving one of the key patents for the sealed nickel/cadmium battery.

A detailed survey of the history of maintenance-free lead-acid and nickel/cadmium batteries is given in ([19]).

When the first edition of this book was finished in Summer 1992, the nickel/metal hydride battery system was a newcomer to the market, and it was not yet possible to foresee whether it would be successful. In this few years the system has been fully established in the market with a steadily increasing share.

References

([1]) DIN 40729 (May 1985)

([2]) Draft Publication IEC 896/2, Stationary Batteries; General Requirements and Test Methods, Part 2, Valve Regulated Types. (Draft Febr.1990)

([3]) J.Garche, J.Power Sources, 31 (1990) 401

([4]) G.W.Vinal, Storage Batteries, John Wiley & Sons, New York, 1955, p. 2 f

([5]) K.V.Kordesch, Batteries, Marcel Dekker, Inc, New York, 1977, p. 419

([6]) G.W.Vinal, l.c. p.4

([7]) K.R.Bullock in A.J.Salkind (Editor), Proc.Symp. History of Battery Technology, © The Electrochem. Soc. Proc.Vol 87-14, Pennington NJ 1987, p.106

([8]) C.Drucker and A.Finkelstein, Galvanische Elemente und Akkumulatoren, Akad. Verlags-gesellschaft, Leipzig, 1932, p.224

([9]) O.Jache, Ger. Pat 1194 015 (1958)

([10]) H.A.Kiehne, Battery Technology Handbook, Marcel Decker, Inc, New York, 1989, p. 341

([11]) B.K.Mahato and E.C.Laird, Power Sources 5, Edit. D.H.Collins, Academic Press, London (1975), p.23

([12]) McClelland et al., United States Patent, 3,862,861; Jan. 28, 1975, Filed May 1972

([13]) B.Culpin, K.Peters and N.R.Young, Power Sources 9, Edit. J.Thompson, Academic Press, London (1983) p.129

([14]) A.I.Harrison and B.A.Wittey, Power Sources 10, Edit. L.J.Pearce, The Paul Press LTD, London (1985) p. 447

([15]) S.U.Falk and A.J.Salkind, Alkaline Storage Batteries, John Wiley & Sons, New York, 1969, p. 12 ff.

D.Tuomi *in* A.J.Salkind (Edit.), 'History of Battery Technology', © The Electrochem. Soc., Inc. Pennington, NJ, U.S.A., Proceedings Volume 87-14, 1987, p.21

([16]) A.E.Lange, E.Langguth, E.Breuning and A.Dassler, Ger.Pat. 674,829, (1933)

([17]) G.Neumann, Ger.Pat. 975 909 (1948)

([18]) G.Neumann and U.Gottesmann, U.S. Pat. 2,571,927 (1951)

([19]) R.F.Nelson, J. Power Sources, 31 (1990), p. 3

2. Fundamentals

This Chapter presents a brief review of the basic electrochemical laws, required to describe and explain the features of batteries. They can, however, be mentioned only briefly, and the reader should not regard this Chapter as a textbook, but as a survey on the applied basic rules, and as a facility to look them up when later they are applied in the text. For details and derivations he is referred to the textbooks of electrochemistry ([1]) or more fundamental books on batteries ([2]).

2.1 The Electrochemical Cell

Electrochemical energy storage is based on the conversion of chemical energy into electrical energy and vice versa. Consequently, a chemical reaction, the cell reaction, characterises each battery system, and the amount of energy that can be stored is determined by chemical substances of different energy content that represent the charged and discharged states of the reacting substances.

Chemical reactions include the exchange of electrical charges between ions. If this exchange occurs directly between the reacting substances, the released energy is converted predominantly into heat and, to a small extent, into volume energy. To gain electrical energy, the reaction has to be split up into two separate electrode reactions, one that releases electrons, and the other one that absorbs electrons. Then the electron exchange can be transformed into a current that flows through the consumer (or the charging device).

For example, the discharge reaction of the lead-acid battery

$$\boxed{Pb + PbO_2 + 2{\cdot}H_2SO_4 \Rightarrow 2{\cdot}PbSO_4 + 2{\cdot}H_2O} \qquad (2.1)$$

has to be split up into negative- and positive-electrode reactions, as shown in Fig. 2.1.

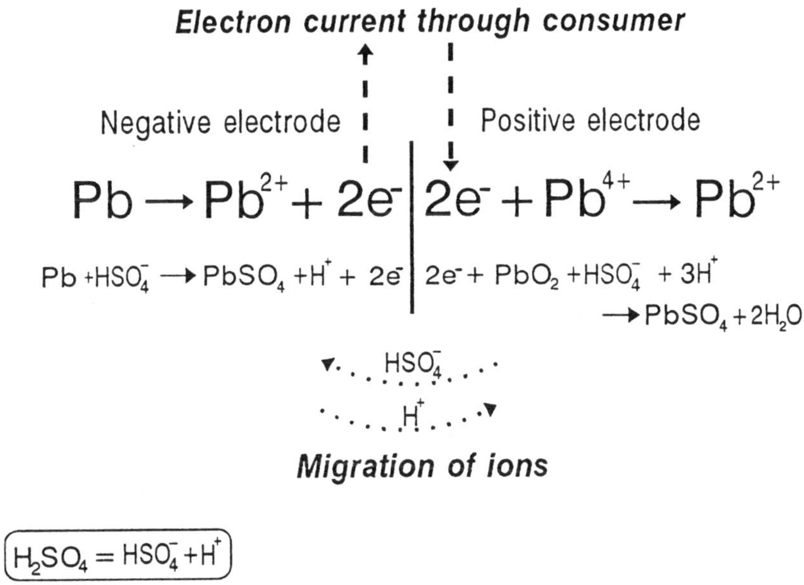

Fig. 2.1 Split-up of the cell reaction into electrode reactions. Example: Discharge reaction of lead-acid batteries (Eq. (2.1)). Per multiple of the cell reaction, two electrons are released at the negative electrode ($Pb \Rightarrow Pb^{2+} + 2 \cdot e^-$). They flow as an electron current through the consuming device, and are absorbed at the positive electrode ($Pb^{4+} + 2 \cdot e^- \Rightarrow Pb^{2+}$). Thus, chemical conversion energy can directly be utilised as electrical energy. (Dissociation of H_2SO_4 cf. Section 4.1.1.1.)

The basic charge transfer reactions that release or absorb electrons are:

$$\boxed{Pb \Rightarrow Pb^{2+} + 2 \cdot e^-} \quad \text{and} \quad \boxed{Pb^{4+} + 2 \cdot e^- \Rightarrow Pb^{2+}} \tag{2.2}$$

At the negative electrode, lead is oxidised to Pb^{2+} ions, at the positive electrode Pb^{4+} ions are reduced to Pb^{2+} ions. Both equations have to be complemented by chemical reactions, as shown in the Figure. So, Pb^{2+} ions form lead sulphate ($PbSO_4$), and Pb^{4+} ions exist only in lead dioxide (PbO_2).

The completed reactions are:

- At the negative electrode, oxidation of lead (Pb) to divalent lead ions (Pb^{2+}) means formation of lead sulphate ($PbSO_4$). Two electrons are released:

$$\boxed{Pb + HSO_4^- \Rightarrow PbSO_4 + H^+ + 2 \cdot e^-} \tag{2.2a}$$

- At the positive electrode, reduction of (Pb^{4+}) to the divalent ion (Pb^{2+}) means conversion of lead dioxide (PbO_2) into lead sulphate ($PbSO_4$). Two electrons are absorbed:

$$\boxed{PbO_2 + HSO_4^- + 3 \cdot H^+ + 2 \cdot e^- \Rightarrow PbSO_4 + 2 \cdot H_2O} \tag{2.2b}$$

The sum of both electrode reactions is the **cell reaction**. For the above example, it is Eq. (2.3):

$$\boxed{Pb + PbO_2 + 2{\cdot}H_2SO_4 \Leftrightarrow 2{\cdot}PbSO_4 + 2{\cdot}H_2O} \qquad (2.3)$$

that equals Eq. (2.1), except the double-pointed arrow which indicates that the reaction can be reversed and the battery recharged.

In Eq. (2.3), Pb and PbO_2 are the primary products at the negative and positive electrodes, respectively. They are called **reactants**. $PbSO_4$ is the **product** of the cell reaction. In the lead-acid battery, used here as an example, lead sulphate is formed as the product at both electrodes since lead is the basic element in both electrodes. In most battery systems, different elements are used in both electrodes (cf., e.g., Section 5.1.1, or Section 6.1).

Such a split-up of electrochemical reactions is achieved in the electrochemical cell, shown in Fig. 2.2: a positive and a negative electrode are immersed in the electrolyte and the reacting substances (the active materials) are (usually) stored within the electrodes, and to some extent also in the electrolyte as H_2SO_4 and H_2O in lead-acid and H_2O in nickel/cadmium batteries.

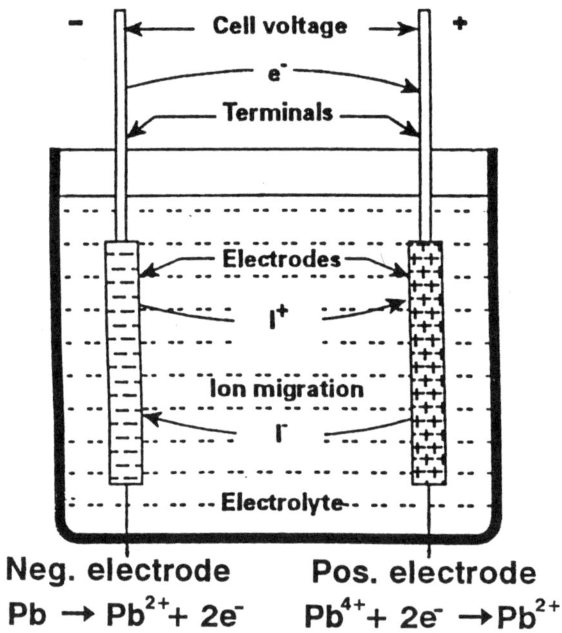

Fig. 2.2 The electrochemical cell and the split-up of the cell reaction Eq. (2.1) (discharge of the lead-acid battery) into the electrode reactions: $Pb \Rightarrow Pb^{2+} + 2.e$ at the negative, and $Pb^{4+} + 2{\cdot}e^- \Rightarrow Pb^{2+}$ at the positive electrode. Here, oxidation of lead (Pb) occurs at the negative electrode, while Pb^{4+} is reduced at the positive electrode.

The split-up of the cell reaction requires a number of structural components and determines the design of the battery. Current-conducting parts like electrode supports, connecting bridges and terminal posts are passive components, required to hold the active material and conduct the current (cf. Section 4.2.2, especially Fig. 4.3). Separators prevent direct electronic contact between the electrodes.

The electrochemical cell is the basic element of each battery, and connection of several cells in series forms a battery. This definition, however, often is not observed in practice, and single cells are frequently also designated as batteries.

Note: *The arrangement shown in Fig. 2.2 resembles an electrolytic capacitor where also two electrodes are separated by the electrolyte. However, charging and discharging of such a capacitor means only charge shifting within the double layer at the electrode/electrolyte interface and within the elctrolyte. Chemical reactions do not occur and the physical structure is not affected. For this reason, a nearly unlimited number of charge/discharge cycles are possible.*

In batteries, such a double layer also exists, and the large surface area of the active material gives rise to a high double-layer capacitance when impedance measurements are made. The real battery capacity, however, is based on chemical reactions. Each charge/discharge cycle changes the physical structure of the electrodes, and these changes inevitably cause ageing processes. For this reason, the number of possible charge/discharge cycles are limited, and performance changes with service live are unavoidable.

2.2 Equilibrium or Thermodynamic Parameters

The laws of thermodynamics generally apply to the state of equilibrium, which means that all reactions are balanced. In the electrochemical cell, these data can only be measured when no current flows through the cell or its electrodes. On account of this balance, the thermodynamic parameters do not depend on the reaction path; they depend only on different energy levels between the final and initial components (the 'products' and the 'reactants' of the electrochemical reaction).

Because of the equilibrium conditions, the laws of thermodynamics describe the possible upper limit of performance data. As soon as current flows through the cell, these data are reduced, owing to kinetic parameters described in Section 2.3.

The energy exchange, connected with electrochemical reactions, is described by three **thermodynamic parameters**:

Enthalpy H,

Free enthalpy (or **Gibbs free energy G**),

Entropy S.

As far as these parameters concern chemical or electrochemical reactions, they actually describe the differences between these parameters before the reaction started and after it was completed. (*The deduction of such differences is illustrated by Eqs. (4.7) and (4.8)*). To indicate their character as differences, they are usually written with the 'Δ' sign and are called:

- **Enthalpy of reaction ΔH**, which represents the amount of energy released or absorbed. ΔH describes the maximum heat generation, provided that the chemical energy is converted into heat by 100%.
- **Free enthalpy of reaction ΔG** also called **change of Gibbs free energy ΔG**, which describes the (maximum) amount of chemical energy that can be converted into electrical or mechanical energy and vice versa (discharge and charge, respectively, cf. the following Note).
- **Entropy of reaction ΔS**, which characterises the energy loss or gain connected with the chemical or electrochemical process. The product T·ΔS, the **reversible heat effect** ($\mathbf{Q_{rev}}$), represents the heat exchange with the surroundings when the process occurs 'reversibly', which means that all equilibria are balanced. This is synonymous with minimal heat loss of the system, which is true only when no current flows through the battery.
  T·ΔS can be positive or negative. In the first case, additional energy is generated by cooling of the environment (Peltier or heat-pump effect) (cf. Section 2.2.4).

Values of these parameters listed in tables are always related to one multiple of the reaction equation.

Note: *The sign of these thermodynamic parameters is positive when the system gains energy and negative when the system loses energy. Consequently, in discharge, the sign of ΔG is negative, because the (electrochemical storage) system delivers energy. When the battery becomes recharged, ΔG is positive.*

Important relations between the three parameters are:

$$\Delta G = \Delta H - T\cdot\Delta S \quad \text{or} \quad \Delta H - \Delta G = T\cdot\Delta S \tag{2.4}$$

with T : temperature in K.

According to this relation, the above mentioned **reversible heat effect T·ΔS** expresses the difference between ΔH and ΔG.

2.2.1 Equilibrium Voltage

ΔG describes the amount of energy that can be converted into electrical energy, which leads to the relation

$$\Delta G = -\, n\cdot F\cdot E^{o} \tag{2.5}$$

Its transposition represents the **equilibrium voltage**:

$$\boxed{E^o = -\frac{\Delta G}{n \cdot F}} \tag{2.6}$$

with n : number of exchanged electronic equivalents (2 in Eq. (2.1))
F : FARADAY constant, equivalent to 96485 As/equiv.
n·F means the amount of electrical charge connected with the reaction.
(1·F = 26.802 Ah; 2·F = 53.604 Ah.)
(*Electrochemical equivalent*)
E^o: equilibrium voltage of the cell (V).
$n \cdot F \cdot E^o$ describes the generated electrical energy (kJ or Ws).

This is the cell voltage under reversible conditions, i.e. when all reactions are balanced. In practice it means when no current flows through the cell.

Note 1: *Eqs. (2.5) and also (2.6) are sometimes written with a 'plus', sometimes with a 'minus' sign. This is a question of convention, because E^o is a voltage difference and the arithmetic sign depends on the convention of how the voltage difference is formed: right hand electrode minus left hand electrode or in reverse order, and on the current sense that has been agreed* (3).

Note 2: *Occasionally the equilibrium voltage is called the open-circuit or rest voltage. Strictly speaking these two terms only mean a voltage without external current flow, and may concern a mixed potential as well (cf. Fig.2.9, p.31). Actually, on account of secondary reactions, the open-circuit or rest potentials in batteries often are mixed potentials, but this is not strictly observed in practical language and they are often also called equilibrium potentials* .

2.2.1.1 Dependence on Concentration of the Reacting Components

Thermodynamic quantities like ΔH and ΔG depend on the concentrations (or better activities) of the reacting components, as far as these components are dissolved. The corresponding relation is:

$$\boxed{\Delta G = \Delta G^{o,s} + R \cdot T \cdot \left(\sum \ln \left[(a_i)^{j_i} \right]_{prod} - \sum \ln \left[(a_i)^{j_i} \right]_{react} \right)} \tag{2.7}$$

with a_i : activity (cf. Note below) of the reacting component i
(approximately the concentration) ($mole \cdot cm^{-3}$),
j_i : number of equivalents of this component which take part in the reaction,
R : molar gas constant for an ideal gas ($R = 8.3145\ J \cdot K^{-1} \cdot mole^{-1}$),
T : temperature (K),
$\Delta G^{o,s}$: standard value when all activities are unity,
react and prod : reactants and products when the reaction equation is written so that it occurs spontaneously (negative ΔG) as in Eq. (2.1).

Note: *The '**activity**' a (in mole/dm^3) describes the 'effective concentration' and is defined by the relation a = f·c, where c represents the concentration (mole/dm^3) and f the **activity coefficient**. Thermodynamic rules that were derived for dilute solutions can be applied for more concentrated solutions, when instead of the concentrations the activities of the reacting substances are used. The activity coefficient f approaches the value one for dilute solutions, but f can be very different from one in concentrated solutions (in certain solutions f attains values even > 1 (cf. [4])), but this does not concern the concentrations used with batteries). The introduction of the activity means that the considered substance is less effective (or more effective in the case f > 1) with respect to the thermodynamic rules than would be expected from its actual concentration. (R·T·lnf represents the energy difference between the ideal and the actual state).*

In the following, Eq. (2.7) is applied as an example to the reaction equation of the lead-acid battery. Under consideration that sulphuric acid (H_2SO_4) mainly dissociates into H^+ and HSO_4^- (cf. Section 4.1.1.1 and Eq. (4.6)):

$$\textbf{Reactants}\ \{Pb + PbO_2 + 2\cdot H^+ + 2\cdot HSO_4^-\} \Rightarrow \textbf{products}\ \{2\ PbSO_4 + 2\ H_2O\} \quad (2.8)$$

Application of Eq. (2.7) to this reaction, using the free enthalpies of formation for the considered substances, results in:

$$\begin{aligned}\Delta G = \{&2\cdot\Delta G^s(PbSO_4) + 2\cdot(\Delta G^s(H_2O) + RT\cdot\ln(a_{H_2O}))\}\\ &-\{\Delta G^s(Pb) + \Delta G^s(PbO_2) + 2\cdot(\Delta G^s(H^+) + RT\cdot\ln(a_{H^+})\\ &+ 2\cdot(\Delta G^s(HSO_4^-) + RT\cdot\ln(a_{HSO_4^-})\end{aligned} \quad (2.9)$$

(The concentrations of pure condensed phases, like Pb, PbO_2, and $PbSO_4$ do not appear in this equation).

When the standard values are combined in the term $\Delta G^{o,s}$

$$\begin{aligned}\Delta G^{o,s} = &2\cdot\Delta G^s(PbSO_4) + 2\cdot\Delta G^s(H_2O)\\ &- \Delta G^s(Pb) - \Delta G^s(PbO_2) - 2\cdot\Delta G^s(H^+) - 2\cdot\Delta G^s(HSO_4^-)\end{aligned} \quad (2.10)$$

Eq. (2.9) can be written:

$$\Delta G = \Delta G^{o,s} + RT\cdot(2\cdot\ln(a_{H_2O}) - 2\cdot\ln(a_{H^+}) - 2\cdot\ln(a_{HSO_4^-})) \quad (2.11)$$

or

$$\Delta G = \Delta G^{o,s} + R\cdot T\cdot\ln\left(\frac{a_{H_2O}^2}{a_{H^+}^2\cdot a_{HSO_4^-}^2}\right) \quad (2.12)$$

When inserted into Eq. (2.6), the equilibrium cell voltage results:

$$E^o = E^{o,s} + \frac{R\cdot T}{n\cdot F}\cdot\ln\frac{a_{H^+}^2\cdot a_{HSO_4^-}^2}{a_{H_2O}^2} \quad (2.13)$$

(The negative sign in Eq. (2.6) is now positive, because numerator and denominator have changed places between the Eqs. (2.12) and (2.13)).

This Eq. is simplified for 25 °C (298.2 K) under consideration that

ln(..) = 2.303 log(..)

R = 8.3145 J/(K·mole) = 8.3145 Ws/(K·mole),

F = 96485 As/equiv.,

thus R·T/F = 0.02569 V· equiv.·mole^{-1}

or 2.303·R·T/F = 0.0592 V·equiv.·mole^{-1}

$$E^{o} = E^{o,s} + \frac{0.0592}{n} \cdot \log \frac{a^2_{H^+} \cdot a^2_{HSO_4^-}}{a^2_{H_2O}} \quad V \qquad (2.14)$$

According to Eq. (2.2), the number of equivalents per multiple of Eq. (2.3) is n = 2, and so the resulting **Nernst equation** that describes the concentration dependence of the equilibrium voltage for lead-acid batteries is:

$$E^{o} = E^{o,s} + 0.0296 \cdot \log \frac{a^2_{H^+} \cdot a^2_{HSO_4^-}}{a^2_{H_2O}} \quad V \qquad (2.15)$$

with $E^{o,s}$ = 1,931 V (cf. Section 4.1); as *pure condensed phases PbO_2, $PbSO_4$ and Pb are implied in $E^{o,s}$ but do not appear in the logarithmic expression.*

2.2.1.2 Temperature Coefficient

The thermodynamic data also determine the **temperature coefficient** of the equilibrium cell voltage or electrode potential according to the relation:

$$\frac{dE^{o}}{dT} = \frac{\Delta S}{n \cdot F} \qquad (2.16)$$

because of the thermodynamic relations

$$\frac{dE^{o}}{dT} = -\frac{1}{n \cdot F} \cdot \frac{\partial(\Delta G)}{\partial T} \quad \text{and} \quad \frac{\partial(\Delta G)}{\partial T} = -\Delta S \qquad (2.17)$$

Temperature coefficient and equilibrium voltage or potential are often listed for the standard temperature of 25 °C (= 298.2 K) ([5]). The equilibrium cell voltage at any temperature (T) is

$$E^{o}(T) = E^{o}(298.2\ K) + (dE^{o}/dT) \cdot (T - 298.2) \qquad (2.18)$$

assuming the temperature coefficient to be constant for the considered temperature range. This assumption represents a rough approximation, but for the considered battery systems, the temperature coefficient of the equilibrium voltage is so small that, anyway, its influence can usually be neglected.

To evaluate the equilibrium potential in a wider temperature range, de Bethune (6) introduced the **isothermal temperature coefficient** $(dE^o/dT)_{iso}$. Its variation with temperature is described by the so called **second temperature coefficient** $(d^2E^o/dT^2)_{iso}$ which is also listed in the mentioned tables.

With the aid of both coefficients, the equilibrium potential $E^o(T)$ for a temperature T against a standard hydrogen electrode at the same temperature can be calculated according to

$$E^o(T) = E^o(298.2°K)+(d\,E^o/dT)_{iso}\cdot(T-298)+1/2\cdot(d^2E^o/dT^2)_{iso}\cdot(T-298.2)^2 \quad (2.19)$$

The temperature coefficient used in Eq. (2.18) is called the **thermal temperature coefficient** $(dE^o/dT)_{th}$ in the above mentioned tables.

2.2.2 Single-Electrode Potentials

Thermodynamic calculations are always based on a cell, and the derived voltage means the voltage difference between two electrodes (Fig. 2.2). The voltage difference between the electrode and the electrolyte, the 'absolute potential', cannot exactly be measured, since potential differences can only be measured between two electronic conductors (7).

When the Nernst equation concerns single-electrode potentials, it means the cell voltage between this electrode and a reference electrode (cf. Section 2.2.3). To meet the sign convention used here (Eq. (2.6)) the electrode reaction has to be written so that negative current flows from the electrode into the electrolyte (cathodic current, cf. Note below), when the cell voltage is chosen as the difference between right hand electrode minus left hand electrode (as in Fig. 2.2). For the lead-acid battery, the desired (cathodic) current direction at the positive electrode is the discharge

$$Pb^{4+} + 2\cdot e^- \Rightarrow Pb^{2+} \quad (2.20)$$

or in its completed form

$$PbO_2 + H_2SO_4 + 2\cdot H^+ + 2\cdot e^- \Rightarrow PbSO_4 + 2\cdot H_2O \quad (2.21)$$

The corresponding Nernst equation ($H_2SO_4 \Leftrightarrow 2\cdot H^+ + HSO_4^-$) that describes the equilibrium potential of the PbO_2 electrode, is

$$E_{PbO_2/PbSO_4} = E^{o,s}_{PbO_2/PbSO_4} + \frac{R\cdot T}{2\cdot F}\cdot \ln\frac{a^3_{H^+}\cdot a_{HSO_4^-}}{a^2_{H_2O}} \quad (2.22)$$

To evaluate the potential referred to the hydrogen electrode, Eq. (2.21) has to be completed by the corresponding reaction (Eq. (2.30)):

$$PbO_2 + H_2SO_4 + 2\cdot H^+ + 2\cdot e^- \Rightarrow PbSO_4 + 2\cdot H_2O$$

$$H_2 \Rightarrow 2\cdot H^+ + 2\cdot e^-$$

$$PbO_2 + H_2SO_4 + H_2 \Rightarrow PbSO_4 + 2\cdot H_2O \quad (2.23)$$

with the corresponding Nernst equation:

$$E_{h(PbO_2/PbSO_4)} = E^{o,s}_{h(PbO_2/PbSO_4)} + \frac{R \cdot T}{2 \cdot F} \cdot \ln \frac{a_{H^+} \cdot a_{HSO_4^-} \cdot a_{H_2}}{a^2_{H_2O}} \tag{2.24}$$

that results in the curve 'PbO$_2$/PbSO$_4$ electrode' in Fig. 2.3. (a_{H_2} *can be replaced by the hydrogen gas pressure as shown in Section 2.2.3.1 to achieve the conditions of the standard hydrogen electrode*). E_h denotes the electrode potential referred to the hydrogen electrode in the same solution (HESS) (cf. Section 2.2.3.1).

The same result is achieved, when the two Nernst equations Eq. (2.22) and Eq. (2.31) are combined by subtraction:

$$E_{h(PbO_2/PbSO_4)} = E_{PbO_2/PbSO_4} - E_{H^+/H_2} \tag{2.25}$$

At the negative electrode, a negative current flows into the electrolyte (cathodic current) when the electrode is charged

$$PbSO_4 + 2 \cdot H^+ + 2 \cdot e^- \Rightarrow Pb + H_2SO_4 \tag{2.26}$$

and the corresponding Nernst equation that describes the equilibrium potential of the negative electrode, the Pb/PbSO$_4$ electrode, is

$$E_{Pb/PbSO_4} = E^o_{Pb/PbSO_4} + \frac{R \cdot T}{2 \cdot F} \cdot \ln \frac{a^2_{H^+}}{a_{H^+} \cdot a_{HSO_4^-}} \tag{2.27}$$

Combination with Eq. (2.31) gives the Nernst equation of the negative electrode referred to the hydrogen electrode, as shown above for the positive electrode:

$$E_{h(Pb/PbSO_4)} = E^{o,s}_{h(Pb/PbSO_4)} + \frac{R \cdot T}{2 \cdot F} \cdot \ln \frac{a_{H_2}}{a_{H^+} \cdot a_{HSO_4^-}} \tag{2.28}$$

which leads to the curve 'Pb/PbSO$_4$ curve' in Fig. 2.3.

The cell voltage is also derived by combination of the two Eqs. (2.22) and (2.27) by subtraction $E_{PbO_2/PbSO_4}$ - $E_{Pb/PbSO_4}$:

$$\begin{aligned} E &= E_{PbO_2/PbSO_4} - E_{Pb/PbSO_4} \\ &= E^{o,s}_{PbO_2/PbSO_4} - E^{o}_{Pb/PbSO_4} + \frac{0.0592}{2} \log \frac{a^2_{H^+} \cdot a^2_{HSO_4^-}}{a^2_{H_2O}} \end{aligned} \tag{2.29}$$

and corresponds to Eq. (2.15) with $E^o_{PbO_2/PbSO_4} - E^o_{Pb/PbSO_4} = E^{o,s}$.

2.2.3 Reference Electrodes

The potential of a single electrode must always be referred to another electrode, and the term **electrode potential** actually means a voltage difference measured with the aid of a reference electrode. To get a basis for the electrode-potential scale, the zero point was arbitrarily equated with the potential of the **standard hydrogen electrode (SHE)**, a hydrogen electrode in specified conditions at 25 °C. For some general remarks cf. ([8]).

2.2.3.1 The Standard Hydrogen Electrode

The standard hydrogen electrode is specified as an electrode (usually platinised platinum) that is immersed in acidic solution with the H^+ ion activity of 1 mole/dm^3, and surrounded by hydrogen bubbles with an H_2 pressure of 1 atm (= 101325 Pa).

The reaction equation of the hydrogen electrode is

$$\boxed{H_2 \Leftrightarrow 2 \cdot H^+ + 2 \cdot e^-} \tag{2.30}$$

and the corresponding Nernst Equation

$$\boxed{E = E^{o,s}_{H^+/H_2} + \frac{R \cdot T}{2 \cdot F} \ln \frac{a^2_{H^+}}{a_{H_2}}} \tag{2.31}$$

When the the H_2 activity is replaced by hydrogen pressure, and the concentration a_{H^+} is substituted by the hydrogen gas pressure on account of its solubility (cf. Eq. (2.120)), and the rule $\ln(..)^2 = 2 \cdot \ln(..)$ is applied, Eq. (2.31) can be converted into

$$\boxed{E = E^{o}_{H^+/p_{H_2}} + \frac{R \cdot T}{2 \cdot F} \ln \frac{a^2_{H^+}}{p_{H_2}}} \quad \text{or} \quad \boxed{E = E^{o}_{H^+/p_{H_2}} + \frac{R \cdot T}{F} \cdot \ln \frac{a_{H^+}}{\sqrt{p_{H_2}}}} \tag{2.32}$$

Under the above mentioned standard conditions, the logarithmic expression is zero and the definition of the standard hydrogen electrode means that at 25 °C

$$\boxed{E^{o}_{H^+/p_{H_2}} = 0} \tag{2.33}$$

The temperature coefficient of the hydrogen electrode is $+0.871$ mV/°C .

Temperature coefficients of electrode potentials are often listed as $(dE_o/dT)_{th}$, which means the temperature coefficient against the standard hydrogen electrode at 25 °C, or as $(dE_o/dT)_{isoth}$, which represents the temperature coefficient referred to the standard hydrogen electrode at the same temperature as the other electrode (cf. Eq. (2.19)).

Mainly in alkaline electrolytes, electrode potentials often are referred to the hydrogen electrode in the solution under consideration (= **hydrogen electrode in same solution, HESS**), because nickel and cadmium electrodes as well as oxygen evolution follow the same dependence on H^+ ion activity a_{H^+} (a_{H^+} is also called the **acidity**).

To express the acidity of a solution, usually the pH value is used with the relation pH = -log(a_{H^+}); i.e. pH stands for the negative logarithm of the activity of the H^+ ions. According to this definition, the activity of 1 mole/dm^3 equals pH = 0. Using the pH value, Eq. (2.31) can be written (under consideration that ln(..) = 2.303 log(..)):

$$E = -2.303 \cdot \frac{R \cdot T}{F} \cdot \left\{ pH + \frac{1}{2} \cdot \log(p_{H_2}) \right\} \quad \text{with } -2.303 \cdot \frac{R \cdot T}{F} = -0.0592\ \text{V} \qquad (2.34)$$

So the potential of the hydrogen electrode is shifted by -0.0592 volts when the H^+ concentration is reduced by one decade.

Table 2.1 shows calculated values for the pH value and the hydrogen electrode potential.

KOH concentration Weight%	mole/dm^3	Density g/cm^3 (20 °C)	pH	Hydrog.el.pot. E_o/V ref.to S.H.E.
0.56	0.099	1.003	12.9	-0.762
1.04	0.199	1.008	13.2	-0.778
2.39	0.450	1.020	13.5	-0.798
5.37	0.995	1.047	13.9	-0.821
7.06	1.34	1.063	14.0	-0.828
10.41	1.995	1.094	14.2	-0.842
14.80	2.995	1.131	14.5	-0.873
22.07	4.49	1.210	14.9	-0.880
28.91	6.58	1.280	15.3	-0.907
38.13	9.35	1.380	15.9	-0.947
41.05	10.30	1.410	16.2	-0.955
43.98	11.30	1.445	16.3	-0.966

Table 2.1 Hydrogen electrode potential referred to the standard hydrogen electrode for various KOH concentrations. pH values and electrode potentials calculated with the aid of activity coefficients published in ([9]).

According to general usage, in this book E_o means the equilibrium potential referred to the standard hydrogen electrode, while the potential referred to the hydrogen electrode in the same solution is denoted by E_h.

Hydrogen reference electrodes are not used in battery practice. In one respect they are rather troublesome; a further obstacle is the risk of contamination of the battery electrodes by hydrogen overvoltage reducing noble metals, because the hydrogen electrode is based on such metals as platinum or palladium ([10]).

2.2.3.2 Reference Electrodes in Lead-Acid Battery Practice

The most commonly used reference electrode in vented lead-acid batteries, the **cadmium electrode,** consists of a stick of cadmium metal immersed directly into the sulphuric acid. This electrode does not attain an equilibrium potential according to Eq. (2.6); instead, a mixed potential (cf. Section 2.3.3) is established by two reactions opposing each other, namely the (corrosive) dissolution of cadmium and hydrogen evolution according to:

$$Cd \Rightarrow Cd^{2+} + 2 \cdot e^- \qquad (2.35)$$

$$2 \cdot H^+ + 2 \cdot e^- \Rightarrow H_2 \qquad (2.36)$$

$$Cd + 2 \cdot H^+ \Rightarrow Cd^{2+} + H_2 \qquad (2.37)$$

The result is a fairly stable potential, slightly increasing with acid density as shown in Fig. 2.3. For the usual acid density range in charged batteries its value is close to -0.42 V, referred to the standard hydrogen electrode. Thus, 0.42 V has to be added to convert the standard-electrode potential values into such referred to the cadmium electrode (cadmium readings). Table 2.2 gives examples. One disadvantage of the cadmium reference electrode is that it gradually dissolves according to Eq. (2.37). Consequently, this reference electrode cannot be used for continuous measurements over extended periods of time.

The potential of the cadmium electrode is slightly below the equilibrium potential of the negative electrode. So, the negative-electrode potential is positive referred to cadmium. But, when the negative electrode is polarised, for example during overcharging, this sign is changed.

The positive electrode, the **$PbO_2/PbSO_4$ electrode**, is sometimes used as a reference electrode. Usually a small positive electrode is inserted as an auxiliary electrode into the cell. Gradual self-discharge is the main disadvantage that requires special precautions such as regular recharge. Furthermore, when the electrode has been charged, it can take some days, until the stable potential is reached again, due to small deviations from the stoichiometry just after overcharging ($PbO_{(2+\delta)}$).

For accurate and continuous potential measurements also in valve-regulated batteries, the **mercury/mercurous sulphate reference electrode** is applied. The potential of this electrode is based on a true equilibrium reaction. It depends to some degree on acid concentration as shown in Fig. 2.3. Usually, it consists of a small glass cylinder with a platinum glass-seal contact. This contact is covered by mercury that again is covered by mercurous sulphate (Hg_2SO_4) mixed with mercury to prevent conversion into $HgSO_4$ (disproportionation). The whole is enclosed by a further glass cylinder, filled with dilute sulphuric acid, and equipped with a porous disc that allows the flow of ionic current but largely prevents mixture between the interior and exterior electrolyte. (Different acid concentrations across that connection generate small deviations, because a diffusion potential is generated.)

When this reference electrode is filled with acid in the usual range of density applied for batteries (1.24 to 1.30 g/cm^3), its potential results between 0.63 and 0.64 V referred to the standard hydrogen electrode.

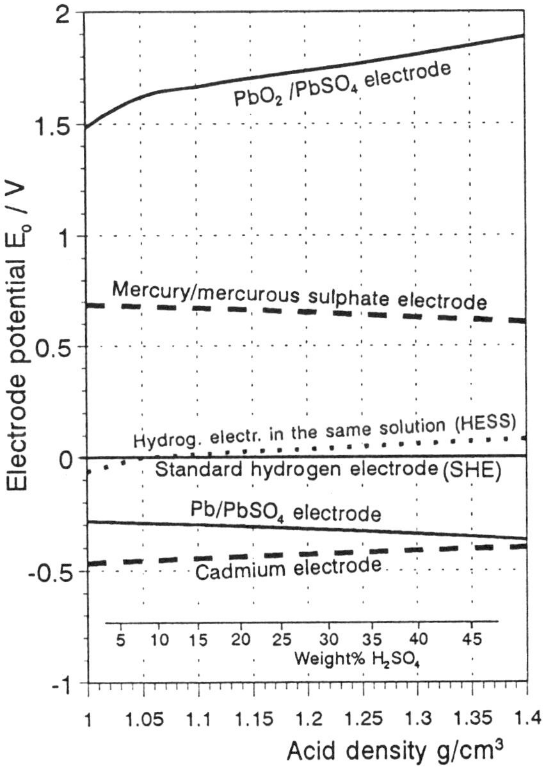

Fig. 2.3 Lead-acid battery: electrode potential versus acid concentration for positive and negative electrodes and the various reference electrodes (20°C).

Table 2.2 shows two examples for the equilibrium potential of the positive and the negative electrodes versus the various reference electrodes, using the above mentioned approximate values for their potentials.

Acid. density	Cell voltage	Potent. vs. SHE Table (4.2)		Cadmium Readings (- 0.42 V)		$Hg/HgS0_4$-Readings (0.63 V)	
		Pos.	Neg.	Pos.	Neg.	Pos.	Neg.
g/cm³	V	V	V	V	V	V	V
1.22	2.070	1.748	-0.322	2.17	+0.098	1.12	-0.952
1.28	2.126	1.790	-0.336	2.21	+0.084	1.16	-0966

Table 2.2 Eqilibrium cell voltage and equilibrium electrode potentials of positive and negative electrodes at 20 °C and their conversion into cadmium and $Hg/HgSO_4$ readings (cf. Table 4.2, p.103).

In Fig. 2.3 the potential of the hydrogen electrode in the same solution is only shown for comparison. It is not used as a reference electrode in lead-acid battery practice, as mentioned above.

2.2.3.3 Reference Electrodes in Nickel/Cadmium Battery Practice

In vented nickel/cadmium batteries, two reference electrodes are usually applied: the cadmium/cadmium hydroxide reference electrode ($Cd/Cd(OH)_2$), which corresponds to the open-circuit potential of the negative electrode itself, and the mercury/mercuric oxide reference electrode (Hg/HgO).

The **cadmium reference electrode** which is identical with the negative electrode of the nickel/cadmium battery, forms a genuine equilibrium according to the reaction equation

$$\boxed{Cd + 2\cdot OH^- \Leftrightarrow Cd(OH)_2 + 2\cdot e^-} \qquad (2.38)$$

Its equilibrium potential referred to the standard-hydrogen electrode (SHE) is

$$\boxed{E_o = E_o^{o,s} - 0.0592 \log a_{OH^-}} \qquad (2.39)$$

with the standard potential

$$\boxed{E_o^{o,s} = - 0.809 \text{ V}} \qquad (2.40)$$

This potential is 20 mV above the potential of the hydrogen electrode and remains so fairly independent of the KOH concentration ([11]).

A serious disadvantage of this reference electrode is its sensitivity to oxidation. For this reason, its usage is limited.

The **mercury/mercuric oxide reference electrode** (Hg/HgO electrode) also forms a stable equilibrium potential in alkaline electrolyte. Mercurous oxide (Hg_2O) does not exist as a chemical compound, so there are no disturbing effects due to a variable valence. The construction and design parameters are described in detail by Falk and Salkind ([12]). Fig. 2.4 shows the potential versus standard hydrogen electrode.

The potential of the Hg/HgO electrode depends in the same way on the pH value as does the hydrogen electrode: it decreases by 0.0592 V per pH value increase. The broken line in Fig. 2.4 shows this decay and deviations that are observed at higher KOH concentrations. These deviations, however, are small, and in a fairly good approximation the potential of a hydrogen electrode in the same solution (HESS) is always 0.93 V below the Hg/HgO potential, while that of the oxygen electrode is 0.3 V above the Hg/HgO potential.

Fig. 2.5 shows deviations between Hg/HgO electrode and HESS potential, observed at varying electrolyte concentration and temperature. In this Figure, the potential scale is considerably enlarged and covers only 15 mV. So the deviations are small. Furthermore, measurements reported by different authors concur quite well, which shows the high reliability of the Hg/HgO reference electrode.

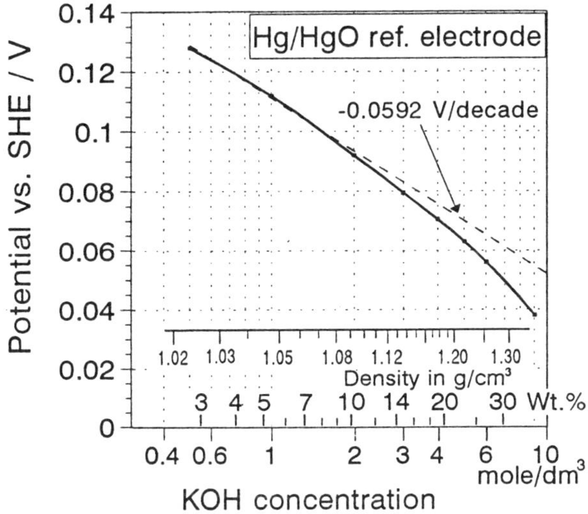

Fig. 2.4 Equilibrium potential of the mercury/mercuric oxide electrode referred to standard hydrogen electrode potential vs electrolyte concentration.

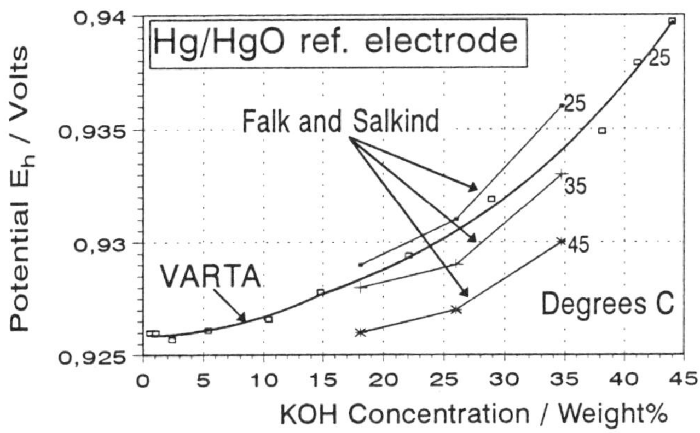

Fig. 2.5 Potential of the Hg/HgO electrode against a hydrogen electrode in the same solution (HESS). Data reported by Falk and Salkind ([12]) and VARTA coincide quite well.

2.2.4 The Reversible Heat Effect

The reversible heat effect, introduced by Eq. (2.4),

$$\boxed{Q_{rev} = T \cdot \Delta S} \tag{2.41}$$

represents the unavoidable heat absorption or heat emission connected with electrochemical reactions. It is related to the thermodynamic (equilibrium) parameters of the concerned reaction, and is strictly connected with the amount of material (in electrochemical equivalents) that reacts. Thus, the reversible heat effect does not depend on discharge or recharge rates. When the cell reaction is reversed, the reversible heat effect is reversed too, which means it gets the opposite sign. Thus, energy loss in one direction means energy gain when the reaction is reversed, and so the effect is 'reversible'.

During discharging or charging of batteries, the reversible heat effect can only be observed at low rates. In general, it is covered up by the larger Joule effect.

ΔS is given per multiple of the cell reaction and usually listed in Joule·K^{-1} = Ws·K^{-1}. So, Q_{rev} is given in Ws. The reversible heat effect per time unit can be related to current flow, because each multiple of the reaction requires the current amount n·F:

$$\boxed{\frac{dQ_{rev}}{dt} = \frac{Q_{rev}}{n \cdot F} \cdot i} \quad / \text{W} \tag{2.42}$$

with n : number of exchanged electrons

F : Faraday constant (= 96485 As/equivalent)

i : current in A

Q_{rev}/nF has the dimension 'V'. So it is equivalent to a voltage, although it is not a voltage that can be measured, but can be used for caloric calculations, as shown below.

2.2.4.1 Calorific Voltage

For caloric evaluations it is convenient to combine E^0 with Q_{rev}/nF. The difference

$$\boxed{E_{cal} = E^0 - \frac{Q_{rev}}{n \cdot F}} \quad / \text{V} \tag{2.43}$$

is called 'calorific voltage' (in German 'Heizwertspannung'). E_{cal} is a hypothetical voltage that includes the reversible heat effect, and is used instead of the equilibrium voltage for caloric calculations (cf. Eq. (2.79).

E_{cal} is equivalent to ΔH, the enthalpy of reaction, as becomes apparent when Eq. (2.6) and Eq. (2.4) are inserted in Eq. (2.43):

$$\boxed{E_{cal} = -\frac{\Delta G}{n \cdot F} - \frac{T \cdot \Delta S}{n \cdot F} = -\frac{\Delta H}{n \cdot F}} \tag{2.44}$$

2.3 Current Flow, Kinetic Parameters

When current flows, the cell reaction must happen at a corresponding rate. For every delivered ampere-second, a corresponding number of electron exchanges must have occurred. This means that at the electrodes the elementary charging or discharging processes, as shown in Eq. (2.2), must have happened $62.4 \cdot 10^{18}/n$ times ($62.4 \cdot 10^{18}$ = the reciprocal of one elemental charge of $1.602 \cdot 10^{-19}$ As per electron). To achieve this current flow, additional energy is required, which intensifies the electron and ion flow in the desired direction. This additional energy finds its expression in **overvoltages**, i.e. deviations from the equilibrium voltage (sometimes denoted as 'irreversible entropy loss' $T \cdot \Delta S_{irr}$). Furthermore, current flow through conducting elements causes **ohmic voltage drops**. Both mean irreversible energy loss and corresponding heat generation, caused by current flow (cf. Section 2.4.1).

As a result, the voltage under current flow E is reduced or increased compared to the equilibrium value E^o. The difference $E^o - E$, when measured as cell-voltage deviation, comprises the overvoltage of the electrochemical reactions and ohmic voltage drops, caused by the electronic as well as the ionic currents in the conducting parts. The sum of all the voltage drops caused by the current flow is called **polarisation**, i.e.

$$\boxed{\text{polarisation} = \text{overvoltage} + \text{ohmic voltage drops}} \qquad (2.45)$$

The quantity determined in practice is always polarisation. Overvoltage can be only separated by special electrochemical methods.

Usually, the reaction path consists of a number of reaction steps that precede or follow the actual charge transfer step, as shown in Fig. 2.6.

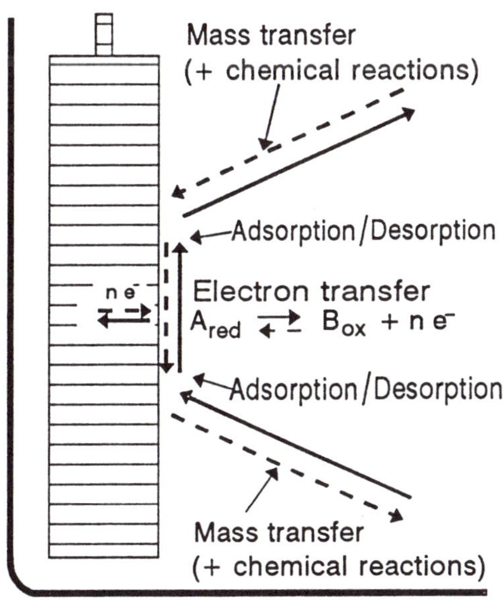

Fig. 2.6 Course of an electrochemical reaction. Charge transfer often can only occur with adsorbed species, then adsorption/desorption steps are included. Furthermore, chemical reactions may precede or follow the electron transfer step.

The rate of each of these reaction steps is determined by **kinetic parameters,** such as exchange-current density, diffusion coefficient or transport numbers. The slowest partial step of this chain is decisive for the rate of the overall reaction. As a consequence, overvoltages, or even limitations of the reaction rate, often are not caused by the electron transfer step itself, but by preceding or following steps.

Transport of the reacting species, as indicated in Fig. 2.6, is achieved by two mechanisms: diffusion and migration.

When only one-dimensional transport is assumed, mass transfer is given:

$$N_j = \frac{i_j}{n \cdot F} = -D_j \frac{\partial c_j}{\partial \xi} + \frac{i \cdot t_j}{z_j \cdot F} \tag{2.46}$$

with: N_j : flux of species j in mole·cm^{-2}
i_j/nF : current equivalent
c_j : concentration of species j in mole·cm^{-3}
$\partial c_j / \partial \xi$: concentration gradient in mole·cm^{-4}
D : diffusion coefficient in cm^2·s^{-1}
t : transference number
z_j : valence number (charges per ion j)
ξ : diffusion direction (cm)

Addend 1 describes transport by diffusion; addend 2 that by migration.

2.3.1 Diffusion

Diffusion depends on the concentration gradient (Fick's first law) (cf. Eq. (2.118), p.50). It always levels out different concentrations. For simplification, often a linear concentration gradient can be assumed as an approximation. Then the first addend in Eq. (2.46) can be written as

$$\frac{i_{j,D}}{n \cdot F} = D_j \frac{c_j - c_{j,o}}{d} \quad \text{or} \quad i_{j,D} = \frac{n \cdot F \cdot D_j}{d}(c_j - c_{j,o}) \tag{2.47}$$

with $c_j - c_{j,o}$: (negative) concentration gradient of species j
d : Thickness of the diffusion layer

So the diffusion current is proportional to the concentration gradient in the diffusion layer, divided by the thickness of this layer.

Diffusion can limit the reaction rate when the mass transport by diffusion precedes the electron transfer step. If the concentration c_j attains the value zero at the electrode surface, all the arriving ions or molecules are charged or discharged immediately, and further increase of the current by increased overvoltage is not possible (cf. Fig. 2.8). Thus the current cannot exceed a certain value, the **limiting current**, that can be derived from Eq. (2.47) with $c_j = 0$

$$i_{lim,j} = -D_j \cdot \frac{n \cdot F}{d} \cdot c_{j,o} \tag{2.48}$$

Limiting currents, caused by diffusion, play an important role in connection with the internal oxygen cycle (cf. Chapter 8).

2.3.2 Migration, Transference Numbers

Addend 2 in Eq. (2.46), describes the movement of ions under the influence of an electric field, which means transport by the electric current and is called **Migration**. Migration causes concentration gradients. The **transference number** t in Eq. (2.46) denotes the share of the total current that is carried by the corresponding ionic species. In a binary electrolyte, dissociated into A^+ and B^-, the transference numbers are connected by the relation

$$\boxed{t_+ + t_- = 1} \tag{2.49}$$

Transference numbers depend on concentration of the ion and on temperature. In binary-salt solutions they are fairly close to 0.5, which means that both ion species share more or less the same ion conductivity. Larger deviations are observed in acids and bases on account of the much higher ion mobility of H^+ and OH^- ions. So the values for the battery electrolytes sulphuric acid (dissociated into H^+ and HSO_4^-) and potassium hydroxide are

Sulphuric acid	$t_+ = t_{H^+} = 0.9$	$t_- = t_{SO_4^-} = 0.$
Potassium hydroxide	$t_+ = t_{K^+} = 0.22$	$t_- = t_{OH^-} = 0.78$

Table 2.3 Transference numbers. For sulphuric acid, given in ([13]) for diluted solutions, but also true for concentrations used in batteries, according to experiments and calculations. For potassium hydroxide true for a wide concentration range ([14]).

So, in sulphuric acid 90% of ionic current are carried by protons (H^+) and only 10% by the negative ion (HSO_4^-), while in potassium hydroxide about 78% of the current is carried by OH^- ions.

The transference numbers can be referred to the corresponding equivalent conductivities:

$$\boxed{t_+ = 1 - t_- = \frac{\Lambda_+}{\Lambda_+ + \Lambda_-}} \tag{2.50}$$

Λ denotes the equivalent conductivity ($\Omega^{1} \cdot cm^2$ per equivalent).

2.3.3 Electron Transfer

The electron transfer reaction denotes the central reaction step where electric charge is exchanged (cf. Fig. 2.6). Current flow affords additional forces because of an energy barrier that has to be surmounted by electrons. The required additional energy is called '**activation energy**' and the dependence of reaction rates is expressed by the **Arrhenius equation** that reads:

$$k = k_o \cdot \exp\left(-\frac{E_A}{R \cdot T}\right) \tag{2.51}$$

with: k: Reaction constant

E_A: Activation energy ($J \cdot mole^{-1}$)

As an approximation, E_A often is treated like a constant. Actually it depends on temperature.

In electrode reactions, n·E·F is the driving force, and the corresponding relation is:

$$i = k' \cdot c_j \cdot \exp\left(\frac{n \cdot F}{R \cdot T} E\right) \tag{2.52}$$

k' includes the 'equivalence factor' n·F, E is the electrode potential, and c_j the concentration of the reacting substances that releases or absorbs electrons.

Electron transfer, however, does not occur in only one direction: the reverse reaction is possible as well, and the balance between both depends on electrode potential. Thus, Eq. (2.52) has to be completed:

$$i = k_+ \cdot c_{red} \cdot \exp\left(\frac{\alpha \cdot n \cdot F}{R \cdot T} E\right) - k_- \cdot c_{ox} \cdot \exp\left(-\frac{(1-\alpha) \cdot n \cdot F}{R \cdot T} E\right) \tag{2.53}$$

where addend 1 describes the anodic reaction (e.g. $Pb \Rightarrow Pb^{2+}$), addend 2 its reversal. α denotes the transference factor (usually ≈ 0.5) that denotes how 'symmetrically' the reaction and its reversal depend on electrode potential (difference in activation energies). n is the number of charges, and c_{red}, c_{ox} are reduced and oxidised states of the reactants. The resulting current $i = i_+ + i_-$ is represented in Fig. 2.7 by the solid curve as the combination of anodic and cathodic current/voltage curves.

Such an electron transfer also occurs during equilibrium at equal rates in both directions as a **dynamic equilibrium**. The **Equilibrium voltage** (E^o) is determined by the point at which the forward and reverse reactions are equally fast. For the lead electrode, described by Eq. (2.2), dynamic equilibrium means that lead atoms (Pb) are transformed into Pb^{2+} ions at the same rate as the reverse reaction occurs, although no current flow is observed through the electrode, and the Equation must be written with a double arrow:

$$Pb \Leftrightarrow Pb^{2+} + 2 \cdot e^- \tag{2.54}$$

At a dynamic equilibrium, current flow occurs in both directions, but neither appears externally nor causes changes in the electrode. *(A similar situation is observed with the mixed potential, but this does cause gradual changes and therefore is no equilibrium, cf. Section 2.3.5).*

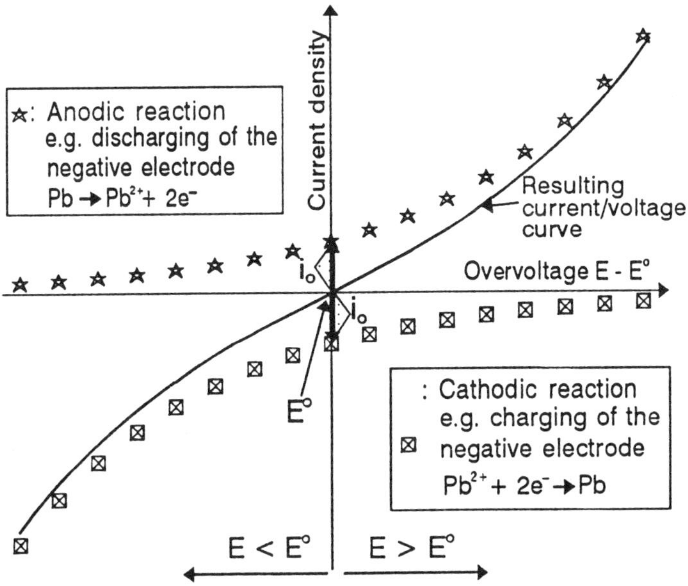

Fig. 2.7 The current/voltage curve. The horizontal axis represents the electrode potential related to the equilibrium potential E^o; the vertical axis the current density, which is synonymous to the reaction rate. The left part of Eq. (2.2) is inserted as an example.

According to Eq. (2.53) anodic and cathodic currents are:

$$
\begin{aligned}
i_+ &= k_+ \cdot c_{red} \cdot \exp\left(\frac{\alpha \cdot n \cdot F}{R \cdot T} E\right) \\
i_- &= -k_- \cdot c_{ox} \cdot \exp\left(-\frac{(1-\alpha) \cdot n \cdot F}{R \cdot T} E\right)
\end{aligned}
\tag{2.55}
$$

At the equilibrium potential $E = E^o$, the current flow in both directions is balanced $i_+(0) = -i_-(0) = i_0$; consequently

$$
\begin{aligned}
i_0 &= i_+(0) = k_+ \cdot c_{red} \cdot \exp\left(\frac{\alpha \cdot n \cdot F}{R \cdot T} E^o\right) \\
-i_0 &= i_-(0) = -k_- \cdot c_{ox} \cdot \exp\left(-\frac{(1-\alpha) \cdot n \cdot F}{R \cdot T} E^o\right)
\end{aligned}
\tag{2.56}
$$

where i_o denotes the **exchange-current density** that characterises the dynamic equilibrium. It is an important kinetic parameter. High exchange-current density means that the equilibrium potential is rather stable, while low exchange-current density indicates that the electrode potential will be polarised (shifted), even when quite small current densities flow through the electrode. In the lead electrode, for example, the exchange-current density of the charge reaction is of the order of 10^{-5} A/cm^2, while the exchange-current density for hydrogen evolution is of the order of 10^{-13} A/cm^2 (cf.([15])).

Eqs. 2.48 and 2.49 can be combined by division

$$\frac{i_+}{i_+(0)} = \cdot \exp\left(\frac{\alpha \cdot n \cdot F}{R \cdot T} \cdot (E - E^o)\right) = \frac{i_+}{i_0}$$
$$\frac{i_-}{i_-(0)} = \cdot \exp\left(-\frac{(1-\alpha) \cdot n \cdot F}{R \cdot T} \cdot (E - E^o)\right) = \frac{i_-}{-i_0} \tag{2.57}$$

with $i = i_+ + i_-$ follows the important current/voltage relation

$$i = i_o\left[\exp\left(\frac{\alpha \cdot n \cdot F}{R \cdot T}(E - E^o)\right) - \exp\left(-\frac{(1-\alpha) \cdot n \cdot F}{R \cdot T}(E - E^o)\right)\right] \tag{2.58}$$

that describes the **current/voltage** or **current/potential curve**, as shown in Fig. 2.7.

The voltage deviation from equilibrium E - E^o is called **overvoltage** (in this case **transference overvoltage**). The exponential relation between current and voltage means that the current increase might be enormous when the overvoltage exceeds certain values.

Overvoltage is often denoted with the letter η which leads to the usual form of Eq. (2.58):

$$i = i_o\left[\exp\left(\frac{\alpha \cdot n \cdot F}{R \cdot T}\eta\right) - \exp\left(-\frac{(1-\alpha) \cdot n \cdot F}{R \cdot T}\eta\right)\right] \tag{2.59}$$

2.3.3.1 TAFEL Lines

If the potential is shifted far enough from the equilibrium value, the reverse reaction can be neglected in Eq. (2.59). So the resulting current/voltage curve in Fig. 2.7 becomes a simple exponential function

$$i = i_o \cdot \exp\left[\frac{\alpha \cdot n \cdot F}{R \cdot T} \cdot (E - E^o)\right] \quad \text{or} \quad i = i_o \cdot \exp\left[\frac{\alpha \cdot n \cdot F}{R \cdot T} \cdot \eta\right] \tag{2.60}$$

This equation can be re-arranged into

$$\eta = \frac{R \cdot T}{\alpha \cdot n \cdot F} \cdot \ln(|i|) - \frac{R \cdot T}{\alpha \cdot n \cdot F} \cdot \ln(|i_0|) \tag{2.61}$$

Eq. (2.61) can be written in a form known as the 'TAFEL equation'. (*J.Tafel was the first to describe this relation in connection with hydrogen overvoltage measurements on noble metals ([16]).*)

$$\eta = a + b \cdot \log(|i|) \tag{2.62}$$

The curves represented by Eq. (2.62) can be linearised when plotted semilogarithmically (cf. Fig. 2.8) and the resulting linear sections of the current/voltage curve are frequently called the **TAFEL lines**. The constant b represents the slope of the TAFEL line and means the potential difference that causes a current increase of one decade (cf. Section 7.2.1). TAFEL lines are important tools when reactions are considered that occur at high overvoltages. Gassing reactions or grid corrosion in lead-acid batteries are such examples.

Fig. 2.8 shows current/voltage curves plotted semilogarithmically, i.e. the current on a logarithmic scale, the voltage (the electrode potential) on a linear scale. In this way, the current rise, which frequently covers several orders of magnitude, can be shown better. A further advantage is that linear sections of the curves, the TAFEL lines, are formed which easily can be handled for quantitative evaluations (cf. Section 7.2.1).

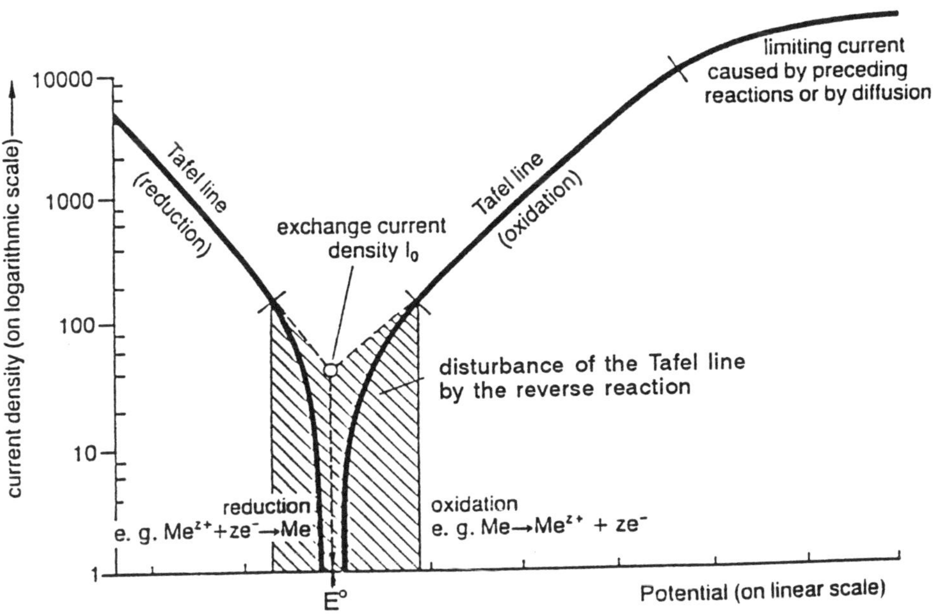

Fig. 2.8 Current/voltage curve on a semilogarithmic scale. Because logarithms of negative numbers are not possible, the reverse reaction must be plotted in absolute values of the the current, as in the left part.

Two sections are observed that are linearised in the semilogarithmic plot; these are the mentioned TAFEL lines. The exchange-current density i_0 is represented by the intersection of the extrapolated TAFEL lines. The equilibrium potential is the projection of this intersection point onto the horizontal axis (the potential axis).

Close to the equilibrium value, forward and reverse reactions have reaction orders of the same order of magnitude. This causes the deviation from straight lines, and the consequence that the simplification of Eq. (2.59) is not admissible within this voltage range.

On the right hand in Fig. 2.8, the curve finally bends into a horizontal line. The transport rate of the reacting substance to the electrode then determines a **limiting-current density**, which cannot be exceeded with overvoltage increase.

2.3.4 Concentration Overvoltage

Electrochemical reactions, chemical reactions as well as transport processes that precede or follow the charge/discharge step, lead to changes in the concentration of the reacting substances at the electrode surface and thereby may change the current/voltage curves. The overall voltage deviation from the equilibrium voltage is the sum of these partial overvoltages.

The influence of concentration is given by the Nernst equation (e.g. Eqs. 2.14 and 2.15). Division of the Nernst equations concerning the actual activity a_j and the equilibrium value a_j^o of a reacting species j, results in the **concentration overvoltage**

$$E - E^o = \eta_{conc.} = \frac{R \cdot T}{n \cdot F} \ln\left(\frac{a_j}{a_j^o}\right) \tag{2.63}$$

Concentration overvoltage often is caused by a number of reaction steps.

2.3.4.1 Limiting Currents

When diffusion processes or reactions precede the transfer step, it can happen that the concentration of the reacting species falls to zero, and causes a limiting current, as has been mentioned in Section 2.3. Then the current can no longer be increased by potential increase, but remains constant as indicated by the horizontal section of the current/voltage line, in the right hand part of Fig. 2.8.

If diffusion of the reacting species to the electrode surface is the slowest partial step, the overvoltage current relation can be deduced. Division of Eq. (2.47) by Eq. (2.48) results in

$$\frac{i_j}{i_{lim,j}} = -\frac{c_j - c_{j,o}}{c_{j,o}} = 1 - \frac{c_j}{c_{j,o}} \quad \text{or} \quad \frac{c_j}{c_{j,o}} = 1 - \frac{i_j}{i_{lim,j}} \tag{2.64}$$

Insertion into Eq. (2.63) gives the diffusion overvoltage:

$$(E - E^o)_{diff} = \eta_{diff} = \frac{R \cdot T}{n \cdot F} \cdot \ln\left(1 - \frac{i}{i_{lim}}\right) \tag{2.65}$$

with i_{lim}= limiting-current density

As long as $i_{lim} >> i$ (supply of reactants is fast compared to the transfer rate), the value of Eq. (2.65) is small and its contribution to the overall overvoltage might be negligible. But when i approaches the limiting current i_{lim}, the increasing diffusion overvoltage dominates the overall situation and the current increase is gradually reduced to zero even though the electrode potential continues to increase. The reason is that processes are now rate-limiting that are not accelerated by electrode potential.

2.3.5 Mixed Potentials

A **mixed potential** is characterised by two electrochemical reactions that occur simultaneously at one electrode and compensate each other. So, it appears like an equilibrium, although it is not. Fig. 2.9 shows the situation for the self-discharge of the lead electrode that forms a couple with hydrogen evolution.

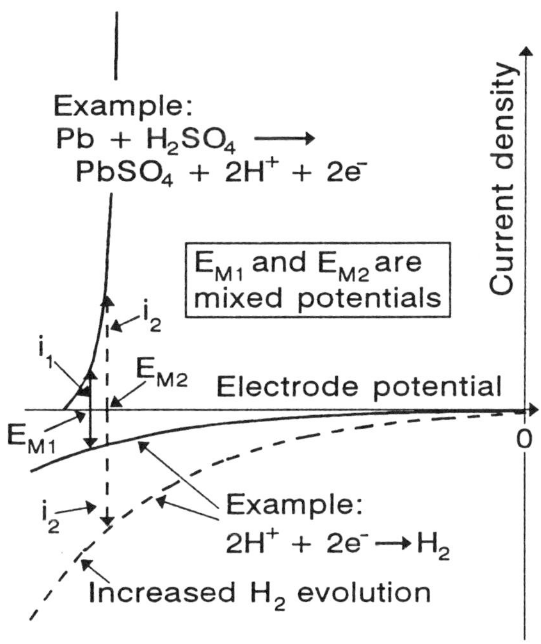

Fig. 2.9 Open-circuit potential as a mixed potential between discharge of the lead electrode and hydrogen evolution with the result of self-discharge. E_M: mixed potential. i_1, i_2: rate of self-discharge. The dashed curve represents reduced hydrogen overvoltage.

The fairly steep current/voltage curve pointed upwards represents the discharge of the lead electrode. The fast rise of this curve indicates the low overvoltage that characterises this reaction. High discharge rates can be achieved at low overvoltage or polarisation values. The high overvoltage, marking the hydrogen-evolution reaction, is expressed by the gradual rise of the corresponding current curve. At open-circuit voltage, discharge reaction and hydrogen evolution have to balance each other, because no current flows through the electrode. The result of this balance is the mixed potential E_M in Fig. 2.9. As for an equilibrium potential, no external current appears, although electrochemical reactions occur. But the mixed potential is not an equilibrium potential, because the couple is formed by two different reactions that cannot form a genuine equilibrium. Rather, a continuous change of the active material in the electrode is associated with the mixed potential, namely the gradual conversion of metallic lead into lead sulphate.

The position of the mixed potential is largely determined by the faster of the two reactions, while the rate of the reaction, the self-discharge in this example, is determined by the slower reaction. This is illustrated by comparison of the continuous and broken curves in Fig. 2.9. The continuous curve represents a fairly high hydrogen overvoltage; the broken curve depicts the case when hydrogen can be evolved more easily. The position of E_M is only slightly changed between the two examples, but the rate of the self-discharge, the current i_2, grows to a multiple of i_1 in the second case.

On account of such coupling with secondary reactions, a number of battery electrodes form mixed potentials at open-circuit.

2.3.6 Temperature Influence

The kinetic parameters depend on temperature as do the rates of chemical reactions. This dependence is described by the **ARRHENIUS equation**, which already has been introduced in Section 2.3.3 in connection with the term 'activation energy':

$$\boxed{k = k_o \cdot \exp\left(-\frac{E_A}{R \cdot T}\right)} \tag{2.66}$$

with: k: Reaction constant

E_A: Activation energy

R : molar gas constant (8.3143 J·mole^{-1}·K^{-1})

E_A actually depends on temperature, but often can be treated like a constant.

The logarithmic form of Eq (2.66) reads

$$\boxed{\ln(k) = -\frac{E_A}{R \cdot T} + \ln(k_o)} \quad \text{or} \quad \boxed{\ln(k) = -\frac{E_A}{R} \cdot \frac{1}{T} + \ln(k_o)} \tag{2.67}$$

On account of this relation, the temperature dependence of kinetic parameters can often be linearised, when the logarithm of the reaction rate is plotted against 1/T, which is often called an **ARRHENIUS plot** (for examples, cf. ([17])).

Very often, the approximation holds true that a temperature increase of 10 K (ten degrees Centigrade) doubles the reaction rate. In electrochemical reactions, this means that the equivalent currents are doubled, which denotes a quite strong temperature dependence. A temperature increase of 20 K means a current increase by a factor of 4; a rise in temperature of 30 K corresponds to a factor of 8.

This relation can be expressed by

$$\frac{k(T+\Delta T)}{k(T)} = 2^{(\Delta T/10)} \tag{2.68}$$

with k: reaction rate (mole/sec) which might be expressed as a current
T: temperature in K

It can be established when the quotient of $k_1(T_1)$ and $k_2(T_2)$ is formed with the aid of Eq. (2.67), and gives:

$$\ln(k_1 / k_2) = \frac{E_A}{R} \cdot \left(\frac{1}{T_2} - \frac{1}{T_1} \right) \tag{2.69}$$

If $k_1 = k(T + \Delta T)$ and $k_2 = k(T)$ are inserted, and Eq. (2.69) is combined with Eq. (2.68) the result is:

$$\ln(k_{(t+\Delta T)} / k_T) = \frac{E_A}{R} \cdot \left(\frac{1}{T_T} - \frac{1}{T_{T+\Delta T}} \right) = \frac{\Delta T}{10} \cdot \ln(2) \tag{2.70}$$

or

$$\ln(2) = \frac{E_A}{R} \cdot \frac{10}{T \cdot (T+10)} \tag{2.71}$$

This means, for normal room temperature (T = 300 K,) that the activation energy E_A must be in the range of 50 kJ/mole (or 10 kcal/mole) which, as a matter of experience, is approximated for most reactions. But this relation also indicates that E_A depends on temperature, and that Eq. (2.68) only can represent a rough approximation.

2.4 Heat Effects

The reversible heat effect, connected with the thermodynamic data, was described in Section 2.2.4. It is strictly linked to the amount of converted material, and represents loss or gain of energy, because it is reversible. Current flow causes also heat effects, but they always mean energy loss (they are irreversible), and the generated heat corresponds to the energy gap that is caused by deviation from equilibrium. In practice, heat effects by current flow usually exceed the reversible heat effect by far, especially at high current rates, because then the deviation from equilibrium increases, but also when secondary reactions, like water decomposition or the internal oxygen cycle, occur that generate much heat (cf. Sections 4.5.2, 5.4.2, and 6.6.2).

2.4.1 Current-Related Heat Effects (The Joule Effect)

Current flow through any conducting object generates heat proportional to the voltage drop caused by the current itself according to:

$$\boxed{dQ_{Joule}/dt = \Delta E \cdot i} \qquad (2.72)$$

with Q_j: generated heat (Joule effect) [J]
t: time [s]
ΔE: voltage drop caused by the current [V]
i: current [A]

This heat is called the **Joule effect**; it always means loss of energy.
Note: *Strictly speaking, the negative absolute value should be used in Eq. (2.72) for consistency with the arithmetical sign of the thermodynamic parameters (lost energy always has the negative sign).*

In an electrochemical cell, the voltage drop caused by the current is represented by the difference between the cell voltage under current flow (E) and the open-circuit cell voltage (E^o). Then the Joule effect reads according to Eq. (2.72):

$$\boxed{dQ_{Joule}/dt = (E - E^o) \cdot i} \qquad (2.73)$$

Note: *Strictly speaking Eq. (2.73) describes a composite parameter of*

- *the 'genuine' Joule effect $dQ/dt = i^2 \cdot R$, which describes the heat generation by current flow (i) through the ohmic resistance (R) of the conducting elements and the electrolyte,*

and

- *the irreversible heat effect $T \cdot \Delta S_{irreversible}$, which causes voltage drops between the electrode surface and the electrolyte on account of reaction hindrance. This additional heat is caused by an additional entropy of reaction ($\Delta S_{irreversible}$) which cannot be recouped when the reaction is reversed.*

Both effects, however, cannot be separated in electrochemical cells in normal conditions. For this reason, the sum of both is usually called the Joule effect, as in this book.

The difference $E - E^o$, when measured as cell-voltage deviation, comprises the overvoltage and ohmic voltage drops, corresponding to Eq. (2.45) in Section 2.3:

$$\boxed{\text{polarisation} = \text{overvoltage} + \text{ohmic voltage drops}}$$

For heat effects it does not matter that in practice polarisation is always determined, since **heat generation is proportional to polarisation**.

The Joule effect can be described as the energy generated per unit of time:

$$\boxed{\frac{dQ_{Joule}}{dt} = (E - E^o) \cdot i} \quad / \text{W} \qquad (2.74)$$

or in its integrated form for a period (t in hours)

$$\boxed{Q_{Joule} = \int_0^t \left\{ \left(E - E^o \right) \cdot i \right\} dt} \quad / \text{Wh} \qquad (2.75)$$

$E - E^o$ does not remain constant during charging or discharging, because it is related to the internal resistance, that usually increases with proceeding discharge, because of the often observed lower conductivity of the discharged products. In lead-acid batteries, conductivity of the electrolyte decreases with discharge, and acid dilution influences E^o (cf. Fig. 4.2, p.101).

2.4.2 Heat Generation in Total

Summation of the Joule effect and the reversible heat effect (Eq. (2.41)) gives the total heat generated in the cell or the battery, which means

$$\boxed{Q_{total} = Q_{Joule} + Q_{rev}} \qquad / \text{ Wh} \tag{2.76}$$

as energy, e.g. Wh, or as work per time unit:

$$\boxed{\frac{dQ_{total}}{dt} = \frac{dQ_{Joule}}{dt} + \frac{dQ_{rev}}{dt}} \qquad / \text{ W} \tag{2.77}$$

Depending on the sign of dQ_{rev}/dt, the total energy generation may be larger or smaller than the Joule effect.

According to Eq. (2.74), dQ_{Joule}/dt can be substituted by $(E - E^o)\cdot i$ and gives:

$$\boxed{\frac{dQ_{total}}{dt} = (E - E^o)\cdot i + \frac{dQ_{rev}}{dt}} \tag{2.78}$$

Substitution of dQ_{rev}/dt through Eq. (2.42)and application of Eq. (2.43) results in the simple relation that is convenient for heat calculations:

$$\boxed{\frac{dQ_{total}}{dt} = \left(E - E_{cal}\right)\cdot i} \tag{2.79}$$

Strictly speaking, Eq. (2.79) should have the negative sign according to thermodynamic parameters, since the Joule effect is always lost energy, as mentioned in connection with Eq. (2.72). To overcome this difficulty, $Q_{gen} = -Q_{total}$ *is introduced in Section 2.5.1.1.*

2.5 Heating of the Battery

While a battery is being charged or discharged, the heat generation caused by the flowing current raises the temperature until balance is achieved between heat generation in the cell and heat dissipation to the environment (cf. Fig. 2.10). This balance depends on both the rate of heat generation and the possibility of heat dissipation.

2.5.1 Heating Effects, Heat Capacity

The changes of battery temperature during discharge (dT/dt) can be estimated by heat generation values, as compiled in Tables 4.9 and 4.10 (p. 122 and 123, respectively):

$$\frac{dT}{dt} = \frac{1}{C_{Batt.}} \cdot \left(\frac{dQ_{gen}}{dt} - \frac{dQ_{diss}}{dt} \right) \tag{2.80}$$

with dQ_{gen}/dt : generated energy per unit of time

dQ_{diss}/dt : dissipated energy per unit of time

(Here, Q_{gen} is positive, when energy is generated. Thus $Q_{gen} = - Q_{total}$ in Tables 4.9 and 4.10)

and the heat capacity of the battery $C_{Batt.}$ (in kJ/K) defined by

$$\sum \left[m(i) \cdot C_p(i) \right] = C_{Batt.} \tag{2.81}$$

with m(i): component i in kg

(i): the components in the battery

$C_p(i)$: specific heat of component (i) in kJ/(kg·K)

The **heat capacity** C_{batt} - the sum $\Sigma[m(i) \cdot C_p(i)]$ in $J \cdot K^{-1}$ - of batteries with aqueous electrolyte is mainly determined by the electrolyte content, because the heat capacity of the aqueous electrolytes is large, compared to the heat capacity of the other cell components (cf. Tables 4.5 and 5.5, p.101 and 186, respectively).

Heat effects are more significant in maintenance-free batteries compared to the vented design for two reasons:

- $m(i) \cdot C_p(i))$ in J/K is comparatively small on account of the limited electrolyte content,
- all the electrical energy employed in the internal oxygen cycle is converted into heat within the battery (cf. Fig. 4.13, p.125).

2.5.1.1 Adiabatic Heat Effects

A fairly simple approximation can be derived on the assumption that heat dissipation does not occur and all the generated heat remains within the battery. This is approximated to by a high-rate discharge, where the discharging period is too short for noticeable heat dissipation. Adiabatic heating can also occur (to a certain extent) in batteries in thermally insulated packages or housings.

Adiabatic heating means that $dQ_{diss.}$ in Eq. (2.80) becomes zero, and the heating rate is:

$$\frac{dT}{dt} = \frac{1}{C_{Batt.}} \cdot \frac{dQ_{gen}}{dt} \quad \text{in °C/time unit,} \tag{2.82}$$

with ΔT : Temperature increase during charge or discharge in °C

Q_{gen} : Energy generation in total ($Q_{gen} = - Q_{total}$)

$C_{batt.}$: Heat capacity according to Eq. (2.81)

or, when the integrated form is used

$$\boxed{\Delta T = \frac{1}{C_{Batt.}} \cdot Q_{gen}} \tag{2.83}$$

2.5.2 Heat Dissipation

Heat is carried away from the battery as soon as its temperature exceeds the temperature of the environment.

The heat emission is split between heat radiation and heat transport by air convection mainly at the walls of the battery container. The amount of heat carried away by these processes is increased with a growing difference between the temperature of the wall and the temperature of the environment. On account of limited thermal conductivity of the container material, the temperature inside the battery can considerably exceed that of the outer surface, especially when heat is generated at a high rate. Under normal operational conditions, however, this temperature difference is small.

The heat emission depends on

- heat transport through the container wall to the outer surface;
- heat loss by radiation;
- heat loss caused by flow of the air around the battery.

Fig. 2.10 illustrates the heat flow out of the battery.

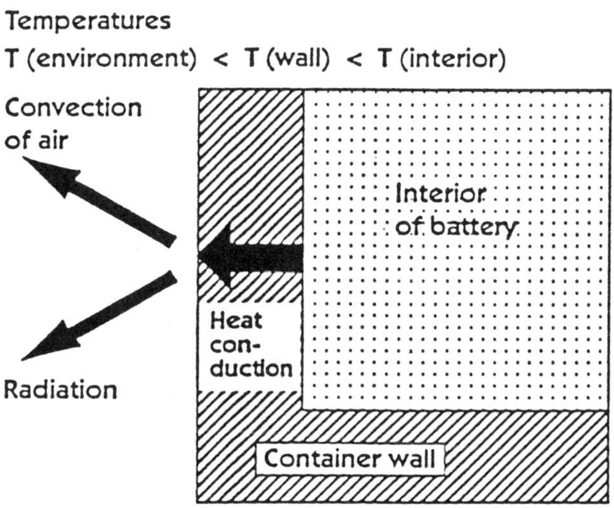

Fig. 2.10 Heat flow out of the battery.

As mentioned above, the temperature drop across the container wall usually is small compared to the temperature difference between container surface and the environment (cf. Figs. 2.11 and 2.12). For this reason, the surface temperature of the container can be used for the surveillance of the battery temperature. Furthermore, cooling through the side walls is quite effective, especially by means of circulated water (cf. Fig. 10.32, p.352).

Heat dissipation (Q_{diss}) increases with a growing temperature difference between the battery and its surrounding atmosphere, as will be shown below. A stable situation is reached at a certain ΔT_{bal}, when heat generation is balanced by heat dissipation:

$$\boxed{\frac{dQ_{gen}}{dt} = \frac{dQ_{diss}}{dt}} \tag{2.84}$$

2.5.2.1 Heat Flow Through Container Walls

Heat flow through the container wall is determined by heat conductivity and thickness of the container material.

Material	Heat conductance $J \cdot s^{-1} \cdot m^{-1} \cdot K^{-1}$
SAN	0.174
Polystyrene	0.15 - 0.16
PVC	0.16
Polypropylene (PP)	0.22 *
PP, reinforced	0.4
Polyamide	0.2 - 0.3
Glass	1.0 - 1.3
Iron	80
Nickel	81

Table 2.4 Heat conductance (λ in Eq. (2.85)) of container materials. * Values of foamed PP are slightly lower than those of the compact version.

The heat flow is described by

$$\boxed{\frac{dQ}{dt} = f \cdot \lambda \cdot \frac{\Delta T}{d}} \tag{2.85}$$

i.e. heat flow (per m^2) is proportional to the surface area f (m^2), the specific heat conductance λ ($J \cdot s^{-1} \cdot m^{-1} \cdot K^{-1}$) and the temperature difference ΔT across the wall. Heat flow is reduced with increasing thickness d of the container wall.

Table 2.2 shows that heat conductance is similar for most plastics materials. But the heat conductance of metals, as used for the housings of sealed nickel/cadmium batteries and nickel/hydride batteries, is about three orders of magnitude higher than that of plastics materials. This means that the temperature drop across the container wall can be neglected for such batteries.

2.5.2.2 Heat Emission by Radiation

Usually, the major part of the heat generated in the battery is emitted by radiation. According to the law of Stefan-Boltzmann:

$$dQ/dt = \varepsilon \cdot \sigma \cdot T^4 \qquad (2.86)$$

The heat-flow density dQ/dt (i.e. the heat per unit area and time) transported by radiation is proportional to the fourth power of the absolute temperature T (in K). ε is the Stefan-Boltzmann constant ($5.67 \cdot 10^{-8}$ $W \cdot m^{-2} \cdot K^{-4}$), and σ is the emission ratio of the material with respect to an ideal emitter, the so called black body (= ca. 0.95 for usual container materials).

The law of Stefan-Boltzmann means that any body emits heat according to the fourth power of its temperature while it receives heat according to the fourth power of the temperature of the surroundings. There is no net heat flow between bodies having the same temperature. Heat flow always happens from the warmer to the colder body. The net heat-flow density (by radiation) between two bodies A, B is

$$dQ/dt = \varepsilon \cdot \sigma \cdot (\, T(A)^4 - T(B)^4 \,) \qquad (2.87)$$

This is also true when one of the 'bodies' is the surroundings.

The curve 'Radiation' in Fig. 2.11 is calculated with Eq. (2.87) and 293 K assumed for T(B), i.e. surroundings at room temperature. It shows that heat dissipation by radiation can be approximated for small temperature differences from environmental temperature by:

$$dQ/dt = 5 - 6 \quad W \cdot m^{-2} \cdot K^{-1} \qquad (2.88)$$

which means that a battery emits by radiation about 5-6 watts per m^2 of container surface for each K (or °C) of difference between container-surface and (lower) environmental temperature.

Radiation is quite effective in regard to emission of heat (cf. Fig. 2.11). So, a battery should be installed in a way that restrictions for radiation do not exist. If the battery wall is facing another surface, the temperature of this surface is relevant. Consequently, no net heat flow occurs between the facing walls of battery blocks having the same temperature. If there is a hot unit nearby, e.g. a rectifier, it will heat up the battery by radiation.

2.5.2.3 Heat Dissipation by Air Convection

Heat transfer by flow of air is complex, and empirical rules only describe the heat transport by air convection.

In free convection of air at the outer vertical surfaces of stationary batteries of the usual size, the approximations ([18]):

$$\boxed{dQ/dt = 1.4 \cdot (T_{wall}-T_{air})^{4/3} \quad W \cdot m^{-2}} \qquad \text{if } h^3 \cdot (T_{wall}-T_{air}) > 0.2\ m^3\ deg \tag{2.89}$$

(tall cells and larger ΔT)

or

$$\boxed{dQ/dt = 1.3\ h^{-1/5} \cdot (T_{wall}-T_{air})^{5/4} \quad W \cdot m^{-2}} \qquad \text{if } h^3 \cdot (T_{wall}-T_{air}) < 0.2\ m^3\ deg \tag{2.90}$$

(low profile cells and small ΔT)

can be used. The temperatures T_{wall} and T_{air} are measured in °C. The height h of the container (in m) is important for the convection flow.

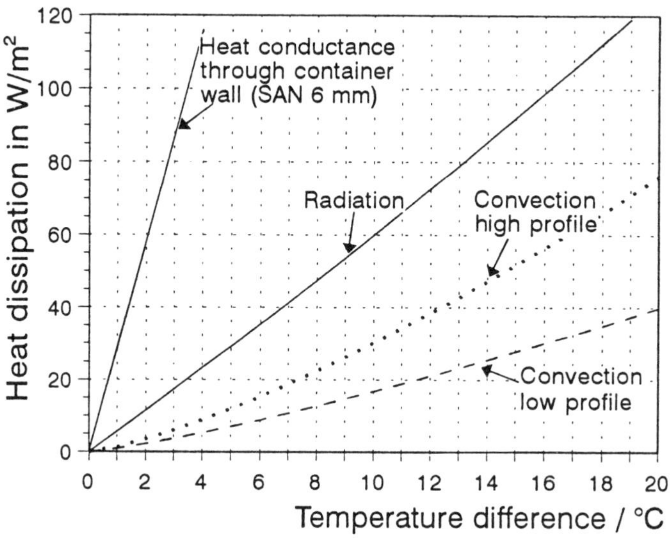

Fig. 2.11 Calculated heat flow. Heat conductance according to Eq. (2.85) with the assumption: d = 6 mm of SAN. Heat radiation according to Eq. (2.81). Heat transport by air convection according to Eqs. 2.80 and 2.81. Assumed battery height 0.2 and 0.6 metres, respectively.

The curves 'Convection' in Fig. 2.11 are the result of Eqs. (2.89) and (2.90) for high and low profile cells of the height 0.6 and 0.2 m respectively. According to the above equations a rough approximation for heat dissipation by air convection is:

$$\boxed{dQ/dt \approx 3 - 4\ W \cdot m^{-2} \cdot K^{-1}} \tag{2.91}$$

for the taller cell design, according to Eq. (2.89), and

$$\boxed{dQ/dt \approx 2 - 3\ W \cdot m^{-2} \cdot K^{-1}} \tag{2.92}$$

for the low profile cell design according to Eq. (2.90).

These figures hold only for free air convection. This means that these relations are true only when neither a flow hindrance nor a forced flow (e.g. in a battery cabinet or by a fan) are influencing air convection. *(Heat transport by air can be increased markedly by strong convection (turbulent flow) ([19]))*. To permit free convection, a minimum distance of 1 cm is required between facing walls. Corresponding spacing of battery blocks should always be observed.

Fig. 2.11 shows that heat transfer by air convection is smaller than heat transfer by radiation if no restrictions exist for either. This indicates the importance of energy dissipation by radiation even at room temperature (which often is underestimated). Consequently, the battery should be installed so as to secure uniform radiation conditions for all cells as far as possible, and heated surfaces in the neighbourhood (e.g. from rectifiers) must be well shielded, otherwise the battery will be heated by radiation from these surfaces. Fig. 2.11 states furthermore, that heat conductance through the container wall occurs much faster than heat transfer away from the container surface by radiation and air convection. So, the temperature drop across the container wall is much smaller than the temperature drop between the container and the air, and the container surface reflects fairly well the temperature of the interior of the cell.

Fig. 2.12 is based on the same calculations as is Fig. 2.11, but shows the temperature drop across the container wall and the total temperature drop versus dissipated energy. The total temperature drop includes both radiation and air convection. The difference between the two formulas, applied for air convection is small, because heat dissipation by air convection is of minor importance as a cooling effect, if not increased by forced air flow.

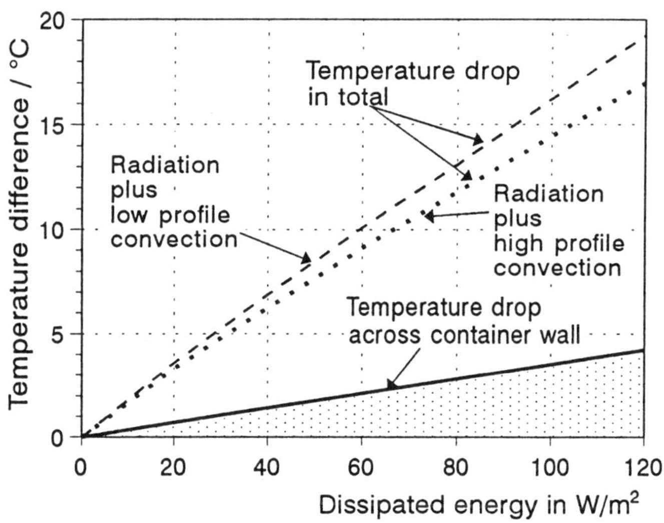

Fig. 2.12 Temperature difference caused by heat dissipation. Based on the same calculation as Fig. 2.11.

Heat dissipation is proportional to the available surface area, which usually means the side walls and cover of the battery, or the corresponding areas of the cell in a spaced single cell. Heat generation, on the other hand, is proportional to the capacity, which corresponds more or less to the volume of the battery. Assuming that the shape of the cell or battery approximates to a cube, heat generation is proportional to the linear dimension raised to the third power, while at the surface heat dissipation increases only in proportion to this dimension to the second power. This means that the ratio between surface area and volume (≈ capacity) decreases with battery size. For this reason, heating effects are more critical in large batteries.

2.5.2.4 Heat Conduction via Bottom and Connectors

In most cases, the heat transfer through the bottom area or through the connecting wires (despite their high thermal conductivity) does not influence the equilibrium temperature of a battery and can be neglected with respect to radiation and air convection. The thermal conductivity of the ground in most cases is low; floors in cabinets are heated from below. Most of the cables connect cells or blocks of similar temperature, which means that there is no net heat transfer between them. Furthermore, the heat transfer of the connectors to the air is reduced by the electrical insulation.

But direct cooling of conducting elements like poles and connectors can be applied with some success, owing to the high heat conductance of metallic parts as described in the following Section.

2.5.3 Forced Cooling

The above described heat dissipation usually is sufficient to control the temperature of stationary batteries. Cabinets with installed batteries sometimes are equipped with fans for forced ventilation to avoid accumulation of hydrogen. Such a forced air flow simultaneously increases heat dissipation by air convection. In traction applications, forced cooling is required when the battery is subject to repeated charging/discharging cycles that do not leave enough time for cooling. Electric powered road vehicles are typical examples. The large batteries in conventional submarines mostly are also equipped with cooling systems ([20]).

Forced cooling is applied for two reasons:

1. Increase of heat dissipation.
2. Equalisation of the temperature within the battery.

The second item often is the more important one to maintain a uniform state of all cells, especially for batteries that contain a great number of cells.

In submarine batteries, often the terminals and bus-bars are formed as thick-walled tubes of copper through which water flows. The considerable high heat conductivity of the grids is used for heat transport (especially when the negative grid contains a copper core, cf. Section 4.8). But also convection of the electrolyte contributes to achieve uniform cell temperature, and in these tall cells, convection of the electrolyte usually is supported by forced flow according to Fig. 7.11 (p.258). Large lead-acid batteries with liquid electrolyte that powered busses ([21]) were equipped with cooling coils which, in combination with forced convection, reduced and equalised the temperature of the cells.

In batteries with immobilised electrolyte, heat transport by convection of the electrolyte is not possible, and, as a consequence, the above methods that cool the top part of the cell are less effective. Most cooling systems for such batteries use heat transport through the side walls of the cells. A rather simple method is forced air flow through adequate spacing of the cells (cf. Fig. 12.4, p.402). Liquid cooling mediums, usually water, mostly stream through pockets of plastic, that are pressed against the side walls of the cells (cf. Fig. 10.32, p.352).

2.6 Aqueous Electrolyte, its Interactions with the Surrounding Atmosphere

The three battery systems treated in this book are based on aqueous electrolyte, a mixture of sulphuric acid (H_2SO_4) or potassium hydroxide (KOH) with water (H_2O). Direct loss of electrolyte by spilling or leakage of the cell container can be neglected because maintenance-free batteries do not contain liquid electrolyte, on account of its immobilisation (cf. Section 4.6.4). For this reason, electrolyte cannot be spilled from intact batteries and, even if the container shows leakage, only a few droplets of electrolyte can leave the cell.

The electrolyte, however, is not only present as a liquid, but:

- Electrochemical decomposition of water generates oxygen and hydrogen (O_2 and H_2) that might have to escape from the cell.
- A certain amount of hydrogen and oxygen is dissolved in the electrolyte.
- The aqueous component evaporates and a certain vapour pressure is always connected with the electrolyte.
- Fumes can be formed and, as an aerosol, carried out of the cell by a gas stream.

So, interaction can exist between the electrolyte and the surrounding atmosphere by gases, water vapour, and fumes. All three mean water loss, that affects performance of the battery when a certain level is exceeded. Another result of such interaction can be oxygen intake that discharges the negative electrode and so reduces the available capacity (cf. Sections 4.6.4 and 5.5.3).

For nickel/cadmium and nickel/hydride batteries of virtually sealed design, and encapsulated in a sealed metallic housing, in normal operational conditions gas and vapour escape are not relevant, and the above interactions are restricted to the interior gas space of the cell. Even if such batteries are equipped with resealing valves (cf. the Note in Section 1.1.1), that only open in an emergency, water loss can be neglected as long as the battery is operated more or less properly.

Interaction with the surrounding atmosphere is unavoidable in valve-regulated lead-acid batteries: hydrogen evolution cannot be completely suppressed and periodic gas escape occurs on account of increasing internal pressure. For this reason, the valve must open from time to time even in normal operational conditions (cf. Section 8.3). Furthermore, these batteries are usually encapsulated in plastic containers that are permeable to some extent to water vapour and gases. Owing to the limited electrolyte content and the fact that water that has been lost cannot be replaced, attention must be paid to these processes, because water loss is one of the major causes of capacity decrease with this battery design ([22]).

Processes of water loss by evaporation or permeation take place also in conventional lead-acid and nickel/cadmium batteries with liquid electrolyte. But, compared to electrolytic water decomposition with this battery design, the additional water loss is small and usually not noticed. Oxygen intake is also not relevant in flooded batteries, since the slow transport of dissolved oxygen protects the negative electrode from oxidation (cf. Section 2.6.3.1).

2.6.1 Electrochemical Water Decomposition

Water decomposition means the split-up of water into hydrogen and oxygen according to:

$$\boxed{H_2O \Rightarrow H_2 + \tfrac{1}{2}\cdot O_2} \qquad (2.93)$$

which is the combination of two electrochemical reactions:

The **oxidation of oxygen** according to

$$\boxed{O^{2-} \Rightarrow \tfrac{1}{2}\cdot O_2 + 2\cdot e^-} \qquad (2.94)$$

which reads in the completed form

$$\boxed{H_2O \Rightarrow \tfrac{1}{2}\cdot O_2 + 2\cdot H^+ + 2\cdot e^-} \quad \text{for acidic medium} \qquad (2.95)$$

or

$$\boxed{2\cdot OH^- \Rightarrow \tfrac{1}{2}\cdot O_2 + H_2O + 2\cdot e^-} \quad \text{for alkaline medium,} \qquad (2.96)$$

and the **reduction of hydrogen** which means hydrogen evolution on the negative electrode and can be written as:

$$\boxed{2\cdot H^+ + 2\cdot e^- \Rightarrow H_2} \quad \text{for acidic medium} \qquad (2.97)$$

or

$$\boxed{2\cdot H_2O + 2\cdot e^- \Rightarrow 2\cdot OH^- + H_2} \quad \text{for alkaline medium.} \qquad (2.98)$$

2.6.1.1 Thermodynamic Data

The thermodynamic data for the water decomposition reaction (Eq. (2.93)) according to the definitions in Section 2.2 are given in Table 2.5.

The enthalpy and the free enthalpy of formation are zero under standard conditions for hydrogen (H_2) and oxygen (O_2), since both substances are pure elements. As a consequence, according to Table 2.5, the values for enthalpy and free enthalpy of reaction (ΔH and ΔG) of water decomposition are:

Enthalpy

$$\boxed{\Delta H = 285.8 \text{ kJ}} \tag{2.99}$$

and free enthalpy

$$\boxed{\Delta G = 237.2 \text{ kJ}} \tag{2.100}$$

referred to each multiple of Eq. (2.93). The sign is positive because the system gains energy when the water is decomposed. The equivalent amount of energy is released when the hydrogen and oxygen are recombined.

Substance	Molar weight g/mole	Enthalpy of formation (H) kJ/mole	Free enthalpy of formation (G) kJ/mole
H_2	2.016	0	0
O_2	32.00	0	0
H_2O	18.02	-285.8	-237.2

Table 2.5 Molar weights and thermodynamic data for the chemical compounds of Eq. (2.93); Elements in standard conditions; $p = 1$ atm ($= 1.013 \cdot 10^5$ Pa), $T = 298.16$ K ([23]).

2.6.1.2 Equilibrium Voltage

Inserting Eq. (2.100) into Eq. (2.6) results in the equilibrium voltage of the water decomposition (1 J = 1 Ws = 1 V·As)

$$\boxed{E^o = 1.229 \text{ V}} \tag{2.101}$$

When the cell voltage exceeds this value, water is decomposed according to Eq. (2.93).

Note: *To get the relation $\Delta G = - n \cdot F \cdot E^o$, Eq. (2.93) would have to be reversed with the result that ΔG and ΔH would be negative. The reason is, that water formation ($H_2 + \frac{1}{2}O_2 \Rightarrow H_2O$) is the spontaneous running reaction, while water decomposition requires energy.*

2.6.1.3 Dependence on Concentrations and Pressures

The dependence of the equilibrium potential on the concentration of reacting components is given by the Nernst equation according to Section 2.2.1.1. When Eq. (2.95) is written in the cathodic direction, it has to be reversed and the result is the equilibrium voltage of the oxygen electrode in acid medium:

$$\boxed{E^o_{O_2,acid} = E^{o,s}_{O_2,acid} + \frac{R \cdot T}{2F} \ln \frac{a^{1/2}_{O_2} \cdot a^2_{H^+}}{a_{H_2O}}} \tag{2.102}$$

and the equilibrium voltage in alkaline medium derived from Eq. (2.96) is:

$$E^{o}_{O_2,alkaline} = E^{o.s}_{O_2,alkaline} + \frac{R \cdot T}{2F} \ln \frac{a^{1/2}_{O_2} \cdot a_{H_2O}}{a_{OH^-}} \quad (2.103)$$

The activity of O_2 can be substituted by the gas pressure according to HENRY'S law (Section 2.6.3). Under consideration that ln = 2.303·log, and that at 25 °C 2.303·RT/F = 0.0592 V, the equilibrium potential in acidic medium in V is:

$$E^{o}_{O_2,acid} = E^{o,s}_{O_2,acid} + \frac{0.0592}{4} \cdot \log(p_{O_2}) + 0.0592 \cdot \log(a_{H^+}) - \frac{0.0592}{2} \cdot \log(a_{H_2O}) \quad (2.104)$$

with the standard value

$$E^{o,s}_{O_2,acid} = 1.229 \text{ V} \quad (2.104a)$$

and in alkaline medium

$$E^{o}_{O_2,alkaline} = E^{o,s}_{O_2,alkaline} + \frac{0.0592}{4} \cdot \log(p_{O_2}) - 0.0592 \cdot \log(a_{OH^-}) + \frac{0.0592}{2} \cdot \log(a_{H_2O}) \quad (2.105)$$

with the standard value

$$E^{o,s}_{O_2,alkaline} = 0.401 \text{ V} \quad (2.105a)$$

Eqs. (2.104) and (2.105) show that the equilibrium potential is increased by 14.8 mV when the oxygen pressure is increased by one decade i.e. the influence of oxygen pressure is weak. The acidity of the solution can be expressed as a pH value: pH = $-\log(a_{H^+})$, cf. Section 2.2.3.1), and in both equations a decrease by 59.2 mV is observed per pH value.

The dependence of the hydrogen electrode potential on concentration has already been mentioned in Section 2.2.3.1. The Nernst equation corresponding to Eq. (2.97) is given by Eq. (2.31) for acidic medium:

$$E^{o}_{H^+/H_2} = E^{o,s}_{H^+/H_2} + \frac{R \cdot T}{2 \cdot F} \cdot \ln \frac{a^2_{H^+}}{a_{H_2}} \quad (2.106)$$

and that corresponding to Eq. (2.98) for alkaline medium:

$$E^{o}_{OH^-/H_2} = E^{o,s}_{OH^-/H_2} + \frac{R \cdot T}{2 \cdot F} \cdot \ln \frac{a^2_{H_2O}}{a^2_{OH^-} \cdot a_{H_2}} \quad (2.107)$$

When the above substitutions are correspondingly applied, both equations can be written as:

$$E^{o}_{H^+/H_2} = E^{o,s}_{H^+/H_2} - \frac{0.0592}{2}\log(p_{H_2}) + 0.0592\cdot\log(a_{H^+}) \quad V \tag{2.108}$$

with the standard value zero (Eq. (2.33), p.16), and

$$E^{o}_{OH^-/H_2} = E^{o,s}_{OH^-/H_2} - \frac{0.0592}{2}\cdot\log(p_{H_2}) - 0.0592\cdot\log(a_{OH^-}) + 0.0592\cdot\log(a_{H_2O}) \quad V \tag{2.109}$$

with the standard value (a_{OH^-} = 1 mole/dm^3 or pH = 14)

$$E^{o,s}_{OH^-/H_2} = -0.829 \;\; V \tag{2.109a}$$

Eqs. (2.108) and (2.109) are valid for 25 °C. They show that with increasing hydrogen pressure the equilibrium potential is reduced by 29.6 mV per decade of pressure. Per increase of the pH value by one unit, the equilibrium potential is reduced by 59.2 mV.

2.6.1.4 Reversible Heat Effect, Calorific Voltage

The difference between ΔH and ΔG, equals the reversible heat effect (cf. Eq. (2.41), p.21):

$$Q_{rev} = T\cdot\Delta S = \Delta H - \Delta G = 48.6 \text{ kJ} \tag{2.110}$$

This is 20.5% of the free reaction enthalpy, and the positive sign means that the system gains this energy in addition to the value of ΔG. In other words, 20.5% of additional energy is required to decompose water according to Eq. (2.93). When the reaction is reversed (recombination of water from hydrogen and oxygen), these 48.6 kJ are generated as additional heat.

According to Eq. (2.42), Q_{rev} can be expressed as a (fictive) voltage

$$\frac{Q_{rev}}{2\cdot F} = 0.252 \;\; V \tag{2.111}$$

and the calorific voltage of water decomposition is according to Eq. (2.43)

$$E_{cal} = 1.229 + 0.252 = 1.481 \;\; V \tag{2.112}$$

As indicated in Eq. (2.44), E_{cal} can also be derived from the enthalpy of reaction ΔH:

$$E_{cal} = \frac{285.8\cdot 10^3}{2\cdot 96485} = 1.481 \quad V \tag{2.113}$$

When water is decomposed by electrolysis, the energy content of the gas equals $n\cdot F\cdot E_{cal}$, which is 20.5% more than $n\cdot F\cdot E^o$. In the reverse case, when hydrogen and oxygen are recombined to water, the amount of released heat corresponds to $n\cdot F\cdot E_{cal}$.

2.6.1.5 Weight and Volume Equivalents

The relations between current and amount of water decomposed or the equivalent volumes of gas generated can be evaluated from Eq. (2.93), using the FARADAY constant (F = 96485 As/equiv. = 26.802 Ah/equiv.). The number of exchanged electrons is 2; consequently, 53.60 Ah are required for each multiple of the molar weights in Table 2.5. This means:

1 Ah decomposes 0.3361 g H_2O and generates 0.0376 g H_2 + 0.2985 g O_2	(2.114)

respectively

1 g H_2O is the equivalent of 2.975 Ah	(2.115)

For most practical purposes, the approximation is sufficient

3 Ah are equivalent to 1 g H_2O or 1 Ah decomposes 1/3 g H_2O	(2.116)

The volume of the generated gas depends on temperature and pressure. In Table 2.6, the volumes of hydrogen, oxygen and gas in total, equivalent to 1 Ah of electricity are listed. These equivalents refer to a pressure of 1 atm (1013 mbar or $1.013 \cdot 10^5$ Pa). The figures are calculated assuming that the two gases oxygen and hydrogen behave like ideal gases, an approximation that in general is admissible. The same figures are valid in cm^3 for 1 mAh. The second column in Table 2.6 contains the figures for the molar volume of an ideal gas.

Temperature	Molar Volume	Equivalents for 1 Ah		
		Hydrogen	Oxygen	Gas in total
°C	dm^3	dm^3	dm^3	dm^3
0	22.41	0.4182	0.2091	0.6272
10	23.23	0.4335	0.2167	0.6502
20	24.06	0.4488	0.2244	0.6732
25	24.47	0.4564	0.2281	0.6846
30	24.88	0.4641	0.2320	0.6961
40	25.70	0.4794	0.2397	0.7191
50	26.52	0.4947	0.2473	0.7420

Table 2.6 Molar volume of an ideal gas and equivalents of hydrogen, oxygen and the gas in total for the water decomposition reaction related to 1 Ah. (Pressure 1 atm = 1013 hPa.)

In Table 2.7, evolution rates in dm^3 per hour, day, and month are listed for hydrogen, oxygen and gas in total, referred to 1 A.

Temp	Equivalents for 1 A								
	Hydrogen evolution			oxygen evolution			Total gas evolution		
°C	dm^3/h	dm^3/d	dm^3/M	dm^3/h	dm^3/d	dm^3/M	dm^3/h	dm^3/d	dm^3/M
0	0.4182	10.04	301.1	0.2091	5.018	150.5	0.6272	15.05	451.6
10	0.4335	10.40	312.1	0.2167	5.212	156.0	0.6502	15.60	468.1
20	0.4488	10.77	323.1	0.2244	5.385	161.6	0.6732	16.16	484.7
25	0.4564	10.95	328.6	0.2282	5.477	164.3	0.6846	16.43	492.9
30	0.4641	11.14	334.1	0.2320	5.569	167.1	0.6961	16.71	501.2
40	0.4794	11.51	345.2	0.2397	5.753	172.6	0.7191	17.26	517.7
50	0.4947	11.87	356.2	0.2473	5.936	178.1	0.7420	17.81	534.3

Table 2.7 Equivalents of hydrogen, oxygen and the total gas evolution per hour, day and month (M) related to 1 A. Pressure 1 atm = 1013·hPa (1 hPa = 1 mbar).

2.6.2 Gas Exchange with the Surrounding Atmosphere

Gas exchange with the surrounding atmosphere presupposes direct connection between the interior of the cell and the surrounding atmosphere. Such a connection is the vent of vented batteries. Valve-regulated and sealed batteries allow solely outlet of gas when the internal pressure exceeds a given limit. Unwanted gas transport into the battery can here only be caused by leakages, a faulty valve, for example, or inadequate welding or sealing of the container.

Gas transport occurs in two ways:

- Gas flow caused by a pressure difference.
- Gas transport by diffusion with no pressure difference between the interior and the surrounding atmosphere. The gas as a whole does not flow, only the molecules of its components migrate.

2.6.2.1 Gas Flow Through an Opening

The gas flow through an opening is (approximately) described by Poiseuille's law (in German called 'Hagen-Poiseuille'sches Gesetz'):

$$\frac{\Delta V}{\Delta t} = \frac{\pi \cdot r^4}{8 \cdot \eta \cdot l} \cdot \Delta p \quad cm^3/s \tag{2.117}$$

with: ΔV : streaming volume in cm^3

Δt : time period in seconds

r : radius of the (cylindrical) opening in cm

η : viscosity: air: $1.87 \cdot 10^{-9}$, H_2 : $0.886 \cdot 10^{-9}$, O_2 : $2.03 \cdot 10^{-9}$ $N \cdot s/cm^2$

stoichiometric mixture of H_2 and O_2 : $1.92 \cdot 10^{-9}$ $N \cdot s/cm^2$

all figures at 20 °C ([24])

l : length of the flow route
Δp : pressure difference in N/cm^2 (= 10^{-4}·Pa)
($1\ N/m^2 = 1\ Pa = 10^{-5}$ bar; 100 Pa = 1 hPa = 1 mbar)

Poiseuille's law applies strictly only to a viscous flow through a cylindrical opening under the condition that the length of this opening is large compared to its diameter (cf., e.g., (25)). For this reason, the following calculations based on this rule can only be approximate.

2.6.2.2 Gas Transport by Diffusion

Diffusion has been introduced by Eq. (2.46) as a current equivalent. The basic relation for one-dimensional diffusion (Fick's first law) is:

$$\boxed{dn = -D \cdot f \cdot \frac{dc}{dx} \cdot dt} \qquad (2.118)$$

with: n : Number of moles which are transported
D : Diffusion coefficient:
$0.25\ cm^2 \cdot s^{-1}$ for water vapour in air
0.18 and $0.63\ cm^2 \cdot s^{-1}$ for O_2 and H_2 resp. in air
f : Diffusion cross sectional area (area of opening) in cm^2
dc/dx : Concentration gradient in mole/cm^3 per cm
dt : Period of time in seconds

For approximate calculations, a linear concentration gradient can be assumed, and Eq. (2.118) simplified to:

$$\boxed{\frac{dn}{dt} \approx -D \cdot f \cdot \frac{c - c_0}{x - x_0}} \quad \text{mole/s} \qquad (2.119)$$

2.6.2.3 Comparison Between Gas Flow and Diffusion

To give an impression of the capability of both transport mechanisms, Tables 2.8 and 2.9 show results calculated from Eqs. (2.117) and (2.119) respectively, for openings of the same diameter and an assumed length of 5 mm, which is about the thickness of container walls.

Table 2.8 emphasises the enormous influence of the size of the opening on gas flow. The reason is that the radius is raised to the power four in Eq. (2.117). The current equivalent shows also, how critically even a small leakage can affect gas exchange between the interior of a cell and the surrounding atmosphere as soon as a pressure difference develops on account of water decomposition or other effects as temperature variation or changes of atmospheric pressure (cf. Section 10.8.6).

At the assumed pressure difference of 1000 Pa (= 10 mbar), which is small compared to the usual opening pressures of valves (cf. Section 10.6), a leak, corresponding to a cylindrical diameter of 0.1 mm, causes oxygen and hydrogen escape of 0.39 and 0.9 dm^3 per hour, corresponding to a water decomposition current of about 6 and 7 A, respectively. These values show that even a small leakage would equalise pressure variations to a large extent by outlet or inlet of gas, and that a small opening in a valve is sufficient to release a large volume of gas within a short period.

A leak allows gas transport in both directions. As a consequence, air can flow through the leak into the cell when the internal pressure is lower than that of the surroundings. The oxygen content of the air causes chemical discharge of the negative electrode (cf. Sections 4.4 and 5.3), which affects further inlet of gas. But air that leaks into the cell contains only 21% of oxygen, and nitrogen that simultaneously enters the cell causes a gradual increase of the internal pressure until pressure balance is reached. This limits chemical discharge by direct gas inlet (but not the effect of diffusion, described below).

Note: *The presence of nitrogen in a battery that had been on float charge for a certain period, indicates air intake by a leakage.*

1	2	3	4	5	6
Radius	Area	Oxygen outlet equiv.		Hydrogen outlet equiv.	
of opening		when 100% of intern. atm.		when 100% of intern. atm.	
mm	mm^2	cm^3/s	mA	cm^3/s	mA
0.01	0.00031	0.341	0.760	0.886	1.97
0.05	0.0079	213	474	554	1235
0.1	0.0314	3411	7600	8865	19754

Table 2.8 Oxygen and hydrogen flow through a circular shaped opening calculated from Eq. (2.117) for the pressure difference $\Delta p = 1000$ Pa = 10 mbar. Current equivalent at 20 °C. Length of opening 0.5 cm.

Figures for transport by diffusion for the same diameters of a leak are listed in Table 2.9 . As mentioned above, gas is not actually flowing in this case, and the volumes (cm^3/h) in columns 3 and 5 are not volumes but equivalents of the transported oxygen or hydrogen.

The values in columns 3 and 4 of Table 2.9 are calculated with the diffusion coefficients given in Column 5 of Table 2.10 and under the assumption that the interior gas phase does not contain oxygen, since oxygen is consumed by chemical discharge of the negative electrode. The values in columns 5 and 6 are based on the fact that the interior gas phase only consists of hydrogen, a situation that approximates the state of the battery during normal float charging (cf. Section 13.1.1).

Comparison between transport rates in Tables 2.8 and 2.9 shows that the diffusion process is slower than gas flow by orders of magnitude. However, the above-mentioned effect that the gas flow gradually comes to a standstill on account of inert nitrogen gas accumulating in the interior atmosphere does not apply to diffusion. The partial O_2 pressure remains zero as long as the oxygen is consumed by the negative electrode, and even a current of only 1 mA means 8.7 Ah of discharge per year, when flowing continually.

1	2	3	4	5	6
Radius	Area	Oxygen inlet equiv. (21% of 1 atmosphere)		Hydrogen outlet equiv. 100% of the intern. atm	
of opening					
mm	mm^2	cm^3/h	mA	cm^3/h	mA
0.01	0.00031	0.0009	0.0041	0.015	0.017
0.05	0.0079	0.022	0.102	0.38	0.42
0.1	0.0314	0.092	0.41	1.52	1.7
0.5	0.785	2.29	10.21	38.2	42.5

Table 2.9 Diffusion transport rates from Eq. (2.118).
Equivalents in cm^3/h at 20 °C and current equivalents:
in columns 3 and 4 with the interior O_2 partial pressure zero;
in columns 5 and 6 with the exterior H_2 partial pressure zero.
Length of diffusion opening 0.5 cm (as in Table 2.2).

In battery practice, gas transport caused by pressure differences and by diffusion usually occur in parallel, because pressure differences between the interior of the battery and the surrounding atmosphere are caused by temperature changes as well as by air pressure variations.

2.6.3 Solubility of Gases

The concentration of a gas in the gaseous phase, treated as an ideal gas, is given by:

$$c = \frac{n}{V} = \frac{p}{R \cdot T} \quad \text{mole/cm}^3 \tag{2.120}$$

with: n : number of moles
V: volume in cm^3
R: gas constant ($8.3413\ J{\cdot}K^{-1}{\cdot}mole$)
p: pressure in 10^{-4} Pa ($= N/cm^2$)
T : temperature in K

For slightly soluble gases, which do not react with the solvent, the solubility is nearly proportional to the partial pressure (HENRY'S law). For the individual gas, this proportionality is described by the BUNSEN absorption coefficient:

$$\alpha = \frac{\text{moles per volume unit of solution}}{\text{moles per volume unit of gas at 1 atm pressure}} \tag{2.121}$$

The characteristic parameters are listed in Table 2.10. These values are rounded, but they are adequate for estimation of the gas-transport rate. Actually, solubility and diffusion coefficient depend on the concentration of solvent and temperature.

1	2	3	4	5	6
Gas	Solubility as BUNSEN coefficient. α	Diffusion coeff. in solution D_{liq} $cm^2 \cdot s^{-1}$	Diffusion coeff. times solubility $\alpha \cdot D_{liq}$ $cm^2 \cdot s^{-1}$	Diffus. coeff. in air D_{air} $cm^2 \cdot s^{-1}$	Relation $D_{air}/\alpha \cdot D_{liq}$
In 5 N H_2SO_4					
O_2/ H_2	0.02	$2 \cdot 10^{-5}$	$4 \cdot 10^{-7}$	0.18	$4.5 \cdot 10^{6}$
In 1 N KOH					
O_2	0.02	$4 \cdot 10^{-5}$	$8 \cdot 10^{-7}$	0.63	$7.9 \cdot 10^{6}$
H_2	0.01	$4 \cdot 10^{-5}$	$4 \cdot 10^{-7}$	0.63	$1,5 \cdot 10^{7}$

Table 2.10 Solubility and diffusion coefficients of oxygen and hydrogen (approximate values) in solution and in air at 25 °C (cf., e.g., ([26])).

To compare the diffusion rates within the liquid to those in the gas filled space, $\alpha \cdot D_{liq.}$ and D_{air} have to be inserted into Eq. (2.118) instead of D. From the figures in Table 2.10, it follows that in similar conditions the diffusion rate in the gas phase is about 10^7 times faster than that in the liquid.

2.6.3.1 Gas-Transport Rate Through the Liquid Phase

The rate of the gas transport through the electrolyte can be estimated from the figures above. At 25 °C, the molar volume of an ideal gas is 24 470 cm^3 (Table 2.6). Consequently, the concentration in the gas phase is

$$\boxed{c = 4.1 \cdot 10^{-5} \quad \text{moles per cm}^3} \tag{2.122}$$

when the space above the liquid is 100% filled by the gas under consideration at 1 atm.

With the above BUNSEN coefficient, the oxygen concentration in sulphuric acid is

$$\boxed{c_{liq.} = \alpha \cdot c \approx 8.2 \cdot 10^{-7} \text{ moles per cm}^3} \tag{2.123}$$

With the assumption $c_o = 0$ (no oxygen at the surface of the negative electrode) and $D = 2 \cdot 10^{-5}$, Eq. (2.119) reads

$$\boxed{\frac{dn}{dt} = -\frac{f}{x} \cdot 1.6 \cdot 10^{-12}} \quad \text{moles per second} \tag{2.124}$$

for the transport rate of dissolved oxygen through area f over distance x, when the solution is saturated to 1 atm at x, and free from oxygen at distance $x_o = 0$.

According to Eqs. 7.15 and 7.16, (p.263) one mole of dissolved O_2 is equivalent to $4 \cdot F$ of electric charge or 107.21 Ah or 385 920 As (cf. Eq. (2.7)). When this figure is inserted into Eq. (2.124) the result is the oxygen-diffusion current:

$$\boxed{\frac{dn}{dt} = i_{O_2} = -\frac{f}{x} \cdot 6.2 \cdot 10^{-7}} \quad / \text{A} \tag{2.125}$$

For lead-acid batteries, a geometric surface area f of 1000 cm^2 can be assumed for 100 Ah of capacity as a rough approximation. If the distance x between the plates is 0.3 cm, the resulting current density that characterises the internal oxygen cycle (cf. Chapter 8) in batteries with liquid electrolyte, is

$$\boxed{i_{O_2} \approx 2 \;\; mA / 100 \; Ah} \qquad (2.126)$$

In nickel/cadmium batteries, the electrodes usually are narrow spaced, but the results would be in the same range. This rough estimation shows clearly that the transport rate by dissolved oxygen is extremely slow.

For this reason, the gas transport for the internal oxygen cycle, shown in Fig. 8.1, must take place largely through the gaseous phase to take advantage of the about 10^7 times faster transport rate. Only then, are acceptable overcharge rates to be reached.

Nevertheless, the transport rate of the dissolved oxygen plays an important role, even when the electrolyte is immobilised. Oxygen always has to penetrate an electrolyte film, the diffusion layer, at the negative-electrode surface, although oxygen transport occurs largely through the gaseous phase. This film of electrolyte is required, because protons (H^+ ions) have to migrate to the place of reaction, to neutralise the generated O^{2-} or OH^- ions (cf. Eqns. 7.15 and 7.16, p.263). Otherwise the negative space charge, caused by the generated O^{2-} and OH^- ions, would soon bring the reaction to a standstill. This electrolyte film, however, is thin and the diffusion length in the range of 10 µm. When this value is inserted for 'x' in Eq. (2.125), an oxygen-diffusion current of about 6 mA/cm^2 results. Actually, in 'dry' batteries maximum rates of the internal oxygen cycle in the order of 10 A/100 Ah are observed, as is shown in Fig. 4.19, p.140.

2.6.3.2 Pressure Dependence of the Diffusion Rate

The diffusion rate of oxygen in sulphuric acid and in alkaline electrolyte is directly proportional to the concentration of the dissolved gas (Eq. (2.118)). This concentration again depends on the partial pressure of the gas above the liquid because of the proportionality given by the BUNSEN coefficient (Eq. (2.123)). If this diffusion rate is the limiting step of the overall reaction, a direct proportionality between reaction rate and internal gas pressure is to be expected.

This applies when the diffusion through the wetting film at the electrode surface is the rate determining factor. For example, the maximum rate of the internal oxygen cycle is proportional to the internal gas pressure when O_2 diffusion through the wetting layer of the negative electrode is rate determining. This is important for nickel/cadmium batteries that are overcharged with a fairly high constant currents (cf. Section 13.2.1). Furthermore, the self-discharge rate of nickel/hydrogen batteries depends on the internal H_2 pressure, since the diffusion of hydrogen through the wetting layer of the positive electrode is the rate-determining step (cf. Fig. 8.4, p.274).

2.6.4 Water Vapour Exchange with the Surrounding Atmosphere

Sulphuric acid and potassium hydroxide are aqueous solutions, and have a water vapour pressure that has to be balanced by water concentration in the surrounding atmosphere.

Otherwise, water escapes from the solution by evaporation, or water is absorbed from the air, until a balance is achieved.

2.6.4.1 Vapour Pressure

The water content in the atmosphere can be expressed as water vapour concentration (g/cm^3) or as partial pressure of water vapour. The relation between water-vapour concentration and partial pressure, is well approximated when the water vapour is treated as an ideal gas:

$$\boxed{p = \frac{n}{V} \cdot R \cdot T} \qquad /Pa \tag{2.127}$$

with p : Partial pressure of water vapour in Pa
(1 atm = 1.013 bar = 1013 mbar = $1.013 \cdot 10^5$ Pa = 1013 hPa)
n : Number of moles
V : Volume $/dm^3$
R : molar gas constant
($R = 8.314 \cdot 10^3\ dm^3 \cdot Pa \cdot K^{-1} \cdot mole^{-1}$)
T : Temperature (K)

2.6.4.2 Temperature Dependence of the Vapour Pressure

The temperature dependence of the vapour pressure is described by the CLAPEYRON-CLAUSIUS equation:

$$\boxed{\frac{dp}{dT} = \frac{Q_V}{\Delta V \cdot T}} \tag{2.128}$$

with: Q_V: Heat of evaporation (per mole)
ΔV: Volume difference between the gaseous and liquid phases (V_G - V_L) (also per mole).
V_L can usually be neglected, because $V_L << V_G$
(for Water at 25 °C: $V_L = 18\ cm^3$; $V_G = 24470\ cm^3$)

Note: *In the German literature, this equation is usually called CLAUSIUS-CLAPEYRON equation. Clapeyron published the equation in 1834, while Clausius later presented the exact proof.*

For dilute solutions, the values for Q_V and ΔV are similar to those of water. This results in the same temperature dependence of the vapour pressure for water and the electrolytes used in batteries, at least for not too-high concentrations (cf. Figs. 2.14 and 2.15).

Table 2.11 shows vapour pressure of water at different temperatures in different units, together with the corresponding concentrations of water vapour per volume. These values represent the equilibrium state, the maximum possible content of water in the air (100% humidity, cf. below). The two bottom lines represent the quotient n/V in Eq. (2.127).

Temp.°C	0	10	20	30	40	50	60	80	100
Vapour pressure									
mm Hg	4.58	9.21	17.53	31.82	55.32	92.51	149.4	355.1	760
mbar	6.11	12.3	23.37	42.41	73.75	123.3	199.2	473.5	1013.2
Pa	611	1227	2337	4241	7375	12334	19916	47353	101320
H_2O concentration									
$\cdot 10^{-4}$ mole/dm^3	2.68	5.21	9.58	16.82	28.32	45.91	71.90	161.3	326.6
$\cdot 10^{-2}$ g/dm^3	0.484	0.939	1.72	3.03	5.10	8.27	12.95	29.1	58.8

Table 2.11 Water vapour pressure (equilibrium or saturated values) at different temperatures in various units. These values represent the maximum water content at the given temperature (100% relative humidity). The two bottom lines show the corresponding H_2O concentration in the air in mole/dm^3 and g/dm^3. 1 mm Hg = 1 Torr = 133.3 Pa, or 1 Pa = 0.00750 Torr.

2.6.4.3 Humidity

The water content of the open atmosphere does not usually reach the saturation value but remains considerably below this maximum. This is described as 'humidity', defined as absolute or relative humidity (cf., e.g., ([27])):

- The **absolute humidity** is the mass of water vapour in the unit of volume, usually measured as grams per cubic metre, or expressed in terms of partial pressure of the water vapour.
- The **relative humidity** is the ratio of the quantity of water vapour in the atmosphere to the quantity that would saturate at the existing temperature. It is the ratio of the partial pressure of water vapour present to the partial pressure of saturated water vapour at the same temperature, or the corresponding ratio of the concentrations.

In the following, the term 'humidity' is always used with the meaning 'relative humidity'.

As an example, values for the concentration in mole per dm^3 and grams per dm^3 as well as partial pressure in mm Hg and pascals are listed in Table 2.12 for different percentages of relative humidity at 20 °C. (The values for 100% rel. humidity correspond with those for 20 °C in Table 2.11.)

Normally, 100% relative humidity is only reached in clouds or in fog. Humid air has about 80% relative humidity, desert climate less than 20% . In normal European weather conditions, the relative humidity is in the range between 40 and 70% .

2.6.4.4 Humidity Changes with Temperature

Humidity changes considerably when air is heated or cooled. Heated air can get very dry, if there is no possibility to absorb water. In battery practice, this can happen when air is heated up in an electrical device before it flows around the battery.

% Rel. humidity	100	80	70	60	50	40	30	20	10
Vapour pressure									
mm Hg	17.53	14.02	12.27	10.51	8.76	7.01	5.25	3.51	1.75
mbar	23.37	18.7	16.36	14.02	11.68	9.35	7.01	4.67	2.34
Pa	2337	1870	1636	1402	1168	934.8	701.1	467.4	233.8
H_2O concentration (= absolute humidity)									
$\cdot 10^{-4}$ mole/dm^3	9.59	7.67	6.71	5.57	4.79	3.83	2.87	1.92	0.959
$\cdot 10^{-2}$ g/dm^3	1.72	1.38	1.20	1.04	0.864	0.691	0.518	0.346	0.173

Table 2.12 Relative humidity and corresponding values for the vapour pressure of water in various units. The two bottom lines show the corresponding H_2O concentration in air in mole/dm^3 and g/dm^3 (absolute humidity). All values referred to 20 °C.

Fig. 2.13 illustrates the effect of temperature upon humidity. The curve shows the decrease or increase of relative humidity when air with a humidity of 40% at 20 °C is heated or cooled. A temperature increase by 10 °C reduces the relative humidity to about 20%, while a temperature decrease causes a corresponding increase of relative humidity. In battery practice, heating is the critical point, which may occur unnoticed by an unfavourable conception of the ventilation system. So, even in a moderate climate zone, the battery can be surrounded by 'desert atmosphere' and so exposed to water loss by evaporation (cf. Section 2.6.5).

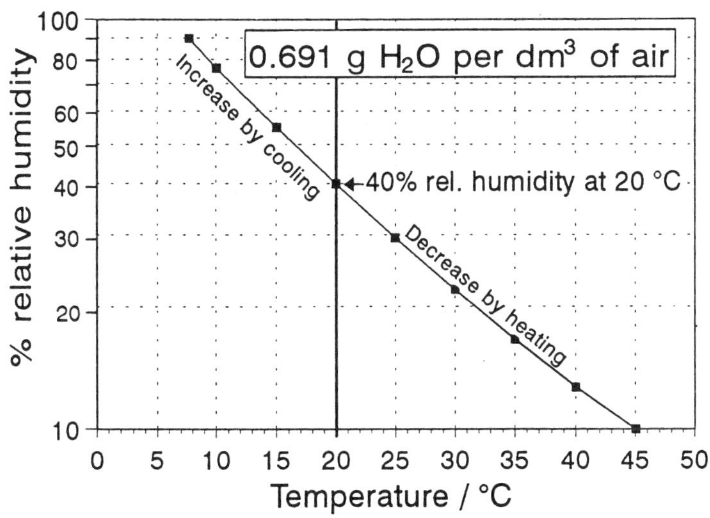

Fig. 2.13 Relative-humidity changes with temperature. The curve represents humidities at various temperatures for air that contains 0.691 g/dm^3 of water (40% rel. humidity at 20 °C according to Table 2.12).

2.6.4.5 Vapour Pressure of Sulphuric Acid

Sulphuric acid as a whole is not vaporised in normal conditions. The boiling point of H_2SO_4 at atmospheric pressure is 338 °C, but it cannot be reached, because of earlier decomposition, and so H_2SO_4 can only be distilled in a vacuum. In the normal operational conditions of batteries, the partial pressure of H_2SO_4 is so small that evaporation of H_2SO_4 can be neglected. But equilibrium is established between the water content of the diluted acid and the vapour content in the air.

The curve in Fig. 2.14 shows the vapour pressure that is in equilibrium with dilute sulphuric acid. This vapour pressure decreases with increasing acid concentration. The value for zero H_2SO_4 concentration is identical to the vapour pressure of water. The horizontal lines represent different steps of water vapour pressure expressed as steps of relative humidity referred to 20 °C (Table 2.12).

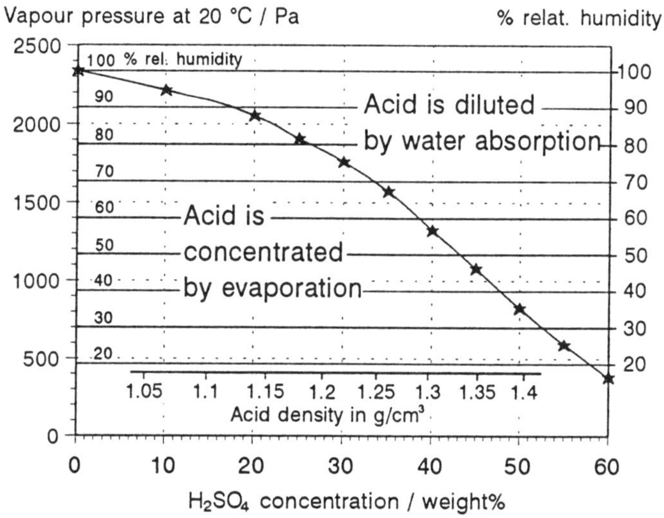

Fig. 2.14 Vapour pressure of sulphuric acid at 20 °C. The horizontal lines represent relative humidities (right hand scale). Up to about 40 weight% of H_2SO_4, the relation between the curve and these lines is almost independent of temperature.

As long as the vapour pressure of the acid is less than that of the surrounding air, the sulphuric acid absorbs water and becomes diluted. For this reason, concentrated sulphuric acid can be used as a drying medium.

Water evaporates when the vapour pressure of the acid is greater than that of the air, and the concentration of the acid increases until the vapour pressure of the air balances that of the acid.

The concentration finally attained depends on the relative humidity of the atmosphere, and is given in Fig. 2.14 by the intersection of the vapour pressure curve with the corresponding humidity level. In normal European weather conditions, with relative humidities between 40 and 70%, the equilibrium acid concentration reaches between

35% and 50% sulphuric acid by weight, which corresponds to an acid density between 1.28 and 1.4 g/cm^3. In a very dry climate, the acid becomes more concentrated.

In concentrations up to 40% sulphuric acid by weight, no difference is observed between the temperature dependence of the vapour pressure of the acid and the corresponding vapour pressure of water (the relative humidity), as mentioned with Eq. (2.128) . This is quite important, because it means that the ratio between the vapour pressure of the acid and the relative humidity remains constant for all temperatures. In other words, the same relative humidity always balances a given acid concentration, fairly independently of temperature.

Fig. 2.14 indicates that, even in moderate climatic conditions, sulphuric acid concentrates up to a density of 1.3 g/cm^3 (40% by weight). Such acid concentrations approach or even exceed the upper limits tolerated for a lead-acid battery. But they are not completely different from the concentrations applied for batteries. As a consequence, the driving force for diffusion, according to Eq. (2.119, p.50) is small, namely the gradient between the vapour pressure of the acid within the battery and the surrounding vapour pressure corresponding to the humidity of the air. As a consequence, serious evaporation losses are not to be expected in normal operating conditions.

2.6.4.6 Vapour Pressure of Aqueous Potassium Hydroxide

A similar relation to that of sulphuric acid exists for the electrolyte of nickel/cadmium and nickel/metal hydride batteries, as shown in Fig. 2.15. The usual concentrations of between 20 and 32% of potassium hydroxide by weight (cf. Section 5.5) are balanced by relative humidities in the range of 55 and 77%.

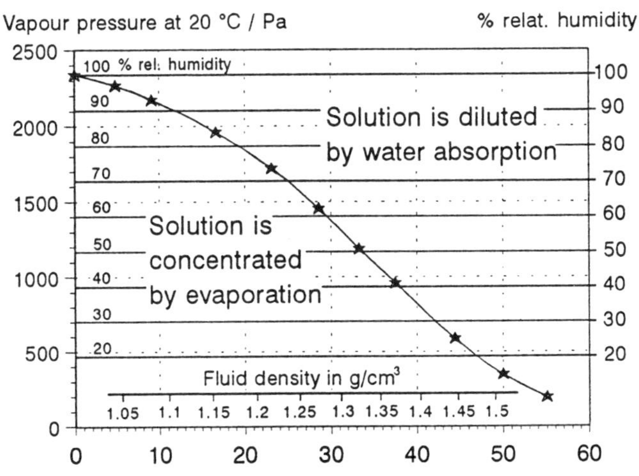

Fig. 2.15 Vapour pressure of aqueous potassium hydroxide solution ([28]). The horizontal lines represent the relative humidities of the atmosphere. All values at 25 °C. (1 Pa = 0.01 mbar = 0.0075 mmHg = $1.450 \cdot 10^{-4}$ psi; 1psi = 68.97mbar.)

For sealed nickel/cadmium batteries, water loss by evaporation is of no relevance, because:

- The sealed design is really sealed, and water vapour cannot be carried out of the cell by gas that leaves the cell.
- Sealed nickel/cadmium batteries are usually encapsulated by a metallic housing that does not allow the permeation of water or water vapour.

2.6.5 Water Vapour Loss of Batteries by Evaporation

Water vapour exchange with the surrounding atmosphere can cause water loss or, under special conditions, water gain, by various mechanisms:

- **Saturation of electrolysis gases** always means water loss, because the gas, generated inside the battery, automatically becomes saturated with water vapour. This water loss depends directly on the amount of gas that leaves the battery.
- **Water loss through the container walls** is caused by permeation of water vapour (which also means diffusion) through the container walls and can be critical for valve-regulated lead-acid batteries in unfavourable operating conditions (low humidity of the surrounding atmosphere).
- **Water loss by air-lift pumps** only concerns flooded low-maintenance lead-acid batteries (Section 7.5.1.2), and is caused by water-vapour absorption by the air used for forced flow of the electrolyte. The lost amount of water can be large when too much air is used (ineffective mixing methods). Effective airlift pumps and a minimised flow of air are ways to reduce water loss. Moistening of the air before it flows through the pump is another possibility.

All these kinds of water loss can be neglected in conventional lead-acid batteries, when compared to water loss by electrolysis, and therefore are of no interest, particularly since they can be compensated by refilling. In valve-regulated lead-acid batteries, however, each kind of water loss is critical. For this reason, environmental conditions are more important for the operation of valve-regulated lead-acid batteries than for vented types.

2.6.5.1 Saturation of Hydrogen Gas with Water Vapour

The gas generated in the battery becomes saturated with water vapour. When the gas leaves the cell, a corresponding amount of water is carried away with the gas. This amount is proportional to the partial pressure of the water vapour compared to the partial pressure of the gas. The vapour pressure of sulphuric acid in the usual concentration and between 15 and 40 °C is between 1000 and 5000 Pa (= 10 to 50 mbar). Consequently, the vapour pressure is between 1 and 5% of the total gas pressure (1 atmosphere). Accordingly, the additional water loss caused by saturation of the generated gas, is at most about 5% of the water loss by electrolytic decomposition, and so not noticed in practice.

In conventional stationary lead-acid batteries with liquid electrolyte, the float current is 10 - 30 mA/100 Ah, which corresponds to 7 - 21 cm^3 per hour of the gas mixture ($H_2 + 0.5 \cdot O_2$). This is a water loss of between 2.5 and 7.3 g per month and per 100 Ah of nominal capacity. In valve-regulated lead-acid batteries in float conditions, mostly hydrogen gas leaves the battery, and the hydrogen loss is in the range 0.5 to 2 cm^3/h/100 Ah n.c. This corresponds to 0.3 to 2 grams of water loss per month and per 100 Ah, or approximately 0.4 to 1.4% of the water content per year.

In both cases, the water loss due to saturation of the generated (and escaped) gas amounts to only 1 to 5% of these figures and so can be neglected.

2.6.5.2 Water Loss Through Container Walls

This effect, caused by the diffusion of water vapour or gaseous oxygen through the container wall, concerns mainly the lead-acid battery, because of plastic containers generally used for this battery design. Sealed nickel/cadmium batteries are mostly enclosed in metallic containers, but, if plastics are employed, these batteries are subject to similar problems. Plastic seals in metallic containers are of minor importance for water loss, because of their small surface area.

Plastics materials are permeable to most gases including water vapour, and the permeability usually increases with temperature. Table 2.13 shows the permeability of water vapour, oxygen and hydrogen through different container materials used for lead-acid and nickel/cadmium batteries.

Water vapour permeability is most important for packaging materials, especially in the food industry. For this reason, the methods of determination are standardised for thin foils ([29]). Usually the amount of water lost by permeation through the foil is determined by weight loss when the environmental atmosphere has less than 2% relative humidity.

	Permeability (per day) referred. to a 100 µm thick foil		
	Water vapour $g \cdot m^{-2} \cdot d^{-1}$	Oxygen (gas) $cm^3 \cdot m^{-2} \cdot d^{-1} \cdot bar^{-1}$	Hydrogen (gas) $cm^3 \cdot m^{-2} \cdot d^{-1} \cdot bar^{-1}$
Polyethylene	2 - 2.5	1900 - 4500	6000 - 13000
Polypropylene	2 - 5	2500 - 4800	7000 - 16000
PVC	11 - 19	100 - 300	2000 - 4000
ABS	27 - 33	400 - 900	
Polystyrene	28	470	
Polyamide	70 - 120	200 - 400	1000 - 2000

Table 2.13 Permeability of water vapour, oxygen and hydrogen ([30]) These figures are determined for thin foils. For container wall materials they are to be regarded as only approximate values for comparison.

Values as in Table 2.13 depend largely on the history and the manufacturing data of the material. They relate to thin foils and cannot be applied readily to container walls of usual thickness. But Vaccaro and Klatte ([31]) observed similar figures for container wall thickness. They calculated, for the material with the highest permeation rate, namely ABS, that 10% of the water content would be lost within 5.3 years at 25 °C and within 1.33 years at 50 °C, with 20% relative humidity assumed for the environmental atmosphere. Although 50 °C and 20% relative humidity are extreme conditions, the evaluation shows that water loss through the container wall may be critical in unfavourable conditions. Attention has to be paid to the environmental conditions. Possible water loss depends mainly on those conditions. Countermeasures are metallic housings or metallised container walls ([32]), coating the container with a water vapour resistant paint ([33]) or a combination of suitable plastic materials ([34]).

The permeation of hydrogen is critical only in nickel/metal hydride batteries, because it means loss of active material in the negative electrode. Thus these batteries cannot be encapsulated in plastic container materials. In lead-acid and nickel/cadmium batteries, hydrogen, generated as a gas, has to escape from the battery because there is no corresponding 'recombination cycle' as for oxygen (cf. Chapter 8). Thus, hydrogen has to escape from the cell anyway and permeation through the container wall has no relevance. But oxygen that escapes from the cell means water loss, because it is no longer available for the internal oxygen cycle, and an equivalent amount of hydrogen has to be generated instead at the negative electrode.

Water and gas permeation are based on diffusion according to Eq. (2.118, p.50). The driving force is the concentration gradient of the water vapour or of the concerned gas between the interior of the battery and the surrounding atmosphere. As an approximation, a linear concentration gradient dc/dx across the container wall can be assumed, given by the difference $c_{interior} - c_{outside}$:

$$\frac{dc}{dx} = \frac{c_{interior} - c_{outside}}{d} \tag{2.129}$$

with: $c_{interior}$: concentration of water vapour inside the container
$c_{outside}$: concentration of water vapour outside the container
both as $mole/cm^3$ or g/cm^3
d : thickness of the wall (cm)

When water vapour is treated as an ideal gas, concentration can be converted into partial pressure according to Eq. (2.122, p.53) :

$$c = \frac{n}{V} = \frac{p}{R \cdot T} \quad \text{or} \quad p = R \cdot T \cdot c \tag{2.130}$$

with: p: partial pressure (hPa = mbar)
n : number of moles
V : volume (cm^3)
R : gas constant (8,314 $J \cdot K^{-1} \cdot mole^{-1}$ or. 8,315 $cm^3 \cdot hPa \cdot K^{-1} \cdot mole^{-1}$)
T : temperature (K)

Eq. (2.129) is so converted into:

$$\frac{dn}{dt} = -\frac{D \cdot f}{R \cdot T \cdot d}(p_{interior} - p_{outside}) \tag{2.131}$$

with: $p_{interior}$ and $p_{outside}$ are partial vapour pressure inside and outside the container.
f : surface area of the container wall, facing the surrounding
d : thickness of the container wall

With the pressure difference Δp:

$$\Delta p = p_{interior} - p_{outside} \tag{2.132}$$

and combination of $D/(R \cdot T)$ to D_{perm} the diffusion equation is written:

$$\frac{dn}{dt} = -\frac{D_{perm} \cdot f}{d} \cdot \Delta p \tag{2.133}$$

with: D_{perm} : permeation coefficient that can be derived from Table 2.13
Δp : partial pressure difference (water vapour or gas)

In normal European weather conditions, the permeability of the container wall is of minor importance on account of the moderate humidities. Sometimes, these processes may even be helpful in stabilising the acid concentration when the relative humidity outside the battery is high enough. The overriding process, however, is additional water loss, as pointed out in Figs. 2.14 and 2.15.

Water loss is determined not only by the water vapour that leaves the battery. As mentioned above, any loss of oxygen gas through the container walls results also in a corresponding loss of water, because an equivalent amount of hydrogen has to be produced on the negative electrode instead of oxygen reduction. Table 2.13 shows that polystyrene and ABS perform rather well in this respect.

2.6.6 Formation of Fumes

Electrolyte fumes are formed in batteries with liquid electrolyte when rising gas bubbles burst at the surface of the electrolyte. Fine droplets of sulphuric acid or potassium hydroxide are sprayed as an aerosol. If the particles are fine enough, they can float in the gas and may be carried away with the gas stream leaving the cell. For this reason, electrolyte fumes are mainly observed when a flooded battery is heavily overcharged.

The amount of electrolyte, lost as fumes, can be neglected. Nevertheless, even traces of sprayed electrolyte can be a nuisance, because the droplets are deposited in the area surrounding the battery opening.

- Sulphuric acid remains there because the acid itself is not vaporised. Rather its concentration is governed by the relative humidity, as described in Section 2.6.4 . Corrosion occurs if the acid fumes reach metallic surfaces; furthermore, the wet surface of the cells gets dirty by absorbing dust, and may reduce insulation resistance to ground.

- Potassium hydroxide is not stable when exposed to air, but is gradually converted into carbonate according to

$$\boxed{KOH + CO_2 \Rightarrow KHCO_3} \quad (2.134)$$

The generated carbonate forms a white crust of crystals on the cell which may affect the insulation resistance within the battery.

In maintenance-free batteries, liquid electrolyte is not present; consequently, gas bubbles cannot rise in the liquid and burst at its surface, and so the generation of fumes is reduced. Furthermore, there is no continual gas stream leaving the battery to carry fumes. Thus, formation of electrolyte fumes can be neglected with these batteries.

Formation of carbonate crystals is occasionally observed with sealed nickel/ cadmium and nickel/hydride batteries after prolonged service. These precipitations, however, are not caused by fumes, but by electrolyte creeping through the seal.

References

([1]) G.Kortüm and J.O'M Bockris 'Textbook of Electrochemistry', Elsevier Publishing Co., New York, Amsterdam, London, 1951

K.J.Vetter, 'Electrochemical Kinetics', Academic Press, New York, London, 1967; Original German Edition: Springer-Verlag, Berlin, 1961

J.O'M.Bockris and D.M.Drazic, 'Electro-Chemical Science', Taylor & Francis Ltd, London, 1972

A.J.Bard and L.R.Faulkner, ‘Electrochemical Methods - Fundamentals and Applications, © John Wiley & Sons, Inc., New York, 1980

F.Beck and K.Euler, 'Elektrochemische Energiespeicher', VDE-Verlag, Berlin, 1984 (in German)

T.R.Crompton, 'Battery Reference Book', Butterworth & Co, London, 1990

J.S.Newman, ‘Electrochemical Systems’. Second Edition, Prentice-Hall, Inc., Englewood Cliffs, 1993

([2]) H.Bode, 'Lead Acid Batteries', John Wiley & Sons, New York, 1977

S.U.Falk and A.J.Salkind, 'Alkaline Storage Batteries', John Wiley & Sons, New York, P.C.Milner and U.B.Thomas, 'The Nickel Cadmium Cell', *in* 'Advances in Electrochemistry and Electrochemical Engineering, Vol.5', John Wiley & Sons, New York, 1967, p.1

([3]) K.Vetter 'Electrochemical Kinetics', Academic Press, New York, London, 1967, p. 13

([4]) R.A.Robinson and R.H.Stokes, Electrolyte Solutions, Butterworth Scientific Publ.,London 1959

([5]) C.A.Hampel, 'The Encyclopedia of Electrochemistry', Reinhold Publ. Co., New York, 1964, p.414

([6]) A.J.de Bethune and N.A.Swendeman Loud, 'Standard Aqueous Electrode Potentials and Temperature Coefficients at 25 °C', D.A.Hampel, Skokie, Ill., 1964

([7]) A.J.Barn and L.R.Faulkner, loc. cit. ([1]), p.49

(8) J.S.Newman, 'Electrochemical Systems'. Second Edition, Prentice-Hall, Inc., Englewood Cliffs, 1993, p. 120 f.

(9) R.A.Robinson and R.H.Stokes, 'Electrolyte Solutions', Butterworth Scientific Publ.,London, 1959, pp. 494, 504

(10) D.J.G.Ives and J.J.Jantz, Reference Electrodes, Academic Press, New York, London 1961 p.71 .

(11) S.U.Falk and A.J.Salkind,'Alkaline Storage Batteries', John Wiley & Sons, New York, 1969, Table 7.5, p. 534

(12) S.U.Falk and A.J.Salkind,'Alkaline Storage Batterie', John Wiley & Sons, New York, 1969, p.534

(13) J.S.Newman, 'Electrochemical Systems', © Prentice-Hall, Englewood Cliffs, New Jersey, 1991, Table 11.1, p.255

(14) S.U.Falk and A.J.Salkind, 'Alkaline Storage Batteries', John Wiley & Sons, New York, 1969, p. 598

(15) Vetter, loc. cit. (1), Fig.197, p.539

(16) J.Tafel, Z phys. Chemie, **50** (1905), p.641

(17) S.U.Falk and A.J.Salkind, loc. cit.(2), p.556

(18) Textbooks of heat transfer, e.g. F.Hell, 'Grundlagen der Wärmeübertragung', VDI-Verlag GmbH, Düsseldorf ,1979, pp.84

(19) S..W.Eckroad, C.A.Luongo and R.J.Loyd, Proceedings of the Symposium on Stationary Energy Storage, The Electrochemical Society, Pennington, N.J., 1987, p.81

(20) N.Bagshaw, Batteries on Ships, © John Wiley & Sons Ltd. 1982, p. 107 f.

(21) VARTA Battery AG (Editor), 'Bleiakkumulatoren', 11th Ed., VDI-Verlag, Düsseldorf, 1986, p.117 (in German)

(22) F.J.Vaccaro and P.Casson, Proc. 9th Intelec Meeting, Stockholm, © IEEE Comm. Soc., 1987, p. 128

(23) H.Bode, 'Lead-Acid Batteries', John Wiley & Sons, New York, 1977, p.366

(24) D'Ans Lax, 'Taschenbuch für Chemiker und Physiker, Band 1', J.Springer-Verlag, Berlin,1976, pp. 1423, 1428

(25) C.M.van Atta, 'Vacuum Science and Engineering', McGraw-Hill Book Company, New York, 1965, p.26

(26) 'CRC Handbook of Chemistry and Physics' 69th Edition, 1988/89, CRC Press, Boca Raton D-273

(27) ibid, pp. F. 88, F. 102

(28) International Critical Tables Vol.III, McGraw-Hill, New York 1928, p. 373

(29) ASTM Standard E 96-80; DIN 53122

(30) H.Dominghaus, 'Die Kunststoffe und ihre Eigenschaften', 2.Auflage, © VDI-Verlag, Düsseldorf, 1986 (in German)

H.J. Saechtling, 'Kunststoff-Taschenbuch', VDI-Gesellschaft Kunststofftechnik, 24. Ausgabe, © Carl Hauser Verlag, München, 1989, p.399 (in German)

Prospectuses of Chemical Companies: Hoechst AG, BASF

(31) Proc. of the 11th Intelec Meeting, Florence 1989, Paper 6.4

(32) Ch.S.C.Bose, J.B.Baldasty, M.C.Weeks, K.R.Bullock and A.G.Cannone,Proc. 16th Intelec Meeting, Vancouver 1994, © IEEE Communications Soc., p.30

(33) E.g. Japanese Patent Application 61.13555

(34) E.g. German Patent Application 2025 631, disclosure Dec. 16th, 1971

3. Battery Parameters, Definitions

Basically, two types of batteries are distinguished:

- Primary batteries that can only be discharged.
- Secondary batteries that are rechargeable.

The border between the two types is vague. A number of battery systems that are mainly used as primary batteries can also be recharged to a certain extent. The three battery systems treated in this book are definitely secondary batteries that reach high numbers of discharging/charging cycles.

Batteries are made for a great variety of applications. They are characterised by a number of design features and performance parameters that to a certain extent are standardised (cf. Chapter 16). Due to the different requirements, such standards often only concern a special field of application and can differ from those for other applications, even for the same battery system.

3.1 Cells and Batteries

The design principle of the electrochemical cell has been shown in Fig. 2.2, p. 8, with the lead-acid battery as an example. The substances that form the cell reaction often are called the **active material**. In Fig. 2.2, PbO_2 and Pb are the active material in the positive and the negative electrodes, respectively. The term 'active material' may, however, also include additives which do not participate in the cell reaction, like expanders or conducting aids.

When the battery is discharged, the active material in the positive electrode (PbO_2) is reduced ($Pb^{4+} + 2 \cdot e^- \Rightarrow Pb^{2+}$). Such a reducing current flow is also called '**anodic**'. In the negative electrode, the active material is oxidised ($Pb \Rightarrow Pb^{2+} + 2 \cdot e^-$) by a '**cathodic**' current. During charging, these currents are reversed.

Note: *In the field of primary batteries, the positive electrode is usually called 'cathode' while the negative electrode is called 'anode'. Both terms are strictly connected to the direction of current flow through the concerned electrode which in primary batteries is unambiguous, since only discharge occurs. In Leclanché batteries, e.g., through the anode, the zinc electrode, an anodic current flows which is an oxidising current ($Zn \Rightarrow Zn^{2+} + 2\cdot e^-$). Through the cathode, the MnO_2 electrode, a cathodic current flows which is a reducing current, e.g. ($Mn^{4+} + e^- \Rightarrow Mn^{3+}$).*

In secondary batteries, it depends on charge or discharge which of the two electrodes is anode or cathode. Thus, in lead-acid batteries, the negative is the anode during discharging ($Pb \Rightarrow Pb^{2+}$), but the cathode during charging. The opposite applies to the positive electrode: $Pb^{4+} + 2\cdot e^- \Rightarrow Pb^{2+}$ is a cathodic reaction, and the PbO_2 electrode is the cathode during discharging and the anode during charging. Because of this ambiguity, the terms anode and cathode are normally not used with secondary batteries. If they are used, they concern the discharge. Characterisation of current directions by the terms 'anodic' and 'cathodic', however, is usual, and will occasionally also be applied in this book

In Fig. 2.2 only one electrode of each sign is shown. In practice often a number of positive and negative electrodes are switched in parallel within each cell.

A connection of several cells in series forms a **battery**. This difference between cells and batteries is, however, not strictly observed in practice. Single cells are frequently also designated as batteries. Single cells with a complete encapsulation are likewise called '**monocells**'. Mainly in the field of lead-acid batteries, a number of cells are often assembled within one multicell container. Such compact batteries are also called '**block batteries**' or '**monoblocks**'. For many applications, like tools or mobile telephones, nickel/cadmium and nickel/metal hydride batteries are used as **battery packs**, also called **power packs**, that combine a number of cells within a common housing. The cells in such a pack are often selected to have uniform capacity. Otherwise, it may happen that individual cells within the pack are reversed when the battery pack is discharged fairly deeply.

Battery packs, and to some extent also single cells, are often equipped with safety devices, like temperature sensors, thermal fuses, or devices that increase the internal resistance when a specified temperature is exceeded. **Smart batteries** have an incorporated controlling system, based on a processor that provides information in regard to capacity and ageing of the battery, controls proper charging, and prevents overdischarge (cf. Section 14.3).

3.2 Capacity

During discharge, the battery delivers energy, and the amount of this energy is described as the **capacity**. The precise definition of the capacity is important, because it is not an unambiguous quantity, but depends on the conditions in which it is determined.

In general form, the capacity is represented by the equation

$$C_{Wh} = \int_0^t E(t) \cdot I(t) \cdot dt \qquad /Wh \tag{3.1}$$

with E = Voltage / V
I = discharge current / A
t = discharge period / hours

It describes the energy output of the battery. Sometimes, especially in the field of motive power, such measurements are made with constant load (W) which means that voltage decrease with proceeding discharge has to be compensated by increased current.

In the general battery practice, it is more usual to measure only the current output, since its more convenient and sufficient in most cases for comparison. This definition of capacity is

$$C = \int_0^t I(t) \cdot dt \qquad /Ah \tag{3.2}$$

To derive the energy output, the so defined 'Ah capacity' has to be complemented by indications concerning the voltage. Corresponding terms are:

Starting discharge voltage or **Initial discharge voltage**: voltage at the moment the load is applied. For lead-acid batteries, usually defined as the voltage after 10% of the capacity has been discharged, and an initial voltage anomaly, the 'coup de fouet', has been passed (cf. Fig. 3.1 and text below).

End-of-discharge voltage or **final voltage**: voltage that determines the end of the discharge. It is defined by the battery manufacturer, to specify the discharge performance of the battery. But it may also be specified by the battery user, according to the requirements of the application. Any comparison of capacity data must always consider the end-of-discharge voltage.

Average discharge voltage or **mean discharge voltage**: approximate average of the voltage during the whole discharge period (average discharge voltage x current = discharged energy).

Mid point discharge voltage: voltage after 50% of the capacity has been discharged (≈ average voltage).

When the battery is discharged with a constant current, which is mainly done in the laboratory, Eq.(3.2) is simplified to

$$C = I \cdot \Delta t \tag{3.3}$$

Sometimes the results of Eqs. 3.1 and 3.2 or 3.3 are called **Wh capacity** (**C_{Wh}**) and **Ah capacity** (**C_{Ah}**), respectively.

Nominal or rated capacity: The nominal or rated capacity of a battery (C_r in Eq. (3.4) below) is specified by its manufacturer as the standard value that characterises this battery. Usually it is specified for a constant-current discharge at 20 °C or for room temperature. For the various applications, nominal or rated capacities are often referred to different discharge durations, termed as C_{20}, C_{10}, or C_5, respectively. So the rated capacities of starter batteries (SLI) but also those of small portable batteries usually are referred to the 20-h discharge rate. The rated capacity of conventional stationary lead-acid batteries is based on the 10-h discharge rate in Europe and on the 8-h discharge rate in the US. The rated capacities of traction batteries mostly are related to the 5-hour discharge rate, and this applies also to many of the alkaline batteries. But 10-h discharges are also used to specify the rated capacity of nickel/cadmium batteries. (Examples are shown in Chapters 10 to 12.) The end-of-discharge voltage is a further parameter that has to be specified for the rated capacity, as shown in Table 3.2.

Note: *The nominal or standard capacity is an important parameter that characterises the energy content of the battery. Comparison of rated capacities can, however, be misleading when batteries are selected for discharge conditions that are considerably different from those that specify the rated capacity. Then the influence of the various discharge parameters must be considered, and the comparison of capacities under such conditions can have a quite different result.*

Retained capacity: The high-rate discharge is to a large extent determined by rather slow diffusion processes. After an interruption period, required for equalisation, the discharge can be continued at a lower rate or, to a limited extent, at the high rate. The charge removed from the battery during the second step is called the **retained capacity**. In the sequence high-rate/low-rate discharge, the sum (high-rate capacity + retained capacity) equals approximately the normal low-rate capacity of the battery.

3.3 Discharge Parameters

The capacity that can be drawn from a battery under a certain load depends on the rates of reactions that determine the discharge characteristic. For this reason, it depends on the kinetic parameters that influence the rates of the various reaction steps, as has been described in Section 2.3. Thus, the capacity depends to a large extent on the design of the battery. There is a wide variation of such design parameters like plate thickness, plate distance, specific plate surface, and similar parameters.

The discharge parameters, beside the design of a battery, that mainly influence the capacity are:

- **discharge current**,
- **voltage limit**, i.e. the final voltage or end-of-discharge voltage, that has to be specified as described below
- and **temperature**.

Further parameters that also influence the capacity are the state of charge and, to some extent, the previous history of the battery.

Typical discharge curves of a lead-acid battery are shown in Fig. 3.1. Increased current reduces the cell voltage on account of polarisation effects that occur at both electrodes. Thus, at an increased current, the discharge occurs at a lower voltage, and the end-of-discharge voltage is reached earlier. The upper curve in Fig. 3.1 represents a low-rate discharge, the lower curve stands for a high-rate discharge (for actual curves cf. Fig. 4.6, p.113).

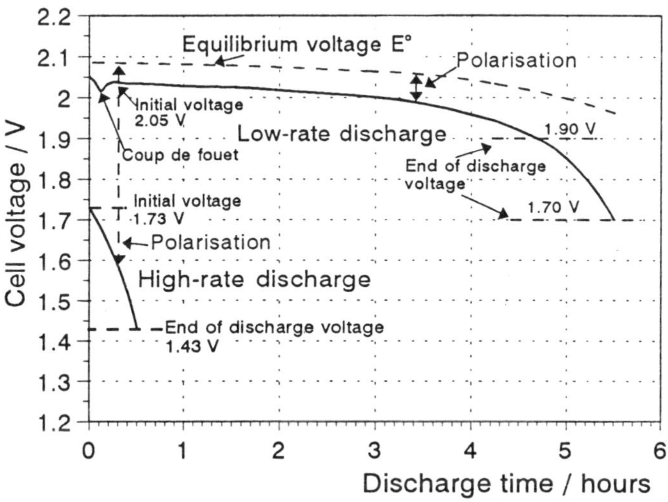

Fig. 3.1 Typical low-rate discharge and high-rate discharge curves of a lead-acid battery. The broken curve denotes the equilibrium voltage (cf. Eq. (2.14), p.13).

The broken curve in Fig. 3.1 represents the equilibrium voltage that gradually decreases with progressing discharge for a lead-acid battery on account of acid dilution. In nickel/cadmium and nickel/metal hydride batteries, E^0 is approximately constant, as indicated, e.g. in Fig. 5.10, p.197. The difference between the actual discharge curves and E^0 is caused by polarisation according to Eq. (2.45), p.23. This difference increases with progressing discharge on account of the gradual increase of the internal resistance (cf. Section 3.6).

The influence of the end-of-discharge voltage is shown in Fig. 3.1 for the low-rate discharge. On account of the slope of the curve, the discharge period would have been limited to about 4.7 hours, if 1.9 V had been chosen as end-of-discharge voltage instead of 1.7 V. A capacity reduction to 85% would be the result.

At high discharge rates, the cell voltage is so far reduced that the end-of-discharge voltage, used for the low-rate discharge, would limit the discharge at the very beginning. As a consequence, in battery practice the end-of-discharge voltage is often reduced with increasing load. In Table 3.2, e.g., the end-of-discharge voltage is reduced step-by-step from 1.85 V/cell to 1.675 V/cell for discharge rates between 10h and 0.25h. But this is not practised uniformly. In the US for example, it is the more general use to specify 1.75 V/cell as end-of-discharge voltage for a wide range of loads and only reduce this value for extremely high loads.

The **coup de fouet** or **initial voltage minimum** (in German 'Spannungssack = 'voltage pocket') is a voltage minimum at the beginning discharge of fully charged lead-acid batteries in the range of 20 mV/cell. It is caused by the positive electrode, and can be ascribed to crystallisation overvoltage ([1]). (*Nickel/cadmium and nickel/metal hydride batteries do not show a corresponding effect.*) The coup de fouet is observed at low and medium discharge rates. On account of this anomaly, the initial voltage reading is usually taken after 10% of the capacity has been discharged, and in discharge curves the first 10% are often not drawn, as in Fig. 3.2. Although the coup de fouet amounts to only about 1% of the cell voltage, it can be important for the battery layout in special applications where a high end-of-discharge voltage is specified. In such conditions, only the flat section of the discharge curve is used, and the coup de fouet can represent the minimum voltage of this section. Then this anomaly has to be considered for the battery layout (cf. Table 3.3).

3.3.1 Influence of Discharge Current, Representation of Data

The influence of the discharge current on capacity has been indicated in Fig. 3.1. In the following, some examples will be shown that simultaneously indicate the various ways that are used to represent the dependency of the capacity on current.

Note: *These Figures are examples that may not show the latest development. This is intentional: any attempt to be completely up to date would be hopeless because of the enormous number of variations and the fast progressing development. For information about the actual situation on the market, the reader can only be referred to the handbooks and brochures of the manufacturers.*

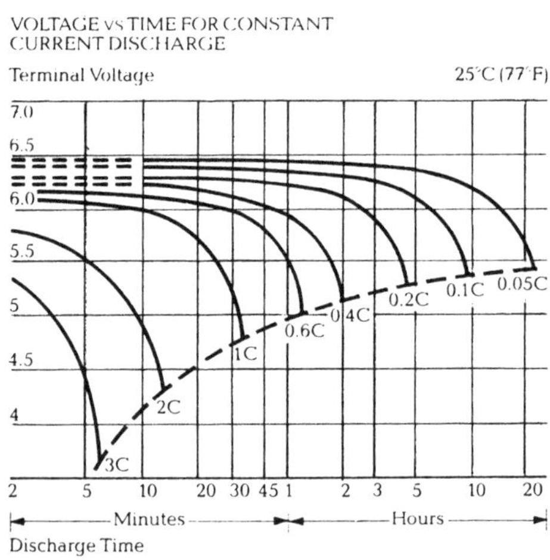

Fig. 3.2 Discharge curves for a maintenance-free lead-acid battery. Discharge time in logarithmic scale. Ratings are referred to multiples of the nominal capacity C_{10} (cf. Eq. (3.4)). The broken sections at the beginning of the curves indicate the range where the coup de fouet is to be expected (cf. Fig. 3.1). (CHLORIDE, POWERSTORE Battery.)

Fig. 3.2 corresponds to Fig. 3.1, except that the horizontal axis is in logarithmic scale. This has the advantage that the small discharge periods are expanded. But the low-rate curves appear flatter than in a linear scale.

Capacity ratings based on currents as multiples of the nominal capacity were originally mainly used for nickel/cadmium batteries. But with the development of valve-regulated lead-acid batteries, this method also became common for these smaller lead-acid batteries. The discharge current can be written as

$$\boxed{i = m \cdot C_r} \qquad (3.4)$$

with i : current in A

m : number that indicates multiple or fraction of C_r

C_r : rated or nominal capacity in Ah, but expressed in A

r : discharge period, the capacity is rated to (10 hours in Fig. 3.2)

The current as a multiple of the capacity is often written as 'CA' or 'C(A)' (cf., e.g., Fig. 3.7) to indicate that 'C' does not mean a capacity in Ah but a corresponding current in A. In regard to the dimensions, Eq. (3.4) is not correct (i in A, C_r in Ah, m without dimensions). But this description of discharging and charging currents is well established and understood.

It has, however, to be considered, that m has no relation to discharge time: m is a pure number. For this reason, discharge rates referred to multiples of capacity according to Eq.(3.4) cannot be converted into discharge times. In Fig. 3.2, e.g., 0.1 C(A) corresponds to the 10-hour rate, since C_{10} is the rated capacity. The discharge rate $1 \cdot$C(A), however, does not correspond to one-tenth of the discharge time (1 hour), but only to about 32 minutes. To evaluate the discharge duration, the discharge curves must be known.

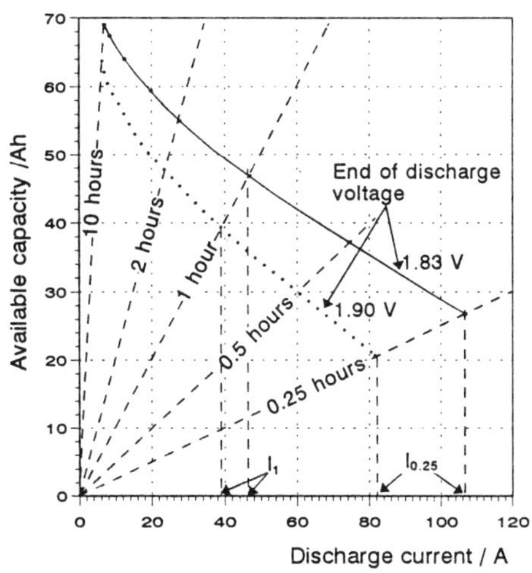

Fig. 3.3 Lead-acid battery. Dependence of the available capacity (or actual capacity) on discharge current for the end-of-discharge voltage 1.83 and 1.90 V/cell. Values listed in Table 3.1. The broken lines correspond to Eq. (3.5).

In the field of lead-acid batteries a different description is often used for charge/discharge currents where the designation of the current expresses the discharge duration. I_{10} and I_5 mean currents that result in 10 and 5 hours of discharge duration.

Fig. 3.3 illustrates how these currents are derived. In this example, data are used from Table 3.3 (Vsb 6904T, end-of-discharge voltage 1.83 and 1.90 V/cell). The rated capacity at the end-of-discharge voltage of 1.83 V/cell, C_{10}, is 69 Ah, and the corresponding current I_{10} is 6.9 A. When this discharge rate is applied, 100% of the rated capacity (= 69 A) can be drawn. Increased discharge rate reduces the available capacity, as shown by the Ah/A curve.

The slopes of the broken lines in the Figure correspond to the simple relation:

$$\frac{\text{Available capacity (Ah)}}{\text{Discharge current (A)}} = \text{Discharge duration (hours)} \tag{3.5}$$

So the 10, 1, and 0.25 hour lines correspond to slopes of 70 Ah/7 A, 70 Ah/70 A, and 30 Ah/120 A, respectively. The intersections of these lines with the measured capacity/current curve determines the available capacities and the discharge currents, as shown for the 1-h and 0.25-h rate.

The solid curve in Fig. 3.3 is based on the end-of-discharge voltage of 1.83 V/cell. The dotted curve indicates the decreased values, when the end-of-discharge voltage is increased to 1.90 V/cell.

The results are compiled in Table 3.1:

End of disc. Volt.	1.83 V/cell		1.90 V/cell	
	Current (A)	Capacity (Ah)	Current (A)	Capacity (Ah)
10-hour rate	6.9 (I_{10})	69	6.3 (I_{10})	63
1-hour rate	47 (I_1)	47	39 (I_1)	39
0.25-hour rate	107 ($I_{0.25}$)	27	82 ($I_{0.25}$)	20.5

Table 3.1 Discharge currents and available capacities according to Fig. 3.3.

Comparison of the values derived for the two end-of-discharge voltages indicates how important it is to consider the specification of this voltage.

For stationary lead-acid batteries, the discharge data are often compiled in tables as shown in Tables 3.2 and 3.3. With the aid of these data, the user can easily select a battery that delivers the required energy when the autonomy period and the minimum voltage are specified.

The data in Table 3.2 show the rated capacities, specified for various discharge durations and end-of-discharge voltages, and the corresponding discharge currents. Individual end-of-discharge voltages are assigned to the different discharge rates, as mentioned with Fig. 3.1.

Electrical Data

Type	Capacity (Ah) hourly						Discharge current (A) hourly						End discharge voltage (V/cell) hourly					
	10	8	3	1	1/3	1/6	10	8	3	1	1/3	1/6	10	8	3	1	1/3	1/6
Vsb 6904 T	68.0	66.0	60.0	49.2	35.6	27.2	6.8	8.25	20.0	49.2	106.8	163.2						
Vsb 6905 T	85.0	82.5	75.0	61.5	44.5	34.0	8.5	10.31	25.0	61.5	133.5	204.0	1.85	1.845	1.825	1.79	1.73	1.675
Vsb 6907 T	119.0	115.5	105.0	86.1	62.3	47.6	11.9	14.44	35.0	86.1	186.9	285.6						
Vsb 6910 T	170.0	165.0	150.0	123.0	89.0	68.0	17.0	20.63	50.0	123.0	267.0	408.0						

Table 3.2 Rated capacities for different discharge duration and end-of discharge voltages, and the corresponding discharge currents. Valve-regulated lead-acid battery (12 volt blocks) for stationary application (VARTA, Vsb-Battery, 1986 design).

Table 3.3 shows 'project planning data' for the same type of battery. Each of these three Tables is assigned to a certain end-of-discharge voltage (1.9, 1.87 and 1.83 V/cell). The top lines of these tables show the discharge duration, and the lines below show the currents that can be drawn from the different battery sizes during this period, before the end-of-discharge voltage is reached. The required size and number of cells for a battery in a standby application can easily be deduced from this Table, when for this application the autonomy period and the minimum voltage are specified.

Project planning data

E_{min} = 1.90V/cell Discharge current in A
6V-Block Batteries

Type	30"	1'	2'	3'	5'	7'	10'	15'	20'	30'	1h	2h	3h	5h	8h	10h
Vsb 6904 T	124.0	124.0	124.0	124.0	114.0	106.0	96.0	82.0	73.6	60.0	39.2	23.8	17.2	11.4	7.6	6.3
Vsb 6905 T	155.0	155.0	155.0	155.0	142.5	132.5	120.0	102.5	92.0	75.0	49.0	29.8	21.5	14.2	9.5	7.9
Vsb 6907 T	217.0	217.0	217.0	217.0	199.5	185.5	168.0	143.5	128.8	105.0	68.6	41.7	30.1	19.9	13.3	11.0
Vsb 6910 T	310.0	310.0	310.0	310.0	285.0	265.0	240.0	205.0	184.0	150.0	98.0	59.5	43.0	28.4	19.0	15.7

E_{min} = 1.87V/cell Discharge current in A
6V-Block Batteries

Type	30"	1'	2'	3'	5'	7'	10'	15'	20'	30'	1h	2h	3h	5h	8h	10h
Vsb 6904 T	152.0	152.0	152.0	146.0	130.0	120.0	110.0	94.0	81.6	67.2	42.8	25.8	18.6	12.0	8.1	6.0
Vsb 6905 T	190.0	190.0	190.0	182.5	162.5	150.0	137.5	117.5	102.0	84.0	53.5	32.3	23.3	15.0	10.1	8.3
Vsb 6907 T	266.0	266.0	266.0	255.5	227.5	210.0	192.5	164.5	142.8	117.6	74.9	45.2	32.6	21.0	14.1	11.6
Vsb 6910 T	380.0	380.0	380.0	365.0	325.0	300.0	275.0	235.0	204.0	168.0	107.0	64.5	46.5	30.0	20.2	16.6

E_{min} = 1.83V/cell Discharge current in A
6V-Block Batteries

Type	30"	1'	2'	3'	5'	7'	10'	15'	20'	30'	1h	2h	3h	5h	8h	10h
Vsb 6904 T	194.0	194.0	184.0	174.0	154.0	140.0	124.0	106.8	92.8	74.4	46.8	27.7	19.8	12.5	8.4	6.9
Vsb 6905 T	242.5	242.5	230.0	217.5	192.5	175.0	155.0	133.5	116.0	93.0	58.5	34.6	24.8	15.6	10.5	8.6
Vsb 6907 T	339.5	339.5	322.0	304.5	269.5	245.0	217.0	186.9	162.4	130.2	81.9	48.4	34.7	21.8	14.7	12.0
Vsb 6910 T	485.0	485.0	460.0	435.0	385.0	350.0	310.0	267.0	232.0	186.0	117.0	69.2	49.6	31.2	21.0	17.2

Table 3.3 The relation between discharge duration and discharge current for three different end-of-discharge voltages. Extract of the project data list for the same battery whose standard performance data are compiled in Table 3.2.

For very short discharge durations, some of the current values in Table 3.3 remain unchanged. This is caused by the coup de fouet, described in Fig. 3.1. Under a high load and at a high end-of-discharge voltage, only a short part of the (nearly horizontal) discharge curve can be used. Under such circumstances the coup de fouet represents the lowest point of the discharge curve and limits further current increase. With decreasing end-of-discharge voltage, the coup de fouet loses its importance since now the end of the more declining discharge curve represents the lowest point of the discharge curve. For this reason, the range with constant-current values is extended to 3 minute discharges at 1.90 V/cell end-of-discharge voltage, while for 1.83 V/cell as final voltage, this range ends at 1 minute discharges.

Nickel/cadmium, nickel/metal hydride batteries: Fig. 3.4 shows discharge curves for sealed nickel/cadmium button cells in comparison to corresponding nickel/metal hydride batteries. The horizontal axis shows the delivered capacity as a percentage of the nominal or rated capacity. According to Eq. (3.4), the (constant) discharge current is given in multiples of the rated capacity C_r which for this cell type is C_5 and the end-of-discharge voltage is specified between 1.1 and 0.9 V/cell, depending on discharge current. The horizontal axis shows the available capacity as a percentage of C_5.

Fig. 3.4 shows the similarity of the discharge behaviour between Ni/Cd and Ni/MeH batteries. Only at high loads, the capacity reduction of nickel/metal hydride batteries is slightly higher since the desorption rate of hydrogen becomes limiting. (C_5 is of course higher for the nickel/metal hydride battery as shown in Fig. 6.6, p.229.)

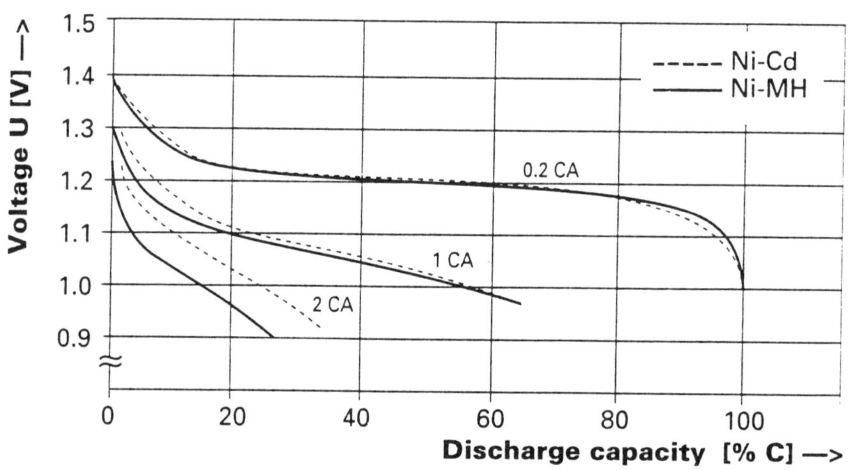

Fig. 3.4 Discharge curves for button cells (nickel/cadmium and nickel/metal hydride). Voltage referred to percentage of the rated capacity (C_5). Current ratings in multiples of the rated capacity according to Eq. (3.4). (VARTA ([2]).)

The great influence of battery design on discharge behaviour can be seen by comparison of Fig. 3.4 with Fig. 5.10, p.197. The latter also shows discharge curves for a sealed nickel/cadmium battery, but for a cylindrical cell with sintered electrodes. For this battery type the discharge currents are related to the 10-hour capacity ($C_r = C_{10} =$ 10 Ah). 100 x I_{10} in Fig. 5.10 approximately corresponds to 5 CA in Fig. 3.4. The superior high-rate performance of the battery with sintered electrode is obvious.

A further way to present discharge data is shown in Fig. 3.5. Here discharge duration is plotted versus the discharge current in a double logarithmic scale. The resulting lines allow an easy selection of the appropriate battery size when load and discharge duration are specified.

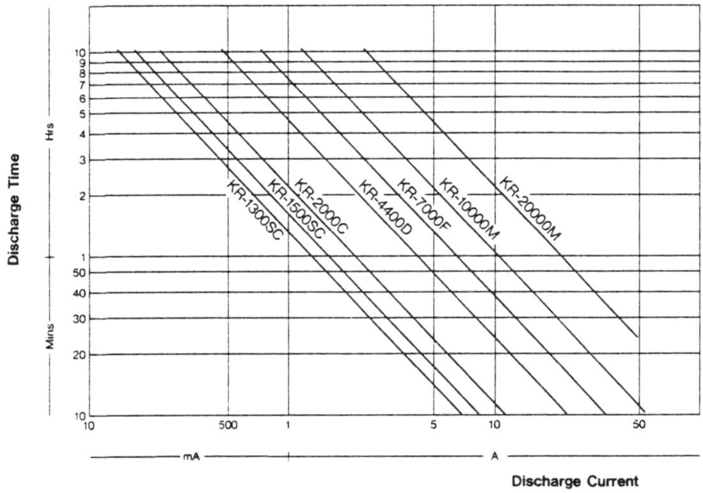

Fig. 3.5 Discharge curves for cylindrical sealed nickel/cadmium batteries. Discharge time vs. discharge current in double logarithmic scales. (SANYO [3].)

3.3.2 Temperature Influence on Capacity

Fig. 3.6 shows an example for the influence of the temperature on capacity.

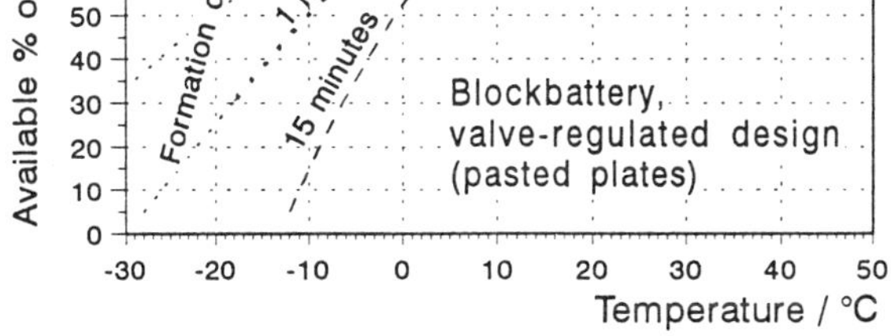

Fig. 3.6 Valve-regulated lead-acid battery. Temperature influence on capacity at various discharge rates (VARTA).

The four curves in Fig. 3.6 are given as percentage of their rated values at 20 °C. Actually, these rated values decrease with increasing current, as shown in Fig. 3.7. But the plot of Fig. 3.6 shows more clearly the increasing influence at higher discharge currents. The slope of the 15 min. discharge curve with decreasing temperature is much steeper than that of the 20-hour discharge curve. The reason is that hindrance of the kinetics gains more importance at high loads. At temperatures above 20 °C, a stronger increase of the capacities at the higher discharge rates is observed, compared to their rated values, since now the kinetic hindrance is reduced and a larger part of capacity is available, that at room temperature remains unused.

A general problem for lead-acid batteries at minus temperatures is the formation of ice, which raises the internal resistance and causes increased polarisation. On account of the higher utilisation at low discharge rates, and the consequently more diluted acid, the critical point is reached for the low-rate discharge at a higher temperature compared to high-rate discharge. At very high discharge rates, a further complication can be ice formation within the pore system of the active material, since there may not be enough time for equalisation by diffusion. To counteract ice formation, occasionally a higher concentrated electrolyte (arctic electrolyte) is used for batteries that continuously are operated at low temperatures.

Fig. 3.7 shows also the influence of the temperature on capacity. Here the curves are referred to the rated capacity at the 20-hour discharge rate. (The current I = 0.05 $C^{(A)}$ discharges 100%).

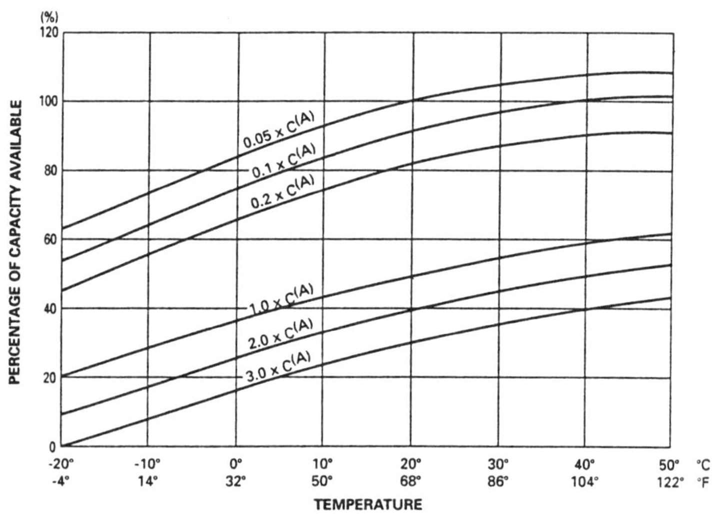

Fig. 3.7 Valve-regulated lead-acid batteries. Effect of temperature and discharge current on capacity. Curves referred to the 20-hour discharge capacity. Yuasa NP series ([4]).

In **nickel/cadmium batteries**, the capacity also is remarkably reduced with decreasing temperature. Although the electrolyte is not diluted, its conductivity is reduced significantly at low temperatures (cf. Fig. 5.14, p.206), and the kinetic parameters of the reactions also depend on temperature.

Fig. 3.8 shows the influence of temperature on the available capacity of nickel/cadmium batteries.

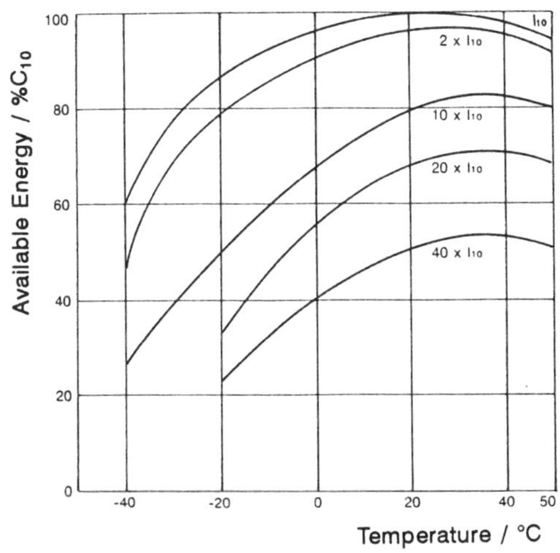

Fig. 3.8 Sealed nickel/cadmium batteries. Available energy in % C_{10} of cylindrical sintered cells as a function of temperature. Charging at the nominal rate at room temperature. VARTA RS Type ([5]).

Fig. 3.8 shows that the available capacity rapidly falls with temperature. Since the electrolyte is not diluted during discharge, ice formation, as described in Fig. 3.6, is not relevant for nickel/cadmium batteries.

In Fig. 3.8 a decrease of capacity can also be observed at temperatures above 30 °C. This is caused by the instability of the higher nickel hydroxides at elevated temperatures.

Nickel/metal hydride batteries: The effect of temperature on the capacity is similar to the corresponding effect on nickel/cadmium batteries. At high temperatures the instability of the nickel hydroxide results is the same decrease of capacity as mentioned with Fig. 3.8. At low temperatures, the reduced desorption rate of the hydrogen increases the polarisation of the negative electrode. For this reason, the capacity decrease at low temperatures is greater, compared to that of nickel/cadmium batteries.

Fig. 3.9 shows discharge curves for nickel/metal hydride batteries at various temperatures.

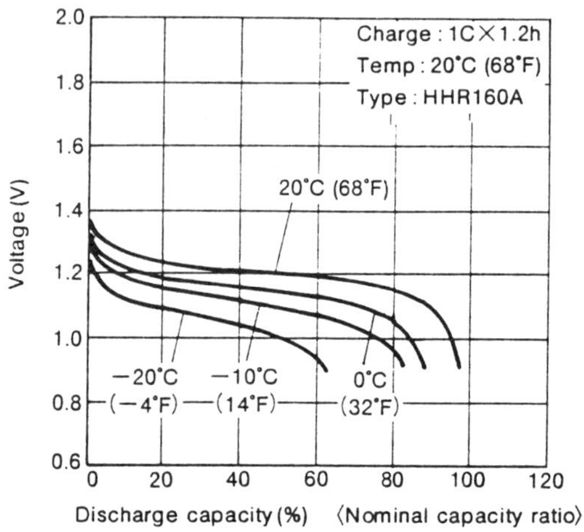

Fig. 3.9 Nickel/metal hydride batteries. Discharge curves at various temperatures. Discharge at $1C_5$. Panasonic ([6]).

Note: *In industrial standards ([7]) approximations are often used to correct capacity readings to the standardised reference temperature with the temperature coefficient δ:*

$$C_a = \frac{C}{1+\delta(\Theta-20)} \tag{3.6}$$

with C *: capacity measured at temperature* Θ

C_a *: actual capacity at 20 °C*

Θ *: temperature in degrees Celsius*

δ *= 0.006 for most lead-acid batteries unless otherwise specified by the manufacturer*

The factor $\delta = 0.006$ *means a temperature coefficient of 0.6% per degree. Comparison with the above Figures shows that Eq. (3.6) applies to only a small temperature range.*

3.4 Deep Discharge

'Deep discharge' means that the battery is discharged beyond the end-of-discharge voltage specified by the manufacturer. Thus, an excessive capacity is drawn from the battery. Deep discharging is disadvantageous because the active material is worked more, especially when deep discharges occur repeatedly, and the consequences are earlier degradation and loss of capacity. Even with normal end-of-discharge voltages the possible number of cycles decreases markedly with increased depth of discharge. Figs. 3.10 and 3.14 show such relations between cycle numbers and depth of discharge for sealed nickel/cadmium and lead-acid batteries, respectively. When the discharge is extended beyond its normal end, the increased stress of the active material can dramatically reduce the number of cycles or cause premature failure.

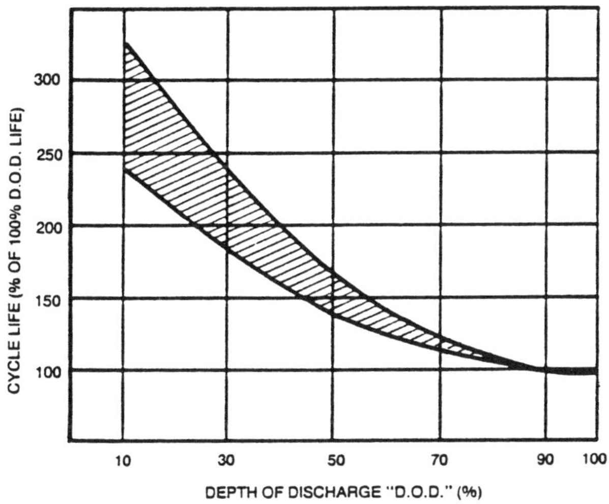

Fig. 3.10 Sealed nickel/cadmium batteries. Effect of depth of discharge on cycle life (from ([8])), cf. also Fig. 3.14.

Deep discharge is mainly a problem of lead-acid batteries. For nickel/cadmium and nickel/metal hydride batteries, occasional deep discharging are not critical. Prolonged reversal may cause short-circuits in Ni/Cd batteries, because of cadmium dendrites which then grow at the positive electrode, and in Ni/MeH Batteries oxidation of the hydrogen-storing alloy can be caused. Furthermore, in both systems stabilising additives such as cobalt oxide can be reduced in the positive electrodes at too low an electrode potential.

3.4.1 Deep Discharge of Lead-Acid Batteries

In lead-acid batteries, deep discharges are critical for both electrodes:

- In the solution mechanism (Section 4.3) the active material is converted extensively and its original structure is lost when lead (Pb) and lead dioxide (PbO_2) are dissolved in the electrolyte and precipitated as a new chemical compound, lead sulphate ($PbSO_4$). Figs. 10.1 to 10.3 (p.314 and 315) show the different appearance of these substances. The original structure of the active material is largely lost, when this conversion is carried too far.

  This is especially critical for the positive electrode, because the conducting bridges between the lead dioxide particles are degraded and the particles may partially lose contact between themselves and with the grid, which means that they can no longer participate in the electrochemical reactions (cf. Section 10.1.1.1).

The mechanical strength of the active material is also weakened when the bonds between the particles are destroyed and the tendency to material loss (shedding) is increased.

- For the negative electrode, deep discharges are critical because of the danger of reversal. Expanding additives of the active material may then be lost by oxidation.
- The danger that lead dendrites grow during the subsequent recharge is increased in lead-acid batteries by the fact that most of the acid is consumed during deep discharge, and the residual electrolyte may become diluted. In such a solution, the solubility of lead is increased markedly and thereby so is the tendency to form thin fibres (dendrites) that surround the separator and form short-circuits when the battery is recharged later. As a remedy, often sodium sulphate (Na_2SO_4) is added to the electrolyte to reduce the solubility of sulphate when the battery is discharged.

In battery practice, the definition of deep discharge is fairly vague:

- In the field of traction batteries, a discharge that draws more than 80% of the nominal capacity is already called a 'deep discharge'.
- In stationary batteries, the usual understanding of 'deep discharge' is that they have been discharged beyond the specified end-of-discharge voltage and individual cells may even have been reversed.

The reason for the different interpretations is that a repetition of the 'deep discharge' has more serious consequences. With traction batteries, the discharge is repeated each cycle and the effect of 'deep discharges' is additive. Stationary batteries, on the other hand, are discharged only rarely and 'deep discharges' are also rare events. These sporadic 'deep discharges' can be much more intensive, before damage of the battery becomes evident.

Valve-regulated lead-acid batteries are less sensitive to deep discharges. The immobilisation of the electrolyte reduces the ion mobility because convection of the liquid is not possible. This retards the dilution of the acid between the plates and prevents concentration profiles caused by gravity (stratification). The development of short-circuits is therefore considerably retarded. This effect especially applies to gelled electrolyte on account of the higher immobilisation (cf. Section 4.6.4). So valve-regulated lead-acid batteries can be stored in a deep-discharged state for more than one month without being harmed ([9]). But this advantage must not be confused with an ability to resist repeated deep discharges. They age the active material as described above.

The described effects of deep discharges indicate that a 'deep discharge' cannot be characterised sufficiently by a too-low final voltage, but is a question of the grade of conversion that the active material undergoes. At high loads, low end-of-discharge voltages may be reached, mainly on account of acid depletion within the pores of the active material, but only a limited part of the active material is converted. Then the structure of the electrode material is changed only superficially. On the other hand, discharge at a small current may cause excessive conversion of the active material, even at fairly normal end-of-discharge voltages. To summarise:

- High-rate discharge to low end voltages does not harm the lead-acid battery unless heat evolution is too high or the current-connecting parts are overheated.
- Complete discharges at low current densities with excessive utilisation of the active material are detrimental.

Repetition of deep discharges is another critical factor. Each deep discharge aggravates the situation with the result that a large number of less critical 'deep discharges' cause the same damage as a small number of severe ones.

3.5 Charge Parameters

The charging process basically represents the reversal of the discharge. It starts as soon as the voltage of the cell attains a value just above the equilibrium voltage E^0. In practice, however, higher voltages have to be applied to achieve full charge within a desired period and for all cells in the series connection of a battery. But with increased cell voltage, the secondary reactions of water decomposition gain more significance as parasitic reactions. Charging methods therefore always have to compromise between charging at a rather low rate with high efficiency, and faster but less effective charging. As a consequence, a great number of different charging methods have been developed, adapted to the type of battery as well as to the requirements of the application. Such charging schedules are discussed in some detail in Chapter 13.

The charging process is mainly influenced by the three parameters:

- **Charging current**
- **Charging voltage**
- **Temperature of the battery**

3.5.1 Charge Acceptance, Charge-Current Efficiency

'Charge acceptance' or 'charge-current efficiency' describes the share of the current, that actually is 'accepted' by the battery and could be retrieved by subsequent discharging. It is determined by the balance between the kinetic parameters of the charging reaction and the secondary processes that simultaneously occur. Charge acceptance is not a constant parameter, but depends largely on the state of charge.

As an example, typical curves for charging a lead-acid battery by a combination of constant current and constant voltage are shown in Fig. 3.11. The charging current is limited to 40 A which means ≈4.2 hour discharge duration for the rated capacity C_{10} = 200 Ah in this example.

The initial charge acceptance of the discharged battery is high, and the charging rate is limited by the maximum current, supplied by the charger. During this period, the final cell voltage is not yet reached.

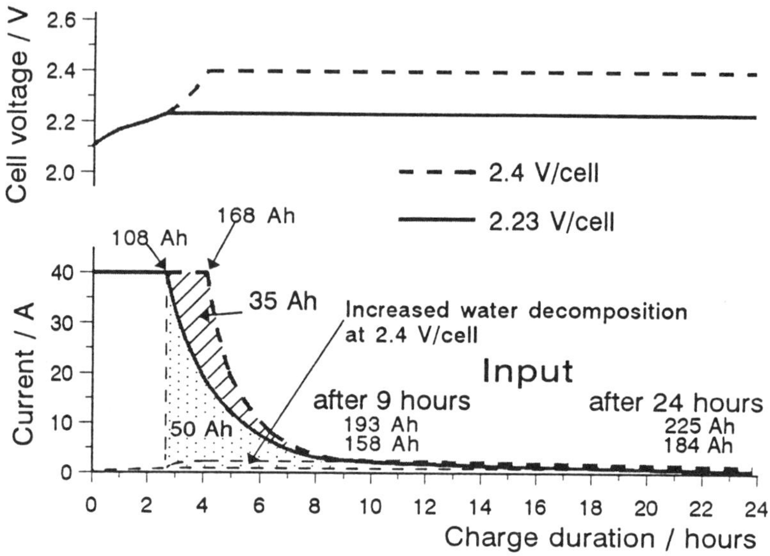

Fig. 3.11 Constant-current / constant-voltage recharge of a vented lead-acid battery after complete discharge. (200 Ah had been drawn before recharging.) (VARTA bloc 428, $C_{10} = 200$ Ah.)

After 2.7 hours (input 108 Ah ≈ 54% of the preceding discharge) the cell voltage of 2.23 V is reached. When this voltage is maintained, the current starts to decrease since charge acceptance of the battery is reduced more and more. Finally the current rate approximates to the value that remains when the battery is fully charged and only the secondary reactions, hydrogen and oxygen evolution, occur. This is the 'float current' that is observed when the battery is continuously operated at that voltage. In the example of Fig. 3.11 it would be about 60 mA (30 mA/100 Ah).

Between 2.7 and 9 hours of charging time, about 50 Ah are recharged, as indicated by the dotted area underneath the decreasing current curve. So, after 9 hours 158 Ah = 79% is recharged at 2.23 V/cell.

When 2.4 V/cell is applied as charging voltage, it takes 4.2 hours until this level is reached (input 168 Ah ≈ 84% of the preceding discharge). But the current decay is steeper with 2.4 V/cell, and when both current curves meet after about 9 hours, the gain in input for the higher charging voltage is only about 35 Ah. Then 203 Ah are 'recharged' in total. So the battery is nearly fully charged after 9 hours, even under the consideration that the charge acceptance is somewhat reduced, since water decomposition occurs about 7 times faster at 2.4 V/cell. (*0.17 V increases the current by 0.85 logarithmic decade, cf. Fig. 7.4, p.241.*) Thus, the parasitic current that decomposes water is increased to about 0.4 A (0.2 A/100 Ah). During the whole charging time (24 hours) at 2.23 and 2.4 V/cell, the water loss is equivalent to 1.2 and 8 Ah, respectively. Thus, after 9 hours 0.45 and 3 Ah are lost by water decomposition at 2.23 and 2.4 V, respectively. The increased water loss is the reason that increased charge voltage normally is only used when a certain level of recharge has to be reached within limited time.

In Fig. 3.11, the battery is not fully charged after 24 hours when 2.23 V/cell are applied (≈183 Ah = 91% is recharged), while complete charge is reached at 2.4 V/cell after about 10 hours.

Fig. 3.11 illustrates the strong influence of charge acceptance. At the beginning, a discharged battery can be recharged very fast. But it takes a certain time to get the battery fully charged, and the possibility to shorten this process by voltage increase is limited.

The example for lead-acid batteries applies also to the other battery systems. In general, charge acceptance is high for batteries designed for heavy load currents (large electrode surface area, low internal resistance). For this reason, charge acceptance is comparatively high for nickel/cadmium and nickel/metal hydride batteries. They can be recharged with high currents, but only under special precautions, as shown in Section 13.2.2, to avoid unwanted gassing or heat effects.

Charge acceptance is also influenced by temperature. The reduced reaction rates at low temperatures require increased polarisation, and at a given voltage a smaller current flow results. Sufficient charge acceptance at low temperatures is important for starter batteries in motor cars to keep the batteries charged in winter time.

3.5.2 Efficiency of the Charging Process

On account of the secondary reactions, the charging process never reaches 100% efficiency. The efficiency actually achieved, depends on the above described parameters. Often the **charging time** is an important parameter for the decision as to which charging method is used to reach both complete recharge and a satisfactory efficiency. This efficiency is expressed either by the

$$\text{Ampere-hour efficiency } \eta_{Ah} = \frac{\text{Available capacity in Ah}}{\text{Required charge input in Ah}} \tag{3.7}$$

or its reciprocal, the

$$\text{Ah charging factor } \alpha_{Ah} = \frac{\text{Required charge input in Ah}}{\text{Available capacity in Ah}} \tag{3.8}$$

For many applications the energy delivered by the battery is the more important parameter. The corresponding relations are

$$\text{Watt-hour efficiency } \eta_{Wh} = \frac{\text{Available energy (Wh)}}{\text{Required input in Wh}} \tag{3.9}$$

or its reciprocal, the

$$\text{Watt-hour charging factor } \alpha_{Wh} = \frac{\text{Required input in Wh}}{\text{Available energy (Wh)}} \tag{3.10}$$

On account of the voltage drop between charging and discharging, the watt-hour efficiency is always smaller, and the watt-hour charging factor larger, than the corresponding terms referred to Ah.

Battery system	η_{Ah}	α_{Ah}	η_{Wh}	α_{Wh}
Valve-regulated lead-acid	0.8 - 0.94	1.07 - 1.2	0.65 - 0.8	1.2 - 1.5
Sealed nickel/cadmium and nickel/metal hydride batteries	0.7 - 0.85	1.2 - 1.4	0.6 - 0.75	1.4 - 1.65

Table 3.4 Ah and Wh charging efficiencies (η_{Ah} and η_{Wh} resp.) and the Ah and Wh charging factors (α_{Ah} and α_{Wh} resp.) achieved in favourable conditions.

The charging factor is occasionally also called the charging coefficient.

Most charging methods represent combinations of several charging steps, where either the current or the voltage is determined. Occasionally, continuous charging with constant current or constant voltage is also applied (cf. Chapter 13).

Note 1: *Charging efficiency and charging factor are appropriate parameters only for charging processes limited in time. They lose any meaning with prolonged overcharging, e.g. during float charging, when only the secondary reactions are maintained by the current and 'charging' no longer occurs. The charging factor would then exceed all limits and the efficiency approach zero.*

Note 2: *All four parameters in Eqs.(3.7) to (3.10) depend not only on the rate of discharge and charge, but also on the depth of discharge when the charging process starts and on the degree of recharge finally reached, because of the decreasing charge acceptance of the battery with increasing state of charge.*

As a result, a high efficiency (a low charging factor) can be reached when, after a fairly deep discharge, the battery is recharged only partially, since nearly all the current actually is recharged. The charging efficiency is necessarily lower when the battery is recharged completely, since at the end the charging process, most of the current is used for water decomposition or the internal oxygen cycle.

Quoted values as in Table 3.4 usually refer to the complete discharge at its nominal rate and a suitable charging schedule that achieves minimum overcharge but ensures complete recharge.

3.6 Internal Resistance

The internal resistance of a battery is often quoted as a characteristic parameter. A battery that is suitable for high-rate discharges must have a low internal resistance, otherwise the voltage drop caused by the current would limit the discharge too early (cf. Fig. 3.1, p.71). Another field of application that makes use of the internal resistance is conductance measurements that during recent years attracted great attention as a tool to determine the state of valve-regulated lead-acid batteries in stationary applications (cf. Section 3.6.4).

The meaning of the term 'internal resistance' has to be considered with some caution, because it is not a simple ohmic resistance.

- The charge-transfer step or the transfer reaction are characterised by the current/voltage curves or current/potential curves, discussed in Section 2.3, which do not lead to a linear relation between voltage and current. Consequently, the internal resistance, determined for a battery, is not an ohmic resistance, but depends on the current.
- The current/voltage curves are not symmetrical for the forward and backward reactions (cf. Fig. 2.8, p.29). As a consequence, the internal resistance usually is different for the discharging and charging reactions. Values quoted for the internal resistance in general concern the discharge of the battery.
- The internal resistance depends on the state of charge of the battery.
- The internal resistance is increased at reduced temperature owing to reduced conductivity and retarded kinetic parameters.

For these reasons, the values quoted for the internal resistance of a battery can only be approximations that depend on the measuring method.

A part of the internal resistance is the grid resistance, described in Section 10.2.1.

3.6.1 The Direct-Current Method

The direct-current method is often applied in battery practice. With this method, the internal resistance (the d.c. resistance) of the battery is determined by measuring the terminal voltage at various loads. The battery is loaded with the current i_1 for a few seconds and the voltage E_1 is measured. Then the current is increased to the value i_2 and the battery voltage is reduced to the value E_2.

The R_i can be calculated according to the formula:

$$\boxed{R_i = \frac{E_1 - E_2}{i_2 - i_1} = \frac{\Delta E}{\Delta i}} \qquad (3.11)$$

Eq. (3.11) represents the usual method to determine an ohmic resistance, but in batteries the result depends on the currents i_1 and i_2 that were applied, as well as on the difference Δi between them. To achieve values for R_i that are relevant to practical applications, the currents i_1 and i_2 usually are specified and adapted to different battery types (cf. Tables 3.6 and 3.7, p.90).

The internal resistance R_i, determined according to Eq. (3.11), comprises ohmic resistance within the electrodes and the electrolyte as well as overvoltage at the phase boundaries between the electrodes and the electrolyte. Eq. (3.11) can only be applied for the discharge reaction, because it implies that the overvoltage is comparatively small compared to the ohmic voltage drop. This is true for the discharge.

The overvoltage increases with proceeding charge, and especially the high overvoltages for hydrogen and oxygen evolution, would lead to high internal resistance that would depend so strongly on current, that a simple linear relationship as in Eq. (3.11) would make no sense.

3.6.1.1 Short-Circuit Current

Short-circuit current flows when the two terminals of a cell or battery are (inadvertently) connected to each other. The short-circuit current is determined by the internal resistance of the cell or battery (cf. also Section 4.5.3.4).

The short-circuit current represents a dynamic parameter that decreases quickly with proceeding discharge. Published values are always referred to the charged battery. The short-circuit current is (approximately) determined according to Eq. (3.11) by extrapolating the resulting line to the voltage zero.

The short-circuit current represents the maximum current that could be supplied by the battery and could flow through the consuming device for a short period of time. The value of the short-circuit current helps in estimating the size of fuses to operate with the battery.

3.6.2 Alternating-Current Method; Cell Impedance; Cell Conductance

When an alternating current of sine-wave shape and the frequency ω flows through the battery the resulting complex resistance is composed of three components:

	Resistance	Conductance
Ohmic Resistance	R	1/R
Inductance	$i\omega L$	$1/i\omega L$
Capacitance	$1/i\omega C_{DL}$	$i\omega C_{DL}$

Table 3.5 Components of the alternating resistance. i denotes the imaginary unit, C_{DL} is caused by charge shifting within the double layer at the phase boundary between the active material and the electrolyte.

The resulting resistance Z can be written as:

$$Z = \sqrt{R^2 + \left(i\omega L + \frac{1}{i\omega C_{DL}}\right)^2} \qquad 3.12$$

This is a complex resistance that depends on frequency.

The inductance of batteries is very small (in the order of mH or even nH) and can usually be neglected. The double layer capacitance is quite high on account of the large surface area of the active material (cf. Chapter 9). At a usual value of 10 μF per cm^2 of the true surface area, it amounts to the order of magnitude of Farads per Ah of battery capacity. (*Often it cannot precisely be separated from the capacity caused by the electrochemical reactions*.)

Alternating-current methods are often applied in electrochemical research to determine the kinetic parameters of a reaction (cf., e.g., ([10])). The method is based on a comparison of the measured impedance values with those calculated for an equivalent-circuit diagram. Measurements at low frequencies can be particularly related to diffusion processes and concentration overvoltages. Analysis of complete impedance measurements is rather difficult and requires sophisticated calculation programs that have to be adapted to the special measurement. One reason is that no universal equivalent-circuit diagram is available to describe the response of an electrode. The situation is complicated if the method is applied to complete cells or even batteries, because the influence of the positive and negative electrodes and all the other components can hardly be separated.

For this reason, impedance measurements in battery practice are usually limited to measurements with fixed frequencies. Often 1 kHz is applied, and the real component of the complex resistance is used as 'impedance' or 'conductance' since at these high frequencies mainly ohmic resistances are registered on account of the high values of C_{DL}. An advantage of the a.c. method is that the battery is not discharged by the measurement.

Fig. 3.12 shows discharge curves of a lead-acid battery together with the internal resistance curves.

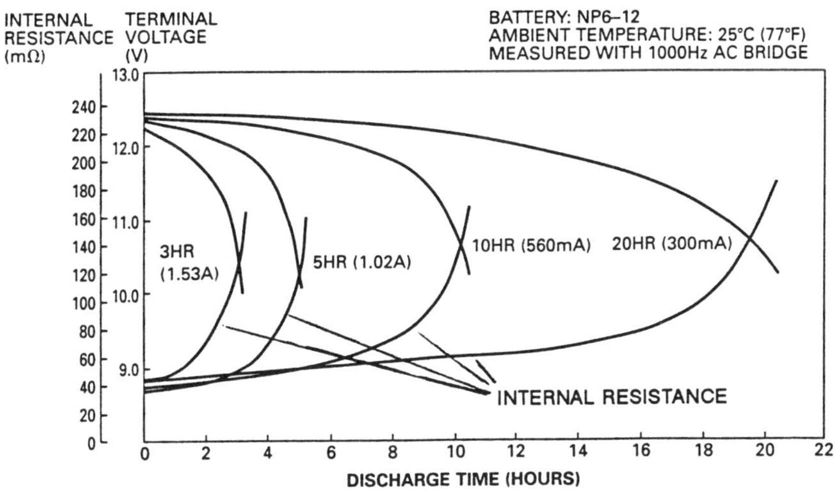

Fig. 3.12 Valve-regulated lead-acid battery. Discharge curves and internal resistance. (YUASA NP Series (4).)

With progressing discharge an increase of the internal resistance is observed. It is partly caused by acid dilution, but mostly by the decreasing conductivity of the active material.

3.6.3 Data in Practice, Comparison of Both Methods

Values of the internal resistance for various valve-regulated lead-acid batteries are listed in Table 3.6, all for one cell and 100 Ah of nominal capacity.

	Cylindrical	Prismatic cells		
	Pasted plates	Pasted plates	Pasted plates	Tubular plates
	Absorbent type	Absorbent type	Gel type	Gel type
	2.5 - 5 Ah	10 - 100 Ah	50 - 100 Ah	100 - 3000 Ah
mΩ/100 Ah	0.1 - 0.3	0.3 - 0.7	0.5 - 1.2	1 - 3

Table 3.6 Internal resistance of valve-regulated lead-acid batteries, all referred to 100 Ah of nominal capacity.

The differences in Table 3.6 depend partly on the fact that with gelled electrolyte a microporous separator has to be used to prevent penetration of lead dendrites, causing a slight increase of internal resistance. Another factor influencing the internal resistance is the different plate size, because the resistance of the plate grows with size, when the other parameters (plate thickness, grid structure) remain unchanged (cf. effective resistance of plates in Section 10.2.1).

Table 3.7 shows data for the internal resistance of nickel/cadmium batteries that have been derived by both methods.

	Button cells			Cylindrical cells					Prismatic
						High rate		High rate	
Capacity	30 mAh	250 mAh	1000 mAh	250 mAh	1.2 Ah	1.2 Ah	4 Ah	4 Ah	15 Ah
mΩ (d.c.)	2200	300	50	58	19	12	7	4.5	2.5
mΩ (a.c.)	740	150	14	24	10	10	3.75	3.75	3.5

Table 3.7 Internal resistance of sealed Ni/Cd batteries (from [11]). Cell condition: fully charged. (d.c.): values determined by Eq. (2.28) with: i_1 = 0.1 C(A); i_2 = 2.1 C(A) for button cells i_1 = 0.2 C(A); i_2 = 4.2 C(A) for cylindrical and prismatic cells. (a.c.): alternating-current method with 1000 Hz.

The d.c. resistance is determined according to Eq. (3.11). To compensate at least partly for the influence of the load current. Different voltage steps are applied for different battery types. As can be seen in the caption of Table 3.7, the load is smaller for button cells because of their higher internal resistance.

The a.c. values are determined with 1000 Hz alternating current. Comparison shows that there is no direct proportionality between the two methods.

The values depend also on the size of the battery, which further limits the comparability of such values.

The values in Table 3.7 demonstrate the difference between batteries with pressed-powder electrodes and those with sintered ones. For the comparable cell size of

250 mA, the internal resistance of a button cell is five times that of a cell with sintered electrodes.

The differences between the standard and high-rate versions of cylindrical cells (19:12 and 7:4.5) are caused by constructional improvements, mainly of the current-conducting parts.

Conversion of the values listed in Table 3.7 to 100 Ah nominal capacity leads, for the cylindrical cells, to about 0.2 mΩ/100 Ah. Comparison with the values for lead-acid batteries, listed in Table 11.1, illustrates the comparatively low resistance of sealed nickel/cadmium batteries.

3.6.4 Conductance Measurements as a Controlling Tool

The special situation of stationary valve-regulated lead-acid batteries is that the usual controlling tools for vented batteries are not available, like visual inspection of the plates and the acid level in transparent containers, and determination of the acid density. A situation of uncertainty as to the performance and reliability of such batteries was the consequence when these batteries were introduced on a large scale, and this the more, since expectations in regard to service life were often not fulfilled.

Additional information can be supplied by conductance measurements to discover increased resistance caused by progressive corrosion of the conducting elements or by increased water loss ([12]). This possibility has been systematically examined by large practice tests in combination with the development of suitable measuring devices ([13]).

Fig. 3.13 shows an example, illustrating the capacity/conductance correlation plot for a string of a 48 V battery of 1000 Ah cells. The capacity was measured at the 2.5 hour rate to an end-of-discharge voltage of 1.80 V/cell. The Figure shows the wide range of this battery that obviously was in a poor condition. The good relation between the measured conductance and the remaining capacity is also obvious. The four segments A, B, C and D belong to the so called 'box-score analysis' that was introduced as a tool to simplify the check for larger batteries that consist of a number of parallel strings. With the aid of capacity and conductivity measurements, the critical conductivity is determined for one string. Then bad cells are selected alone by conductivity measurements in the remaining strings ([14]). In Fig. 3.13, the horizontal line indicates 80% of the rated capacity. Its intersection with the regression line suggests the critical value of conductance. In this example it amounts to 2.25 kMhos (ksiemens) and is represented by the vertical line. The so formed 'box' A contains the cells that have capacities above 80% of C_r and conductivies above the critical value, while 'box' B contains cells with insufficient capacity and a too-low conductivity. The 'boxes' C and D are empty since in this example the correspondence between measured values and regression line is excellent. Otherwise, the relation between the number of cells in the boxes A and B and those in the boxed C and D can be used to estimate the accuracy to detect 'bad cells' within a battery by this method (cf. the Table in ([14])).

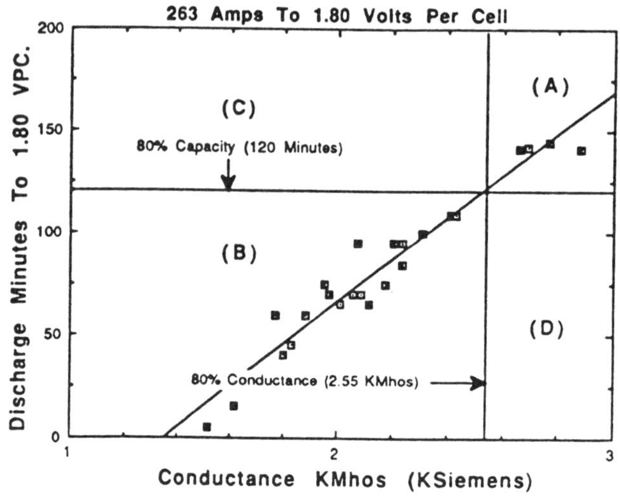

Fig. 3.13 Discharge capacity versus conductance of 24 VRLA 1000 Ah cells. (Fig. 1 in ([14]).)

Fig. 3.13 shows a very clear example of a battery that contains a larger number of cells that totally fail during discharge. Often, however, the correlation between conductivity and still available battery capacity is less significant and influenced by a great number of parameters ([15]).

During the recent years, the conductance technology has been improved to be used also as an on-line method ([16]).

3.7 Service Life, Life Expectancy

The basic processes in a battery are connected with material changes and transport of substances. As a consequence, ageing processes are inherent in batteries. This is a fundamental difference from electronic modules where only charges are displaced while the physical structure remains unaltered.

Thus the service life of a battery is necessarily limited and the life expectancy of batteries an important question. But this life expectancy depends not only on battery design, but largely on the material stress the battery is subjected to; or, in other words, life expectancy depends largely on the operational conditions. For this reason, statements about battery life can only be rather vague.

Fig. 3.14 is an example that illustrates the enormous influence of the operational conditions on service life of a battery.

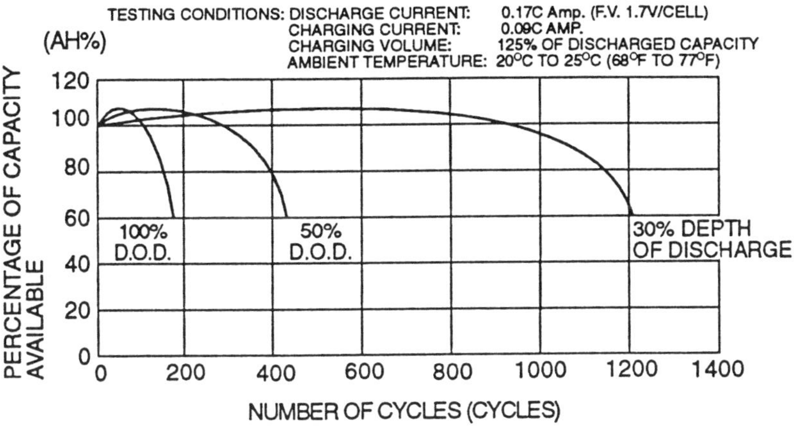

Fig. 3.14 Cycle service life in relation to depth of discharge of a valve-regulated lead-acid battery. YUASA NP-series (4).

Fig. 3.14 concerns a series of rather small batteries, designed for 5 years of float charging. The Figure shows how much service life can be reduced when such a battery is subjected to discharges, especially when deep discharges are applied.

Service life is judged differently, depending on the application of the battery:

- For standby batteries, which are only rarely discharged, service life expressed in calendar time is the important parameter.
- For other applications, like motive power or tools, service life is characterised by the possible number of discharge/charge cycles.

Often the battery application is a mixture between extended charging periods and more or less deep discharges. This applies for example to many household appliances, like electric shavers, where life expectancy includes both float-charging time and a certain cycle load.

For most stationary and motive power applications, the definition is accepted:

- **The end of service life is determined as the point at which the actual capacity of the battery is reduced to 80% of its rated capacity**.

  This definition includes the necessity to oversize the battery, initially to about 125% of the calculated capacity, when full capacity is required for the whole service life.

But there are also applications that use different definitions of the end of life e.g. for starter batteries in motor cars, the end of a cycle test is reached when the capacity falls below 60% of its original value. This end-of-life criterion is also used in Fig. 3.14.

The ageing of batteries is described to some detail in Section 15.2.

3.7.1 Solar Applications

Energy supply by photovoltaic systems is of special interest, because solar energy is converted directly into electric energy. But solar energy is not always available, and the main problem is the necessity for electrochemical energy storage, which means a rather expensive battery, when larger amounts of energy have to be stored. This applies especially in climate zones where the insolation is markedly different in summer and winter, so that the daily cycles are superimposed by a gradual discharge of the battery during the winter. As a consequence, the battery is more and more discharged during the winter and only gradually recharged when the days are getting longer.

On the whole, a battery for voltaic application has to be tough. It cannot be recharged as it should be, because the conditions for charging do not depend on the demand of the battery, but on the energy available. Due to the above mentioned season problems, batteries in solar applications often have to remain in a partly or even largely discharged state for prolonged periods, while excessive overcharging may occur in summer. High environmental temperatures present additional problems in the sun-belt regions of the world.

Many battery manufacturers offer special solar-battery designs, mainly vented lead-acid; but often usual designs are recommended including valve-regulated lead-acid and sealed nickel/cadmium batteries for smaller photovoltaic devices (a survey is given in ([17])). Vented lead-acid batteries are preferred on account of their high charge efficiency and their comparatively low price. For consumer applications like caravans automotive batteries are often used to reduce cost.

Larger photovoltaic installations have usually been sponsored by public funds, for example by the European Commission ([18]). Extensive experience has been gained from these installations, and repeatedly the result was that the difficult operational conditions, usual in solar applications, often cause premature failure of the battery ([19]).

References

([1]) D.Berndt and E.Voss, Batteries 2, Proc. 4th Intern. Symp. Brighton 1964, Pergamon Press, Oxford, 1965

([2]) VARTA, Ni-MH Button Cells, Sales Program and Technical Handbook, VARTA Batterie AG, Daimlerstraße 1, D-73479 Ellwangen/Jagst, Germany

([3]) SANYO, Cadnica, Rechargeable Nickel Cadmium Batteries, Engineering Databook SANYO Electroc Co,. Ltd, Osaka, Japan

([4]) YUASA, NP Valve Regulated Lead Acid Battery Manual, Yuasa Battery Sales Ltd., Hawksworth Industrial Estate, Swindon, Wiltshire SNW 1EG, U.K.

([5]) VARTA Batterie AG (Editor) 'Sealed Nickel Cadmium Batteries', © VDI Verlag GmbH, Düsseldorf, 1982, Fig. 5/1. p. 96

([6]) Panasonic, 'Nickel Metal Hydride Batteries' Technical Handbook, Matsushita Battery Industrial Co, Ltd, Kanagawa 253, Japan

(7) IEC 896-1, First Edition 1987, Part 1, p. 17 (Stationary Batteries); also DIN 43 539, Teil 4 (August 1984)

IEC Standard, Publication 254-1, 1983, Part 1, p. 15 (Traction Batteries); also DIN 43 539, Teil 3, December 1980

(8) Gates Energy Products (Editor), 'Sealed Rechargeable Batteries, Application Manual', Gates Energy Products Inc., Gainesville, Florida, 1989, Fig.3-70, p.89

(9) SONNENSCHEIN (Editor), 'Handbook for Rechargeable Sealed Lead-Acid Batteries for Military Applications', p, 1-005, Akkumulatorenfabrik Sonnenschein GmbH, D-W-6470 Büdingen

(10) M.Sluyters-Rehbach and J.H.Sluyters *in* A.Bard (Editor), 'Electroanalytical Chemistry, Vol.4',©Marcel Dekker, Inc., New York, 1970, p. 1

A short survey in: D.Linden (Editor), Handbook of Batteries, Second Edition, © 1995, Mc Graw Hill, Inc., p. 2.25

(11) VARTA Sealed NC-Accumulators, Sales Programme and Handbook (40 121 e/0190)', VARTA Batterie AG, Am Leineufer 31, D-3000 Hannover, p.59

(12) F.J.Vaccaro and P. Casson, Proc. 9th Intelec Meeting, 1987, Stockholm, © IEEE Publishing Inc., p.128

(13) D.O.Feder, T.G.Croda, K.S.Champlin, S.J.McShane, and M.J.Hlavac, Journal Power Sources, **40** (1992), p. 235

D.O.Feder and M.J.Hlavac, Proc. 16th Intelec Meeting, Vancouver 1994, © IEEE Communications Soc., p. 282

(14) D.O.Feder and M.J.Hlavac, Batteries International, Issue 18, January 1994, p. 60

(15) R.Heron, Al McFadden, and J.Dunn, Proc. 16th Intelec Meeting, Vancouver 1994, © IEEE Communications Soc., p. 270

(16) M.Hawkins, Proc. 16th Intelec Meeting, Vancouver 1994, © IEEE Communications Soc., p. 263

M.J.Hlavac and D.O.Feder, Proc. 17th Intelec Meeting, The Hague 1995, © IEEE Communications Soc., p.284

(17) D.W.H.Lambert, Batteries International, Issue 5, October 1990, p. 24

(18) Batteries International, Issue 5, October 1990, p.20

(19) M.Whitehead, Batteries International, Issue 5, October 1990, p.30

4. The Lead-Acid Battery System

The lead-acid battery has always been the most important rechargeable electrochemical storage system, maintaining its prime position unchallenged now for more than a century. This seems surprising, because the fundamental data concerning the amount of energy that can be stored are rather modest on account of the high weight of the reacting substances. However, there are many other reasons that contribute to the outstanding position of this system, e.g.

- The reactants are solids of low solubility, and the reactions are highly reversible.
- The reacting compounds are well defined, and there are no intermediate states of oxidation. As a consequence, any voltage above the open-circuit voltage results in complete charge, and equalising charges are not required when the battery is float charged in standby applications.
- The electronic conductivity in the electrodes is high enough; conducting additives are not required.
- The cell voltage of 2 V is fairly high, and so a comparable small number of cells is sufficient for a certain battery voltage.
- Despite the high potential of the positive electrode, lead can be used for the conducting elements, owing to its passive properties (cf. Section 4.7).

This list could be extended (cf. ([1])). It indicates, how many chemical and physical properties are required for an efficient electrochemical energy-storage system.

4.1 The Thermodynamic Situation of Lead-Acid Batteries

Fig. 4.1 shows reactions that are possible in lead-acid batteries on account of thermodynamic parameters in relation to electrode potential.

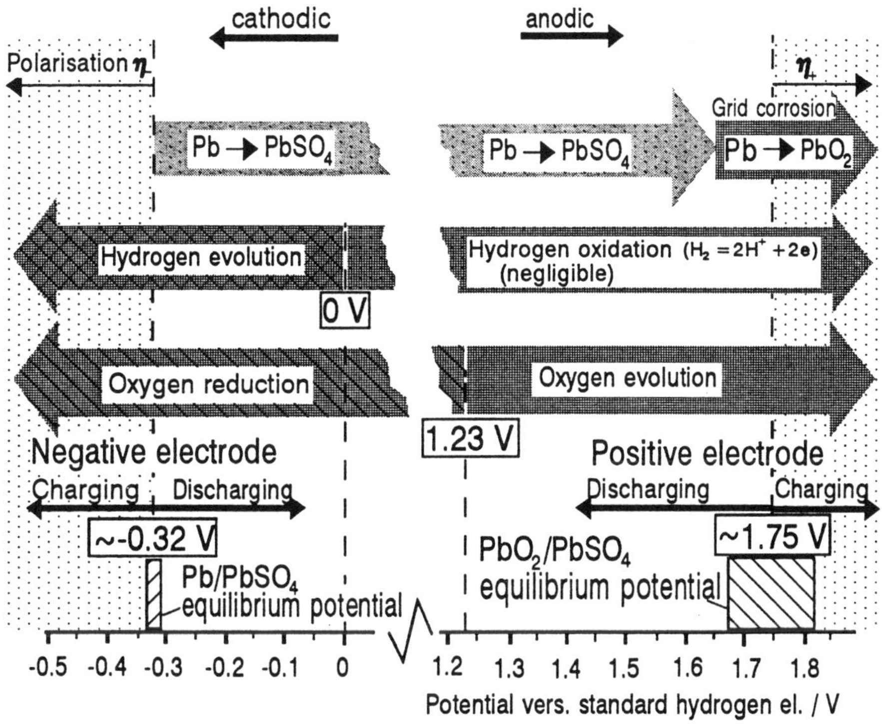

Fig. 4.1 Reactions that occur in lead-acid batteries dependent on electrode potential (thermodynamic situation). Their equilibrium potentials are inserted as enframed numbers. Equilibrium potentials of the charge/discharge reactions ($Pb/PbSO_4$ and $PbSO_4/PbO_2$) are represented by columns, to indicate their dependence on acid concentration. The inserted equilibrium potentials (-0.32 and +1.75 V) of the charge/discharge reactions correspond to an acid density 1.23 g/cm^3. (Table 4.2, p.103).

Besides charging and discharging, a number of secondary reactions are possible, because their equilibrium potentials are within the potential range covered by the cell voltage of the battery.

- **Hydrogen evolution** occurs at an electrode potential below 0 V which is about 0.3 V above the open-circuit potential of the negative electrode.
- **Oxygen evolution** occurs at an electrode potential above 1.23 V which is about 0.5 V below the positive electrode potential.
- **Oxygen reduction**, as the reversal of oxygen evolution, is possible below 1.23 V vs. SHE. Thus, oxygen, evolved at the positive electrode or intruding into the battery from outside, is reduced as soon as it reaches the negative-electrode surface. This reaction is utilised to establish the internal oxygen cycle in valve-regulated batteries. In flooded batteries it is largely hindered by the slow diffusion rate of solved O_2 .

- **Hydrogen oxidation**, as the reversal of hydrogen evolution, is possible for thermodynamic reasons above 0 V vs. SHE, but is hindered at the PbO_2 surface so much, that it can be neglected. This means that gaseous hydrogen, once generated in the battery, cannot be removed by electrochemical oxidation (unless auxiliary electrodes or catalysts are applied), but has to escape from the cell - an important fact for valve-regulated batteries.
- **Lead corrosion** (or **grid corrosion**) is a secondary reaction that finally limits service life. Lead, used as grid material and for conducting elements, is stable only below the open-circuit potential of the negative electrode; at higher potentials it is converted into lead sulphate ($PbSO_4$) or lead dioxide (PbO_2).
  In the positive electrode, a layer of PbO_2, formed by corrosion, protects the underlying lead against spontaneous oxidation, but conducts electronic current, and therefore does not hinder electrochemical processes (cf. Section 4.7). The situation at the phase boundary is, however, not stable, and the corrosion layer penetrates slowly into the bulk of the lead, and a corresponding anodic current is continually required to re-establish the PbO_2 layer (cf. Fig. 4.26, p.153).

According to this thermodynamic situation, the lead-acid battery system is unstable in two respects:

1. Water is decomposed according to Eq. (2.93) into oxygen and hydrogen above 1.23 V/cell which is considerably below the 2 V of open-circuit voltage of the lead-acid battery. Thus, water decomposition cannot be avoided.
2. Lead corrosion to lead dioxide (PbO_2) occurs with all conducting elements connected to the positive electrode.

Thermodynamic data, as shown in Fig. 4.1, however, only indicate whether a reaction is possible at a given electrode potential or not. The actual occurrence of reactions and their rates are characterised by the laws of electrode kinetics, described in Section 2.3 (in regard to water decomposition cf. also Chapter 7).

Fortunately, hydrogen and oxygen generation are very slow reactions, and the lead corrosion rate is also largely reduced by the described passivation effects. So these secondary reactions mostly can be tolerated, but they always have to be considered as unwanted discharge reactions (cf. Section 4.4) that consume water and often limit service life.

4.1.1 The Charge/Discharge Reactions

The basic electrode reactions (already mentioned as examples in Chapter 2) of the lead-acid system are both based on lead (Pb), but on different states of oxidation, namely Pb^{2+} and Pb^{4+}. During discharge,

Pb is oxidised to Pb^{2+} at the negative electrode

$$\boxed{Pb \Leftrightarrow Pb^{2+} + 2 \cdot e^-} \tag{4.1}$$

Pb^{4+} is reduced to Pb^{2+} at the positive electrode

$$\boxed{Pb^{4+} + 2 \cdot e^- \Leftrightarrow Pb^{2+}} \quad (4.2)$$

These equations have to be completed into:

$$\boxed{Pb + H_2SO_4 \Leftrightarrow PbSO_4 + 2 \cdot H^+ + 2 \cdot e^-} \quad \text{(at the neg. electrode)} \quad (4.3)$$

and

$$\boxed{PbO_2 + H_2SO_4 + 2 \cdot H^+ + 2 \cdot e^- \Leftrightarrow PbSO_4 + 2 \cdot H_2O} \quad \text{(at the pos. electrode),} \quad (4.4)$$

because the solubility of the Pb^{2+} ions in sulphuric acid is extremely low ($\approx 10^{-6}$ mole/dm^3), and Pb^{2+} is precipitated as $PbSO_4$. In the positive electrode, Pb^{4+} ions do not occur in the free form, but only in compounds such as lead dioxide (PbO_2).

In detailed examinations, attention has to be paid to the fact that lead dioxide does not always correspond exactly with the stoichiometric formula PbO_2, but some (small) deviations are observed (2) that indicate that the true formula reads $PbO_{1.99}$ (3).

Summation of the two electrode reactions results in the cell reaction:

$$\boxed{Pb + PbO_2 + 2 \cdot H_2SO_4 \Leftrightarrow 2 \cdot PbSO_4 + 2 \cdot H_2O} \quad (4.5)$$

When the lead-acid battery is discharged, lead (Pb) and lead dioxide (PbO_2) are converted into lead sulphate ($PbSO_4$). In other words, during discharge sulphate from the sulphuric acid is absorbed into the electrodes. When the battery is charged, these reactions are reversed.

4.1.1.1 Dissociation of sulphuric acid

In the concentration range applied with batteries, dilute H_2SO_4 is dissociated mainly into H^+ and HSO_4^- ions. Only about 1 per cent of the H_2SO_4 molecules dissociate into $2 \cdot H^+$ and SO_4^{2-}. In consideration of the actual dissociation Eq. (4.5) can be written

$$\boxed{Pb + PbO_2 + 2 \cdot H^+ + 2 \cdot HSO_4^- \Leftrightarrow 2 \cdot PbSO_4 + 2 \cdot H_2O} \quad (4.6)$$

4.2 Thermodynamic Data

The thermodynamic data (cf. Section 2.2) for the substances employed in Eq. 4.5 are displayed in Table 4.1.

When the values for $H^{o,s}$ and $G^{o,s}$ of Table 4.1 are inserted in Eq. (4.6) instead of the chemical symbols, and the sum of the terms on the left hand side of this equation are subtracted from the corresponding sum on the right side the results are:

the enthalpy of the reaction

$$\boxed{\Delta H_s = -2411 + 2052 = -359.4 \text{ kJ}} \quad (4.7)$$

the free enthalpy of the reaction

$$\Delta G_s = -2101 + 1728 = -372.6 \text{ kJ} \tag{4.8}$$

and the reversible heat effect according to Eq. (2.4), p.10

$$T \cdot \Delta S = \Delta H - \Delta G = 13.2 \text{ kJ} \tag{4.9}$$

for each multiple of the cell reaction (cf. Section 2.2).

Substance	Enthalpy of Formation ($H^{o,s}$) in kJ/mole	Free Enthalpy of Formation ($G^{o,s}$) in kJ/mole
Pb	0	0
PbO_2	-277.4	-217.4
$PbSO_4$	-919.9	-813.2
H^+	0	0
H_2SO_4	-814.0	-690.1
HSO_4^-	-887.3	-755.4
SO_4^{2-}	-909.3	-744.0
H_2O	-285.8	-237.2

Table 4.1 Standard values (T = 25 °C) of the thermodynamic data for the chemical compounds of Eqs. (4.5), (4.6) and 4.12 (cf. [4])). (*In older Tables, the energy often is given in calories: 1 cal = 4.187 J.*)

4.2.1 The Equilibrium Voltage

The standard value of the equilibrium voltage of the lead-acid battery results when the value of Eq. (4.8) is inserted into Eq. (2.6), p.11 (1 kJ = 0.2778 Wh; 372.6 kJ = 103.5 Wh):

$$E^{o,s} = 1.931 \quad V$$

This standard value of the equilibrium voltage applies when the activities (effective concentrations) of the soluble components of the reaction, namely H^+, HSO_4^-, and H_2O amount to 1 mole/dm^3 i.e. the quotient

$$\frac{a_{H^+} \cdot a_{HSO_4^-}}{a_{H_2O}} = 1$$

This value is approximated to by the acid concentration of 1.083 mole/dm^3 (≈ 10 wt.% or a density of 1.066 g/cm^3) (cf. Table 4.2 and Note 2 below).

The dependence of the equilibrium voltage on concentration is given by the Nernst equation, Eq. (2.15), p.6:

$$E^o = 1.931 - \frac{R \cdot T}{2 \cdot F} \cdot \log \frac{(a_{H_2O})^2}{(a_{H^+})^2 \cdot (a_{HSO_4^-})^2} = 1.931 + 0.0592 \cdot \log \frac{a_{H^+} \cdot a_{HSO_4^-}}{a_{H_2O}} \quad V \qquad (4.10)$$

According to Eq. (4.10) the equilibrium cell voltage depends only on the acid concentration. It is independent of the amount of lead, lead dioxide or lead sulphate present in the battery, so long as all three substances are available.

The results of Eq.(4.10) are plotted in Fig. 4.2 and compiled in Table 4.2, together with the single-electrode potentials (cf. Section 2.2.2). The calculations for the curve in Fig.4.1, as well as the values for cell voltage and electrode potential, are based on the measurements of Harned and Hamer ([5]) and Covington and co-workers ([6]) (cf. also ([4]), p. 93-95).

In battery practice, the approximation:

$$\text{Equilibrium cell voltage} = \text{acid density (in g/cm}^3 \text{ or kg/dm}^3) + 0.84 \qquad (4.11)$$

is used instead of Eq. (4.10). Fig. 4.2 shows that the calculated curve and the approximate formula coincide quite well.

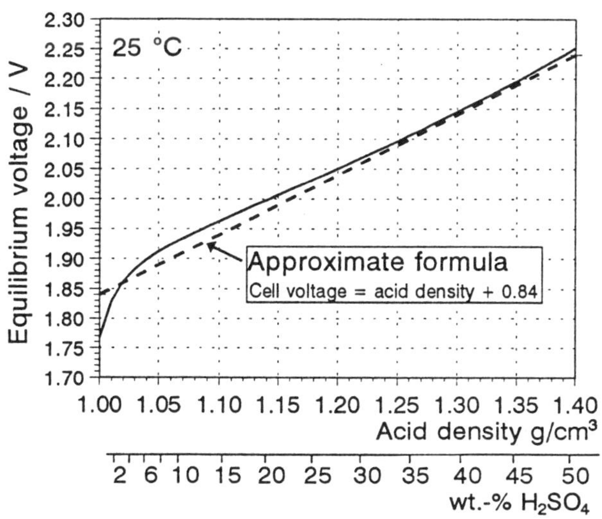

Fig. 4.2 Equilibrium cell voltage of the lead-acid battery referred to acid density and acid concentration in wt.% H_2SO_4. The dashed line represents the approximate formula Eq. (4.11).

Note 1: *Actually not the true equilibrium voltage but only the* ***open-circuit voltage*** *can be measured with lead-acid batteries. Due to the unavoidable secondary reactions of hydrogen and oxygen evolution (cf. Chapter 7), mixed potentials (cf. Section 2.3.5) are established at both electrodes, which are a little different from the true equilibrium potentials. But, owing to high overvoltages of both gas evolution reactions (flat* *current/voltage curves) and small overvoltages of the discharge reactions (steep current/voltage curves), these differences are small and can be neglected.*

Since H_2SO_4 participates in the electrode reaction, the electrolyte becomes more diluted during discharge and is re-concentrated during charge. Furthermore, according to Fig. 4.2, the open-circuit potential depends on acid concentration. Both facts together cause the disadvantage that the discharge voltage does not remain constant even at low discharge rates. Furthermore, the conductivity of the acid varies when the battery is charged or discharged, and during discharge, its freezing point may be reached in cold climates (cf. Section 4.6.2). On the other hand, the acid dilution provides a tool to determine the state of charge with the aid of density measurements.

Note 2: *Sometimes the calculations of equilibrium voltage are based on the assumption of completely dissociated sulphuric acid. Instead of Eq. (4.6) the (hypothetical) equation for the reaction is:*

$$\boxed{Pb + PbO_2 + 4 \cdot H^+ + 2 \cdot SO_4^{2-} \Leftrightarrow 2 \cdot PbSO_4 + 2 \cdot H_2O} \tag{4.12}$$

On account of the different energy content of SO_4^{2-} from that of HSO_4^- the equilibrium voltage attains then the value (ΔG^s = -395.4 kJ = -109.8 Wh; cf. Table 4.1)

$$\boxed{E^{o,s} = 2.049 \quad V} \tag{4.13}$$

This value is as correct as that derived through Eq. (4.6). Both are only referred to different standard acid concentrations: the first one (Eq. (4.6)) is based on the mean activity a_{H^+} and $a_{HSO_4^-}$ = 1 mole/dm^3, while Eq. (4.13) is based on $(a_{H^+})^2$ and $a_{SO_4^{2-}}$ = 1 mole/dm^3, and the equilibrium voltage $E^{o,s}$ corresponds to the quotient

$$\frac{(a_{H^+})^2 \cdot a_{SO_4^{2-}}}{a_{H_2O}} = 1$$

Table 4.2 shows, that this is true for a concentration of about 2.42 mole/dm^3, 20.2 wt.%, or the density of 1.145 g/cm^3.

The Nernst equation then reads

$$\boxed{E^o = 2.049 - \frac{R \cdot T}{2 \cdot F} \cdot \log \frac{(a_{H_2O})^2}{(a_{H^+})^4 \cdot (a_{SO_4^{2-}})^2}} \tag{4.14}$$
$$\boxed{= 2.049 + 0.0592 \cdot \log \frac{(a_{H^+})^2 \cdot a_{SO_4^{2-}}}{a_{H_2O}} \quad V}$$

This equation results also in the curve shown in Fig. 4.2.

⇒ **Table 4.2** Acid-concentration parameters: acid density, H_2SO_4 content and H_2SO_4 concentration in mole/dm^3. T = 25 °C. Cell voltage and electrode potentials referred to the standard hydrogen electrode.

Acid concentration			Cell voltage	Electrode potentials	
Density g/cm^3	H_2SO_4 Weight%	Concentration mole/dm^3	E^0 Volts	V ref. to std. H_2 electrode Positive	Negative
1.01	1.731	0.1783	1.828	1.539	-0.289
1,02	3,242	0.3372	1.862	1.568	-0.294
1.03	4.746	0.4983	1.883	1.590	-0.293
1.04	6.237	0.6613	1.899	1.606	-0.293
1.05	7.704	0.825	1.913	1.619	-0.293
1.06	9.129	0.9865	1.924	1.630	-0.294
1.07	10.56	1.152	1.935	1.640	-0.295
1.08	11.96	1.317	1.945	1.649	-0.296
1.09	13.36	1.484	1.955	1.657	-0.297
1.10	14.73	1.652	1.964	1.665	-0.299
1.11	16.08	1.820	1.973	1.673	-0.300
1.12	17.43	1.990	1.982	1.680	-0.302
1.13	18.76	2.161	1.991	1.687	-0.304
1.14	20.08	2.334	2.000	1.694	-0.305
1.15	21.38	2.507	2.008	1.701	-0.307
1.16	22.67	2.681	2.017	1.708	-0.309
1.17	23.95	2.857	2.026	1.714	-0.311
1.18	25.21	3.033	2.034	1.721	-0.314
1.19	26.47	3.211	2.043	1.727	-0.316
1.20	27.72	3.391	2.052	1.734	-0.318
1.21	28.95	3.572	2.061	1.741	-0.320
1.22	30.18	3.754	2.070	1.748	-0.322
1.23	31.40	3.938	2.079	1.754	-0.325
1.24	32.61	4.123	2.088	1.761	-0.327
1.25	33.82	4.310	2.097	1.768	-0.329
1.26	35.01	4.498	2.107	1.775	-0.331
1.27	36.19	4.686	2.116	1.783	-0.334
1.28	37.36	4.876	2.126	1.790	-0.336
1.29	38.53	5.068	2.136	1.797	-0.338
1.30	39.68	5.259	2.145	1.805	-0.340
1.31	4082	5.452	2.156	1.813	-0.343
1.32	41.95	5.646	2.166	1.821	-0.345
1.33	43.07	5.840	2.176	1.829	-0.374
1.34	44.17	6.035	2.187	1.837	-0.350
1.35	45.26	6.229	2.197	1.845	-0.352
1.36	46.33	6.424	2.208	1.853	-0.355
1.37	47.39	6.620	2.219	1.861	-0.358
1.38	48.45	6.817	2.230	1.869	-0.361
1.39	49.48	7.012	2.241	1.877	-0.364

4.2.1.1 Single Electrode Potentials

The potentials of the negative and the positive electrode, the $Pb/PbSO_4$ and $PbO_2/PbSO_4$ electrode, respectively, have been used as examples in Chapter 2, Section 2.2.2. Eqs. 2.24 and 2.28 are the corresponding Nernst equations that describe the dependence on acid concentration which is shown as curves in Fig. 2.3.

Comparison with Table 4.2 shows, that the standard potentials used in the Eqs. 2.24 and 2.28 referred to standard hydrogen electrode are

$$\boxed{E^{*o}_{Pb/PbSO_4} = -0.295\ \text{V}} \qquad (4.14a)$$

$$\boxed{E^{*o}_{PbO_2/PbSO_4} = 1.636\ \text{V}} \qquad (4.14b)$$

4.2.1.2 Specific Gravity

In the United States, **<u>Specific gravity</u>** is used instead of density in battery practice. The specific gravity is a dimensionless quantity, namely the ratio

$$\boxed{\text{Specific gravity} = \frac{\text{density of the acid}}{\text{density of water}}} \qquad (4.15)$$

with both densities measured at the same temperature.

The specific gravity includes the temperature dependence of the density of water. The differences between specific gravity and density are rather small and can usually be neglected (cf. Table 20 in ([7])).

4.2.1.3 The Temperature Coefficient

The temperature coefficient of the equilibrium voltage (Eq. (2.16), p.13) is closely related to the reversible heat effect $T\cdot\Delta S$. According to Eq. (4.9) $T\cdot\Delta S$ = 13.2 kJ. Division by temperature gives:

$$\boxed{\Delta S = 45\ \text{J}\cdot\text{K}^{-1}} \quad \text{per multiple of the cell reaction } (T = 293.2\ \text{K}). \qquad (4.16)$$

When this value is inserted into Eq. (2.16), it results in

$$\boxed{\frac{dE^o}{dT} = 0.23\ \ \text{mV/K}} \qquad (4.17)$$

This temperature coefficient is strictly valid only for the standard conditions, because the thermodynamic parameters depend on acid concentration as well as on temperature. At low acid concentrations (1 wt.%) the temperature coefficient even attains the negative value of -0.18 mV/K ([8]); cf. also Eq. (2.18), p.6.

The temperature coefficient dE^o/dT is so small over the whole concentration range that it always can be neglected in lead-acid battery practice.

4.2.1.4 The Reversible Heat Effect, Calorific Voltage

According to Eq. (2.4), p.10, the reversible heat effect is given by the difference between the enthalpy ΔH and the free enthalpy ΔG of the reaction. According to Eq. (4.9) it amounts to

$$\boxed{Q_{rev} = T \cdot \Delta S = 372.6 - 359.4 = 13.2 \text{ kJ}} \quad \text{per multiple of the cell reaction.} \tag{4.18}$$

Comparison with the value for ΔG (-372.6 kJ) shows that the reversible heat effect is only 3.5% of the amount of energy converted when the lead-acid battery is discharged or charged.

The positive sign means that additional electrical energy is gained during discharge on account of ($|-\Delta G| > |-\Delta H|$). When the lead-acid battery is charged, the reversible heat effect appears as additional heat. But only when the lead-acid battery is discharged at very low rates (> 10 hours) can slight cooling of the battery be observed. Usually, the reversible heat effect is covered by the Joule effect (cf. Section 4.5).

According to Eq. (2.42), p.22, Q_{rev} can be expressed as a (fictive) voltage (13.2 kJ = 13200 Ws; $2 \cdot F$ = 192970 Ah/mole):

$$\boxed{\frac{Q_{rev}}{2 \cdot F} = 0.068 \text{ V}} \tag{4.19}$$

and the calorific voltage in lead-acid batteries is according to Eq. (2.43)

$$\boxed{E_{cal} = E^{o} - 0.068 \text{ V}} \tag{4.20}$$

This means that the calorific voltage, and thereby heat effects, correspond to a voltage slightly lower than the equilibrium voltage (cf. Figs 4.10 and 4.11, p.121 and 122).

Note: *The above calculated values for the reversible heat effect (Eq. (4.18)), the temperature coefficient (Eq. (4.17)), and the calorific voltage are strictly valid only for the standard conditions; they depend on acid concentration and temperature. But the deviations are rather small and the values so insignificant that the conclusions can be extended to the usual range of acid concentration and temperature (cf. Table 2.36, p. 98, in (4)).*

4.2.2 Weight-Dependent Quantities

The weight that is required to store a certain amount of electrical energy can be evaluated with the aid of the cell reaction shown in Eq. (4.5). The molar weights for the substances are compiled in Table 4.3. This Table also shows utilisation values, derived by required number of moles, and under consideration that the amount of electricity converted per multiple of Eq. (4.5) is

$$\boxed{2 \cdot F = 192{,}970 \text{ As} = 53.61 \text{ Ah}} \tag{4.21}$$

The utilisation figures in Table 4.3, of course, represent maximum values, because they are true only if the considered substances were converted completely (100% utilisation of the active material).

Substance	Molar weight	Required quantities based on Eq. (4.5); 2·F = 53.6 Ah	Utilisation values	
	g	g/Ah	Ah/kg	Ah/lb
Pb	207.2	3.865	258.7	117.3
PbO_2	239.2	4.462	224.1	101.7
$PbSO_4$	303.3	5.657	176.8	80.18
H_2SO_4	98.08	3.659	273.2	123.9
H_2O	18.02	0.6722	1487	674.3

Table 4.3 Molar weights, of the components in Eq. (4.5), required quantities and (maximum) utilisation values derived from thermodynamic data.

Considering that one mole of Pb and PbO_2 and two moles of H_2SO_4 are required for Eq. (4.5), the weight of the required substances is 642.6 g in total.

As shown in Fig. (4.2), E^0 depends on acid concentration, and so does $2{\cdot}F{\cdot}E^0$, the amount of energy converted per multiple of the cell reaction. When the (theoretical) maximum energy output is referred to the standard voltage of 2 V, the result is 2 times 53.6 = 107 Wh. Consequently, the number of watt-hours that can be stored per kg lead-acid battery, **the specific energy**, is:

$$\frac{107}{0.6426} = 166 \quad \text{Wh/kg} \quad = 75.7 \quad \text{Wh/lb} \tag{4.22}$$

This figure represents the theoretical maximum at the standard E^0 of 2 V. It can never be realised in battery practice, because it is unavoidably reduced by a number of factors, illustrated in Fig. 4.3.

From the weight of the substances participating in the cell reaction (Eq. (4.5)), it can be calculated that 6 kg 'theoretical lead-acid battery' are required for the storage of 1 kWh, with its reciprocal, the specific energy of 166 Wh/kg (*9.345 multiples of the cell reaction are required to deliver 1 kWh (9.345 x 0.107), and 9.345 x 0.6426 = 6 kg*). The technically usable system, however, requires diluted acid, since concentrated sulphuric acid has no conductivity and would convert the active material very quickly into lead sulphate. (*2 H_2SO_4 means 196.2 g per multiple of the cell reaction or 1.83 kg per kWh. Dilution to 34% means addition of 3.6 kg of water. Now, 9.6 kg are required per kWh and the specific energy is reduced to 104 Wh/kg.*) Furthermore, there must be excess acid, otherwise the discharge would lead to pure water, which would also cause unwanted chemical reactions, and recharging would not be possible, because of the lack of conductivity. In Fig. 4.3 an excess of 15% electrolyte is assumed, and so the actually required active materials contribute 11.5 kg/kWh and limit the specific energy to 87 Wh/kg.

Much dead weight results from only partial utilisation of the active material (25 to 60%, as indicated in Fig. 4.3), since the conductivity of the active material decreases with advancing discharge and, in addition, parts of the still charged active material (lead as well as lead dioxide) are buried under insulating lead sulphate (cf. Fig. 4.5).

Furthermore, the splitting of the cell reaction into the two electrode reactions (illustrated by Fig. 2.1, p.7) considerably reduces the storage figures, on account of the weight contributed by grids, current-conducting elements, separators and containers. Thus, in actual lead-acid batteries, not 6 but around 30 kg are required to store 1 kWh of energy.

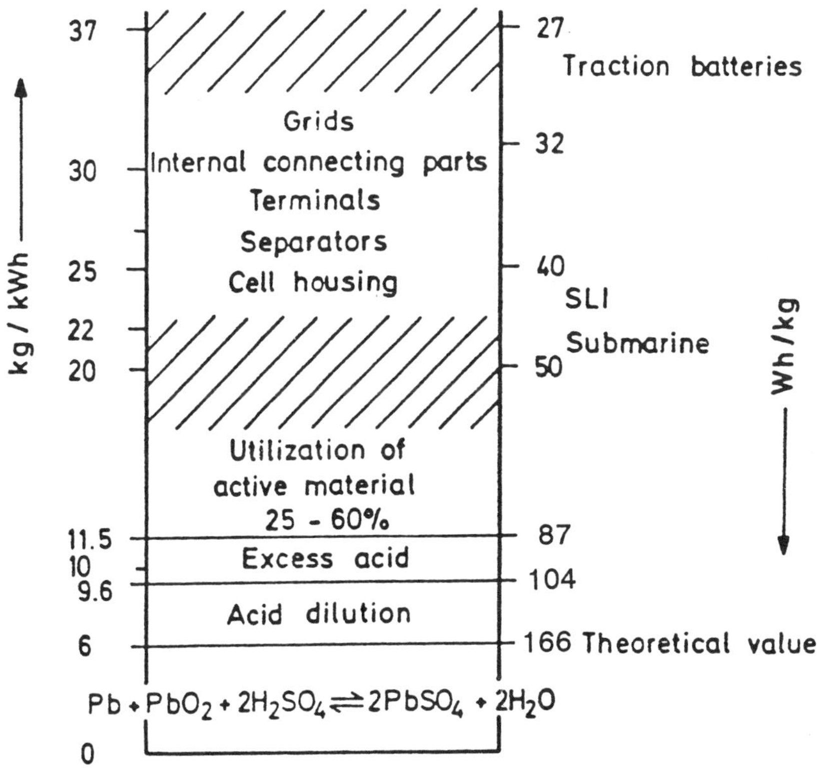

Fig. 4.3 Decrease of the stored energy per unit weight in actual lead-acid batteries compared with the theoretical limit of 166 Wh/kg (referred to the (standard) cell voltage of 2 V. Left hand scale: weight per kWh of stored energy required by the different components. Right hand scale: reciprocal values (Wh/kg).

Some utilisation figures of various lead-acid battery types are compiled in Table 4.4. Such figures are, of course, rough approximations. They depend widely on the specific battery design, and on the discharge conditions, which are rather different in varying applications, as has been pointed out in Section 3.3. Despite these limitations, they indicate characteristic features. The highest values of specific energy (Wh/kg) are reached for submarine and starter batteries, for very different reasons. The submarine battery consists of very large cells (up to 15000 Ah of nominal capacity (cf. ([9])), which shifts the relation between active material and passive parts (connectors, pole bridges) in favour of the active material. Furthermore, the number of discharges is moderate and low-density material can be used in the positive electrode. The automotive or starter battery, on the other hand, is specially designed for high efficiency and the characteristic situation of the car battery, that charging and overcharge periods are rather short compared to service life, and that the battery is rarely discharged in normal service conditions.

	Energy Content of Cells or Batteries			Utilisation Figures of the Active Material	
	Specific. Energy		Energy.dens.	pos.(PbO_2)	neg. (Pb)
Type of Battery	Wh/kg	Ah/kg	Wh/dm^3	Ah/kg	Ah/kg
Starter. SLI (20-h rate)	30 to 45	15 to 24*	75 to 100	100 to 130	130 to 160
Submarine (10-h rate)	35 to 55	18 to 29	90 to 105	100 to 110	110 to 130
Traction (cell, 5-h rate)	25 to 32	13 to 17	60 to 100	60 to 90	80 to 100
Stationary conventional	20 to 28	10 to 15	35 to 60	60 to 100	80 to 100
Bell-Syst. Round-Cell	18	9.5	14	65	90
Planté	11	6	23	90	80 to 100
Stationary, valve-regul. (60 Ah)	20 to 31	10 to 16	45 to 85	66 to 90	80 to 140

Table 4.4 Some examples of energy content per cell weight and volume as well as utilisation figures of the active material for different designs of lead-acid batteries, flooded as well as valve-regulated types. The values for traction and stationary batteries of flooded design are based on cell sizes of 300 to 600 Ah , SLI ≈ 80 Ah, and submarine ≈10,000 Ah. Utilisation figures referred to nominal capacity (5-hour rate for traction, 8- or 10-hour rate for stationary). Energy density values (Wh/dm^3) are rough approximations since they are determined to a large extent by the specific cell design. * per cell.

Traction batteries have to perform charge/discharge cycles more or less continuously. Reduced utilisation of the active material is a prerequisite to withstand this kind of service. For this reason, utilisation values are moderate. *For complete comparison, it has to be taken into account that these values would be reduced to the range of 20 Wh/kg for a complete battery with the tray weight included.*

In stationary batteries, the utilisation figures of the active material are similar to those for traction batteries. The reduced data for energy content are mainly caused by oversized grid weights in the positive plate to withstand corrosion attack during float charge (cf. Section 4.7) with the consequence of thicker electrodes. The rather low Wh/dm^3 values are mainly a consequence of excess electrolyte, partly to achieve a more constant discharge voltage (less concentration change) and partly to provide electrolyte reserve to compensate for water loss (cf. Chapter 6). The relatively low figures for the Bell Syst. Cell ([10]), which has been designed for more than 30 years of service, show how much these data are influenced by special design parameters.

The rather low Wh/kg figure for Planté batteries ([11]) is caused by the heavy weight of the core in the positive plate, cast from pure lead.

The values for valve-regulated lead-acid batteries, shown in Table 4.4, indicate rather moderate utilisation factors and specific energy contents. These lower data are partly caused by the smaller size of these batteries. Furthermore, the utilisation of the active material is slightly reduced on account of the immobilised electrolyte. On the other hand, excess volume is neither required above the plates for surplus acid, nor below the electrodes as mud space. This is the main reason for the fairly high energy density (Wh/dm^3) of valve-regulated batteries.

4.2.3 Heat Capacity (Specific Heat Content)

Temperature response of the battery on heat generation or heat loss is determined by the heat capacity of the battery, which can be calculated by adding up the heat capacities for all the components, given by the products of their weight times their specific heat, according to Eq. (2.81), p.36.

For the acid, this addition is not quite correct, because the enthalpy of the mixture between H_2SO_4 and H_2O differs from the added values of both components on account of the mixing energy (cf. ([12])). Fig. 4.4 shows this difference.

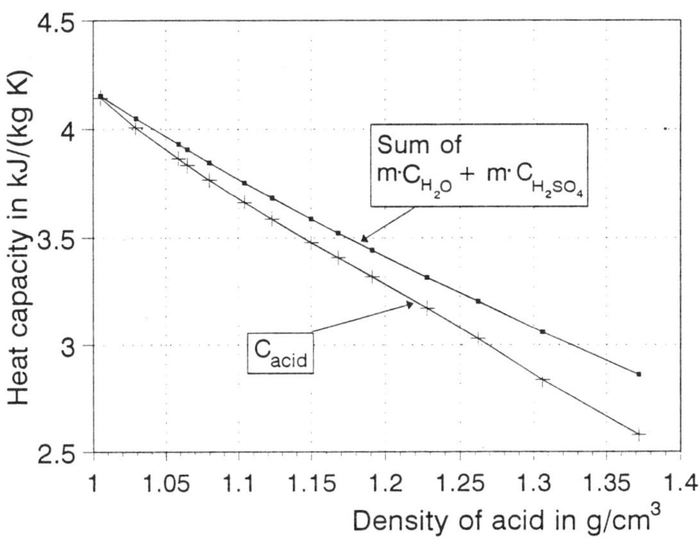

Fig. 4.4 Heat capacity of aqueous solution of H_2SO_4 (c_{acid}) (Table 2.22 in ([4])). Addition of the individual heat capacities for H_2O and H_2SO_4 leads to slightly higher figures (upper curve). m represents the percentages of weight for the two components.

In Table 4.5, values are compiled for active materials and cell components. The large difference between the heat-capacity values of the solid substances and those of the electrolyte indicate that the heat capacity of a battery is largely determined by the acid content. Some examples of materials that are used for separators and containers are added. According to their rather high heat capacity, they may influence the heat capacity of the whole battery markedly, especially in small batteries (cf. Table 4.6).

Substance	Molar Weight g	Heat Capacity J/(K·mole)	J/(K·kg)
Pb	207.2	26.46	127.6
PbO_2	239.2	64.68	270.1
$PbSO_4$	303.3	103.2	340.2
H_2O	18.02	75.45	4187
H_2SO_4, 1.24 g/cm^3			3100
H_2SO_4, 1.28 g/cm^3			2800
Polypropylene			2100
Polystyrene			1200
Glass			800

Table 4.5 Heat capacity of the different components in lead-acid batteries.

In Table 4.6, the heat capacities of some batteries are compiled as examples for the different types of lead-acid battery.

Type of Battery	Heat Capacity kJ/K kg	kJ/(k-100 Ah)	Percentage caused by the Electrolyte
Starter (flooded SLI)	0.94 to 1.2	30 to 37*	75% to 85%
Traction (600 Ah, tubul.)	1.05	7	76%
Stationary (600 Ah, past.)	1.15	8	82%
Stationary (600 Ah, tubul.)	1.18	10	80%
Valve-regulated	0.75 to 1.0	6 to 8	64% to 75%

Table 4.6 Typical values for the heat capacity of various lead-acid batteries, and the share caused by the electrolyte. * per 12 V battery.

Table 4.6 shows, that the heat capacity does not vary too much for the different types. For batteries of similar design, it is mainly determined by the acid percentage of the battery weight. This mainly causes the differences between the kJ/K·kg values of starter and valve-regulated types. Table 4.6 shows, furthermore, that for valve-regulated designs both values are comparatively low: the heat capacity itself as well as the share assigned to the electrolyte content, since the first depends on the latter.

4.3 Kinetic Effects

In the lead-acid battery, the reactions at both electrodes include the dissolved state, which means that the reacting species are dissolved in the course of the reaction. The new chemical compounds formed during the reaction are precipitated again as solid matter. Fig. 4.5 illustrates the course of the reaction. According to the values in Table 4.3, about 8 grams of substance (Pb and PbO_2) have to be dissolved per Ah.

During discharging, lead ions (Pb^{2+}) are dissolved at the negative electrode. A corresponding number of electrons is removed from the electrode as negative charge. The solubility of the Pb^{2+} ions is, however, limited to about 10^{-6} mole/dm^3 in the presence of HSO_4^- or SO_4^{2-} ions (sulphuric acid). As a consequence, the dissolved ions form lead sulphate ($PbSO_4$) which is precipitated on the electrode surface immediately after the dissolution process, mostly within the pore system.

The discharging reaction at the positive electrode proceeds in a similar manner: bivalent lead ions (Pb^{2+}) are formed by tetravalent lead ions (Pb^{4+}) acquiring two electrons. These dissolve and immediately form lead sulphate ($PbSO_4$). In addition, water is formed at the positive electrode during discharging, because oxygen ions (O^{2-}) are also released from the lead dioxide (PbO_2) that combine with the protons (H^+) of the dilute sulphuric acid to form H_2O molecules.

When charging the battery, these reactions occur in the opposite direction, as indicated by the double-line arrows in Fig. 4.5. Then lead (Pb) and lead dioxide (PbO_2) are formed from lead sulphate ($PbSO_4$).

The electrochemical reaction, the transfer step, can only take place where electrons can be supplied or removed, which means that this conversion is not possible on the surface of the lead sulphate, as lead sulphate does not conduct electric current. For this reason, the Pb^{2+} ions must be dissolved and transported by diffusion to the conductive electrode surface (lead or lead dioxide).

The course of the reaction is complex during both charging and discharging. All substances involved in the reaction must temporarily be dissolved in the electrolyte and reformed as new chemical compounds. As mentioned above, about 8 g of active material are dissolved and precipitated again, for each ampere hour drawn from the battery or recharged. This process can be repeated quite often, since the decisive parameters, solubility and dissolution rate, are well matched in the lead-acid battery system. The chemical conversions occur close to each other, and most of the material transport takes place in the micrometer range. Nevertheless, transport processes often limit the performance of the lead-acid battery, and a gradual disintegration of the active material is observed after a certain number of charge/discharge cycles.

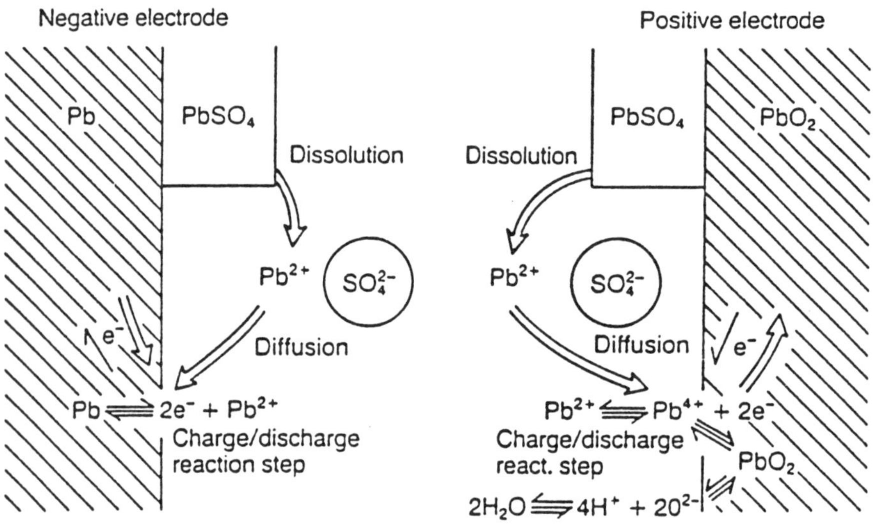

Fig. 4.5 Reaction steps in the lead-acid battery. Double-lined arrows mark the charging reaction. The figure illustrates the solution mechanism.

One consequence of limiting transport processes is the reduction of the amount of energy with increasing load, shown, e.g., in Fig. 4.6. The Figure shows discharge curves referred to the amount of Ah drawn from the battery. The dashed line on the top represents the equilibrium voltage corresponding to the acid concentration according to the Nernst equation (Eq. (4.10), p.101). If all the partial-reaction steps were fast enough, i.e. if no kinetic deceleration occurred, the increased discharge rate would cause only a voltage drop and correspondingly shift the dashed line in parallel to lower voltages. Because of the low internal resistance, which is in the range of 1 mΩ for the cell considered, only a slight deviation would be expected, except near to the end of discharge, where a certain increase is caused by the decreasing conductivity of the more diluted acid (cf. Fig. 4.21, p.144).

Fig. 4.6 shows that the discharge voltage decreases markedly with increasing load. Beyond that, not only a considerable voltage drop can be observed with increasing discharge current, but also a growing decline of the curves. So, with increasing load, the dischargeable share of the capacity is reduced, even when the end-of-discharge voltage is reduced. The decreased utilisation of the active material is caused by limiting diffusion rates. Mainly acid depletion at the electrode surface increases the voltage drop and reduces the rate of the reaction. Furthermore, some of the undischarged material may be buried underneath the growing $PbSO_4$ layer. Because of high concentrations of Pb^{2+} ions, this layer grows very fast at high loads, resulting in a thin but compact covering layer that prevents further discharge very early.

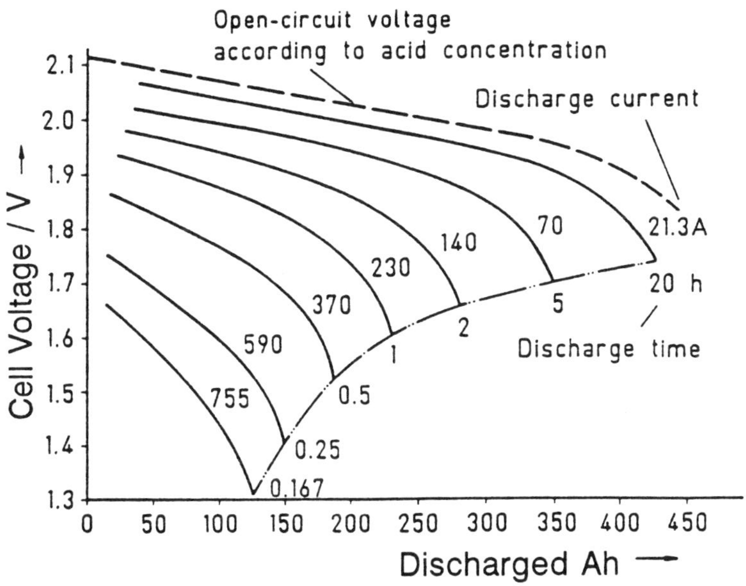

Fig. 4.6 Discharge curves referred to the drawn amount of Ah. Flooded traction cell with tubular plates, 350 Ah (5-hour rate).

The impact of the kinetic parameters on the charging process was mentioned in connection with Fig. 3.11, p.84. During charging, $PbSO_4$ crystals must be dissolved and the ions have to migrate to the lead or lead dioxide surface, where electrons are available. These dissolution and diffusion steps proceed at certain rates and cannot be accelerated by increased electrode potentials. Both chemical reactions that precede the transfer step cause limiting currents as discussed in connection with Fig. 2.8, p.29. These limiting currents are reduced more and more when the battery approaches the fully charged state, because the amount of chargeable material vanishes. As a consequence, the charge acceptance of a discharged lead-acid battery, which initially is very high, is reduced more and more towards the end of charge (cf. Fig. 3.11). So the duration of charging can only slightly be reduced by increased voltage, while a number of disadvantages are connected with increased cell voltage. For these reasons, charging at increased voltage is recommended for lead-acid batteries only for limited charging periods (cf. Chapter 13).

4.3.1 Migration Effects

Migration of ions that leads to the transference numbers has been described in Section 2.3.2 and values for sulphuric acid are shown in Table 2.3, p.25. Migration effects usually are not noticed, since they are equalised when the battery is discharged and charged again, and during float charging, diffusion largely compensates migration, so that only small concentration differences result.

When the lead-acid battery is discharged, the small transference number of HSO_4^- is beneficial, because HSO_4^- ions are carried away from the positive electrode during discharge, as indicated in Fig. 2.1, p.7. Due to their small transference number, migration of HSO_4^- ions corresponds to only one tenth of the flowing current. For example, per multiple of the cell reaction (53.6 Ah), HSO_4^- ions are carried away equivalent to 5.4 Ah. This ion transport means additional reduction of acid concentration, besides consumption of SO_4^{2-} ions by the discharge reaction and acid dilution on account of water generation. The small transference number reduces this unfavourable effect. Nevertheless, migration contributes to the more critical discharge performance of positive electrodes in lead-acid batteries.

A significant effect of migration is observed in block batteries, when a leakage occurs between two adjacent cells, causing a short-circuit through the electrolyte. Fig. 4.7 illustrates the effect:

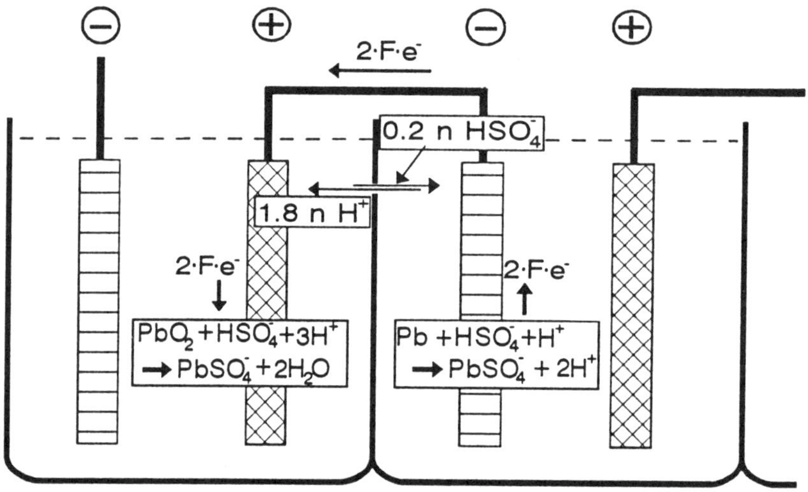

Fig. 4.7 Lead-acid battery in block design with a leakage in the partition of two adjacent cells. Changes of acid concentration in the concerned cells on account of migration.

The intercell leakage, as an 'ionic short-circuit', causes continual discharge of the two adjacent cells. A current flux equivalent to $2\cdot F\cdot e^-$ gives rise to the transport of 0.2 mole HSO_4^- from the cell on the left into that on the right hand side, and 1.8 mol H^+ ions in the reverse direction. Table 4.7 shows the balance after a current flux equivalent to 5 multiples of the reaction equation.

The figures in Table 4.7 show that the intercell leakage transfers 1 mole = 98 g H_2SO_4 from the left to the right hand cell, when 5 multiples of the reaction, i.e. 5x53.6 = 268 Ah are discharged by the intercell leakage. This seems to be a large amount of current. However, the intercell leakage continuously draws current, and 268 Ah would be reached within one year by a current of 30 mA. A corresponding electrolyte resistance would be 66 $\Omega (R = 2\ V/0.03\ A)$, and, with the specific resistance of 1 $\Omega \cdot cm$, this corresponds with a leakage area in the range of 1 mm^2 when 0.5 cm is assumed as the thickness of the partition wall - a plausible size of such a leakage.

Such an intercell leakage would not be noticed, if the battery is on float charge or is charged regularly, because the gradual discharge is compensated again and again (under corresponding water loss by overcharging the counter electrodes in the concerned cells). But recharge does not compensate for the acid transfer between the cells, and diffusion, that equalises concentration gradients, is such a slow process, that its influence can be neglected in these circumstances ($D_{HSO_4^-} \approx 1 \cdot 10^{-5}\ cm^2/s$), due to the small area of the leakage. So, after a certain period of float charging, different concentrations of the acid are observed in the adjacent cells.

	Left hand cell	Right hand cell
	Reaction: $5 \cdot PbO_2 \Rightarrow 5 \cdot PbSO_4$ consumed: $5 \cdot H_2SO_4$ left: $10 \cdot H^+$ and $10 \cdot O^{2-}$	Reaction: $5 \cdot Pb \Rightarrow 5 \cdot PbSO_4$ consumed: $5 \cdot H_2SO_4$ left: $10 \cdot H^+$
	Migration gained: $9 \cdot H^+$; lost: 1 HSO_4^- 'left' $1 \cdot H^+$ (from H_2SO_4)	Migration lost: $9 \cdot H^+$; gained: $1 \cdot HSO_4^-$
Result	lost: $6 \cdot H_2SO_4$; gained: $10 \cdot H_2O$	lost: $4 \cdot H_2SO_4$
After recharge	lost: $1 \cdot H_2SO_4$	gained: $1 \cdot H_2SO_4$

Table 4.7 Concentration changes in the two cells in Fig. 4.7 on account of the leakage, after a current flux of $10 \cdot F \cdot e^-$ (268 Ah) and a corresponding recharge. Recharge only compensates for the losses by discharge within each cell.

4.4 Self-Discharge, Secondary Reactions

Self-discharge of batteries can be caused by electronic short-circuits. Contact between current-conducting elements can be the reason. Such a contact can be caused by the growth of the positive electrode on account of corrosion (cf. Section 4.7.5) when the enlarged positive electrode gets in touch with the negative bus-bar. Electronic contact can also be caused by lead dendrites that grow on the negative-electrode surface and can penetrate the separator, if there are holes or too large pores.

Self-discharge can also be caused by impurities dissolved in the electrolyte, like iron, that form a redox system. Iron (Fe) is solved as Fe^{2+} or Fe^{3+}. When Fe^{2+} ions get in touch with the positive electrode they are oxidised to Fe^{3+} ($Fe^{2+} + e^- \Rightarrow Fe^{3+}$). Subsequently, when Fe^{3+} reaches the negative electrode, it is reduced to Fe^{2+}. Both reactions are repeated again and again and form a 'shuttle' that gradually discharges both electrodes (compare the nitrate shuttle in Section 5.3).

Another kind of self-discharge concerns only the positive or the negative electrode. It is caused by the thermodynamic situation, shown in Fig. 4.1. The unavoidable secondary reactions cause mixed potentials at both electrodes that give rise to gradual self-discharge of the concerned electrode.

4.4.1 Self-Discharge Reactions at the Negative Electrode

One example that concerns the negative electrode has been shown in Fig. 2.9, p.31. It is formed by a mixed potential between the two reactions:

Discharge of the negative electrode $Pb \Rightarrow Pb^{2+} + 2 \cdot e^-$

and hydrogen evolution $2 \cdot H^+ + 2 \cdot e^- \Rightarrow H_2$

complemented by $H_2SO_4 \Rightarrow 2 \cdot H^+ + SO_4^{2-}$

with the result $\boxed{Pb + H_2SO_4 \Rightarrow H_2 + PbSO_4}$ (4.23)

and occurs in the potential range between ≈ -0.32 V and 0 V (cf. Fig. 4.1) where Pb is no longer stable and H^+ ions can be reduced. In this couple of reactions, hydrogen evolution is the slow reaction that determines the rate of discharge, as is shown in Fig. 2.9. Thus, the rate of this self-discharge mechanism is determined by hydrogen evolution which again is strongly influenced by foreign metals that reduce hydrogen overvoltage (cf. Fig. 7.7, p.250). Furthermore, self-discharge by hydrogen evolution depends on the state of charge, because it only occurs at the charged surface. For this reason, the rate of self-discharge by H_2 evolution decreases with progressing discharge (cf. Table 4.8, and also Section 7.2.3).

Another self-discharge reaction of the negative electrode is caused by oxygen reduction, that, according to Fig. 4.1, can occur in the potential range between -0.32 and 1.23 V. The two opposite reactions are

Oxidation of lead $Pb \Rightarrow Pb^{2+} + 2 \cdot e^-$

and reduction of oxygen $1/2 \cdot O_2 + 2 \cdot e^- \Rightarrow O^{2-}$

with the result $\boxed{Pb + 1/2\, O_2 \Rightarrow PbO}$ (4.24)

This reaction happens spontaneously as soon as lead comes in contact with oxygen. It covers any piece of lead exposed to air with a dull grey layer of lead oxide. Water acts as a catalyst for this oxidation reaction.

Note: *The above equation for oxygen reduction is a purely formal reaction. Actually, O^{2-} ions can exist only in a crystal lattice, like an oxide. In aqueous solutions their concentration is so low that noticeable reaction rates could never be reached (activity $a_{O^{2-}} < 10^{-50}$ and $< 10^{-22}$ mole/dm^3 in acid and alkaline solution, respectively). Furthermore, the exchange of two electrons does not happen in one step. Oxygen reduction therefore occurs via a sequence of reaction steps: in neutral and alkaline solution such steps are $O_2 + H_2O + 2 \cdot e^- \Rightarrow H_2O_2$ and $H_2O_2 + 2 \cdot e^- \Rightarrow 2 \cdot OH^-$* (cf. Section 7.5.3.2). *At a dry surface, the reaction (4.24) is a slow solid state reaction including adsorbed atoms.*

Negative plates of the lead-acid battery are sensitive to oxidation, owing to their large surface area. On account of the catalytic activity of water, wet negative electrodes are most endangered by this oxidation. The reaction is connected with an appreciable heat generation ($\Delta H = -219$ kJ), and is well known in the battery manufacturing process.

Special methods have to be applied to prepare fully charged, dry charged lead electrodes, that can be stored in the charged state for years when they are dry (and are kept dry). As long as lead is available at the surface, the oxidation rate is mainly determined by the amount of oxygen that reaches the surface. With proceeding oxidation, this rate is more and more reduced on account of the diminishing lead surface exposed to oxidation. When the surface is kept dry, only a thin layer of oxide is formed, which protects the underlying material. But this protection is destroyed when the plate gets wet, because of the catalytic effectiveness of water, and in unfavourable conditions the oxidation of the active material proceeds apace as long as lead is available at the surface, and charged negative plates may be oxidised so far, that up to 50% of their capacity is lost. (cf. dry charged electrodes, Section 10.7.1).

Lead oxide (PbO) reacts very quickly with sulphuric acid according to

$$\boxed{PbO + H_2SO_4 \Rightarrow PbSO_4 + H_2O} \qquad (4.25)$$

Thus in the presence of sulphuric acid, the two reactions of Eqs.(4.24) and (4.25) cause self-discharge of the negative electrode, with lead sulphate ($PbSO_4$) as the final product. Eq. (4.25) produces a considerable amount of heat ($\Delta H = -173$ kJ).

On account of this heat generation, the extent of negative-plate oxidation (formation of PbO) can be estimated after storage of the above mentioned dry charged lead-acid batteries by the heat, evolved when the battery is filled with acid.

In flooded batteries, the extremely low diffusion rate of dissolved oxygen hinders the access of oxygen to the electrode surface so far, that this reaction normally is insignificant (cf. Section 2.6.3.1). But in valve-regulated lead-acid batteries, self-discharge by oxygen intake can gain importance, because of the direct access of oxygen to the negative-electrode surface on account of the immobilised electrolyte. This self-discharge can have two causes:

- Admission of air to the interior of the battery on account of a leak or a defective valve.
- Unbalanced float conditions, e.g. on account of a too-high hydrogen evolution rate; i.e. the float current cannot compensate for both the internal oxygen cycle and the hydrogen evolution (cf. Section 8, esp. Fig. 8.15). In this case, part of the oxygen generated at the positive electrode cannot be reduced at the negative electrode and leads to oxidation, which means chemical self-discharge.

4.4.2 Self-Discharge Reactions at the Positive Electrode

According to Fig. 4.1, the situation of the positive electrode corresponds to that of the negative insofar as oxygen evolution starts below the $PbSO_4/PbO_2$ equilibrium potential. So, a mixed potential exists that causes gradual self-discharge:

Discharge of the positive electrode

$$PbO_2 + H_2SO_4 + 2 \cdot H^+ + 2 \cdot e^- \Rightarrow PbSO_4 + 2 \cdot H_2O$$

and oxygen evolution

$$H_2O \Rightarrow \tfrac{1}{2}O_2 + 2 \cdot H^+ + 2 \cdot e^-$$

with the result

$$\boxed{PbO_2 + H_2SO_4 \Rightarrow \tfrac{1}{2}O_2 + PbSO_4 + H_2O} \quad (4.26)$$

Lead corrosion is another reaction that induces self-discharge of the positive electrode on account of the following couple of reactions:

Discharge of the positive electrode

$$PbO_2 + H_2SO_4 + 2 \cdot H^+ + 2 \cdot e^- \Rightarrow PbSO_4 + 2 \cdot H_2O$$

and grid corrosion

$$Pb + H_2O \Rightarrow PbO + 2 \cdot H^+ + 2 \cdot e^-$$

with the result

$$\boxed{PbO_2 + Pb + H_2SO_4 \Rightarrow PbO + PbSO_4 + H_2O} \quad (4.27)$$

The formed lead oxide (PbO) is converted into lead sulphate ($PbSO_4$), according to Eq. (4.25), as soon as it gets in touch with sulphuric acid (cf. Section 4.7.1). So the final result of these reactions would be:

$$\boxed{PbO_2 + Pb + 2 \cdot H_2SO_4 \Rightarrow 2 \cdot PbSO_4 + 2 \cdot H_2O} \quad (4.28)$$

This corrosion reaction gradually destroys the passivating layer of PbO_2 that protects the underlying grid material when the battery is stored at open circuit. For this reason, freshening charges are required from time to time to rebuild the PbO_2 layer.

4.4.3 Quantities of Self-Discharge

The described self-discharge reactions concern the individual electrodes. The resultant self-discharge rates of positive and negative electrode are independent from each other. The electrode with the higher self-discharge determines the charge loss of the cell (which, however, may be masked by oversizing of the concerned electrode). Table 4.8 summarises these rates:

In the positive electrode, grid corrosion dominates self-discharge at open-circuit voltage and can be further increased when the passivating layer at the grid surface is destabilised by advanced corrosion (cf. Section 4.7.1). (*The situation at the positive grid surface is more stable about 40 to 80 mV above the $PbSO_4/PbO_2$ equilibrium voltage, where grid corrosion reaches its minimum, cf. Fig. 7.5, p.242*). During storage, grid corrosion can cause irreparable damage when freshening charges are not made at regular intervals.

Self-discharge of cells or batteries is usually determined by the hydrogen evolution rate at the negative electrode. The values in Table 4.8 concern modern stationary or valve-regulated batteries. In batteries with medium or high antimony grid alloys (cf. Section 4.7), hydrogen-evolution rates start at about twice this volume and increase with service life to a multiple. Hydrogen evolution is, of course, only possible at the lead surface. For this reason, the rate of hydrogen evolution gradually decreases with progressing discharge of the electrode.

Affected element	Discharging mechanism
Positive electrode	Oxygen generation (Eq. (4.26)) in the range of 0.5 mA/100 Ah, which means ≈ 0.4% of nom. cap. per month (starting at higher values) Corrosion of the positive grid (Eq. (4.28)) equivalent to about 1 mA/100 Ah at open-circuit voltage and intact passivation layer *.
Negative electrode	Hydrogen evolution in the range of 1 to 4 mA/100 Ah ◊ which means 1 to 3% of nom. cap. per month (gradually decreasing)
Cell (both electrodes)	Usually, determined by the self-discharge of the negative electrode, can be masked by oversized negative electrodes (no reaction between both electrodes)

Table 4.8 Self-discharge mechanisms and their rates in normal conditions. The values are, of course, only rough approximations. * The corrosion rate depends to some extent on alloy composition and is increased with high antimonial alloys. ◊ Values for modern stationary batteries, influencing parameters cf. Section 7.2.

The rates of hydrogen and oxygen evolution as well as grid corrosion are approximately doubled per 10°C of temperature increase (cf. Section 2.3.6). For this reason, self-discharge is largely dependent on storage temperature as shown in Fig. 4.8.

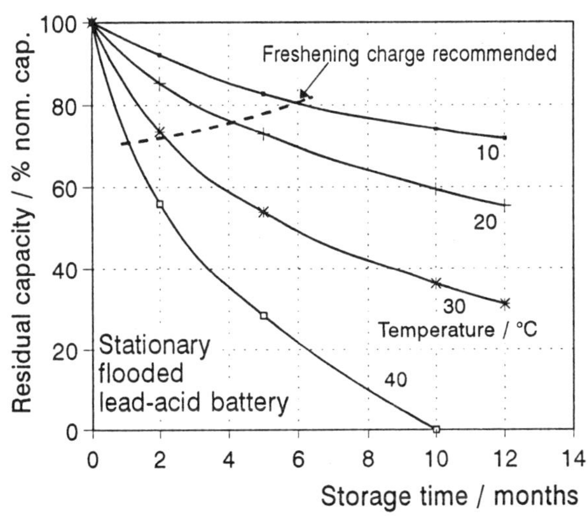

Fig. 4.8 Self-discharge of vented stationary lead-acid batteries (VARTA OPzS) at different temperatures.

Refresher charges, as indicated in Fig. 4.8 usually are constant-current charges with a small current (e.g. I_{200} for 24 hours).

Self-discharge also depends on battery design. In valve-regulated lead-acid batteries, the situation is relieved to some extent by the fact that their design automatically reduces self-discharge and minimises grid corrosion. Nevertheless, they also suffer gradual self-discharge and require refresher charges during long term storage. Fig. 4.9 shows corresponding curves.

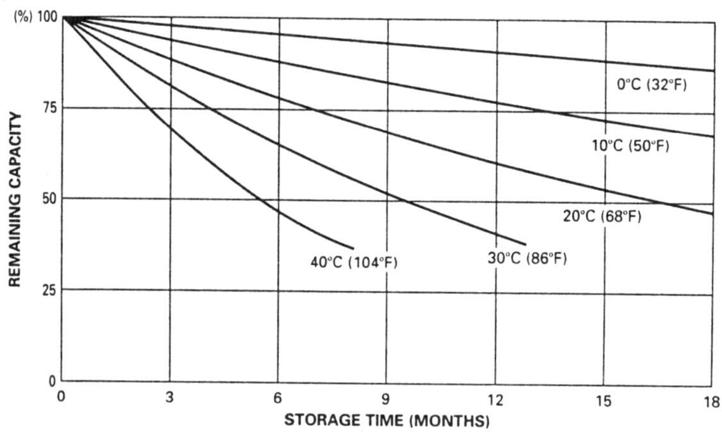

Fig. 4.9 Self-discharge of valve-regulated lead-acid batteries at different temperatures. (YUASA NP series ([13]).)

4.5 Heat Effects

The two heat effects that have to be considered are:

1. The reversible heat effect, described in Section 4.2.1.4, that depends only on the amount of converted active material and can be expressed by means of the calorific voltage E_{cal}.
2. The Joule effect that is proportional to the voltage drop caused by current flow and depends largely on load current and the specific design of the battery (its internal resistance).

4.5.1 Heat Generation During Discharge

Heat generation during discharge is demonstrated for two examples in the following Figures. Fig. 4.10 shows discharge curves for a vented battery with tubular positive plates with inserted curves for equilibrium and calorific voltage, E^o and E_{cal} respectively.

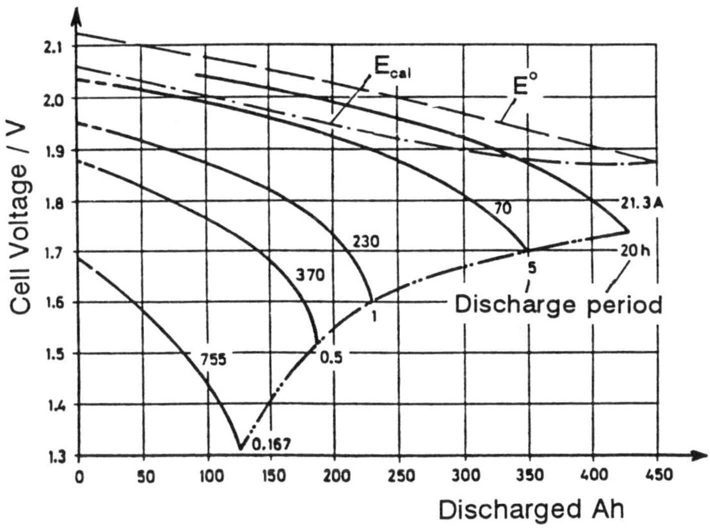

Fig. 4.10 Discharge curves and equilibrium voltage E^0 of a vented lead-acid battery with tubular plates (VARTA OPzS). Nominal capacity 350 Ah (5-hour rate). Numbers at the curves are discharge current and discharge period, respectively. The dash-point curve is the calorific voltage according to Eq. (4.20), but under consideration of the concentration dependence of $T{\cdot}\Delta S$ and thereby the resulting voltage difference $T{\cdot}\Delta S/n{\cdot}F$, given in ([8]). Thus, especially at the end of discharge, the simple approximation of Eq. (4.20), p.105, does no longer apply.

The heat effects and the resulting total heat generation are compiled in Table 4.9 . The rate of total heat generation Q_{total}/dt in the last line of Table 4.9 is the average, namely the quotient of the generated energy and the discharge period. Strictly speaking this quotient changes with discharge time and increases considerably towards the end of discharge, on account of the increasing internal resistance (and the corresponding Joule effect). More exact calculations would afford an evaluation of dQ_{total}/dt for each point of the discharge from Eq. (2.79), p.35. But in view of the approximate character of the following considerations, this average value will be sufficient.

A positive Q_{total} results for the 20-hour discharge rate; i.e. the reversible heat effect exceeds the Joule effect and a slight cooling effect remains (heat-pump effect). But already at the 5-hour rate, the situation is reversed and, with further increasing discharge current, the dominance of the Joule effect is more and more pronounced. This is because of the higher internal resistance of the fairly thick tubular plates; and the larger size of the battery.

This increase of the Joule effect with increasing discharge current is especially obvious in Line 4 of Table 4.9. The energy loss by heat generation due to the Joule effect is only 3% at the 20-hour discharge rate, but more than one-third of the discharged energy at the 10-minute discharge rate. This increase is indicated in Fig. 4.10 by the marked reduction of discharge voltage, which causes a corresponding difference $E - E^0$.

1	Discharge period	20 h	5 h	1 h	0.5 h	10 min
2	Discharge current / A	21.3	70	230	370	755
3	Discharged energy / Wh	834	680	423	329	192
4	Q_{Joule} / Wh	-23	-41	-54	-61	-71
	= % of discharged energy	3%	6%	13%	18.5%	37%
5	Q_{rev} / Wh	29	26	19	15	9
	= % of discharged energy	3.5%	3.8%	4.4%	4.5%	4.7%
6	Q_{total} / Wh	+6	-15	-35	-46	-62
	referred to 100 Ah NC	+1.7	-4.2	-10	-13	-18
	= % of discharged energy	-0.7%	2%	8.1%	14%	32%
7	$Q_{total}/\Delta t$ / W (average)	+0.3	-3.0	-35	-92	-372

Table 4.9 Discharge parameters and values for Q_{Joule} and Q_{total} (always per cell) taken from Fig. 4.10 by graphical integration. Negative sign means generated energy but loss of energy from the system in consistency with thermodynamic rules (cf. Notes in Sections 2.4.1 and 2.4.2). Vented lead-acid battery with tubular plates. Nominal capacity 350 Ah (5-hour rate). $Q_{total}/\Delta t$: average value for the whole discharge duration Δt.

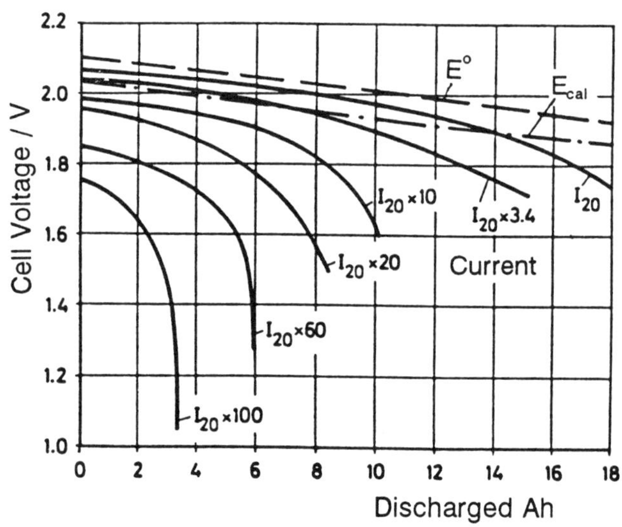

Fig. 4.11 Discharge curves of a valve-regulated block battery with gelled electrolyte (12 V; 18 Ah; SONNENSCHEIN A400 based on prospectus data ([14])). Equilibrium voltage (broken curve) calculated from acid content. The dash-point curve is the calorific voltage based on values as in Fig. 4.10.

Fig. 4.11 shows discharge curves of a valve-regulated lead-acid 12 V block battery of 18 Ah nominal capacity with gelled electrolyte. This battery is a 'long life design' with positive plates of nearly 3 mm thickness. The nominal capacity is based on the 20-hour discharge rate, and the discharge currents are expressed as multiples of this load (see Fig.3.2).

The values from this Figure are listed in Table 4.10. The result is similar to that in Table 4.9. At the lowest discharge rate, slight cooling results. Compared to Table 4.9, the percentage of discharged energy, lost by heat generation, is lower on account of the smaller internal resistance of this battery.

1	Discharge current	I_{20}	I_{20}x3.4	I_{20}x10	I_{20}x20	I_{20}x60	I_{20}x100
	Discharge period	20 h	4.96 h	1.12 h	27 min	6.6 min	2.2 min
2	Discharge. current / A	0.9	3.06	9	18	54	90
3	Discharged. energy / Wh	40.8	33.9	22.5	17.9	12	6.5
4	Q_{Joule} / Wh	-1.11	-1.71	-1.79	-2.07	-2.06	-1.61
	= % of discharged energy	2.7%	5%	8%	11.5%	17%	25%
5	Q_{rev} / Wh	1.8	1.2	0.81	0.61	0.45	0.25
	= % of discharged energy	4.3%	3.5%	3.6%	3.4%	3.7%	3.8%
6	Q_{total} / Wh	+0.7	-0.53	-0.98	-1.45	-1.60	-1.36
	referred to 100 Ah NC.	3.9	-2,94	-5.4	-8.05	-8.9	-7.6
	= % of discharged energy	-1.6%	1.6%	4.3%	8.1%	13.24%	20.9%
7	$Q_{total}/\Delta t$ / W (average)	0.03	-0.11	-0.87	-3.12	-14.7	-36.2

Table 4.10 Discharge parameters and values for Joule effect (Q_{Joule}), reversible heat effect (Q_{rev}) and resulting total heat generation. (Q_{total}) taken from Fig. 4.11 by graphical integration (always per cell). $Q_{total}/\Delta t$: average value for the whole discharge duration Δt. Line 7 is the average value, which means the energy generated in total divided by the discharge duration. Valve-regulated lead-acid battery, 12 V block, nominal capacity 18 Ah (20-hour rate).

At the 20-hour discharge, the positive Q_{rev} exceeds the (negative) Joule effect and causes a positive Q_{total}, which means a slight decrease of the battery temperature at this rate. But at the 2.2-minute discharge, 21% of the discharged energy is lost as heat, mainly caused by the Joule effect.

Line 7 is (as in Table 4.9) the average rate of energy generation. At low discharge rates, these values are small, because the Joule effect Q_{Joule}, and the reversible heat effect Q_{rev}, more or less balance each other. But, with increasing load, heat generation grows faster and faster and reaches large values at high discharge rates, though only for a short period.

Fig. 4.12 illustrates the situation given by the values in Table 4.10. Reversible heat effect, Joule effect and heat in total (Lines 5, 4 and 6 respectively in the table) are represented by the columns. The reversible heat effect always amounts to the same percent of discharged energy, because it is linked to the amount of active material that is converted. With increased load, the Joule effect gains more and more weight and increasingly determines the heat generation in total. Nevertheless, even during the 2.2 min discharge, heat output is reduced from 25% of the discharged energy to 21% due to the cooling, caused by the reversible heat effect.

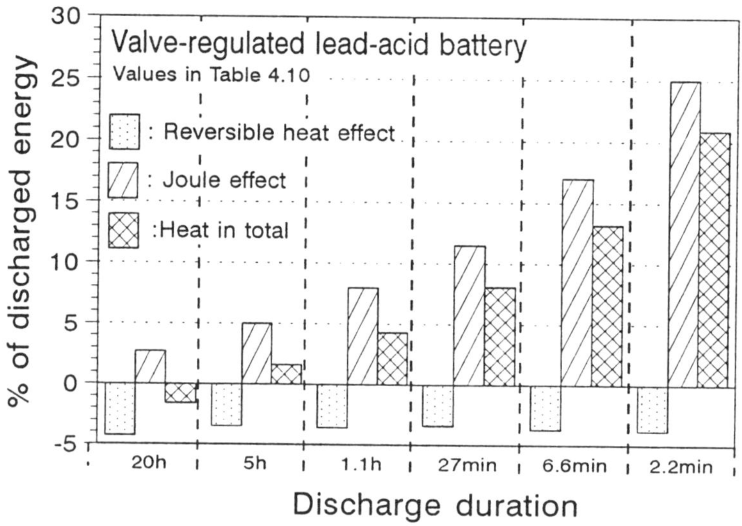

Fig. 4.12 Heat generation in the valve-regulated lead-acid battery, used as example in Fig. 4.11, referred to the discharged amount of energy. Values in the Lines 4,5 and 6 of Table 4.10. *Compared to Table 4.10, the signs of* Q_{rev}, Q_{joule} *and* Q_{total} *are reversed, to correspond with the usual practice that in heat calculations generated heat is positive which is expressed by* $Q_{gen} = -Q_{total}$ *in Equations 2.82 and 2.83, p.36.*

Comparison of the data in Tables 4.9 and 4.10 shows that lead-acid batteries of different design have similar heat generation at low discharge rates. But, with increasing discharge current, differences of internal resistance in the various designs become more and more evident (cf. also Table 3.7, p.90).

Fig. 4.13 is a comparison between both batteries. It shows that heat generation does not differ too much, although these batteries are of quite different design. The vented one is a large stationary battery with 8 mm thick tubular plates, designed for medium loads with discharge duration of more than one hour; the valve-regulated battery is a typical high-load battery for short autonomy periods as they are required for UP systems.

Heat generation is similar in both batteries, despite their different design. This is due to the small heat generation in lead-acid batteries during discharge, which is further reduced by the reversible heat effect that does not depend on battery design.

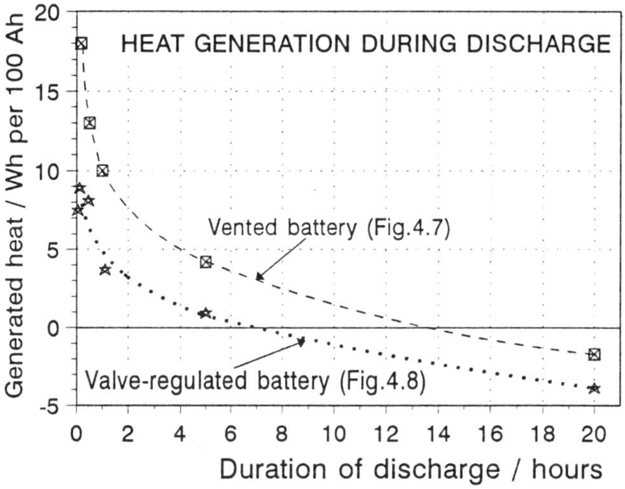

Fig. 4.13 Comparison of the heat generation in the above samples of a vented and a valve-regulated lead-acid battery (Figs. 4.10 and 4.11). (Line 7 in Tables 4.9 and 4.10; 'total heat generation'.)

4.5.2 Heat Generation During Charging

When the battery is charged or overcharged, heat generation is mainly determined by the secondary reactions, water decomposition in vented batteries, and internal oxygen cycle in the valve-regulated design. For this reason, it is important to know how the current is split up between charging and the secondary reactions. Fig. 4.14 is a comparison of heat generation by the various reactions.

The hatched or cross hatched areas in Fig. 4.14 show the amount of heat, contributed by the different reactions when equal current is consumed by each of them.

For the charging reaction, most of the supplied electrical energy $(\Delta G/nF)\cdot i = E^o\cdot i$ is used for the electrochemical reaction, as represented by the dotted area. Heat generation, corresponds to $(E - E_{cal})\cdot i$, and, since E_{cal} is lower than E^o, the reversible heat effect (hatched area Q_{rev}/nF) contributes additional heat. Thus, heat generation corresponds to $0.24\cdot i$ in W, when E and E_{cal} are given in V and i in A. ($E_{cal} = 0.07$ V, according to Eq. (4.20).)

For water decomposition, E_{cal} is higher than E^o (Eq. (2.112), p.47). So during charging, heat generation is reduced by the reversible heat effect that acts as a cooling effect. This means, that more heat is carried away with the evolved gas mixture than corespondents to the equilibrium voltage: an advantage for vented batteries, because heat generation during overcharging is so reduced by 20%. On the other hand, the difference between the actual cell voltage is larger compared to the charging reaction, and heat generation corresponds in this example to $(2.25 - 1.48)\cdot i = 0.77\cdot i$ (in W), or more than 3 times the heat generation of the charging reaction.

The special situation of the internal oxygen cycle in respect to thermodynamic data is that $E^0 = 0$ and also $E_{cal} = 0$. So, all the electrical energy, employed for this reaction, is converted into heat, as represented by the double hatched area in Fig. 4.14. Under these conditions, heat generation is about 3 and 9 times larger, compared to water decomposition and the charging reaction, respectively.

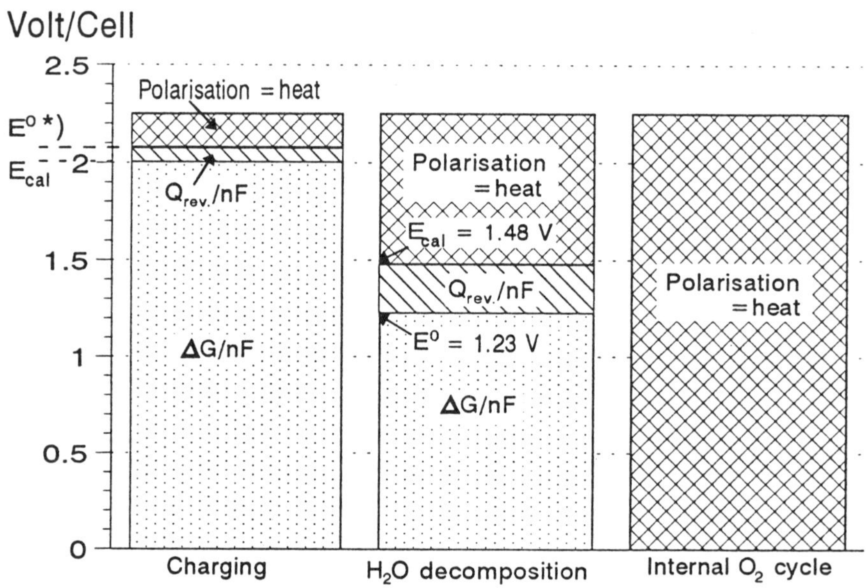

*) depends on acid density ($E^\circ = d + 0.84$); here $d = 1.24$ g/cm^3, i.e $E^\circ = 2.08$ V

Fig. 4.14 Heat generation during charging or overcharging at 2.25 V/cell. Voltage equivalents times current results in heat generation.

Fig. 4.14 shows that heat generation during charge is mainly connected to the secondary reactions. During float charging, when charging of the battery has been completed, water decomposition or the internal oxygen cycle are the heat sources in vented and valve-regulated batteries, respectively.

Owing to these great differences, the heat generation in the battery is decisively affected by the distribution of the charging current to the three reactions. Fig. 4.15 shows how this distribution changes with increasing charge in a normal charging/discharging cycle of a vented lead-acid battery in a traction application.

The voltage curve is shown at the top of the Figure.

- A current-limited initial step of charging is followed by a constant-voltage period at 2.4 V/cell. Equalising charging up to 2.65 V/cell is the final step of the charging schedule.
- Discharge is assumed at constant current ($I_5 = 20$ A/100 Ah). The broken line shows the equilibrium voltage, while the full line represents the actual discharge voltage.

The middle of the Figure shows the current and its distribution to charging, water decomposition and discharging.

- During the initial stage, practically only charging occurs. Only when the voltage approaches the 2.4 V level, does the onset of water decomposition become noticeable. The broken horizontal line marks the average current during the 2.4 V step.
- As long as the cell voltage remains at 2.4 V, gas evolution is maintained at a roughly constant rate (assuming that the potentials of the positive and negative electrodes do not change too much).
- During the equalising step, nearly all the current is used for water decomposition on account of the progressively reduced charge acceptance.
- During the discharge, water decomposition is neglected.

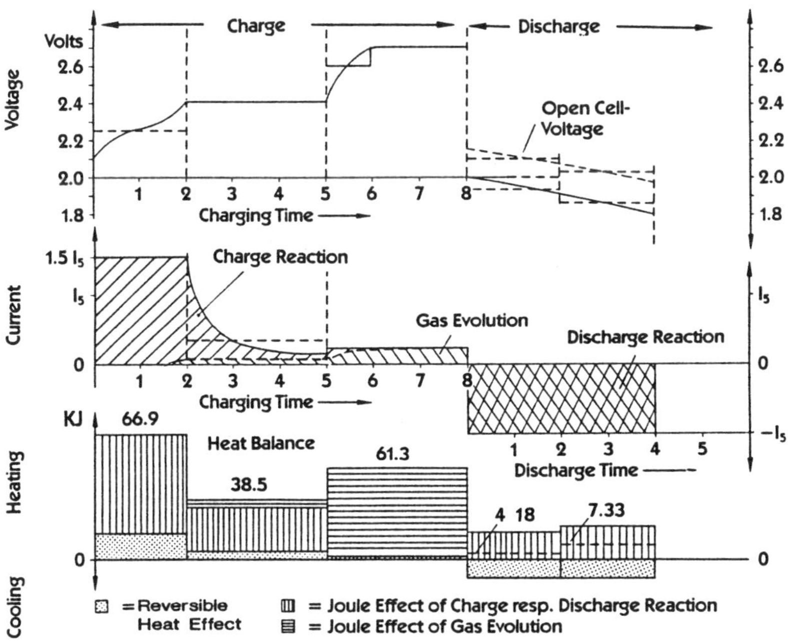

Fig. 4.15 Charging/discharging cycle of a vented traction battery. Lead-acid with tubular positive plates (VARTA PzS), 500 Ah. Heat-generation values referred to 100 Ah of N.C. The figures in the bottom part represent heat generation in total. The sum of the whole charging period amounts to 166 kJ/100 Ah.

At the bottom of Fig. 4.15, the heat generation is drawn as blocks that represent average values for the corresponding sections of the charging/discharging process. The distribution between reversible heat effect, charging and water decomposition is marked by different patterns of the areas concerned. The value above each block is the total heat generation in kJ.

Note: *According to thermodynamic conventions, heat generation means energy loss of the system and is a negative quantity, while the cooling effect indicates energy gain. Strictly speaking heating and cooling should be plotted reversely.*

Fig. 4.15 illustrates characteristic features of the charging process.

- During the first stage of the charging process, gas evolution can be neglected. The main heat is generated by the Joule effect, on account of the fairly high current. But the reversible heat effect also contributes noticeably to heat generation, on account of the converted active material. (40 Ah ≈ 85 Wh is charged during this period, which means a reversible heat effect of about 3 Wh = 11 kJ.)
- When 2.4 V is reached, the current is reduced and, as a consequence, the Joule effect and the reversible heat effect are reduced too. The now approximately constant gas evolution causes a corresponding heat generation.
- During the equalising step, gas evolution (required for mixing of the electrolyte, cf. Section 4.6.3) dominates. On account of the large difference between the actual charging voltage and the calorific voltage, heat generation is considerable, though the current is rather small.

The situation during discharge is as shown in Figs. 4.10 and 4.11. Owing to the small overvoltage, heat generation is also small, and is further reduced by the reversible heat effect.

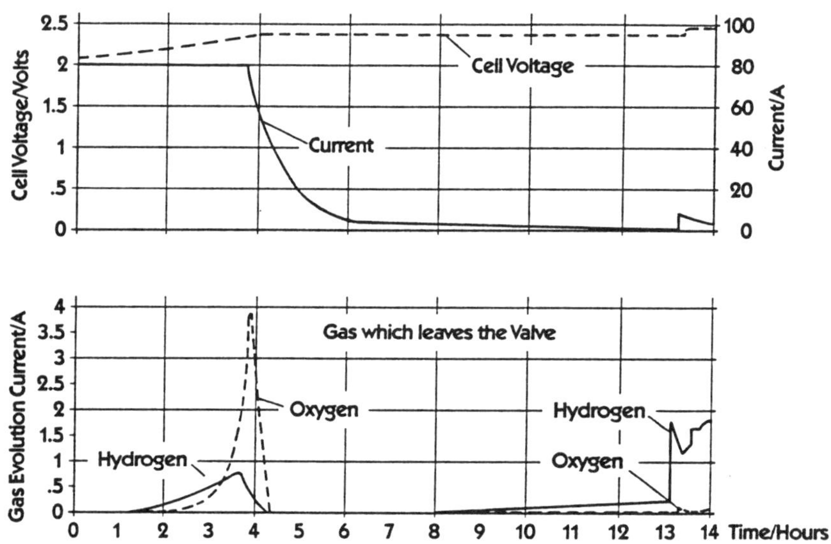

Fig. 4.16 Charging characteristic for a valve-regulated lead-acid traction battery. Upper part: cell voltage and current. Bottom part: gas escape from the valve. (Battery type: VARTA carat, 400 Ah N.C.; gelled electrolyte.)

Fig. 4.15 concerns a vented battery. The charging behaviour of a valve regulated type is shown in Fig. 4.16. The upper part in Fig. 4.16 shows cell voltage and current; the bottom part the evolution rate (in current equivalents) of the gas leaving the cell through the valve.

The initial charging current is 80 A, the 5-hour rate (I_5). At the end of the initial stage of the charging process, a certain amount of oxygen escapes through the valve together with a small amount of hydrogen. The reason for this gas escape is that both electrodes reach a state of polarisation that causes gas evolution while the charging current is still very high. So, at the negative electrode, an increasing amount of hydrogen is generated that has to escape through the valve. The inferior charge acceptance of the positive electrode causes a corresponding high rate of oxygen generation that exceeds the maximum rate of the internal oxygen cycle. So the gas cannot reach the surface of the negative electrode fast enough (cf. Chapter 8). Neither reduction of the oxygen nor chemical discharge of the negative electrode is possible and so oxygen has to leave the cell.

When the final charging voltage of 2.4 V/cell is reached, the current starts to decrease and gas generation also decreases. Then the internal oxygen cycle is enabled to consume the oxygen in the interior of the battery, and the pressure drops below the opening pressure of the valve. No gas escapes, until after about 8 hours of charging time; hydrogen generation increases the internal pressure again, since the negative electrode now approaches the state of complete charge; and charge acceptance is reduced.

At the end of the charging process, a certain overcharge is required to achieve full charge of all cells within a limited period of time. This is the 1-hour period at slightly increased voltage (+0.12 V) at the end of the charging process.

At this charging period, the negative electrodes have reached nearly 100% of recharge, while the positive electrodes are not yet fully charged due to their lower charge acceptance, that always is less than 95%. So, at the end of the charging process, the negative electrode is overcharged and produces a corresponding amount of hydrogen.

1	2	3	4	5
Heat effect		Wh	kJ	kJ/100 Ah
Joule effect	300·(2.25 - 2.0)	75	270	68
Reversible heat effect	0.035 x 615	21.5	77	19
Water decomposition	3·(2.4 - 1.48)	2.76	9.9	2.5
Internal oxyg. cycle	27·2.4	64.8	233	58
Heat generation in total		164	590	147

Table 4.11 Heat generation during charging estimated from Fig. 4.16 (positive arithmetic sign for generated Energy). Nominal capacity 400 Ah.

Heat-generation values, estimated from Fig. 4.16 are compiled in Table 4.11. In the foregoing discharge 300 Ah had been drawn. During the recharge, shown in Fig. 4.16, 330 Ah flow into the battery. 300Ah can be assumed for recharge (100% recharge). Oxygen and hydrogen escape is equivalent to about 3 Ah. So 27 Ah were used for overcharging (internal oxygen cycle).

Most of the recharge occurs during the first 3.7 hours at a mean voltage of 2.25 V. During the initial stage, the equilibrium voltage is reduced on account of acid dilution. So the ΔE value that determines the Joule effect according to Eq. (2.66) can be estimated as 2.25 - 2 V. The reversible heat effect amounts to 3.5% of the (reversible) charged energy, i.e. $0.035\cdot(2.05\cdot300)$, when 2.05 V is assumed as the average equilibrium voltage. The reversible heat effect equals that in the vented battery. Heat generation by water decomposition is given by the difference $E - E_{cal}$, while all the electrical energy is converted into heat.

Referred to 100 Ah N.C., heat generation in total amounts to 147 kJ which is close to the 166 kJ/100 Ah of the vented battery in Fig. 4.15. The reason is that the internal oxygen cycle is much more effective in heat generation than water decomposition. Owing to this heat generation, the fairly low charge factor of 1.10 produces as much heat as the more intensive overcharging of the vented battery shown in Fig. 4.15.

4.5.3 Heating of Batteries

The gradual warming of a battery above its ambient temperature is the result of heat generation within the battery and heat flow out of the battery in combination with the heat capacity of the battery, as described by Eq. (2.80), p.36. Such heating depends to a large extent on the specific parameters of the battery. For this reason, the following estimations can only have an approximative character, and can only point out typical results.

Two borderline cases will be considered:

1. The maximum temperature difference ΔT between the battery and its surroundings that will be reached under adiabatic conditions according to Eq. (2.83), p.36, i.e. without any heat dissipation.
2. The temperature difference ΔT_{bal} that is required between the container wall and the surrounding atmosphere to establish a balance between generated and dissipated heat (Q_{gen} and Q_{diss} respectively) according to Eq. (2.84), p.38.

In both cases, the temperature of the battery is assumed as uniform; temperature gradients within the battery are neglected.

4.5.3.1 Heating Parameters

In Table 4.12 characteristic parameters are compiled for the two batteries, shown as examples in Figures 4.10 and 4.11: the vented one with tubular plates and the valve-regulated battery.

1	2	3	4	5	6	7
				Values referred to one cell and 100 Ah NC.		
Battery	Spec. energy Wh/kg	Weight kg/kWh	Ah Ah/kg	Weight/cell p. 100 Ah kg/100 Ah	Heat capacity C_{Batt}/cell kJ/(K·100 Ah)	Surface area per cell m^2/100 Ah
Vented (Fig. 4.10)	23.1	43	11.1	9	10.5	0.1
Valve reg. (Fig. 4.11)	30	33	15	7.1	5.8	1/6 block area 0.06

Table 4.12 Characteristic data of the two batteries used as examples in Figs. 4.10 and 4.11. Heat capacity in kJ/K·kg is 1.2 and 0.82 for the vented and valve-regulated example, respectively, in correspondence with Table 4.6. Normally, the specific surface area of the container walls (ref. to 100 Ah) increases with decreasing battery size. Here, the small surface area of the valve-regulated battery is due to the block design (cf. Table 4.13).

Heat capacity depends on the design of the battery; of considerable influence is the share of electrolyte in relation to the weight of the others listed in Table 4.5. So, the rather high heat capacity of the vented battery is mainly due to its rather high electrolyte content.

Heat dissipation is proportional to the available surface area, which usually means the side walls and cover of the cell, or the corresponding areas of the battery, provided they are spaced sufficiently. Heat generation, on the other hand, is proportional to the Ah capacity, which corresponds more or less to the volume of the battery. Assuming that the shape of the cell or battery approximates to a cube, heat generation is proportional to the linear dimension raised to the third power, while at the surface, heat dissipation only increases in proportion to the second power. This means that the ratio between surface area and volume (≈ Ah capacity), and thereby the ratio between heat dissipation and heat generation, depends on the size of the battery. As a rough approximation, this relation is:

Single cells	
Large cells	≈ 0.04 m^2/100 Ah
Medium cells	≈ 0.1 m^2/100 Ah
Small cells	≈ 0.3 m^2/100 Ah
Block batteries	
Average per block	≥ 0.06 m^2/100 Ah
Centre cells	≥ 0.04 m^2/100 Ah

Table 4.13 Specific surface area of single cells and block batteries. These values are, of course, rough approximations that just show the order of magnitude.

In block batteries, assembled of several strings, or in compact batteries of a larger number of cells, like traction batteries in a tray, the surface area per 100 Ah may be considerably reduced, compared to these values (cf., e.g., ([15])).

Note: *On account of the fairly high heat conductivity between cells not separated by insulating or spacing materials, such batteries can be treated as a heat-generating unit and the overall outer surface of the battery considered for heat dissipation.*

4.5.3.2 Adiabatic Heating During Discharge

Table 4.14 shows values of adiabatic temperature increase during discharge for the two examples used above (Figs. 4.10 and 4.11) based on heat capacities shown in Table 4.6.

	1	2	3	4	5	6
1	Vented battery (Fig. 4.10, Table 4.9)			Valve-regul. batt. (Fig.4.11, Table 4.10)		
	Duration of discharge	Q_{total} Wh/100 Ah	ΔT °C	Duration of discharge	Q_{total} Wh/100 Ah	ΔT °C
2	5 h	-4.2	1.5	4.96 h	-2.94	1.8
3	1 h	-10	3.4	1.1 h	-5.4	3.3
4	10 min	-18	6.2	6.7 min.	-8.9	5.6

Table 4.14 Temperature increase of the batteries in Figs. 4.10 and 4.11 during discharge without heat dissipation (adiabatic situation) (based on the values of Line 6 in Tables 4.9 and 4.10 and heat capacities 10 and 6 kJ/100 Ah according to Table 4.6).

The values in Table 4.14 indicate that the difference in heating between the two examples, the medium size vented and the small valve-regulated lead-acid battery is small. The higher heat capacity of the vented type, mainly on account of the large electrolyte content, is compensated by the lower internal resistance of this valve-regulated design.

Table 4.14 shows furthermore that even without any heat dissipation, the batteries are only heated up by about 6 °C in the worst case at the high-rate discharge. The increased heat generation by the Joule effect at high loads is compensated for by the reduction of the available capacity (cf. Fig. 4.6).

4.5.3.3 Heat Dissipation During Discharge

Table 4.15 concerns the same discharges as in Table 4.14, but it shows heat generation per time unit (W) and the temperature difference between the battery and its surroundings, when heat generation balances heat dissipation according to Eq. (2.84), p.38.

Table 4.15 illustrates the strong dependence of heat generation on discharge rate. At the 5-hour rate, the required temperature difference is so small that heat dissipation is of no influence, and so the temperature difference ΔT_{bal} equals ΔT caused by adiabatic heating (Table 4.14).

	1	2	3	4	5	6	7	8
1	Vented battery (Fig. 4.10, Table 4.9)				Valve-regul. batt. (Fig.4.11, Table 4.10)			
	Duration of disch.	dQ_{total}/dt W/100 Ah	Dissip. W/m^2	ΔT_{bal} °C	Duration of disch.	dQ_{total}/dt W/100 Ah	Dissip. W/m^2	ΔT_{bal} °C
2	5 h	-0.85	-8.5	1.5	4.96 h	-0.6	10	1.8
3	1 h	-10	-100	16	1.1 h	-4.8	80	13
4	10 min	-106	-1060	>>	6.7 min.	205	3416	>>

Table 4.15 Heat generation, corresponding heat dissipation and balancing temperature difference between the batteries and the surroundings. Discharge of the batteries in Figs. 4.10 and 4.11, data based on the values of Line 7 in Tables 4.9 and 4.10, and spec. surface areas of 0.1 and 0.06 m^2/100 Ah according to Table 4.13. ΔT_{bal} is taken from Fig. 2.12, p.41.

At the 1-hour discharge rate the balancing temperature of Table 4.15 is not nearly reached by the adiabatic heating process, because most of the generated heat is required for warming up. Here, disc charging time is too short to reach the balancing temperature difference ΔT_{bal} .

For the high-rate discharge, the balancing temperature difference is beyond the scales in Fig. 2.12. But this temperature would never be reached, as the adiabatic warming up in Table 4.14 shows. Discharge duration is far too short.

4.5.3.4 Maximum Heating During Discharge

To estimate possible heating effects at extremely high-rate discharges or even short-circuits, data of a starter battery may be used. The energy content of such a 12 V/ 40 Ah battery is 480 Wh, i.e. about 0.5 kWh. The specific energy of a starter battery is about 35 Wh/kg, and so the weight of the considered battery 14 kg. With the specific heat of 1 kJ·kg^{-1}·K^{-1}, the heat content of the battery is 14 kJ/K.

If the whole energy content were converted into heat (0.5 kWh = 1800 kJ) and heat dissipation were excluded (adiabatic heating), the temperature increase would be

$$\Delta T = \frac{1800}{14} = 129\ °C \tag{4.31}$$

This, however, is a theoretical value. Actually only a small fraction of the energy content can be drawn at such a high rate, mainly because lead sulphate ($PbSO_4$), precipitated at the surface of the active material, forms a barrier that hinders further discharge, and so the residual energy content can only be drawn at quite a low rate.

When the above battery is loaded with 200 A at 50 °C (high temperature cranking test ([16])), the voltage is reduced to 1 V/cell within about 5 minutes on account of the growing internal resistance. Only 42% of the nominal capacity (200 x 5/60 = 17 Ah) are available. With an average discharge voltage of 1.65 V/cell, i.e. a voltage drop of 0.45 V/cell, 200 x 0.45 x 5/60 = 7.5 Wh per cell is generated by the Joule effect or 45 Wh per battery.

The reversible heat effect has to be related to discharge under equilibrium conditions (≈2 V/cell), so 17 Ah of drawn charge correspond to 34 Wh of reversibly drawn capacity, and the reversible heat effect amounts to 34 x 0.035 = 1.2 Wh/cell (3.5% of drawn energy) or 7.2 Wh per battery. It reduces the generated heat to 38 Wh in total.

When heat loss by dissipation is again neglected (adiabatic situation), 38 Wh lead to a temperature increase of

$$\boxed{\Delta T = \frac{38 \cdot 3600}{14 \cdot 1000} = 9.8\ ^{\circ}\mathrm{C}} \tag{4.32}$$

In this cranking test, about 78% of the total energy output is converted by the load resistance. In a true short-circuit, this share will be lower and the current higher. So, when the current is doubled, about twice the voltage drop is to be expected, and four times the energy generation (ΔE and i are doubled). Then, 50% of the energy is converted in the load resistance. Discharge duration, however, would then be markedly reduced, and so the energy generation in total presumably only doubled. When the short-circuit continues, the residual energy is discharged at such a reduced voltage that heat dissipation more or less will balance heat generation.

This estimation shows that lead-acid batteries cannot be heated up to detrimental temperatures by their own discharge. Even when discharged by short-circuit and heat cannot dissipate, less than 30 °C of temperature increase are to be expected.

This, however, does not mean that loads above specified limits may not cause damage. Internal connectors or terminals may be interrupted by melting and the sparks may ignite explosive gas mixtures in the head space of the battery.

4.5.3.5 Heating During Charging

Heat generation during charging is much more significant and even can grow hazardous for three reasons:

1. The charging energy comes from outside sources, and normally is not limited. So charge duration can arbitrarily be extended, and can be even unlimited, as in the case of float charging of stationary batteries.
2. The secondary reactions water decomposition, and especially the internal oxygen cycle, as the main reactions during overcharging, generate much heat.
3. The reversible heat effect contributes a limited but additional amount of heat.

Evaluations of maximum heating effects under adiabatic conditions, as in Section 4.5.4, only make sense for a fixed duration of the charging process. In the continual float charging of stationary batteries, even the smallest heat generation gives rise to unacceptably high battery temperatures under adiabatic conditions, when it occurs over a long enough period.

In Table 4.16 data are compiled that concern the charging process, shown above in the two Figures 4.15 and 4.16, when adiabatic conditions are assumed. Values of the vented battery are taken from Fig. 4.15 (charging section).

1	2	3	4
Battery	Heat evolution during charge kJ/100 Ah	kJ/kg	Temp. increase ΔT °C
Vented (Fig. 4.15)	166	16.6	16.6
Valve-regulated batt. (Fig. 4.16, Table 4.11)	145	14.5	14.5

Table 4.16 Estimation of the temperature increase, when the charging process shown in Figures 4.15 and 4.16 occurs under adiabatic conditions. Assumed approximations for both batteries - Specific weight: 10 kg/100 Ah; Heat capacity 1 kJ/K·kg.

The resulting rise of temperature corresponds to the experience that charging of larger traction batteries of the flooded type causes a temperature increase of about 12 °C. In Table 4.16, the temperature rise is smaller for the valve-regulated battery, since the charging process was optimised. But in general, due to the higher heat generation of the internal oxygen cycle, valve-regulated batteries are in more danger of suffering heat problems. For this reason, manufacturers sometimes limit the size of valve-regulated traction batteries to a maximum energy content.

Under normal float conditions, heat generation is small. Table 4.17 shows estimated values for vented and valve-regulated lead-acid batteries based on usual float currents and the rule that this current is doubled per 10 °C of temperature increase. As specific surface area, the smallest value of Table 4.13 is chosen to estimate the 'worst case'.

1	2	3	4	5
Temp. °C	Float current mA/100 Ah	Generated heat mW/100 Ah	Heat p. 0.04 m^2 W/m^2	ΔT_{bal} K
Vented batteries				
		(2.25 - 1.48)·i		
20°	10 - 30	7.7 - 23	0.2 - 0.6	< 1
40°	40 - 120	31 - 92	0.8 - 2.3	< 1
Valve-regulated batteries				
		2.25·i		
20°	20 - 70	45 - 157	1.2 -- 3.9	≤ 1
40°	80 - 300	180 - 675	4.5 - 16.8	1 - 2.5

Table 4.17 Estimated heat generation and temperature difference against surroundings during float charging. Assumed surface area per 0.04 m^2/100 Ah (cf. Table 4.13). Temperature difference ΔT_{bal} taken from Fig. 2.12.

The values in Table 4.17 correspond with the general experience that temperature rise during float charging can be neglected in vented batteries, even at 40 °C. The same applies to valve-regulated types even at 40 °C: a temperature difference of 2.5 °C between battery and surroundings is sufficient for heat dissipation. This agrees with published experimental experiences, e.g. that temperature rise above ambient is less than 3 °C at 40 °C and 2.27 V/cell (Fig. 2 in ([17])), or that even at 55 °C a marginal temperature rise was observed (Fig. 11 in ([18])). Corresponding values are also shown in Fig. 4.17.

4.5.3.6 Thermal Runaway

Thermal runaway means an unstable state of operation where heat generation increases faster than heat can dissipate ([19]). The result is an increase of the battery temperature that again increases the reaction rates, mainly the rate of the internal oxygen cycle. Thus, heat generation is further increased and the process accelerates itself and can heat up the battery, until it is destroyed. Often, however, this definition of the 'thermal runaway' is not observed, and any battery damage connected with temperature increase is called thermal runaway, although such overheating effects may have quite different reasons as described below in Section 4.5.6.1.

Thermal runaway usually is not relevant in vented lead-acid batteries. It was sometimes observed with aircraft batteries, when the aircraft had been on the ground for some hours in hot climates, especially if the battery was charged at a too high a voltage from the d.c. bus-bars in the craft. Water evaporation finally acts as cooling process, but may cause dangerous dry-out of the battery ([20]).

Thermal runaway is, however, of great importance with the valve-regulated design, where it represents one of the items continually discussed (cf., e.g., ([21]) and Section 13.1.1.2). This has two main reasons:

1. The internal oxygen cycle that characterises valve-regulated batteries, represents a strong heat source, especially when its rate exceeds certain limits.
2. Valve-regulated batteries are often installed under unfavourable conditions.

A typical thermal-runaway situation is shown in Fig. 4.17 . It concerns overcharging of a 2V/300 Ah monoblock battery with 2.45 V/cell and 2.50 V/cell at an ambient temperature (circulated air) of 40 °C. Four thermocouples were inserted, No. 1 and 2 within the plate stack of two adjacent cells, No. 3 and 4 on the container surface.

While the battery was charged at 2.27 V/cell, the difference between the interior and the surface was less than 0.5°C and the temperature rise of the surface about 1 °C above the ambient, which confirms the estimation in Table 4.17. Monitoring of the current showed that this situation was stable; no current increase was observed during this initial period.

When the voltage was raised to 2.45 V/cell, the balance between generated and dissipated energy was disturbed and a gradual rise of the temperature began that further increased when the voltage was raised to 2.50 V/cell. This increase of temperature continued until the test was stopped, and a growing temperature difference between the interior and the container surface was observed. The daily monitoring showed an unstable current increase between 0 and 4% per day during the 2.45 V/cell charging. During the 2.50 V/cell-charging period, a continual increase of the current was observed that reached 4% of current increase per day at the end. So, the typical 'runaway' of the current occurs.

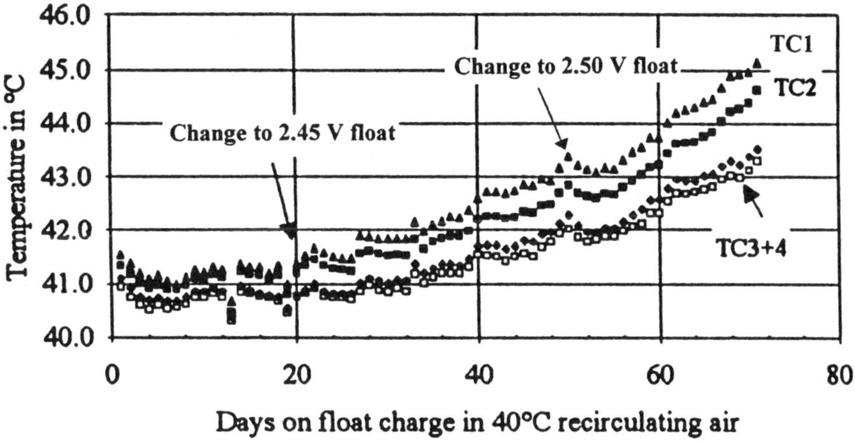

Fig. 4.17 Temperatures of a 2V/300 Ah monoblock battery, valve-regulated with AGM separator. Thermocouples TC1 and TC2 within two cells, TC3 and TC4 at the container surface. Overcharging at 2.27 V/cell during the first 20 days, then 20 days at 2.45 V/cell, and 2.5 V/cell during the rest. (Fig. 15 in ([22]).)

In the example of Fig. 4.17, the incubation time for the thermal runaway was quite slow, despite the high cell voltage and ambient temperature. This time factor is influenced by a number of parameters which may accelerate the runaway, but in most cases, enough time is left to detect such a situation early enough when the battery is inspected regularly.

One important parameter for thermal runaway is the maximum rate of the internal oxygen cycle, since it determines the maximum heat generation that is possible. It is mainly a question of oxygen transport rate and depends on battery design (plate spacing, separation), kind of electrolyte immobilisation and residual water content. The latter two are not independent from each other, as is shown Figs. 4.18 and 4.19.

The kind of electrolyte immobilisation, in combination with the gradual, but unavoidable water loss, affects the internal oxygen cycle on account of 'pore-size competition' between separators and active materials ([23]):

- In absorbent glass-mat batteries, water loss occurs mainly within the separator, because the average pore sizes in the separator are larger than in the active material of positive and negative electrodes. Thus, electrolyte is retained within the electrodes by capillary forces. As a result, much open space is formed within the separator that allows fast transport of oxygen, and so the maximum rate of the internal oxygen cycle can be increased to values of more than 10 A per 100 Ah of nominal capacity (cf. Figs. 4.18 and 4.19).
- In batteries with gelled electrolyte, the above described 'capillary competition' does not apply, and water loss is equally distributed between electrodes and separator. In addition, coverage of surface areas by SiO_2 (the gelling substance)

seems to limit the maximum rate of oxygen reduction. As a consequence, the maximum rate of the internal oxygen cycle is limited to about 1.5 A per 100 Ah of nominal capacity. Thus, heat generation by the internal oxygen cycle is more limited with gelled electrolyte.

In general, the maximum rate of the internal oxygen cycle should not be too high, since it is not required in valve-regulated lead-acid batteries, but increases the danger of thermal runaway (24). *This maximum rate should, however, not be confused with the efficiency of the internal oxygen cycle that has to be as high as possible* (cf. Section 8.3.5).

Figures 4.18 and 4.19 illustrate the influence of temperature and overcharge voltage on thermal runaway, and show some further details which are characteristic for such a situation. Both Figures concern overcharging at 2.6 V/cell, which is absolutely abusive for valve-regulated lead-acid batteries, but here was applied to demonstrate thermal runaway. In practice, such cell voltages only can occur when the controlling device of the current supply fails, or when cells within a battery have metallic short-circuits, e.g. by contact between the positive grid and the negative bus-bar, caused by grid growth, and the rest of the battery gets the full rectifier voltage.

Fig. 4.18 reflects the situation of a new battery that has lost only a little of its original water content. Overcharging at 2.6 V/cell and room temperature causes a current of about 0.6 A (per 85 Ah N.C.) The energy input is in the range of 3 W per battery after a short maximum at the beginning, due to a short recharging period. Temperature rise of the battery was 25 °C to 27 °C within 8 hours. But the rate of increase of this curve is reduced with time, an indication that a dramatic development of 'thermal runaway' has not started yet.

The situation in Fig. 4.18 is, however, not stable after the 8 hours of overcharging. Energy input and energy emission are slightly increasing, mainly on account of the increasing cell temperature. A quasi-stable state was reached with this experiment after 30 hours at a cell temperature of 29 °C. If, however, overcharging at this high voltage were continued, the situation would have been destabilised more and more on account of the gradual water loss, which increases the internal oxygen cycle. The final result would be thermal runaway as shown in Fig. 4.19.

The internal oxygen cycle increases slowly to about 0.4 A or about 60% of the overcharging current. This means that the float current exceeds the **maximum rate of the internal oxygen cycle**, and only about 60% of the oxygen, evolved on the positive electrode, can reach the surface of the negative electrode fast enough. As a consequence, the negative electrode is polarised until hydrogen evolution fills the gap between internal oxygen cycle and overcharging current, and the battery behaves in part like a vented one: hydrogen and oxygen escape in a stoichiometric ratio, equivalent to about 0.35 A which means 160 ml H_2 per hour!

For heat balance, this gas escape is important, because two-thirds of the energy, used for gas evolution, leaves the battery together with the gas, as described in Fig. 4.14 ($1.48 \cdot i \approx 0.5$ W). This increases the dissipation of energy markedly; and energy input approaches energy emission after about 5 hours of overcharging. The evolved energy of about 3 W per battery means about 12 W/m^2 and can dissipate according to Fig. 2.12 when the battery is heated to about 1 °C above the surrounding temperature.

For safety reasons, the situation in Fig. 4.18 should be avoided, because hydrogen evolution in these circumstances is about 100-times that of a normally operating valve-regulated lead-acid battery. This may cause too high a hydrogen concentration in the neighbourhood of the battery.

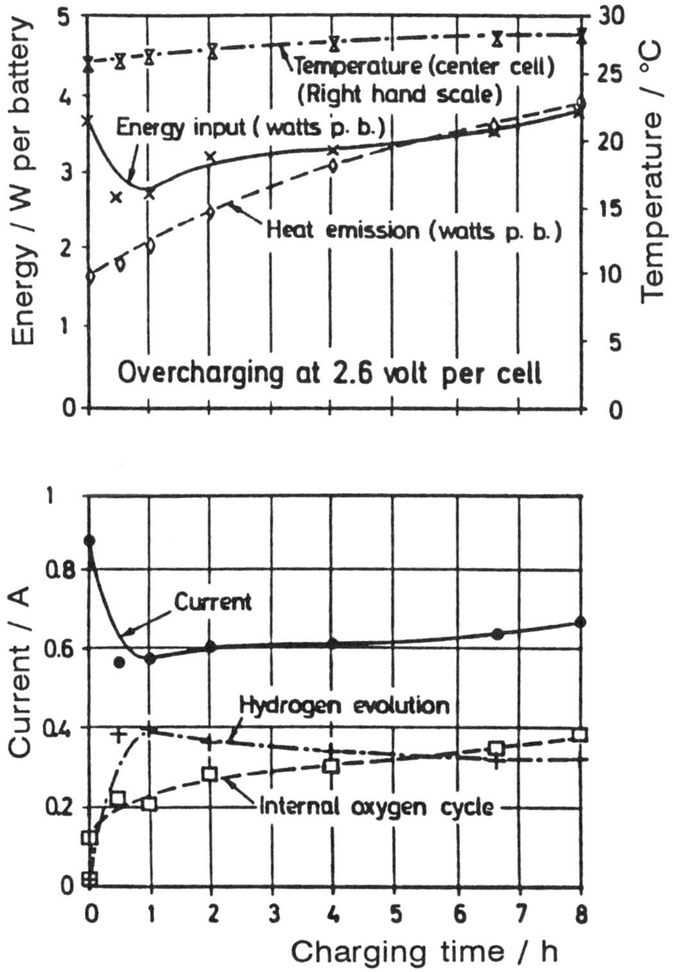

Fig. 4.18 Overcharging of a valve-regulated lead-acid battery with absorbent glass-mat separation at constant voltage (2.6 V/cell). 6 V block battery, 85 Ah of nom. capacity (VARTA solid bloc). Fairly new battery, water loss <5%; starting temperature 25 °C. Instable situation, but not yet a real thermal runaway.

Fig. 4.19 shows a similar overcharging experiment with the same battery, but after a water loss of more than 12% on account of preceding overcharging. A further difference, compared to the experiment in Fig. 4.18, is the higher starting temperature of 50 °C caused by preheating of the battery, while the ambient temperature was 22 °C as in Fig. 4.17. This aggravates the situation because hydrogen- (and oxygen-) evolution rates are increased about 8 times according to the ARRHENIUS relation, given in Eq. (2.68), p.33. On account of the water loss and the subsequent increased open space, the internal oxygen cycle now reaches at maximum about 10 A per battery, as can be seen in the bottom part of the Figure. (*Temperature does not influence this maximum so much, because it is determined by gas diffusion that is less dependent on temperature than the electrochemical reactions.*)

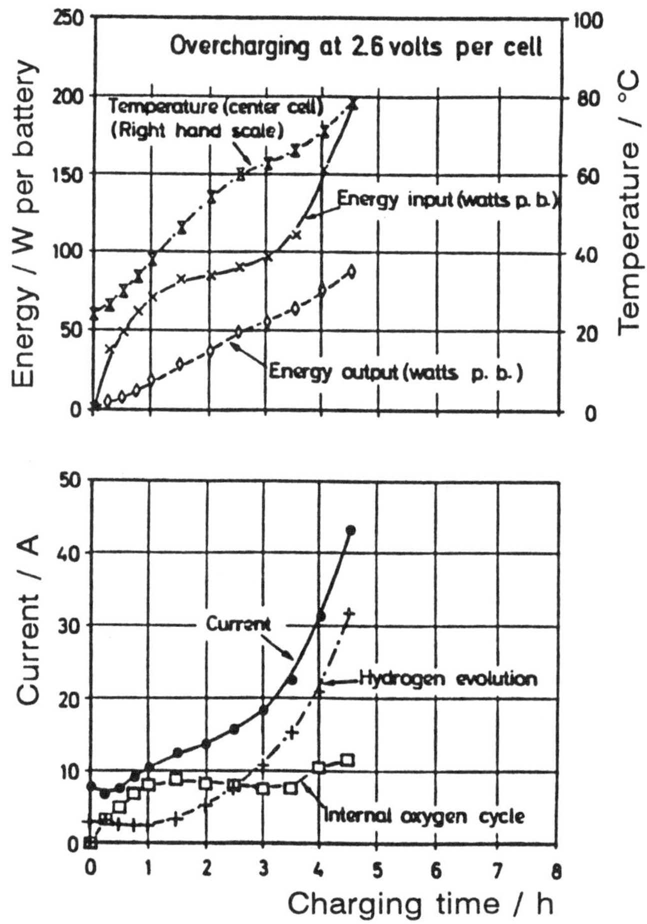

Fig. 4.19 Thermal runaway during overcharging of a valve-regulated lead-acid battery at a constant voltage (2.6 V/cell). Same battery as in Fig. 4.18, but after more than 12% of water was lost on account of overcharging. Furthermore, the temperature at the beginning of overcharge was 50 °C, while the ambient was room temperature.

Now, the high voltage in combination with the high battery temperature causes thermal runaway; the internal oxygen cycle rises rapidly (curve at the bottom of Fig. 4.19) until the maximum value is reached, close to 10 A. This means heat generation by the internal oxygen cycle of 26 W per cell or 78 W per battery (more than 300 W/m^2 referred to the block surface of about 0.25 m^2). A rapid increase of the temperature is the consequence, expressed by the steep slope of the temperature curve in the top of the Figure; this temperature rise again causes a corresponding increase of the current.

Initially, water decomposition into hydrogen and oxygen contributes only a comparatively small portion of heat ((2.6-1.48 V) x hydrogen-evolution rate in current equivalents (A)); but when the current exceeds the maximum rate of the internal oxygen cycle, the gas-evolution rate increases faster and faster and finally generates heat at a comparable rate. This causes the second steep increase of the temperature curve in the upper part of Fig. 4.19.

Fig. 4.19 clearly shows thermal runaway: the gap between energy input and energy output becomes wider and wider, because of the current increase, and the resulting increase of heat generation is faster than the increase of heat dissipation. Thus, the situation deteriorates more and more until the battery is destroyed by thermal damage and evaporation of the residual water, if the current has not been switched off.

Comparison of the three Figures in this Section shows the complex connection between the different parameters. In Fig. 4.18, the increased energy loss with gas escape stabilises the thermal situation of the battery. But, in the more unfavourable conditions shown in Fig. 4.19 (higher starting temperature; dried-out battery), even this reaction contributes to the thermal runaway.

Exact values of voltage or current that cause thermal runaway cannot be given. In float-voltage applications, heat dissipation is the most important parameter. But just this parameter can differ enormously in practice, e.g. when the battery is installed within a small cabinet that is heated by the battery itself, a quasi-adiabatic situation results that favours thermal runaway. This makes it understandable that the danger of thermal runaway is judged very differently. Vaccaro and co-worker (([18]), p. 24) found that 2.35 V/cell for a new and 2.30 V/cell for an aged battery are safe even at 55 °C, while other authors (e.g. ([25])) claim that even 2.27 V/cell is too high at 40 °C.

Valve-regulated batteries with gelled electrolyte are, as mentioned above, less prone to thermal runaway because the internal oxygen cycle never reaches the values observed with fairly dry glass-mat-separated batteries.

Thermal runaway will not occur in normal operational conditions. The danger exists only in abnormal operational conditions like increased voltage and temperature ([26]), cf. also the paragraph 'thermal runaway' in Section 13.1.1.2. The situation is aggravated when the charger can supply high overcharging currents. This applies to most UP-systems where the battery is designed for short autonomous periods of the system. A bridging time of 10 minutes means that, in an emergency, a current of about 200 A is drawn from the battery per 100 Ah of nominal capacity. This load more or less corresponds with the usual input of the alternating-current converter; so the rectifier must be able to supply a corresponding current. This implies that practically no limit exists for the overcharging current, because 200 A represents a factor of 4000 compared to the usual float current of about 50 mA/100 Ah. Batteries in such applications should at least be equipped with sensors that interrupt the current supply when a critical temperature is exceeded, if a monitoring system is not applied.

The most effective remedy against thermal runaway is surveillance of the battery ([27]). The optimum are monitoring systems that continuously control battery parameters including the temperature of pilot cells ([28]). Combination of such sensors with in situ conductance measurements is also a promising method ([29]). Another possibility is to use 'intelligent rectifiers' that regulate the float voltage by temperature sensors and, at an extreme temperature increase, switch down the voltage below the open-circuit voltage, or switch to a small constant current ('trickle charging', cf. Section 13.1.1.2). Thermal runaway is then arrested, but the battery kept at a partial charged state, and so is enabled to supply energy in an emergency ([25]).

4.5.3.7 Overheating

'Overheating effects', described in this Section, differ basically from the above thermal runaway, because they concern only single cells in a longer battery string, and they are not connected to a 'runaway' of the current. Such overheating effects are observed during charging as well as during discharging, especially in battery strings of higher voltage ([30]). They are always caused by faulty cells.

- **During charging**, a significant higher internal resistance of an individual cell causes a voltage drop across this cell that might be a multiple of the average polarisation. When the battery string is long enough, the float current is only slightly reduced. But heat generation in the considered cell is increased, according to the increased Joule effect. So the cell can reach very high temperatures, although the float current remains at its low level. Increased water loss of this cell is the result that aggravates the situation.
  The reasons for such an increased internal resistance are, e.g., dry out, or lost contact between electrodes and the bus-bar due to plate-lug corrosion or poor welding.
- **During discharge**, the situation of such cells is much more endangered, because of the high load that may cause excessive heat generation with heavy gassing and possible evaporation of the residual water in the electrolyte ('whistling valves'). The final stage will be interruption of the current, which in the worst case might cause sparking and explosions.
  During discharge, cells with a significantly reduced capacity are also critical because they reach the end of discharge much too early and are reversed. This also causes heavy gassing and may also destroy the cell as described above.

Such overheating can also happen to vented batteries when plates have lost contact with the bus-bar, and also when water loss has not been replaced or container leakage not noticed. *Fires in remote stations, caused by such overheating, promoted the development of antimony-free batteries for stationary applications in the 1940s (cf. Section 4.7.3).*

4.6 Sulphuric Acid as Active Material

Sulphuric acid is involved in the cell reaction (Eq. 4.5). As a consequence, the acid concentration is reduced when the battery is discharged and is increased again during charge. This causes a number of features, typical for lead-acid batteries:

- The **open-circuit voltage** decreases with proceeding discharge on account of acid dilution (cf. Fig. 4.2, p.101).
- The **conductivity** of the sulphuric acid changes with the state of charge (cf. Fig. 4.20, p.144).
- The **amount of sulphuric acid** (H_2SO_4), required for the reaction, must be stored in a suitable manner to reach the electrode surface fast enough, which means storage between the electrodes. This requires a minimum spacing between the electrodes, which again means that the **internal resistance** of the battery cannot fall below a certain value, determined by this electrolyte layer.
- The **freezing point** of the electrolyte can be reached when the battery is discharged at low temperatures (Fig. 4.21, p.144).
- Another disadvantage of varying acid concentration is **stratification** of the electrolyte, described in Section 4.6.3.
- An advantageous feature, although of minor importance, is the possibility to determine the **actual capacity** of the battery by density readings. Limitations are:

  The amount of acid per cell must be known with some accuracy.

  The acid concentration must not be disturbed by stratification effects. Such stratification usually occurs during charging, unless the electrolyte is mixed by forced flow.

  In valve-regulated lead-acid batteries, advantage cannot be taken of this method, because the acid density cannot be measured and the residual amount of electrolyte is not known, due to water loss.

4.6.1 Conductivity of Sulphuric Acid

On account of the concentration changes during charge/discharge reactions, the conductivity of the sulphuric acid varies, as Fig. 4.20 shows.

The Figure shows that in charged batteries, acid conductivity is close to its maximum, but at deep discharges it can reach low values. Furthermore, conductivity is considerably reduced at low temperatures.

4.6.2 The Freezing Point of Sulphuric Acid

The freezing point of sulphuric acid depends on its concentration. Fig. 4.21 shows the freezing curves. On account of the acid concentration changes during charging and discharging, the freezing point of the electrolyte in a battery depends on the state of charge.

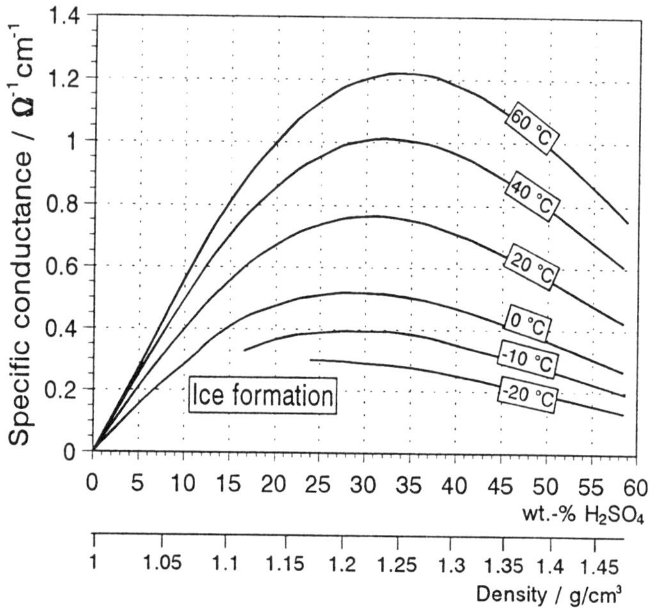

Fig. 4.20 Specific electric conductance of aqueous solution of sulphuric acid and its dependence on temperature. (From ([31]).

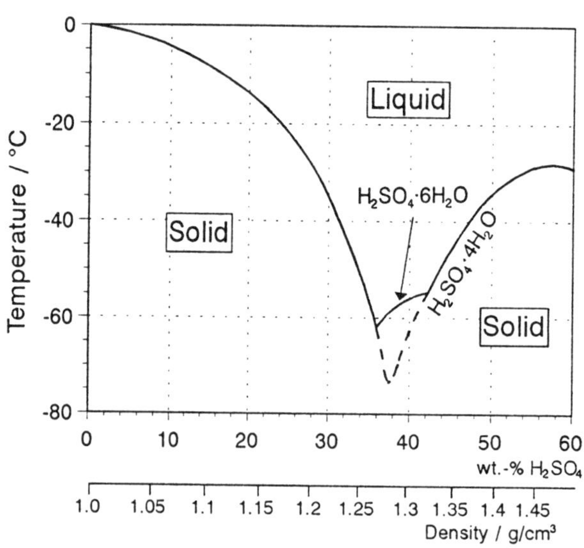

Fig. 4.21 Freezing curves of H_2SO_4/H_2O mixtures.

For the usual acid concentrations in charged batteries, the freezing point is low (-60° to -50 °C), but with progressing dilution it is more and more increased and reaches the zero point when only water is left. In lead-acid batteries, such a complete dilution does not occur, but acid densities of 1.10 kg/dm^3 are normally reached, especially in traction batteries. Fig. 4.21 shows that freezing then occurs at about -8 °C. Thus, ice can be formed when the battery is deeply discharged at low environmental temperatures.

The formation of ice converts the electrolyte into a slurry with rather low conductivity. The internal resistance of the battery is considerably increased, and discharging, but especially recharging, are hindered. *At high-rate discharges at very low temperatures, like the cold-cranking test of starter batteries (minus 18 °C), acid dilution within the pores of the positive electrodes can cause the formation of ice, although the average of the acid concentration is still high. This ice formation ('ice passivation') considerably hinders discharge and reduces the cranking voltage.*

Although the formation of ice does not usually destroy the battery, it may have detrimental effects on the active material by swelling within the pores. To avoid ice formation, lead-acid batteries should only be partly discharged at very low temperatures.

The effect can be compensated to some extent by increased acid concentration ('arctic electrolyte'), which is occasionally applied with vented lead-acid batteries for low-temperature application.

4.6.3 Electrolyte Stratification

Electrolyte stratification is connected with the changes of the specific weight of the sulphuric acid associated with the concentration changes during charge and discharge. In flooded lead-acid batteries, acid stratification may cause severe problems. The immobilisation of the electrolyte, in valve-regulated lead-acid batteries reduces stratification effects to a large extent. A gelled electrolyte practically eliminates them.

Fig. 4.22 illustrates the origin of acid stratification. On the left, the acid concentration is uniform all over the battery. The battery may have been floated for an extended period or the electrolyte been mixed by other means. This will always be the final state when a battery is on float, because the diffusion process always equalises acid concentration. Stratification can only be caused by discharge/charge processes.

In the centre of Fig. 4.22, the battery has been discharged rather deeply. During discharge, the sulphuric acid between the plates is consumed by the electrochemical reaction and natural convection is caused by the fact that the electrolyte between the plates attains a lower specific weight than the electrolyte beside and above the plates. At the end of discharge, the concentration of the electrolyte is uniform again, except for the share localised below the plates, which remains at the original concentration. Diffusion would equalise this concentration difference also, but diffusion is a slow process that takes days or even weeks, depending on the thickness of the concentrated layer.

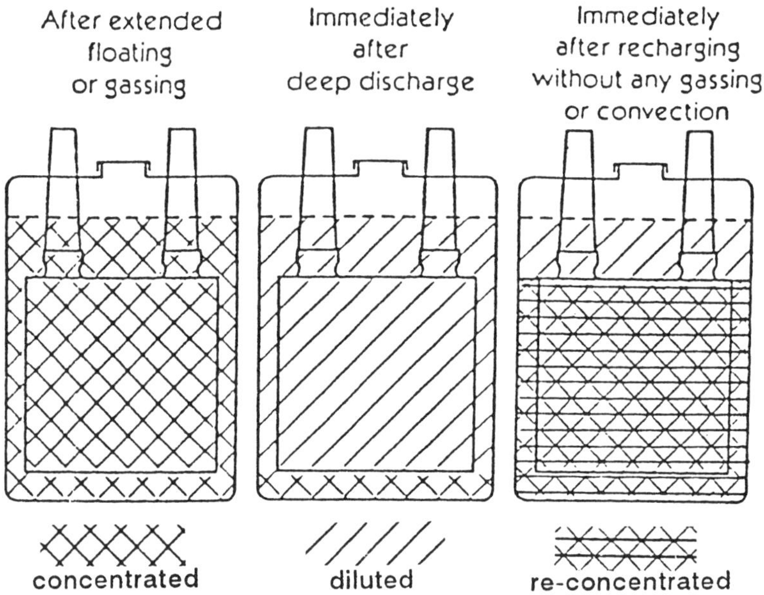

Fig. 4.22 The origin of acid stratification, when the lead-acid battery is discharged and recharged without gassing or other means of mixing the electrolyte.

During charge, sulphuric acid is released from the active material in the plates and the concentration of the acid increases between the plates. Now the highest acid concentration is achieved between the plates, which initiates convection of this acid to the parts beside and underneath the plates. At the end of the charging process, the battery is filled with slightly increased acid concentration to the upper edge of the plates. The acid above the plates remains diluted and there is no reason for convection, because of the low specific weight of this acid. *This situation often irritates maintenance staff when, in vented batteries, acid readings are taken shortly after a capacity test and subsequent recharging (or after refilling with water). In both cases a too low acid density is measured above the plates, although the battery is fully charged. This often induces people to carry out unnecessary overcharging.*

In the usual stationary application, acid stratification causes no problems, because the batteries are discharged only occasionally. There is enough time left for the diffusion to equalise the acid concentration. However, when the battery is running on a cycling schedule without electrolyte mixing, acid stratification causes uneven current distribution, and concentration differences grow between the top and bottom parts of the cell from cycle to cycle, especially in rather tall cells. In flooded cells of 55 cm height, filled with acid of the density 1.24 g/cm^3, severe stratification effects were observed; only after six full cycles, the acid density had decreased to 1.13 g/cm^3 at the top, while a density of 1.31 g/cm^3 was observed at the bottom ([32]). The increasing acid concentration at the bottom may harm the negative plate, which is prone to sulphation when running in too high an acid concentration.

As a remedy against stratification, overcharging by 10 to 20% during each cycle is mostly applied for flooded traction batteries. The gas evolution forms bubbles that mix the electrolyte. A more effective method is forced agitation of the electrolyte with the aid of air-lift pumps that are inserted in each cell (cf. Fig. 7.11, p.258).

4.6.4 Immobilised Electrolyte

In valve-regulated lead-acid batteries, immobilisation of the electrolyte is required to provide open space for the fast transport of oxygen as a gas. Otherwise, the internal oxygen cycle is hindered by the too slow diffusion rate of the dissolved oxygen. In lead-acid batteries, two ways are used for electrolyte immobilisation: absorption of the sulphuric acid by a glass felt, or gelled acid by addition of SiO_2. Both methods immobilise the electrolyte, and a common feature of both techniques is that the battery does not contain liquid electrolyte.

One characteristic of both methods of electrolyte immobilisation is that, during service, the (slight) water loss leads to an increase of the gas channels. The transport of oxygen from the positive to the negative electrode is thereby made easier. The result of this is a slight increase in the float-charging current during the operational life of the batteries due to reduced polarisation of the negative electrode ([22]). This is a system characteristic and is completely normal.

4.6.4.1 Absorption of the electrolyte by a glass mat (AGM)

In lead-acid batteries, absorbent mats (AGM: absorbent glass-mat) of various plastic materials were tried, but without success, until glass fibres with fine enough diameters became available ([33]). Only glass is wetted well enough by sulphuric acid to soak the separators. The fine fibres of these felts (diameters in the μm range) form a pore system that absorbs the electrolyte by capillary forces.

Such microporous glass felts simultaneously fulfil a double function:

- They act as a conventional separator that insulates the electrodes of different polarities from one another and prevents short-circuits between the electrodes, since lead dendrites do not grow from the negative to the positive electrode through the fine pore system of the glass felt.
- They immobilise the electrolyte by effective absorption that fills most of their pores.

In a battery, the dosage of the electrolyte must be kept so low that the absorbent mat separator is not completely saturated with electrolyte. So the required gas channels are formed in the felt. Otherwise, the battery acts, at least partially, like a vented one, until water loss has sufficiently reduced the amount of water.

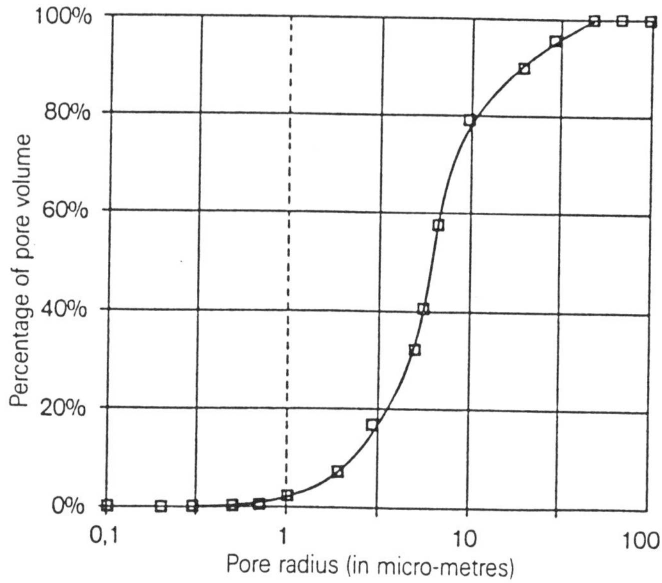

Fig. 4.23 Typical pore size distribution in a fibre separator. The majority of the pore radii are in the 5 to 10 μm range

Fig. 4.23 shows the distribution of pores in such a felt. If the force of gravity did not exist, pores with a radius of more than 10 micrometers would stay empty when the cell is filled with only 80% of its maximum electrolyte content. Gravity forces, however, influence the saturation profile in the separator, which means that at the bottom of the cell large pores are filled with electrolyte, while at the top even smaller pores may remain empty (cf., e.g., ([34])). As a consequence, the saturation of the separator with electrolyte varies with plate height ([35]) and a gravity-caused filling profile is superimposed on the curve of Fig. 4.23 . To keep the influence of gravity small, most manufacturers limit the plate height to less than 300 mm or recommend operation of batteries with taller cells in the side position, to keep the actual separator height in a similar order of magnitude ([36]).

4.6.4.2 Gelled electrolyte by addition of SiO_2

Gelling is another way to immobilise the electrolyte. Attempts were made to gel the sulphuric acid electrolyte in lead-acid batteries by inorganic and organic thickeners as long ago as the last century, to prevent the acid from leaking out of defective cell containers ([37]). Nowadays, highly dispersed mixtures of sulphuric acid with specially prepared silicon dioxide (fumed silica, SiO_2) are used ([38]). These are stable over a long time, even to the powerful oxidising effects of the positive electrode, and are environmentally safe and technically easy to handle.

Sulphuric acid, in the concentrations usually applied in batteries, forms a solid gel when about 6% of SiO_2 are added. It is a thixotropic gel that can be kept highly liquid by intensive agitation, which is taken advantage of when filling the battery. Shrinkage during solidification squeezes electrolyte out of the gel and forms cracks that run through the solid electrolyte and allow fast transport of gaseous oxygen. Real pores, as in a felt, cannot be observed in the gelled electrolyte. But the gelled sulphuric acid behaves in the same way as in a felt, with a pore structure one order of magnitude finer than that of the usual microglass-mat separator. Accordingly, fill profiles caused by gravity are less pronounced, and stratification effects are not observed with gelled electrolyte ([38]) (cf. Section 4.6.4.5).

Gelled electrolyte cannot fulfil the function of a separator, since it does not suppress penetration of lead dendrites. Thus, a conventional separator, as used in vented batteries, has to be inserted between the electrodes.

4.6.4.3 Differences Between Both Immobilisation Techniques

Maintenance-free lead-acid batteries can be constructed using absorbent mat separators as well as gelled electrolyte. Which immobilising method is used, often depends on manufacturing possibilities. Batteries with absorbent glass mats can be made with the usual equipment. On the other hand, in cells with tubular plates, the complicated geometry of the electrode itself makes the gel-binding method more suitable, but also tubular plates with AGM separation are on the market.

Differences, caused by the two immobilisation methods, are of marginal character for most applications, and both types of immobilisation are in wide use. In special applications, however, or in specific circumstances these differences can gain importance.

4.6.4.4 Internal Resistance

As mentioned above, gelled electrolyte requires a conventional separator to prevent short-circuits by growing dendrites. This additional separator causes additional resistance. Thus the internal resistance of comparable batteries is slightly higher with gelled electrolyte, and such batteries are somewhat inferior in the extreme high load range (e.g. car cranking applications, UPS with extreme short autonomous operation periods).

A gradual increase of the internal resistance occurs with both immobilisation techniques during service life on account of the unavoidable loss of water ([39]). With extremely high current loads, this can cause a lower discharge voltage of aged batteries. Batteries of the absorbent mat-separation type are more sensitive in this respect because of the 'pore-size competition', mentioned in connection with thermal runaway (Section 4.5.3.6). It means that the pore sizes in the separator are larger than in the active material of both electrodes. So, any reduction of the electrolyte volume mainly concerns the separator and increases the electrolyte resistance across the separator. As a consequence, a comparable water loss affects the internal resistance of an absorbent glass-mat type more than that of a corresponding battery with gelled electrolyte.

Figs. 4.24 and 4.25 illustrate the effect of water loss on high-load performance. Increased water loss was achieved in these batteries by constant-current overcharging, and the lost volume was determined by weight difference, assuming that oxygen consumption by corrosion was negligible during the fairly short overcharging period.

Fig. 4.24 shows high-rate discharge curves of a valve-regulated lead-acid block-battery before and after removal of 12 wt.% of its water content, and after refilling this amount of water.

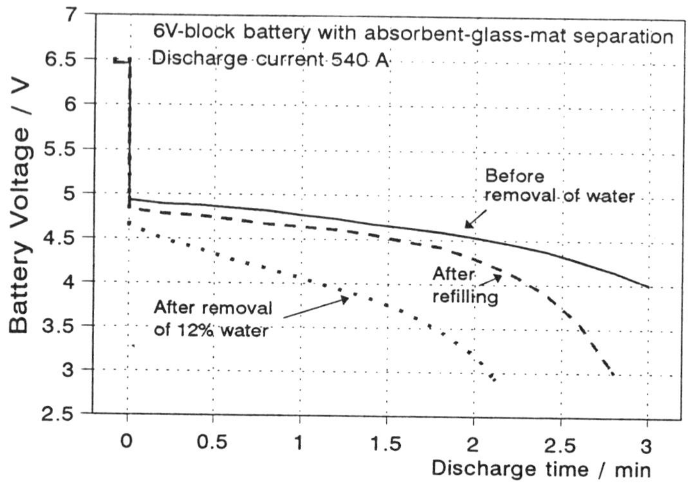

Fig. 4.24 High-load-discharge curves before and after removal of 12 wt. % of water content, and after refilling this amount of water. VARTA solid bloc prototype, 6 V, ≈80 Ah (AGM type). (Fig. 1 in ([23]).)

Due to the increased acid concentration, the open-circuit voltage increased after water loss from 6.46 to 6.5 V/cell, as indicated in the Figure.

After water removal, the increased internal resistance that causes an additional voltage drop on discharge is obvious. This was also shown by impedance measurements at 1 kHz (Table 4.18). The increase of resistance (0.27 mΩ/cell) corresponds fairly well to the observed additional voltage drop. But this increase is higher than would be expected from calculations, assuming that water removal only affected the separation. Obviously, uneven current distribution contributes additional resistance.

	Impedance per battery
Original	1.53 mΩ
After water removal	2.34 mΩ
After refilling	1.75 mΩ

Table 4.18 Impedance values of the charged battery with absorbent glass-mat separation. Measuring frequency 1 kHz.

After refilling it took 100 hours, until the open-circuit voltage of 6.46 V was measured, as an indication that the original acid concentration was approximately reached again. The discharge curve after refilling approximates the original curve fairly closely and shows thereby that the voltage drop was actually caused by water loss.

Fig. 4.25 shows the effect of water removal on high-rate discharge curves of two batteries with gelled electrolyte. (*The different discharge periods in Figs. 4.24 and 4.25 are caused by the different capacity of these prototype batteries.*)

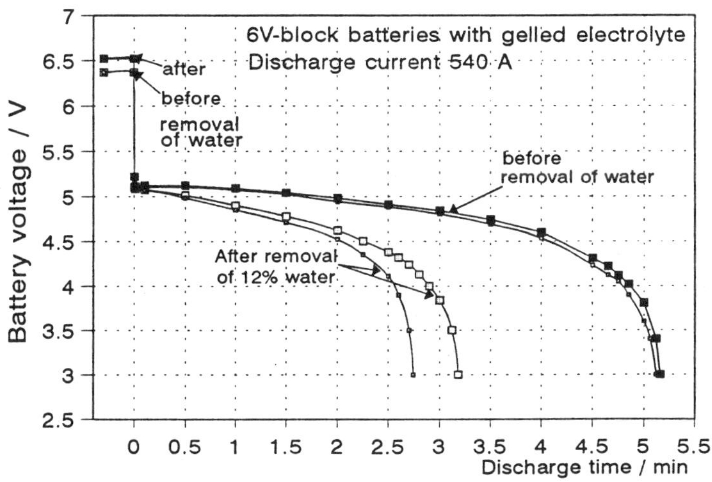

Fig. 4.25 High-load discharge curves before and after removal of 12 wt.% of water content. VARTA solid bloc prototype, 6 V, ≈100 Ah, gelled electrolyte. (Fig. in [23].)

After water removal, an increase of open-circuit voltage is observed, corresponding to the increased acid concentration.

Fig. 4.25 shows that the additional voltage drop, caused by water removal, is smaller at the beginning of discharge, compared to that of the absorbent glass-mat battery in Fig. 4.24. This indicates that the water loss did not mainly occur within the separator, but was distributed between all elements (separators, positive and negative active material). The quite small additional voltage drop is confirmed by results of impedance measurements in Table 4.19.

	Impedance per battery	
	Battery 1	Battery 2
Original	1.45 mΩ	1.42 mΩ
After water removal	1.51 mΩ	1.50 mΩ

Table 4.19 Impedance values of the charged batteries with gelled electrolyte. Measuring frequency 1 kHz.

The discharge curve of the gelled battery descends more steeply after water loss, compared to the corresponding curve in Fig. 4.24. Presumably, this is caused by the reduced amount of electrolyte in the pores of the positive active material and the thereby reduced utilisation factor.

Figures 4.24 and 4.25 show that the effect of increased water loss on high-load performance is different, depending on the method of acid immobilisation. The absorbent glass-mat battery is in danger of failing completely when the end-of-discharge voltage is specified at a high level, because the battery voltage may drop below this value already at the beginning of the discharge. Batteries with gelled electrolyte maintain the initial discharge voltage after a comparable water loss, but they reach only about 60% of their original high-load capacity.

In the absorbent glass-mat battery the extreme defect would have been detected by impedance measurements that are recommended as an additional check for valve-regulated lead-acid batteries ([40]) (cf. Section 3.6.4). Impedance measurements would, however, not indicate the capacity decrease of the gelled-battery design as the small change in Table 4.19 indicates.

Note: *Water loss in AGM separated batteries can cause dramatic capacity loss even at moderate load, when the pre-pressure (that influences the pore size distribution ([41])) on the glass felt is not high enough or is lost on account of bulging of the container. Gradual water loss empties more and more of the pores in the glass felt and so causes a slight shrinkage of the separator. So the contact between the separator and the adjacent electrodes can be lost, at least partially. A corresponding decay of capacity is then observed. Refilling of the lost amount of water restores the original state, because the filling of the pores in the separator by capillary forces causes a slight swelling of the felt ([42]).*

4.6.4.5 Gravity Influence

The influence of gravity is smaller in batteries with gelled electrolyte. It has been mentioned above that the gelled sulphuric acid behaves in the same way as in a felt with a pore structure one order of magnitude finer than that of the usual microglass-mat separator. So the acid is more strongly fixed, and the gravity effects that are observed in a 30 cm high AGM battery would be observed to a similar extent in a 3 m high battery with gelled electrolyte. Convection of the electrolyte, still possible to a certain degree in glass-mat separators, is almost repressed in gelled electrolyte and ion movement is restricted to diffusion.

Thus, the gel prevents electrolyte stratification even during extensive cycle operations. For this reason, comparable block batteries as they are used in wheel-chairs or electric vehicles, reach about 700 cycles with gelled electrolyte, while the AGM version is limited to about 500 cycles. Gel batteries are therefore preferred for such applications.

4.7 Lead as Current-Conducting Material

Corrosion of the positive grid represents a restriction of the lead-acid battery that finally limits the useful life, if no other reasons cause failure before. This corrosion phenomenon is a side-effect of the high cell voltage of this battery system, which implies a high potential of the positive electrode. Metals, usually applied as current conductors, and even noble metals like gold, would be dissolved by oxidation when connected to the positive electrode of the lead-acid battery. In the negative electrode, such metals increase the hydrogen-evolution rate. Thus, foreign metals like copper, nickel or aluminium have thoroughly to be covered by a dense layer of lead which precludes their widespread use as grid material. Only copper is used as inserts in conducting elements (the top lead) and to a certain extent as expanded metal in grids of negative electrodes (cf. Section 4.8).

Lead can be used, because the corrosion itself forms a rather dense passivating layer of lead dioxide that protects the underlying material against fast corrosion (43). Fig. 4.26 illustrates the situation:

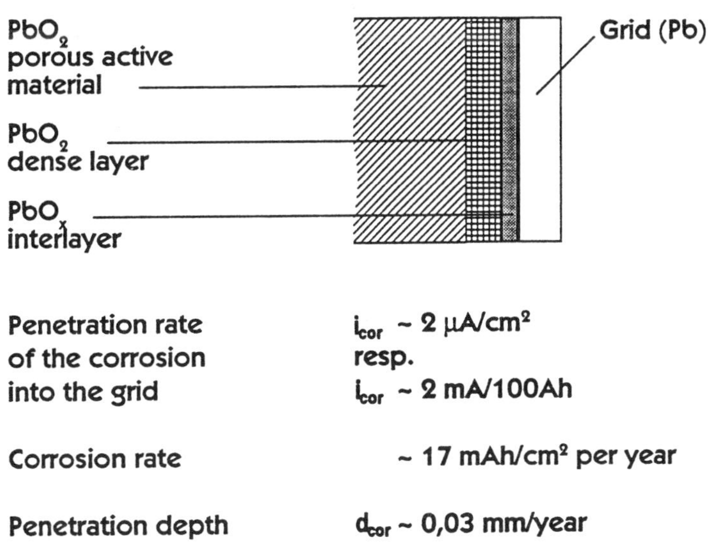

Fig. 4.26 Structure of the corrosion layer at the grid surface. Penetration and corrosion rates approximated for room temperature and normal float voltage (2.23 to 2.25 V/cell).

The active material is represented in Fig. 4.26 by the area on the left; the grid is shown on the right. Underneath the porous lead dioxide that constitutes the active material, a dense layer, also of lead dioxide, covers the grid surface. This layer is formed by corrosion and protects the grid. On account of acid depletion a rather stable oxide layer (mainly of α-PbO_2) is formed (44). However, lead dioxide and lead cannot exist beside each other for thermodynamic reasons, and a thin layer of less oxidised material is always formed between the grid and the lead dioxide (PbO_x in Fig. 4.26) (45). The existence of lead oxide PbO in this layer has been determined; the existence of higher oxidised species is assumed (PbO_x phases cf. (46)), but their structure is not yet known exactly (47).

The PbO_2/PbO_x border slowly penetrates into the grid; however, at a very slow rate as a solid-state reaction. Cracks are formed when the oxide layer exceeds a given thickness, on account of the growth in volume when lead becomes converted into lead dioxide (Table 4.20). Underneath the cracks the corrosion process starts again and again. As a whole, the corrosion proceeds at a fairly constant rate and never comes to a standstill, and a continually flowing anodic current, the **corrosion current** is required to re-establish the corrosion layer (cf. Section 4.7.1).

1	2	3	4
	Molar weight g/mole	Density g/cm^3	Volume ratio related to Pb
Pb	207.2	11.34	1
$PbO_{(red)}$(48)	223.2	9.64	1.26
α-$Pb0_2$	239.2	9.87	1.32
β-PbO_2	239.2	9.3	1.40
$PbSO_4$	303.3	6.29	2.64

Table 4.20 Molar weight (or molar mass), density (49) and volume ratio of corrosion products related to lead.

When the grid material (Pb) is converted into lead dioxide (PbO_2), the basic electrochemical reaction is

$$\boxed{Pb \Rightarrow Pb^{4+} + 4 \cdot e^-} \qquad (4.33)$$

Twice the amount of electricity is required than for the discharge reaction of the negative electrode (Eq. (4.1)), since corrosion includes 4 valences which means 4·F = 107.21 Ah per multiple of Eq. (4.33). Consequently, for the corrosion reaction according to Eq. (4.33) the equivalent values are:

$$\boxed{\frac{207.19}{107.21} = \quad 1.9326 \text{ g Pb/Ah or } 517.4 \text{ Ah/kg Pb}} \qquad (4.34)$$

This figure means that 1 cm^3 of lead (11.34 g; cf. Table 4.20) is equivalent to 5.89 Ah.

A current of 1 $\mu A/cm^2$ means 8760 μAh per year and cm^2. Referred to 5.89 Ah/cm^3, a penetration rate of $1.49 \cdot 10^{-3}$ cm/year results, assuming that the corrosion attack occurs uniformly and progresses at a constant rate. So the value in Fig. 4.26 was derived from the fact that a corrosion current of 2 $\mu A/cm^2$ means a penetration rate of about 0.03 mm/year.

Fig. 4.27 illustrates the consequences for battery practice. The above penetration rate would reduce the cross section of a grid spine in a tubular electrode by about 50% within the usual service life of 15 years. This result is confirmed by field experience and shows that long-life batteries must have a corresponding 'corrosion reserve' in their positive grids.

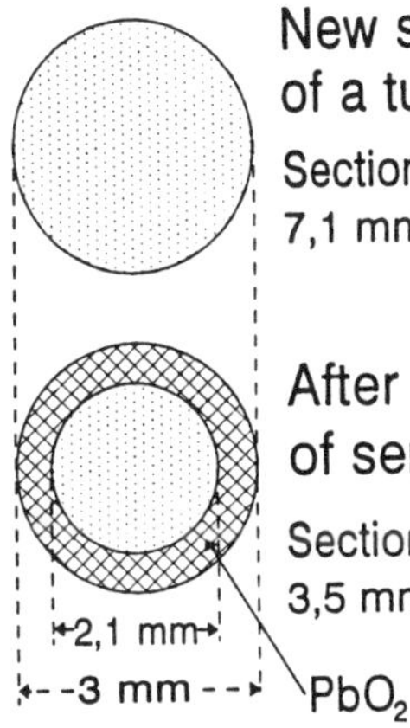

Fig. 4.27 Conversion of grid material into lead dioxide (Pb) by corrosion.
Spine of a positive tubular plate.
New plate: 3 mm diameter means 7.1 mm^2 cross section (πr^2 with r = 1.5 mm).
Aged plate: reduction of r by 0.03 x 15 = 0.45 mm means $\pi r^2 = 3.5$ mm^2.

Since grid material is converted into lead dioxide, a slight increase of the actual capacity is often observed. The reduced cross section in Fig. 4.27 does not affect the performance of such batteries that normally are used for discharge duration in the order of one hour or more. Attention must, however, be paid to batteries that are loaded with high currents, because the conductivity of the grid gains importance with increased current flow. Advanced corrosion of the grid can cause an intolerable voltage drop at high discharge rates.

The 'corrosion capacity of the positive grid' can be estimated from the above figures. The positive grid in lead-acid batteries for stationary and traction applications contains about 10 g of lead per Ah (usually slightly more). This means a positive-grid weight of about 1 kg/100 Ah. With the values of Eq. (4.34), the 'corrosion capacity' is 500 Ah per 100 Ah of battery capacity. The corrosion rate of 1 or 2 mA/100 Ah means 8.76 and 17.52 Ah/100 Ah/year, respectively. Related to the 500 Ah of the total 'corrosion capacity', 2 to 4% of the grid material would be consumed per year under these assumptions.

4.7.1 Corrosion at Open-Circuit Voltage

Corrosion at open-circuit has briefly been treated in Section 4.4.2 as self-discharge of the positive electrode. In connection with Fig. 4.26 it has been mentioned that an anodic current, the **corrosion current**, must flow continuously to stabilise the lead dioxide layer at the grid surface. Then the PbO_x layer remains thin because PbO_x is always converted into PbO_2 by further oxidation.

At open-circuit voltage, no anodic current flow through the positive electrode occurs that can oxidise the PbO (or PbO_x) layer, but the corrosion reaction

$$Pb + PbO_2 \Rightarrow 2 \cdot PbO \tag{4.35}$$

continues between grid and passivating layer. Consequently the PbO_x layer grows between grid and PbO_2.

However, as mentioned in Section 4.4.2, PbO or PbO_x are not stable against sulphuric acid, and react very fast according to

$$PbO + H_2SO_4 \Rightarrow PbSO_4 + H_2O \tag{4.36}$$

as soon as these substances get in contact with sulphuric acid. The protecting lead dioxide layer, shown in Fig. 4.26, would be destroyed by this reaction, and severe grid corrosion is one of the problems that can occur when the battery stands for prolonged periods without any charging. Battery manufacturers therefore recommend recharging a lead-acid battery filled with electrolyte within regular periods that must be shortened when the battery is stored at elevated temperature (cf. Fig. 4.8, p.119).

The situation also applies to valve-regulated lead-acid batteries (cf. Fig. 4.9), but is less critical mainly for two reasons:

1. Fairly corrosion-resistant alloys are employed for the positive electrode.
2. The mobility of the sulphuric acid is reduced to diffusion only (practically no convection).

Thus, prolonged recharging intervals are usually permitted by manufacturers for valve-regulated lead-acid batteries.

4.7.2 Electrochemical Consequences of the Lead Corrosion

The unavoidable corrosion of the positive grid and other parts of the conducting material has far-reaching consequences for the whole battery:

- Grid corrosion finally limits service life of every lead-acid battery, unless other failures occur earlier. A long service life battery must be equipped with a corresponding 'corrosion reserve' of the positive grid, which causes additional weight.
- Grid corrosion also is a water consuming reaction which is of great importance in valve-regulated lead-acid batteries (cf. Section 7.2.2.3).
- The rate of grid corrosion is influenced by the composition of the grid alloy. Selection of appropriate alloying additives is important to reach the desired service life.

- Alloying constituents are released when the grid material is transformed into lead dioxide. If not absorbed by the active material, they are leached out of the positive plate and can reach the negative electrode by diffusion. This explains the immense importance of alloys for lead-acid batteries.

Antimony, for example, is only partially absorbed by the lead dioxide. To some extent it leaves the positive electrode and migrates to the negative ([50]), there increasing the hydrogen-evolution rate. For this reason, the addition of antimony is avoided or at least minimised in valve-regulated lead-acid batteries.

4.7.3 Alloys for Lead-Acid Batteries

The selection of grid alloys must not only consider fabrication requirements (to produce flawless grids), mechanical strength (for handling during production), creep strength (to resist growth during service) and corrosion resistance, but also must pay attention to the fact that the alloy and the alloying constituents may cause electrochemical problems, like increased water decomposition or premature loss of capacity due to passivation effects at the positive grid surface. As it is not possible to match all these requirements at their optimum simultaneously, a number of different grid alloys are in use ([51]). (*A survey on metallurgical properties of lead alloys for batteries is given in* ([52])). In Table 4.19, examples of alloys employed in lead-acid batteries and their characteristic constituents are compiled.

Standard antimony alloys contain 8 to 12% antimony. The composition of these alloys is rather close to the eutectic mixture (11% Sb), which is easy to cast and results in strong grids. Furthermore, a high antimony content stabilises the active material in the positive electrode and improves the cycle performance of the battery ([53]) (cf. Section 10.1.1). For this reason, most traction batteries today contain positive grids with antimony contents between 4 and 11%.

The disadvantage of a high antimony content is the increase of water-decomposition rate with service time. This is caused by antimony released from the positive grid by corrosion and precipitated on the surface of the negative electrode, since hydrogen can be evolved much more easily at the antimony particles than at the lead surface (cf. Section 7.2.3.2).

Low-antimony alloys constitute a possibility to diminish the disadvantages of antimony by a reduction of its content in the positive grid below 2%. The amount of antimony then released becomes so small that antimony contamination of the negative electrode is almost negligible, but stability of capacity during float is better than with antimony-free alloys, and cycling performance is preserved ([54]). This antimony reduction became possible only when grain refining additives were developed. Copper and sulphur are such additives. The most effective additive is selenium. It was introduced for submarine batteries in 1949 ([55]) and its widespread use for stationary and car batteries was initiated by VARTA in the early 1970s ([56]). Nowadays, these 'selenium alloys' are used for SLI and stationary batteries and partly also for low maintenance traction batteries, in the latter with contents below 1% ([57]).

	Kind of alloy	Additives	Field of applications
1	Standard antimony	4 - 11% Sb As, Sn, Cu (Ag)	Traction
2	Low antimony	0.5 - 3.5% Sb Se, Te, S, Cu [1] As, Sn (Ag)	Low maintenance, (stationary, traction) SLI [2] batteries
3	Standard calcium	0.06 - 0.12% Ca 0 to 3% Sn, (Al)	Stationary, vented. & VRLA batteries SLI [2] batteries
4	Low calcium	0.02 - 0.05% Ca 0.3 - 3% Sn; Ag 0.008 - 0.012 Al %	Valve-regulated, SLI [2] batteries (contin. grid manufact.)
5	Lead/tin	0.2- 2% Sn	Valve-regulated batt., conducting elements
6	Pure lead	-	Planté plates, BELL Syst. round cell (BELL Linage 2000)
7	ASTAG, ASTATIN [3]	0.009% As, 0.065 Te, 0.08% Ag Sn (*ASTATIN*)	Submarine batt. stationary batteries (Scandinavia, Italy)
8	Antimony/ cadmium [4]	1.5% Sb; 1.5% Cd	Valve-regulated

Table 4.21 Grid alloys for lead-acid batteries, their characteristic additives and fields of application.
[1]) Se, Te , S, and Cu are added as 'grain refiners'.
[2]) **S**tarting, **L**ighting, **I**gnition means starter battery for motor cars.
[3]) Alloys used by the TUDOR Group in Scandinavia.
[4]) Only used by GNB.

The introduction of **lead/calcium alloys** started in the United States in the 1930s ([58]) and was specified for standby batteries in the Bell telephone service in 1951 (cf. ([59])). Corrosion of the lead/calcium alloy does not affect the electrochemistry of the battery, because calcium is not precipitated at the negative electrode but remains as Ca^{2+} ion in the electrolyte. As a consequence, the hydrogen-evolution rate is low and remains practically unaltered during the whole service life of the battery.

The calcium concentration in these alloys is fairly low, and its observance is aggravated by the tendency of calcium to become oxidised. Initially, this caused great problems that could be markedly reduced by the addition of aluminium in the range

0.01 to 0.03 wt.%. Aluminium is more easily oxidised than calcium and forms an oxide layer that covers the surface of the melt. This layer protects the underlying melt from oxidation and gives several advantages ([60]):

- The calcium in the melt is widely protected against oxidation. As a consequence, the calcium content of the alloy can be kept in a close range and the development of corrosive compounds like Pb_3Ca is avoided.
- The melt is kept clear of suspended oxidised particles that would otherwise remain finely distributed in the melt and give rise to casting problems.

One disadvantage of lead/calcium alloys is grid growth (cf. Section 4.75).

A further demerit of lead/calcium alloys is the reduced stability of the capacity, especially when deep-discharge cycles are performed ([61]). A passivating layer of lead sulphate was often observed between the grid and the active material, (e.g. ([62])) and was assumed to be the reason for capacity loss. More recent investigations ([63]) showed that this capacity loss is a much more general problem, mainly obvious in batteries with antimony-free grid alloys, but depends also on a number of other parameters like the charge current rate and the history of the battery ([64]) (cf. Section 10.1.1). The primary effect is reduced conductivity of the active material, which leads to uneven current distribution. The charge/discharge process is limited more and more to the vicinity of the grid, while the bulk of the active material remains unaltered. As a consequence, the material around the grid is deeply discharged and the lead sulphate layer emerges as a secondary effect.

Tin additions are used in a wide range between 0.02 and 1.2 wt.% for lead/calcium alloys applied for positive grids. The corrosion and grid growth pattern follows the basic principles as that of binary lead/calcium alloys ([65]). Tin addition improves the creep resistance of the grid material ([66]) to some extent, and is helpful in stabilising the active material, including the contact between the grid and the active material ([67]). Some authors assume that tin becomes accumulated at the phase boundary between the grid and the active material ([68]). Experimental results indicate that in a calcium/tin alloy, initially, a tin-passivating layer is established at the grid surface. Later this layer is ameliorated by grain-boundary corrosion caused by the tin content ([69]). This mechanism might explain why antimony-free batteries with calcium/tin alloys often suffer a minimum capacity during the first 20 to 30 cycles, but gain full capacity afterwards in identical operating conditions. In general, tin addition is considered helpful and not critical up to 0.8 wt.% ([51]); but a detrimental influence of tin content above 0.3 wt.% on grid disintegration has also been reported ([70]).

Low-calcium alloys have been introduced during recent years. Their main advantage is reduced corrosion attack, and thereby reduced grid growth. Silver addition is claimed to increase the creep resistance and improve the corrosion behaviour ([71]).

A disadvantage of these alloys is their softness. So they cannot be manufactured by the conventional casting process, because they could not be handled in the usual equipment, like pasting machines. Instead they depend on continuous processes that combine grid manufacture with pasting (cf. Section 10.2.1).

The use of **strontium** instead of calcium seems promising, on account of the general rules for the composition of alloys ([72]). Superior mechanical properties, required by deep-discharge batteries, are expected. Lead-strontium-tin alloys were used in automotive batteries in the USA for a time, but have yet found no widespread use ([73]).

Lead/tin alloys gained much importance as a remedy against corrosion problems of the conducting elements in valve-regulated lead-acid batteries. But they are also applied for grids, where a tin addition of 0.6 wt.% proved to be the optimum ([74]). Lead/tin alloys for conducting elements like the pole bridge may contain fairly high tin additions of up to 3 wt.%.

Pure lead is used for positive Planté plates, the grids for the BELL System round cell (Table 4.21), and its valve-regulated successor, the 'Lineage 2000' ([75]). To improve deep-discharge performance, the addition of small amounts of tin (Sn) is practised to an increasing extent ([76]). So, cylindrical cells of valve-regulated design (Gates ([77])) with punched grids (cf. Figs. 10.24 and 10.25) employ lead for both grids with a small tin addition. Thin grids cannot be made from pure lead by the conventional manufacturing process, because the metal is too soft.

Table 4.19 contains a further alloy, **ASTAG**, used to a small extent for stationary batteries in Scandinavia, where it was introduced by Tudor Sweden, originally for submarine application. A similar alloy, basically corresponding the ASTAG alloy, but using tin as a further additive, is applied under the name ASTATIN ([78]).

The **antimony/cadmium alloy** ([79]) is used by one manufacturer exclusively, namely GNB ([80]) for the 'Absolyte Battery'. It contains 1.5% antimony and 1.5% cadmium. With this alloy, cycling is possible owing to the antimony content. On the other hand, antimony release is rather low because the intermetallic compound SbCd is formed between antimony and cadmium that keeps antimony within the positive plate. The serious disadvantage of this alloy is its high cadmium content, because of the toxicity of cadmium. This does not concern the battery, but requires special treatment for remelting of these batteries when they are recycled after service.

Bismuth is an unwanted component in lead and lead alloys. It is naturally occurring in lead ores and difficult to remove. The bismuth content has often been discussed, and many results are contradictory ([81]). A too-high bismuth content obviously influences the structure of the passivating layer and thereby increases the proneness to corrosion as shown for bismuth contents above 0.7% wt.% ([82]). But, according to general experience, low concentrations of bismuth have neither detrimental nor beneficial effects. For this reason, battery manufacturers usually tolerate bismuth contents up to 0.05 wt.%.

4.7.4 Corrosion-Determining Parameters

Corrosion is a complex reaction, influenced by a number of parameters. Not only ‘external parameters’ like temperature, electrode potential and acid concentration are important, but also ‘internal parameters’ like composition of the corroding alloy, metallographic structure and local distribution of the alloying additives. Exact evaluation of corrosion rates is not possible, but there are a number of general rules that can be combined with empirical experience.

Metallographic structure is determined mainly by the alloy composition but also by the casting conditions. The sometimes confusing results concerning corrosion rates of various alloy compositions mostly are caused by the influence of casting or other manufacturing treatments.

4.7.4.1 Influence of the Electrode Potential

Corrosion is an electrochemical reaction and so it depends on electrode potential. In Fig. 4.28 a number of results on the influence of the electrode potential are compiled.

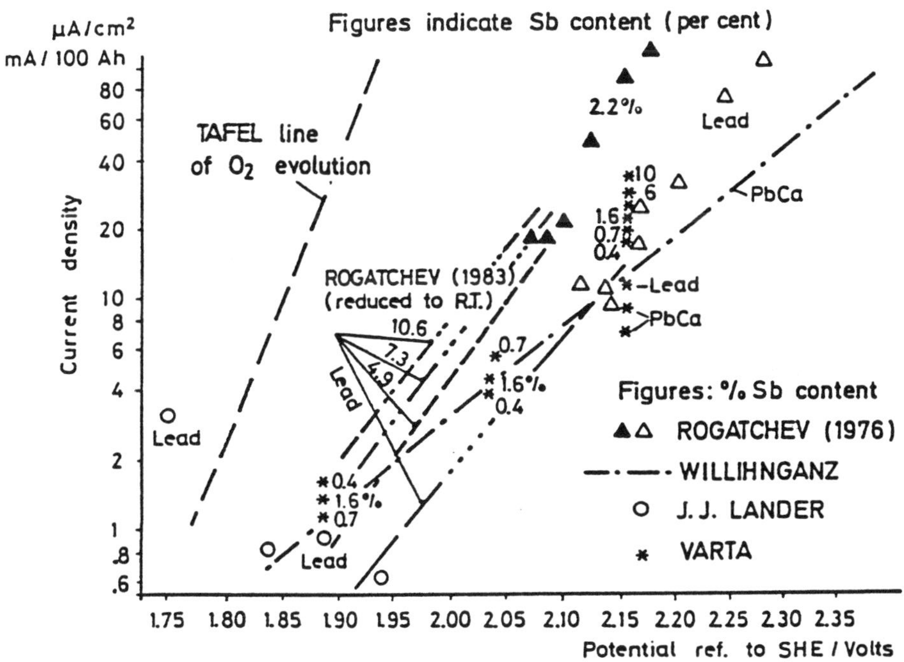

Fig. 4.28 Corrosion rate of lead and lead alloys at constant electrode potential.
The comparison shows the general trend, but should not be overestimated, because test conditions were rather different:
- J.J.Lander ([83]) used cold-rolled pure lead samples;
- Willihnganz ([84]): stationary batteries with lead/calcium alloys;
- Rogatchev (1976) ([85]): special cast samples of lead and Pb, 2.2% Sb, 0.1%As;
- Rogatchev (1983) ([86]): single tubes of tubular plates with special cast spines, corrosion rates determined at 50 °C are reduced by the factor 4 in the Figure;
- VARTA: special cast samples.

Oxygen-evolution curve (1 Decade per 80 mV) is inserted for comparison. (cf. also TAFEL lines in Fig. 7.5, p.242).

The variety of the results is rather wide. But corrosion measurements are difficult to compare, because of many influencing parameters. A major problem is to estimate the true area of the corroding surface, which is changed by the corrosion process itself. Nevertheless, some general rules can be derived:

- There is a strong influence of the electrode potential. Within the different series of measurements, TAFEL lines can be established, but the slope (the b-factor in Eq. (2.62)) varies in the range 160 to 300 mV. This variation indicates that additional parameters influence the results. In general a b-factor of 240 mV is assumed, which means that such an increase of the electrode potential increases the corrosion by one order of magnitude (cf. Fig. 7.5, p.242).
- Corrosion passes through a minimum at about 40 to 80 mV above the $PbSO_4/PbO_2$ equilibrium potential, already observed by Lander ([83]) but stated by many other authors (e.g. ([87,88])). (*It cannot be recognised in Fig. 4.28.*)
- The increase of the corrosion rate with potential increase is smaller than that of oxygen evolution. Thus the increase in the corrosion rate is less than that of oxygen gassing or internal oxygen cycle (= float-current of VRLA batteries).
- Pure lead proves to be most corrosion resistant. Low-calcium alloys and lead/tin alloys also are largely corrosion resistant. With antimony alloys, a slight increase of the corrosion rate is observed proportional to the antimony content. Under a cycle regime, however, reduction of the corrosion rate with increase of antimony content was observed ([89]).

4.7.4.2 Temperature Influence

The temperature dependence of the corrosion reaction corresponds with the general dependence on temperature of chemical reactions, described by the ARRHENIUS equation (cf. Section 2.3.6). This means, that the corrosion rate is approximately doubled per ten degrees of temperature increase (Eq. (2.68), p.33).

4.7.5 Grid Growth

Grid growth is a direct consequence of grid corrosion. It is caused by the volume increase when lead in converted into lead dioxide (> 1.3 according to column 4 in Table 4.20). The result is shear forces at the interface between metal and oxide that try to expand the metal. When the creep resistance of the metal cannot withstand these shear forces, elongation of the spines or grid wires is the consequence. Such elongation can cause short-circuits, when the expanded grid of the positive plate gets in touch with the negative bus-bar. In pasted plates (cf. Fig. 10.4), capacity reduction can be a consequence, since the pellets of active material may lose their contact with the grid.

The situation is aggravated when corrosion does not uniformly progress on the surface, forming a layer as in Figs. 4.26 and 4.27, but penetrates along grain boundaries into the bulk of the grid material. Then the volume increase between Pb and PbO_2 cleaves the grid material, and does not only cause elongation but may even fracture the spine or wire of the grid.

Lead/calcium alloys are susceptible to grid growth for both reasons: their creep resistance is comparably low, even lower than that of 1% antimony alloys ([57]), and they also suffer grain-boundary corrosion. This results in growth of the positive plates, which is often observed in service ([90]). This growth occurs particularly when the concentration of calcium exceeds 0.09 wt.% in positive grids. The reason is a very corrodible compound, Pb_3Ca, is then formed and precipitated at the grain boundaries, causing the above mentioned expanding forces. For this reason, the calcium content in positive grids has to be kept below 0.087 wt.%. This value can be exceeded for negative-grid alloys, because corrosion does not happen at the negative grid in normal circumstances. The use of 'low-calcium alloys' (Line 4 in Table 4.19) is promoted by the wish to reduce corrosion and thereby grid growth.

Since grid growth is connected to grid corrosion, its minimum can be expected at the corrosion minimum. This is a reason for applying a corresponding float voltage. In vented lead/calcium batteries, polarisation of the positive electrodes between 80 and 100 mV above the $PbSO_4/PbO_2$ potential is regarded as most favourable. More recent measurements, however, indicated that this minimum rather can be expected at 40 mV above open-circuit value ([91]). Furthermore all parameters that increase corrosion are unfavourable in respect to grid growth (increased electrode potential, elevated temperature).

Closely related to grid growth is the formation of lead dioxide flakes at the surface of the conducting elements, like post and bus-bars in vented batteries. The above corrosion process occurs also at the surface of these parts, but their cross section prevents elongation. Instead fairly large parts of lead dioxide, the 'flakes', peel off the basic material when the oxide layer exceeds a certain thickness.

4.7.6 Special Types of Corrosion Attack

The corrosion effects described are not limited to the positive grid itself but may occur at all parts connected to the positive grid, such as the bus-bar and post. Here, sometimes, corrosion processes are observed that at the first view seem difficult to understand, such as corrosion problems at terminal sealings and corrosion of current-conducting elements at the negative plate group in valve-regulated batteries. Both are of great importance. The first often causes premature failure of vented batteries, and the second one occured widely as an unexpected experience, when these batteries were new in the market.

4.7.6.1 Crevice Corrosion

A type of corrosion often observed at vented stationary batteries is known as 'Crevice corrosion'. Its mechanism is explained in Fig. 4.29.

The precondition for crevice corrosion is a relatively tight covering around the metallic lead that leaves a crevice into which the acid can penetrate. On the surface of all the lead parts that are exposed to the sulphuric acid, a layer of lead dioxide is formed, as described above. In the crevice, the current required to rebuild this passivating layer causes a voltage drop, which means that along the crevice, the potential of the metal referred to the electrolyte is gradually shifted to less positive values. The situation becomes critical when this potential difference attains values where the passivating layer can no longer be re-established. Then active corrosion starts, and since the growing rate of corrosion requires more current, the voltage drop within the crevice is increased. A dramatically accelerated local corrosion rate results.

Finally, the post seal and even the cover or the battery container may be destroyed, because the corrosion product requires much more volume than the lead, and this corrosion product presses away the sealing compound.

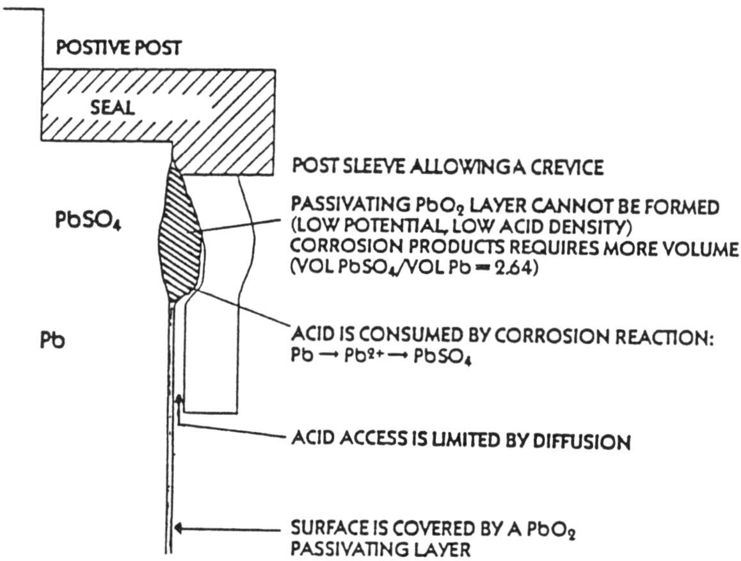

Fig. 4.29 The initiation of crevice corrosion.

Battery manufacturers have made strenuous efforts to overcome the problem of post corrosion, which quite often leads to premature failure of the battery. As a result, a number of different post sealings are offered.

In valve-regulated lead-acid batteries, the post has to be sealed tightly to prevent air admission to the interior of the battery (chemical self-discharge, cf. Section 4.4.2). In view of this sealing, the immobilisation of the sulphuric acid is beneficial, because acid fumes are not present in the free space inside the battery, which could moisten the post. As a consequence, the creep distance for the sulphate ions that are required for the corrosion process is very long.

4.7.6.2 Corrosion Attack at the Negative-Plate Group

In flooded lead-acid batteries, the grid of the negative plate and all parts connected to it, are secured from corrosion attack during float charging on account of 'cathodic corrosion protection'. The potential of the negative electrode is kept below the $Pb/PbSO_4$ equilibrium, i.e. in the region where lead is stable, and Pb^{2+} ions cannot be formed (cf. Fig. 4.1). To ensure this situation, conducting elements, like plate lugs and connecting bus-bars, are immersed under the electrolyte, and only the terminal post is generally above the electrolyte level.

Cathodic corrosion protection also applies to conducting elements above the electrolyte surface, unless their distance to the electrolyte surface is too large. This protection is effected by a wetting film of the electrolyte that creeps up at the surface of these elements, and is continuously renewed by acid fumes. Thus, ionic conductivity is sufficient to keep the potential of this wetting film nearly uniform and very close to that of the bulk of electrolyte. Oxygen that reaches these parts cannot cause corrosion; it can only be reduced at the lead surface according to

$$\boxed{O_2 + 4 \cdot e^- + 4 \cdot H^+ \Rightarrow 2 \cdot H_2O} \qquad (4.37)$$

Oxygen is so reduced and recombines with H^+ ions that originally have been released at the positive electrode by oxygen evolution (cf. Section 7.5.3.2).

When corrosion at properly cast and welded conducting elements in negative electrodes is observed occasionally, it must be assumed that the electrolyte level has been very low for extended periods, such that most of the conducting elements have been above the electrolyte surface. Then, oxygen reduction at the fairly large surface area and the long creep distances within the wetting film, cause a situation, as described below in connection with Fig. 4.30, i.e. uniform potential is no longer retained in the wetting layer and cathodic protection may be lost. This situation is aggravated when the composition and the structure of the metallic parts are not uniform, and local potential differences increase corrosion ([92]).

In valve-regulated batteries, the bus-bar cannot be submerged in the immobilised electrolyte that necessarily ends below these parts. So the wetting film is stretched to larger distances from the electrolyte surface. Furthermore, acid fumes that renew the wetting layer are hardly present. So the uniformity of the wetting film is much more in danger of being disturbed by oxygen reduction. Fig. 4.30 illustrates the situation.

The continual oxygen reduction at the wetted lead parts has the following results:

- Oxygen reduction on the lugs and the connecting bus-bar consumes H^+ ions and causes a current along the wetting film, and thus a voltage drop along the wetting film that displaces the potential in the upper part of the film to more negative values.
- The potential of the metallic parts is uniform. On account of the potential shift along the wetting film, the potential between that film and the metallic surface is increased with the distance from the electrolyte surface (an increase of as much as 200 mV has been observed ([93])). Referred to its wetting film, the potential of metal gets more and more positive, and at a given distance it reaches and even exceeds the $Pb/PbSO_4$ equilibrium value (Zone 2 in Fig. 4.30).

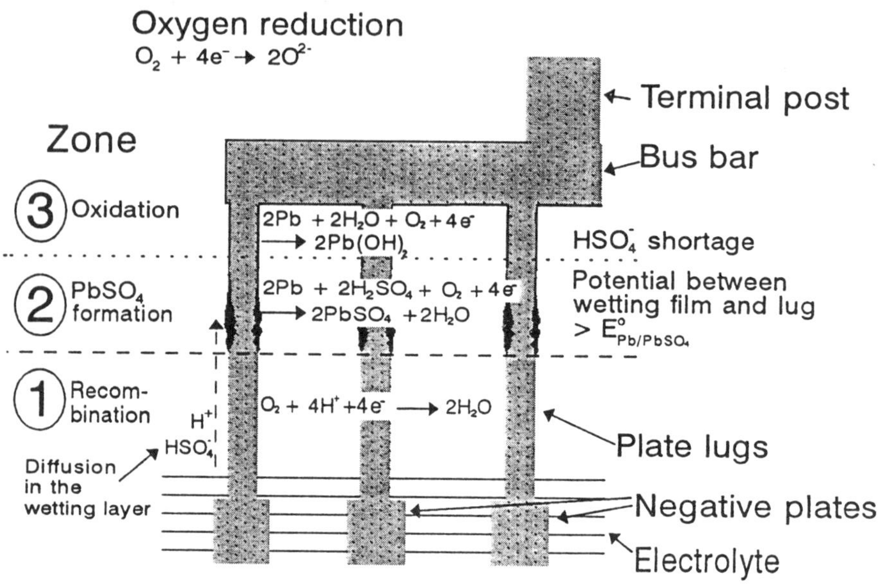

Fig. 4.30 Oxygen reduction at the surface of the conducting elements in VRLA batteries. (Corresponds to Fig. 6 in ([94]) combined with results in ([95]).)

As a result, three different Zones can be observed that are indicated in Fig. 4.30:

Zone 1: The potential of the metal versus its electrolyte film is below the $Pb/PbSO_4$ equilibrium potential. Lead is stable and reduction of oxygen is neutralised by recombination with H^+ ions, according to:

$$\boxed{O_2 + 4 \cdot H^+ + 4 \cdot e^- \Rightarrow 2 \cdot H_2O} \tag{4.38}$$

The required H^+ ions are produced by the reverse reaction at the positive electrode (oxygen generation) and reach the negative by diffusion.

Zone 2 When the potential difference between the metal and its wetting layer reaches the $Pb/PbSO_4$ equilibrium potential or even more positive values, Pb^{2+} ions are formed, and oxygen reduction is now balanced by $PbSO_4$ formation according to:

Oxygen reduction $O_2 + 4 \cdot e^- + 4 \cdot H^+ \Rightarrow 2 \cdot H_2O$

and lead corrosion $2 \cdot Pb + 2 \cdot H_2SO_4 \Rightarrow 2 \cdot PbSO_4 + 4 \cdot H^+ + 4 \cdot e^-$

with the result $\boxed{2 \cdot Pb + O_2 + 2 \cdot H_2SO_4 \Rightarrow 2 \cdot PbSO_4 + 2 \cdot H_2O}$ (4.39)

The situation is aggravated as the acid in the wetting layer at lug and group bar becomes diluted by the generated water. This dilution decreases the conductivity of the wetting layer and increases the voltage drop along the layer. Furthermore, corrosion roughens the surface that again increases current and voltage drop. Thus the progressing corrosion accelerates the process.

Zone 3 in Fig. 4.30 is characterised by a lack of HSO_4^- ions. Lead sulphate formation according to Eq. (4.39) requires that HSO_4^- ions are available. But this applies only up to a distance where this corrosion occurs, since there the HSO_4^- ions are consumed. Above that distance, HSO_4^- ions are no longer are available. So, lead can only be oxidised according to

Oxygen reduction $\quad O_2 + 4\cdot e^- + 4\cdot H^+ \Rightarrow 2\cdot H_2O$

counter reaction $\quad 2\cdot Pb + 2\,H_2O \Rightarrow 2\cdot Pb(OH)_2 + 4\cdot H^+ + 4\cdot e^-$

with the result

$$\boxed{2\cdot Pb + O_2 + 2\cdot H_2O \Rightarrow 2\cdot Pb(OH)_2} \qquad (4.40)$$

Because of the higher vapour pressure of the nearly neutral region compared to the acid electrolyte, drying converts the $Pb(OH)_2$ into PbO and finally stops further oxidation.

The corrosion process is quite complex and depends on a number of parameters including ion-transport rates within the wetting layer. For this reason, position and size of the three zones vary widely, and often corrosion occurs within narrow bounds.

Pavlov and co-workers ([95]) measured the voltage drop between the wetting film and the metal surface at partly immersed electrodes of lead/tin and lead/tin/selenium alloys and observed that 2 to 4 cm above the electrolyte level the critical potential was reached. They also showed the complex influence of selenium addition. Such complex influence is sometimes caused by the structure of the underlying material, e.g., welding areas with their non-uniform structure are endangered, and the described corrosion attack is often observed at the lug/group strap burning area ([96]). Antimony content of the conducting elements promotes the effect ([97]) since antimony enhances oxygen reduction. For this reason, antimony-free alloys, mainly lead/tin alloys ([98]), are employed for the top lead in stationary lead-acid batteries of valve-regulated design in combination with proper welding techniques ([99]).

Such a corrosion mechanism is mainly observed with batteries charged continuously (floated), because then oxygen reduction occurs continuously. Equalisation processes that reduce the effect can take place when the battery is cycled or standing on open-circuit for longer periods. The latter applies for starter (SLI) batteries, but the latent danger is present there as well.

As a remedy, it is recommended, to maintain acidity over the top lead by wrapping these parts with felt, or putting a layer of gelled electrolyte around them ([100]).

4.8 Copper as Conducting Insert and Grid Material

The use of copper as conducting metal is of great interest due to its low electrical resistance, as illustrated by the values in Table 4.22.

Copper, however, is not stable at the potential of the $PbO_2/PbSO_4$ electrode and must be protected thoroughly by a dense layer of lead to prevent contact with the electrolyte. Otherwise, fast corrosion occurs that is dangerous in two ways:

1. The corrosion destroys the conduction path in the positive plate by disintegration of the copper insert.
2. The dissolved copper is precipitated on the surface when it reaches the negative electrode, and increases the hydrogen evolution rate considerably (cf. Fig. 7.7, p.250).

For this reason, copper is mainly used as an insert in parts where it can be covered easily by a thick layer of lead, like terminal posts. Beside the improved conductivity of these parts, a further advantage is the tight and permanent copper/copper connection between such inserts and bolted cell connectors that need no periodic tightening.

Metal or alloy	Electrical conductivy	Electrical resistance	Factor
	$10^6\ \Omega^{-1}\ cm^{-1}$	$10^{-6}\ \Omega\ cm$	
Copper (Cu)	0.57	1.754	1
Lead (Pb)	0.048	21	12
Lead/antimony alloys			
Pb; Sb3 (3% Sb)	0.043	23	13
Pb; Sb5 (5% Sb)	0.04	25	14

Table 4.22 Electrical conductivity and resistance of copper, lead and two lead/ antimony alloys and the factor referred to copper.

In high-load submarine batteries, internal copper connectors are used between twin plate groups within a tall cell to reduce the internal resistance (cf. ([101]), and ([16]), p.306). Those that are connected to the positive electrodes must be protected by an absolutely dense layer of lead sheath that is thick enough to withstand corrosion during service life of the battery .

In the negative electrode, copper is not endangered by corrosion, since its equilibrium potential of 0.337 V vers. SHE is far above the potential of the negative electrode, unless the battery voltage is reversed during too deep discharges. When, however, it is only partly immersed in the electrolyte, and the atmosphere above the electrolyte contains oxygen, a situation results, similar to that shown in Fig. 4.30. The non-immersed part of the copper is covered by a wetting layer. Oxygen reduction causes a current along this layer, and the 'cathodic protection' of the metal is lost when the height above the level of the electrolyte exceeds a few mm ([102]). To prevent this kind of corrosion, copper parts have to be covered by lead, especially in valve-regulated batteries where the electrode is not immersed in the electrolyte and oxygen is present.

Another reason to protect the copper by a dense layer of lead in negative electrodes is the above mentioned increased hydrogen generation at the copper surface. This layer need not be thick, since it is not disintegrated by corrosion, as in the positive electrode. On the contrary, such a layer of lead is even produced on not, or only partly, lead-plated copper grids, when the battery is cycled. During charging, lead is precipitated on the copper surface and gradually forms a layer. This is, however, a slow process and it takes several hundred cycles until the hydrogen-evolution rate is reduced to that of a lead grid ([103]).

Lead-plated copper grids (copper stretch metal, CSM) have been used to a certain extent in vented traction batteries for more than ten years ([104]). They reduce the internal resistance which also results in a more uniform current distribution along the height of the plates. Due to its high conductivity (Table 4.22), the grid requires less volume. A corresponding volume of additional active material can be added, and so a slight increase of capacity is achieved.

Expanded copper metal is also used in other applications. Submarine batteries and large stationary batteries with tall cells are equipped with copper grids in the negative electrodes ([105]).

References

([1]) K.R.Bullock, *in* 'Proc.Symp.Advances in Lead-Acid Batteries', The Electrochemical Society, 1984, p.1

([2]) J.P.Pohl and H.Rickert, 'Electrochemistry of Lead Dioxide', *in* S.Transatti (Edit.), 'Electrodes of Conductive Metallic Oxides, Part A', Elsevier, 1980, p.183

([3]) J.P.Carr and N.A.Hampson, 'The Lead Dioxide Electrode', Chem Rev.,**72** (1972), pp. 679 - 703, esp. p.681

([4]) H.Bode, 'Lead-Acid Batteries', J.Wiley & Sons, New York, 1977, p.366

([5]) H.S.Harned and .W.J.Hamer, J. Amer. Chem. Soc., **57** (1935) pp. 27-33, 33-35

([6]) A.K.Covington, J.V.Dobson and W.K.F.Wynne-Jones, Trans.Far.Soc., **61** (1965), pp. 2050 - 2062

([7]) G.W.Vinal, 'Storage Batteries', John Wiley & Sons, New York, 1955, pp. 120,121

([8]) H.Bode, 'Lead-Acid Batteries', John Wiley & Sons, New York, 1977, Table 2.36, p. 98

([9]) N.E.Bagshaw, 'Batteries on Ships', John Wiley & Sons, New York, 1982, p.117, Table 3-4

([10]) D.E.Koontz, D.O.Feder, L.D.Babusci and H.J.Luer, Bell Syst. Tech. J., **49**, (1970), p.1253 cf. K.v.Kordesch (Editor), 'Batteries', Marcel Dekker Inc., New York, 1977, pp.52 - 54

([11]) cf.K.v. Kordesch, (Editor) 'Batteries', Marcel Dekker, Inc., New York,1977, pp.419-422

([12]) H.Bode, 'Lead-Acid Batteries', John Wiley & Sons, New York, 1977, p 61

([13]) YUASA NP Valve-Regulated Lead Acid Battery Manual, Yuasa Battery Sales (U.K.) Ltd., Hawksworth Industrial Estate, Swindon, Wiltshire SNS 1EG

([14]) ACCUMULATORENFABRIK SONNENSCHEIN GmbH, D-W-6470 Büdingen (Hessen) Thiergarten

([15]) K.Matthes, B.Papp and R.F.Nelson, *in* T.Keily and B.W.Baxter (Editors),'Power Sources 12', Intern. Power Sources Symp. Committee, 1988, p.1, esp. Table 16, p.14

(16) VARTA Batterie AG (Editor), Bleiakkumulatoren, 11th Edition, © VDI-Verlag, Düsseldorf, 1986, p.220

(17) H.D.Thacker, Proc. 14th Intelec Meeting, Washington, DC, 1992, © IEEE Communications Soc., 1992, p.47.

(18) F.J.Vaccaro and R.E.Landwehrle, Proc. 13th International Telecommunications Energy Conference (INTELEC), Kyoto, 1991, © IEEE Communications Soc., 1991, p.20

(19) A.I.Harrison, Proc 13th International Telecommunications Energy Conference (INTELEC), Kyoto, 1991, © IEEE Communications Soc., 1991, p.14 and esp. p.17

(20) Private communication by N.E.Bagshaw

(21) Batteries International, ISSN 0957-9249, Issue 6, January 1991, p.72

(22) R.Frank and H.Giess, Proc. 17th Intelec Meeting, The Hague 1995, © IEEE Communications Soc., p.353

(23) D.Berndt and E Meißner, and E.Rusch, Proc. 15th Intelec Meeting, Paris, 1993, © IEEE Communications Soc., p.139

(24) R.F.Nelson, Proc. 12th Intelec Meeting, Orlando 1990, © IEEE Communications Soc., p.165

(25) K.R.Bullock, D.G.Fend, and P.K.Ng, Proc. 17th Intelec Meeting, The Hague 1995, © IEEE Communications Soc., p.690

(26) D.Berndt and E.Meissner, Proc. 12th Intelec Meeting, Orlando, 1990, © IEEE Communications Soc., p.148

(27) Bellcore, GR-1515-CORE, Generic Requirements for the Detection and Control of Thermal Runaway in VRLA Batteries, © 1994 by Bellcore Customer Relations, 8 Corporate Place, Piscatawayn NJ 08854-4156, USA

(28) D. Berndt, E.Meissner, F.J.Kruger, and R.Luetkehoelter, Proc. 13th Intelec Meeting, Kyoto 1991, © IEEE Communications Soc., p.181

Poulin, R.Héron and D.Mailloux, Proc. 16th Intelec Meeting, Vancouver 1994, © IEEE Communications Soc., p.497

T.Suntio, ibid, p.481

W.Rusch, S.Goebel and R.Luetkehoelter, ibid., p.632

(29) M.J.Hlavac and D.O.Feder, Proc. 17th Intelec Meeting, The Hague 1995, © IEEE Communications Soc., p.284

(30) W.T.Rutledge and R.J.Bowers, Proc. 16th Intelec Meeting, Vancouver 1994, © IEEE Communications Soc., p.168

(31) H.Bode, 'Lead-Acid Batteries', John Wiley & Sons, New York, 1977, p. 74, reprinted by permission of The Electrochemical Society, Inc.

(32) H.Tuphorn, **in** 'Advances in Lead-Acid Batteries', The Electrochemical Soc., Pennington, NJ, 1984, p.441, and esp. p.448, Fig.5

(33) N.E.Bagshaw, Proc. 11th Intelec Meeting, Florence, 1988, © IEEE Communications Soc., U.S.A., Paper 12.4, p. 3

(34) B.Culpin and J.A.Hayman *in* L.J.Pearce (Editor), 'Power Sources 11' (Proc. 15th Intern. Power Sourc. Symp., Brighton, 1966), Intern. Power Sources Symposium Committee, Leatherhead, Surrrey, England, 1986

(35) D.R.Crouch and J.W.Reitz, J. Power Sources, **31** (1990), p. 125, and esp. Fig.4, p.129

(36) E.g. 'CHLORIDE Powersafe VH Range', Instruction Manual edited by CHLORIDE Standby Power, Clifton Junction, Swinton, Manchester M27 2LR, England, 1990

(37) C.Liebenow, Centralblatt für Accumulatoren und Elementenkunde, **1** (4), Febr.1900, pp.67-70

(38) H.Tuphorn, J. Power Sources, **31** (1990), p.57

(39) D.R.Crouch and J.W.Reitz, J. Power Sources, **31** (1990), Fig.2, p.127

(40) D.O.Feder, T.G.Croda, K.S.Champlin, S.J.McShane and M.J.Hlavac, J.Power Sources **40** (1992) p.235

D.O.Feder, J.Hlavac, Proc. 16th Intelec Meeting, Vancouver 1994,

© IEEE Communications Soc., p.282

D.O.Feder, Proc. 17th Intelec Meeting, The Hague 1995, © IEEE Communications Soc., p.22

(41) B.I.Tsenter, Yu.Popowa, R.V.Mustafin, J.Jindra, M.Musolivá and J.Mrha, Journal Power Sources, **39** (1992) p.179

(42) Batteries International, Issue 25, October 1995, p.84

(43) D.Pavlov, Ber. Bunsenges, **71** (1967), p.398

D.Pavlov *in* 'Advances in Lead-Acid Batteries', The Electrochem. Soc., Proc.Vol.84-14, (1984), p.110

(44) P.Ruetschi and R.T.Angstadt, J. Electrochem. Soc., **111** (1964), p.1323

(45) D.Pavlov and Z.Dinev, J. Electrochem. Soc., **127** (1980), p.855

(46) H.Bode, 'Lead-Acid Batteries', John Wiley & Sons, New York,1977, p.18

(47) K.R.Bullock, G.M.Trischan and R.G.Burrow, J. Electrochem. Soc., **130** (1980), p.1283

(48) H.Bode, 'Lead-Acid Batteries', John Wiley & Sons, New York, 1977, p.10

(49) H.Bode, ibid., p.372, Table VII

(50) J.L.Dawson, M.I.Gillibrand and J. Wilkinson, 'Power Sources' 3, Oriel Press, Newcastle upon Tyne, 1971, p.1

(51) N.E.Bagshaw, *in* T.Keily and B.W.Baxter (Editors), 'Power Sources 12' (Proc. 16th Intern. Power Sources Conference, 1988); Intern. Power Sources Sympos. Committee (1988), p.113 (comprehensive survey)

R.D.Prengaman, 'Advances in Lead-Acid Batteries', The Electrochemical Society, 1984, Proc. Vol 84 - 14, p.201

R.D.Prengaman, Batteries International, Issue 13, October 1992, p.24

(52) N.E.Bagshaw, A Battery Technologists Guide to Metallurgical Structures and Alloy Properties, Advanced Lead-Acid Battery Consortium (ALABC), 1994
ILZRO, Inc, 2525Meridian Parkway, Suite 100, P:O Box 12036, Research Triangle Park, NC 27709-2036, USA

(53) E.Y.Weissman *in* K.v.Kordesch (Editor), 'Batteries, Vol.2', Marcel Dekker, Inc., New York, 1977, p.74 ff.

(54) K.A.Lundqvist, Proc. 9th Intelec Meet., Stockholm, 1987, © IEEE Comm. Soc.,1987, p 614, Figs. 5 and 6

(55) H. Waterhouse and R.Willows, B.P.No 622,512 (1949)

(56) H.Borchers, S.C.Nijhawan and W.Scharfenberger, Metall, **28** (1974), p.863

D.Berndt and S.C.Nijhawan, J.Power Sources, **1** (1976/77), p.3

German Patent D.P. 2,151,733 (VARTA Batterie AG)
(57) D.Berndt and H.Franke, Proc.9th Intelec Meeting, Stockholm, 1987, © IEEE Publishing Inc., p.115
(58) H.E.Haring and U.B.Thomas, Trans. Electrochem.Soc., **68** (1935), p.293
(59) D.E.Koontz, D.O.Feder, L.D.Babusch and H.J.Luer, BELL System Technical Journal, **49** (1970), p.1253, and esp. p.1256
(60) R.D.Prengaman, J. Power Sources, **33** (1991), p.13
(61) T.G.Chang, 'Advances in Lead Acid Batteries', The Electrochem. Soc., 1984, p.86
(62) S.Tudor, A. Weisstuch and S.H.Davang, Electrochem. Techn., **3** (1965), p.90
(63) A.Winsel, E.Voss and U.Hullmeine, J. Power Sources, **30** (1990), p.209
(64) E.Meissner and E.Voss, J. Power Sources, **33** (1991), p.231
(65) S.Misra and A.J. Williamson, Proc. 17th Intelec Meeting, The Hague 1995, © IEEE Communications Soc., p.360
(66) N.E.Bagshaw, Proc.11th Intelec Meeting, Florence, 1989, © IEEE Publishing Inc., Paper 12.4
(67) M.N.C.Ijomah, J. Applied Electrochemistry, **18** (1988), p.142
(68) K.Takahashi, N.Hoshihara, H.Yasuda, T.Ishii and H.Jimbo, J. Power Sources, **30** (1990), p.23
(69) R.F.Nelson and D.M.Wisdom, J. Power Sources, **33** (1991), p.165, esp. Fig.10, p.177
(70) E.M.L. Valeriote, A.Heim and M.S.Ho, J Power Sources, **33** (1991), p.187
(71) P.Rao (GNB Inc), US Patent No 5 298 350 (1994)
(72) N.E.Bagshaw, J.Power Sources, **33** (1991), p.3, and esp. p.10
(73) N.E. Bagshaw, J. Power Sources, **53** (1994), p.25
(74) D.M.Wisdom and R.F.Nelson, Proc. 10th Intern. Lead Conference, NICE, May 1990
(75) Batteries International, Vol.8, July 1991, p.24
(76) P.Driver, Batteries International, Issue 6, Jan. 1991, p.17
(77) Gates Energy Products Inc., Warrensburg, Miss., U.S.A.
(78) M.Grossoni and C.Tortora, Proc. Intelec Meeting, Washington, D.C., 1982, © IEEE Publishing Inc., p.237
(79) N.E.Bagshaw, Proc. 3rd Intern. Lead Conference, Venice (Italy) September 1968
N.E.Bagshaw, Proc. 10th Intern. Lead Conference, NICE (France), May 1990
(80) German Patent DE 2457349 C2
(81) M.J.Koop, D.A.J. Rand, B.Culpin, Journal of Power Sources, **45** (1993), p. 365
(82) M.J.Koop, D.F.A.Koch and D.A.J.Rand, J. Power Sources, **34** (1991), p.369
(83) J.J.Lander, J. Electrochem. Soc., **98** (1951), p. 231; **103** (1956), p.1
(84) E.Willihnganz, J. Electrochem. Soc., **126** (1979), p.1511
(85) T.Rogatchev, St.Ruevski and D.Pavlov, J.Appl.Electrochem., **6** (1976), p.33
(86) T.Rogachev, G.Papazov and D.Pavlov, J. Power Sources, **10** (1983), p.291
(87) C.E. Weinlein, J.R. Pierson, D. Marshall, Power Sources 1979, **7**, p. 67
(88) K.R. Bullock, W.H. Tiedemann, J. Electrochem. Soc. 1980, **127**, p.2112

(89) G.Papazov, T.Rogatshov, D.Pavlov, J.Garche and K.Wiesener, J. Power Sources **6** (1981), p.15

(90) W.B.Brecht, D.O.Feder, J.M. McAndrews and A.J.Williamson, Proc.11th Intelec Meeting, Florence, © IEEE Comm.Soc., 1989, Paper 12.2

(91) W.B.Brecht, D.O.Feder, J.M.McAndrews and A.J.Williamson, Proc. 10th Intelec Meeting, San Diego 1988, © IEEE Communications Soc., p.124

(92) S.Hua, Y.Guo and Z Wang, J. Power Sources, **45** (1993) 131

(93) F.J.Vaccaro, J.Rhoades, R.Malley, B.Le, and K.Marion, Proc. 17th Intelec Meeting, The Hague 1995, © IEEE Communications Soc., p.70

(94) D.Berndt, 8th ERA Battery Conference and Exhibition, Solihull, West Midlands, UK, 1994, Paper 2.3.1

(95) D.Pavlov, M.Dimitrov, G.Petkova, H.Giess, and C.Gnehm, J. Electrochem. Soc., **142** (1995), p. 2919

(96) G.Baudo and G.B.Cecchinato, Proc.11th Intelec Meeting, Florence, © IEEE Comm.Soc. 1989, Paper 12.3

(97) S.Warrell, J. Power Sources, **31** (1990), p.36

(98) R.D.Prengaman, Batteries International, Issue 13, October 1992, p. 24

(99) R.D.Prengaman, Proc. 4th Intern. Lead Acid Battery Seminar, San Francisco, ILZRO, 1990, p.19

(100) R.D.Prengaman, cited in Batteries International, Issue July 1990, p.67

(101) H.Lawrenz, Marine-Rundschau, **9** (1980), p.533
cf. N.E.Bagshaw, 'Batteries on Ships', John Wiley & Sons, 1980, p.105

(102) R.Wagner, J. Electrochem. Soc., **143** (1996), p. 139

(103) K.Takahashi, M.Shiomi, M.Tsubota, and K.Yonezu in GS News Technical Report (Japan Storage Battery, C. Ltd), **46**, Issue 1, p. 20 and Issue 2, p.15

(104) R.Kiessling, Journal Power Sources, **19** (1987), p. 147

(105) H.Dominik, K.G.Krämer, and B.Voigt, Proc. 21th IECIC, San Diego, CA, 1986
R.Kiessling, Journal Power Sources, **19** (1987), p. 227

5. The Nickel/Cadmium Battery System

The nickel/cadmium battery is one couple of a larger number of electrodes, working in alkaline electrolyte. Manganese dioxide, silver oxide, zinc, and iron are further members of this species ([1]). Nickel/cadmium batteries have been in industrial production nearly as long as lead-acid batteries. They have gained enormous technical importance as maintenance-free batteries in their sealed version.

Characteristic features of nickel/cadmium batteries are:

- The charge/discharge reactions can frequently be repeated.
- The important fact that the electrolyte only forms the ion conducting medium was already mentioned in the patent by Jungner ([2]). Potassium hydroxide (KOH), and the occasionally used sodium hydroxide (NaOH), are involved in the electrochemical reaction only to an insignificant extent. Thus, the electrodes need not be spaced to provide room for the electrolyte as in the lead-acid battery, and the electrolyte can be kept to a minimum, which results in a battery of low weight and low internal resistance. Since concentration changes during charging or discharging can be neglected, the conductivity of the electrolyte remains fairly constant.

  As further consequences, problems of electrolyte stratification (cf. Section 4.6.3) are of little significance, and the fairly constant electrolyte concentration means a constant low freezing point of the electrolyte (between -30 and -60 °C) (cf. Fig. 5.15, p.206). These features give the reason for the excellent high rate and low temperature performance of nickel/cadmium batteries and their capability of intermediate charging.
- Jungner also recognised the advantage compared to lead-acid batteries that corrosion reactions of the current-conducting parts can be neglected in normal operating conditions (for exceptions cf. Section 5.6). This eliminates all the problems discussed in Section 4.7 for the lead-acid battery.

Note: *When graphite is used as a conducting aid, its gradual conversion into carbonate can be regarded as a corrosion process (cf. Section 11.1.1.1). But graphite is hardly used with sealed nickel/cadmium batteries.*

- Nickel/cadmium batteries can be sealed completely. Such batteries are absolutely independent from their operational position; electrolyte cannot be spilled, and gases do not escape from the battery. Thus, sealed nickel/cadmium batteries are ideal for the cordless power supply of electric or electronic devices, and are in widespread use as small batteries in different cell designs and in a wide range of sizes.

5.1 The Thermodynamic Situation

Fig. 5.1 shows reactions in nickel/cadmium batteries that are possible on account of thermodynamic parameters in relation to the electrode potential.

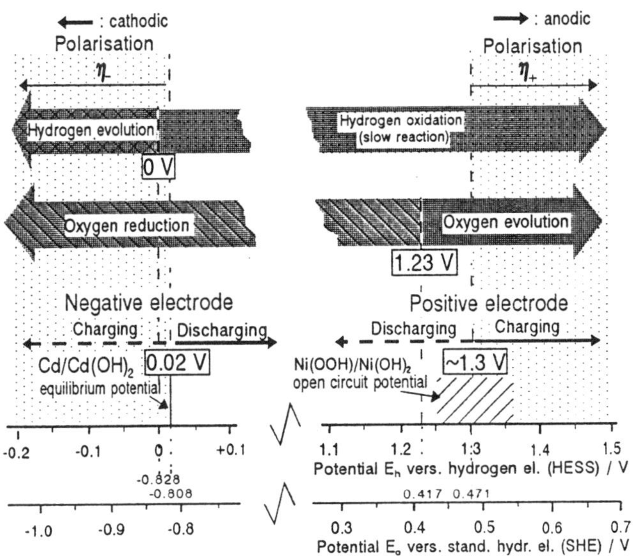

Fig. 5.1 Reactions that occur in nickel/cadmium batteries in dependence on electrode potential (thermodynamic situation). Their equilibrium potentials are inserted as enframed numbers. The charge/discharge reaction of the positive electrode ($NiOOH/Ni(OH)_2$) is not quite reversible and the reacting substances are not precisely defined chemical compounds (cf. Section 5.2.2). Its equilibrium potential is drawn as a column to indicate the uncertainty of this value. Upper scale: electrode potentials referred to the hydrogen electrode in the same solution (HESS). Bottom scale: referred to the standard hydrogen electrode (SHE). The shown relation between both corresponds to 25 °C (cf. Section 2.2.3.1).

In nickel/cadmium batteries, also a number of secondary reactions occur since their equilibrium potentials are within the potential range covered by the cell voltage of the battery, at least during charging.

- **Oxygen evolution** occurs at an electrode potential above 1.23 V which is slightly below the open-circuit potential of the positive electrode. Thus, oxygen evolution cannot be avoided, and the situation is similar to that of the lead dioxide electrode (cf. Fig. 4.1, p.97).
- **Oxygen reduction**, is possible below 1.23 V. As in lead-acid batteries, oxygen is reduced as soon as it reaches the negative-electrode surface, and the internal oxygen cycle in the sealed version is based on this reaction. But also in flooded batteries with narrowly spaced electrodes, e.g. in aircraft batteries, the diffusion rate of dissolved oxygen allows oxygen reduction to a certain degree. Here the slow diffusion rate (cf. Table 2.10, p.53) is partly compensated by the small distance between the electrodes. In such batteries, special measures like cellophane layers are required to hinder the internal oxygen cycle which would act as an internal short-circuit.
- **Hydrogen evolution** occurs at an electrode potential below the equilibrium potential of the cadmium electrode. This is an important difference compared to lead-acid batteries, with the consequence that the cadmium electrode does not suffer self-discharge by hydrogen evolution (cf. Section 4.4.1). Due to the small distance between the H_2/H^+ equilibrium potential and that of the cadmium electrode, hydrogen evolution already occurs when the electrode is only slightly polarised during charging. In sealed nickel/cadmium batteries such a polarisation has to be avoided by a suitable cell layout (cf. 'charge reserve' in Section 8.1.1).
- **Hydrogen oxidation** is very slow also at the nickel hydroxide electrode, although its rate at the nickel surface is increased compared to lead and lead dioxide. As in lead-acid batteries, an effective internal hydrogen cycle is not possible, and gaseous hydrogen, generated in the battery, has to escape from the cell, except small amounts that gradually are oxidised at the nickel surface in the positive electrode.

  Hydrogen oxidation gains importance as a self-discharge reaction especially in nickel/hydrogen batteries where hydrogen is present at high pressures (cf. Section 6.1.1.1).
- **Corrosion** of the nickel or nickel-plated substrate is very slow, and normally can be neglected. Therefore it is not shown in Fig. 5.1.

5.1.1 The Charge/Discharge Reactions

In the charged state, the active material of the negative electrode is cadmium (Cd), that of the positive electrode nickel hydroxide (NiOOH). The latter contains Ni^{3+} ions. When the battery is discharged, cadmium (Cd) is converted (oxidised) into divalent cadmium ions (Cd^{2+}), and trivalent nickel ions (Ni^{3+}) are converted (reduced) into divalent nickel ions (Ni^{2+}). During charge, both reactions are reversed.

For the positive electrode, the reactions in fact are more complicated. Nickel hydroxide is not an exactly defined chemical compound, and not only trivalent nickel ions are present in the charged material, but it can be a mixture between Ni^{2+}, Ni^{3+}, and Ni^{4+}. Furthermore, the nickel hydroxide electrode is not quite reversible. Both will be discussed in Section 5.2.2.1. Thus, the positive electrode in nickel/cadmium batteries does not attain a true equilibrium potential, and thermodynamic data and values based on them are only approximate.

The discharge reaction of the negative electrode is

$$\boxed{Cd \Rightarrow Cd^{2+} + 2 \cdot e^-} \qquad (5.1)$$

which has to be completed as a chemical reaction:

$$\boxed{Cd + 2 \cdot OH^- \Rightarrow Cd(OH)_2 + 2 \cdot e^-} \qquad (5.2)$$

When the positive electrode is discharged, Ni^{3+} ions are reduced to Ni^{2+} according to the (simplified) equation

$$\boxed{Ni^{3+} + e^- \Rightarrow Ni^{2+}} \qquad (5.3)$$

or its completed form

$$\boxed{NiOOH + H_2O + e^- \Rightarrow Ni(OH)_2 + OH^-} \qquad (5.4)$$

When both are summed up, the cell reaction during discharge results that formally can be written as:

$$\boxed{2 \cdot Ni(OOH) + 2 \cdot H_2O + Cd \Leftrightarrow 2 \cdot Ni(OH)_2 + Cd(OH)_2} \qquad (5.5)$$

The above 'formally can be written' means that Eq. 5.5 does not quite correspond with reality (cf. below and Section 5.2.2).

A remarkable feature of Eq. (5.5) is that neither potassium hydroxide (KOH) nor potassium appear in it or in one of the electrode reactions (Eqs.(5.2) and (5.4)). Only H_2O and its dissociation product OH^- ions take part in the electrode reaction. Water, however, is the predominant portion of the electrolyte, because the usually applied concentrations of KOH are in the range between 20 and 32 wt.% (densities at 20 °C between 1.19 and 1.30 g/cm^3).

From Eq. (5.5) it can be deduced that 0.67 g of the water is required per Ah of discharge. (One multiple of that equation includes the current exchange of 2 F = 53.6 Ah and includes 36 g of water (2 moles), thus 36/53.6 = 0.67 g/Ah.)

In vented nickel/cadmium batteries, a minimum of 10 g of electrolyte is contained per Ah of nominal capacity; usually it is much more. The specific weight of 1.20 or 1.22 g/cm^3, means 21.38 or 23.38 wt.% of KOH, respectively. So, 10 g electrolyte of the density 1.20 g/cm^3 contain 2.14 g of KOH and 7.86 g of water.

Complete discharge of the battery consumes 0.67 g/Ah of water, and at the end of discharge the electrolyte contains 2.14 g of KOH but only 7.19 g of water. So, its concentration is increased to $2.14/(2.14+7.19) \cdot 100$ = 22.9% which corresponds to a density of 1.215 g/cm^3. The change in density is only 0.015 g/cm^3, which will not be noticed in practice. This applies also to sealed batteries: despite the lower content of electrolyte, the density change caused by discharge does not exceed 0.02 g/cm^3.

Nevertheless, balance of the liquid within the cell has to be considered. The absorption of electrolyte in the fine pore system of the active material can dry out the separator when the electrolyte in the cell has been limited too stringently.

Note: *Traces of potassium are incorporated, especially in the γ form, the oxidised state (γ-NiOOH, cf. Fig. 5.4), and cause a certain dependence on KOH concentration (cf. Fig. 6 in ([3])), but this is a fairly small effect. It will be described in some detail in Section 5.2.2.*

Eq. (5.5) is sometimes written:

$$2\cdot Ni(OOH)\cdot H_2O + Cd \Leftrightarrow 2\cdot Ni(OH)_2 + Cd(OH)_2 \qquad (5.6)$$

Eq. (5.6) uses the above mentioned hydrated form of the trivalent nickel oxide that includes the water ($NiOOH\cdot H_2O = Ni(OH)_3$).

Thermodynamic data of the charge/discharge reactions are compiled in Table 5.1.

	Substance	Enthalpy ($H^{o,s}$) kJ/mole		Free Enthalpy ($G^{o,s}$) kJ/mole		
		1a	1b	2a	2b	2c
1	Cd	0		0		
2	$Cd(OH)_2$	-557.9 ◊	-571.6•	-470.8 ◊	-482.3•	
3	NiOOH		-391.2•		-321.7•	
4	$Ni(OH)_3$ *	-678.8		-542.2		
5	α-$Ni(OH)_2$				-466.9••	
6	γ-$NiOOH\cdot H_2O$					-679.3••
7	ß-$NiOOH\cdot H_2O$	-676 **	-629.9•	-561 **	-564.4•	-553.7••
8	ß-$Ni(OH)_2$	-537.8 *	-541.8•	-453.5 *	-459.1•	-452.7••
9	H_2O	-285.8		-237.2		

Table 5.1 Thermodynamic data: standard values of enthalpy and free enthalpy (Gibbs free energy) of formation concerning the chemical compounds used in Eq (5.5) and Eq. (5.6). Data from:

• Macdonald and Challingsworth ([4]), Table 1 (different sources)
•• Barnard et al. ([5]), Fig. 8, p.138
◊ Falk/Salkind ([6]), Table 7.1, p.523
* Latimer ([7])
** evaluation based on potential measurements ([8]).

The doubly enframed data are used in calculations in this Chapter, including heat calculations.

On account of the complex situation especially in the positive electrode, the values of the charged nickel hydroxide, especially the highly oxidised γ-NiOOH are not unambiguous and defy simple thermodynamic treatment. Certain discrepancies are also observed for cadmium hydroxide, since $Cd(OH)_2$ forms different modifications (α- and β-$Cd(OH)_2$) with diverging structures and water contents ([9]). To illustrate the situation, some values from different sources are shown in Table 5.1.

The results achieved with the values doubly enframed in Table 5.1 correspond with practical experience and calorimetric measurements. This is partly due to the fact that these thermodynamic data were derived from equilibrium voltage measurements. So the doubly enframed values for $H^{o,s}$ and $G^{o,s}$ for β-$NiOOH \cdot H_2O$ are based on Eq. (5.6) and were evaluated from direct potential measurements and TAFEL lines. They are referred to a mean open-circuit voltage of 1.33 V.

When they, together with the other enframed values, are inserted as the 'reactants' and 'products' into Eq. (5.6), according to Eq. (2.7), p.11, ΔH and ΔG result.

$$\Delta H = 2 \times (-537.8) - 557.9 + 2 \times 676$$

$$\boxed{\Delta H = -1634 + 1352 = -282 \text{ kJ}} \tag{5.7}$$

$$\Delta G = 2 \times (-453.5) - 470.8 + 2 \times 561$$

$$\boxed{\Delta G = -1378 + 1122 = -256 \text{ kJ}} \tag{5.8}$$

and the reversible heat effect according to Eq. (2.4), p.10

$$\boxed{T \cdot \Delta S = \Delta H - \Delta G = -26 \text{ kJ}} \tag{5.9}$$

always for one multiple of the cell reaction (Eqs. (5.5) or (5.6)).

Note: *The approximate character of such calculations must always be kept in mind. Since the hydroxides, especially the NiOOH are not defined uniform chemical compounds the simple difference between 'products' and 'reactants' applied in the Eqs. (5.8) and (5.9) represents only a rough approximation. More exact calculations of thermodynamic data consider that the reacting substances actually are mixtures ([5]).*

5.1.2 Equilibrium or Open-Circuit Voltage

Inserting the value of Eq. (5.8) into Eq. (2.6) results in the equilibrium or open-circuit voltage of the nickel/cadmium battery (1 kJ = 0.2778 Wh):

$$\boxed{E^o = 1.32 \text{ V}} \quad \text{in practice} \quad \boxed{E^o \approx 1.3 \text{ V}} \tag{5.10}$$

The sign '≈' expresses that this potential does not represent a true equilibrium potential and that this value can only be approximated. Actually, values between 1.3 and 1.4 volt are observed after charging, depending on the previous treatment of the cell (cf., e.g., Fig. 3 in ([3]), and also Fig. 5.5 on p.91 in this Chapter). But on open-circuit the cell voltage decreases to less than 1.3 V within about ten days ([10]).

This voltage does not depend on the concentration of the electrolyte, as mentioned in connection with Eq. (5.5), since only water is released or consumed when the battery is charged or discharged, and KOH does not participate in the cell reaction. *Actually, a certain dependence on the KOH concentration is observed (cf. Fig. 6 in (3)) and a small portion of potassium (K^+) is incorporated into the active material of the positive electrode (cf. Section 5.2.2). But this slight influence of the electrolyte concentration cannot be compared to the corresponding effect on the lead-acid battery, discussed in Section 4.2.1. In practice, the concentration of the electrolyte cannot be used to determine the state of charge in nickel/cadmium batteries.*

5.1.2.1 Single-Electrode Potentials

The reactions shown in Fig. 5.1 and their standard potentials are compiled in Table 5.2, together with the temperature coefficient and the isothermal temperature coefficients which allow the calculation of the standard potential at different temperatures according to Eq. (2.19), p.14.

		1	2	3	4	5	6
	Reaction	E_o V	E_h V	E_{HgO} V	dE_o/dT mV/K	dE_o/dT_{iso} mV/K	d^2E_o/dT^2 $(\mu V/K)^2$
1	$NiOOH + H_2O + e^- \Rightarrow Ni(OH)_2 + OH^-$	0.500*)	1.32*)	0.395*)	**	--	--
2	$Cd(OH)_2 + 2 \cdot e^- \Rightarrow Cd + 2 \cdot OH^-$	-0.809	-0.019	-0.913	-0.143	-1.014	--
3	$O_2 + 2 \cdot H_2O + 4 \cdot e^- \Rightarrow 4 \cdot OH^-$	0.400	1.228	0.296	-0.809	-1.68	-6.7
4	$2 \cdot H_2O + 2 \cdot e^- \Rightarrow H_2 + 2 \cdot OH^-$	-0.828	0.0	-0.932	+0.037	-0.834	-7.272
5	$HgO + H_2O + 2 \cdot e^- \Rightarrow Hg + 2 \cdot OH^-$	0.098	0.926	--	-0.249	-1.120	-6.775

Table 5.2 Electrode reactions, electrode potentials and temperature coefficients for the nickel/cadmium battery. $(dE^o/dT)_{iso}$ and d^2E^o/dT^2 for calculations according to Eq. (2.19).
All temperature coefficients from (11)
* Uncertainties of these values are explained in Section 5.2.2
** between -0.5 and -0.64 according to (12).

The equilibrium potentials are based on the different reference electrodes:

E_o (Column 1) is referred to the standard hydrogen electrode (SHE) (Section 2.2.3.1),

E_h (Column 2) to the hydrogen electrode in the same solution (HESS),

E_{HgO} (Column 3) to the mercury/mercuric oxide electrode (Hg/HgO electrode, cf. Section 2.2.3.3).

The values are referred to an OH^- activity of 1 mole/dm^3, corresponding to the pH value pH = 14. According to Table 2.1 (p.17), this corresponds to the density of 1.06 g/cm^3 or 7.06 wt.% of KOH.

The potential of the hydrogen electrode in the same solution is then 14 x 0.059 = 0.826 V below SHE, while the potential of the Hg/HgO electrode is 0.104 V above SHE, as Fig. 2.4 (p.21) shows.

The Nernst equations, according to Eq. (2.13), p.12 for the reactions compiled in Table 5.2 are:

Positive electrode (nickel/nickel hydroxide electrode)

$$E_{Ni^{3+}/Ni^{2+}} = E^o_{Ni^{3+}/Ni^{2+}} + \frac{R \cdot T}{F} \cdot \ln \cdot \frac{a_{H_2O}}{a_{OH^-}} = E^o_{Ni^{3+}/Ni^{2+}} + \frac{R \cdot T}{F} \cdot \ln(a_{H_2O}) - \frac{R \cdot T}{F} \cdot \ln(a_{OH^-}) \quad (5.11)$$

This equation is only correct for the simplified Eq. (5.6). Actually a certain dependence on KOH concentration is observed (cf. Fig. 6 in ([3])).

Negative electrode (cadmium/cadmium hydroxide electrode)

$$E_{Cd/Cd(OH)_2} = E^o_{Cd/Cd(OH)_2} - \frac{R \cdot T}{F} \cdot \ln(a_{OH^-}) \quad (5.12)$$

Oxygen evolution

$$E_{O_2/O^{2-}} = E^o_{O_2/O^{2-}} + \frac{R \cdot T}{4 \cdot F} \cdot \ln(a^2_{H_2O}) - \frac{R \cdot T}{F} \cdot \ln(a_{OH^-}) \quad (5.13)$$

Hydrogen evolution:

$$E_{H^+/H_2} = E^o_{H^+/H_2} + \frac{R \cdot T}{2 \cdot F} \cdot \ln(a_{H_2O}) - \frac{R \cdot T}{F} \cdot \ln(a_{OH^-}) \quad (5.14)$$

Mercury/mercuric oxide reference electrode (Hg/HgO electrode:

$$E_{Hg/HgO} = E^o_{Hg/HgO} + \frac{R \cdot T}{2 \cdot F} \cdot \ln(a_{H_2O}) - \frac{R \cdot T}{F} \cdot \ln(a_{OH^-}) \quad (5.15)$$

Common to all these reactions is the same dependence on OH^- activity that amounts to

$$-\frac{R \cdot T}{F} \cdot \ln(a_{OH^-}) = -0.0592 \cdot \log(a_{OH^-}) \quad \text{V at 25 °C} \quad (5.16)$$

or -0.0592 V per pH·value. This means that the electrode potentials referred to the hydrogen reference electrode (E_h in Table 5.2) or the mercury/mercuric oxide reference electrode (E_{HgO}) are (nearly) independent of the KOH concentration, expressed by the pH value. For this reason, this potential scale is preferred with alkaline electrolyte.

5.1.3 The Reversible Heat Effect, Calorific Voltage

The reversible heat effect is given by Eq. (5.9)

$$\boxed{Q_{rev} = T \cdot \Delta S \approx -26 \text{ kJ}} \tag{5.17}$$

per multiple of the cell reaction. Comparison with the value for ΔG shows that the reversible heat effect is 10.5% of the converted energy when the nickel/cadmium battery is discharged or charged.

The negative sign means that additional heat is generated during discharge. When the nickel/cadmium battery is charged, the reversible heat effect appears as a cooling effect. (*Compared to lead-acid batteries, the reversible heat effect in nickel/cadmium batteries has the opposite sign, and is larger, namely 10.5 instead of 3.5% of the converted energy. Thus, it is more significant in nickel/cadmium batteries and is an additional heat source during discharging.*)

During the charging process, the reversible heat effect causes a remarkable cooling effect as long as heat-generating secondary reactions do not occur (cf. Fig. 5.13, p.201). In this connection, water decomposition can only be critical as a heat source, when the cell voltage considerably exceeds 1.48 V. As a result, vented nickel/cadmium batteries can be charged at quite high rates without suffering heat problems. But in the sealed version, the internal oxygen cycle can cause thermal problems.

On the other hand, attention has to be paid to the heat balance when the battery is discharged with high current. Then the reversible heat effect is generated quickly and has to be added to the Joule effect, which might also be considerable in these circumstances (cf. Table 5.7).

According to Eq. (2.42), Q_{rev} can be expressed as a (fictive) voltage (-26 kJ = -7.22 Wh; $2 \cdot F$ = 192970 Ah):

$$\boxed{\frac{Q_{rev}}{2 \cdot F} = -0.135 \text{ V}} \tag{5.18}$$

and the calorific voltage: in nickel/cadmium batteries is according to Eq. (2.43)

$$\boxed{E_{cal} = E^{o} + 0.14 \text{ V}} \tag{5.19}$$

It means that heat evaluation corresponds to a voltage 0.14 V higher than the equilibrium voltage (cf. Fig. 5.11).

Note: *The above calculated values for the reversible heat effect and the calorific voltage are strictly valid only for the standard conditions. But due to the uncertainties of the equilibrium voltage, dependence on the different parameters like temperature can be neglected.*

5.1.4 The Temperature Coefficient of the Equilibrium Voltage

The temperature coefficient of the equilibrium voltage according to Eq. (2.16), p.13, is closely related to the reversible heat effect. Division of $T \cdot \Delta S$ = -26 kJ by 298.2 K results in:

$$\boxed{\Delta S = -87\ \ J{\cdot}K^{-1}} \tag{5.20}$$

When this value is inserted in Eq. (2.16), p.13, (87 J = 0.024 Wh) under consideration that in the cell reaction (Eq. (5.5)) n = 2 and so nF = 192970 As (=53.6 Ah per multiple of the cell reaction), it results in

$$\boxed{\frac{dE^{o}}{dT} \approx -0.45 \quad mV/K} \tag{5.21}$$

This decrease of the equilibrium voltage of about 0.5 millivolts per degree can usually be neglected in battery practice.

5.1.5 Weight-Dependent Quantities

The weight of the required substances can be evaluated from the cell reaction Eq. (5.5), p.177. However, one has to keep in mind that the composition of the active material in the charged nickel electrode is not known precisely, and thus the resulting values can only be approximate.

The amount of electricity converted per multiple of this equation is

$$\boxed{2{\cdot}F = 192974\ \text{As/cell reaction} = 53.60\ \text{Ah/cell reaction}} \tag{5.22}$$

The molar weights for the substances employed in Eq. (5.5) are compiled in Table 5.3 together with the utilisation figures calculated with the aid of Eq. (5.22). These figures represent the maximum values, because they are true only if the considered substance is converted completely (100% utilisation of the active material).

The maximum energy that can be stored per unit of weight can be calculated from the molar weights. Since two moles of NiOOH and H_2O and one mole of Cd are required for Eq. (5.5), the resulting weight totals to 332.4 g. The energy converted per multiple of the cell reaction is described by the product E^o (≈ 1.33 V) times 53.60 = 71.30 Wh.

Consequently, the number of watt-hours that can be stored per kg of active material in the nickel/cadmium battery is:

$$\boxed{\frac{71.30}{0.3324} = 214.5 \quad Wh/kg \;=\; 97.3 \quad Wh/lb} \tag{5.23}$$

Because the equilibrium voltage is not defined exactly, this value for the maximum energy density can vary between 235 Wh/kg (for E^o = 1.4 V) and 209 Wh/kg (for E^o = 1.3 V). This number represents the theoretical maximum. But, as in lead-acid batteries, it can never be realised, because it is reduced by a number of factors similar to those discussed for the lead-acid battery in connection with Fig. 4.3, p.107.

The utilisation values in Wh/kg of the substances employed in Eq. (5.5), can also be evaluated since 71.3 Wh are converted per multiple of the cell reaction. They correspond to those in the last two columns of Table 5.3 times 1.33.

Substance	Molar weight g/mole	Required quantities based on Eq. (5.5) (2F = 53.6 Ah) g/Ah	Utilisation values Ah/kg	Ah/lb.
NiOOH	91.72	3.421	292.3	132.6
$Ni(OH)_2$	92.71	3.458	289.2	131.2
Cd	112.4	2.731	366.1	166.0
$Cd(OH)_2$	146.4	2.395	417.5	189.3
H_2O	18.02	0.6722	1487	674.3

Table 5.3 Molar weights of the components in Eq. (5.5), required quantities and (maximum) utilisation values derived from thermodynamic data.

Much dead weight in actual batteries results from only partial utilisation of the active material because its conductivity is reduced with advancing discharge. The electric conductivity, especially that of the reduced positive material, the $Ni(OH)_2$, is poor. In batteries with thicker layers of active material, like pocket plates or button cells (cf. Section 11.2), nickel or graphite have to be added as conducting aids which contribute additional weight.

Table 5.4 shows corresponding data of various types of nickel/cadmium batteries and utilisation values of the active material reached in practice.

Type of battery, nominal capacity, and discharge rate (always single cells) Electrode type positive/negative	Energy content: Specific energy Wh/kg	Energy content: Specific energy Ah/kg	Energy content: Energy density Wh/dm^3	Utilisation of active material: NiOOH Ah/kg	Utilisation of active material: Cd Ah/kg
Stationary, 315 Ah, 10 hours pocket/pocket	21	19	36	230	220
Aircraft batt., 40 Ah, 5 hours sintered/sintered	31	28	72	280	250
Sealed round cell, 4 Ah, 5 hours sintered/pressed *	34	31	90	280	220
Sealed button cell, 0.06 Ah, 10 hours pressed/pressed	21	19	63	230	220

Table 5.4 Typical values of energy content per cell weight and volume, as well as utilisation of the active material in nickel/cadmium batteries. The values depend largely on cell size and the specific cell design. The stationary and aircraft batteries are flooded (filled with liquid electrolyte), while the electrolyte in the sealed types is immobilised. * Electrodeposited (cf. 11.2.8). Similar figures are compiled for lead-acid batteries in Table 4.4, p.108.

The first example in Table 5.4 is a stationary battery of flooded design with pocket plates where the active material is encapsulated within perforated nickel sheets. This encasement of the active material provides strong support, but reduces the utilisation considerably. The low energy content per volume (Wh/dm^3) mainly is due to the large space above the plates, for reserve electrolyte and to prevent spilling.

The aircraft battery is also a flooded type, but designed very compactly and equipped with sintered plates. The close contact of the active material to the large surface of the current-conducting sintered support results in quite high utilisation factors. But the weight of the support is fairly high and reduces the Wh/kg figure.

The sealed round cell (cylindrical cell) is also equipped with sintered electrodes. The utilisation factor of the positive active material is somewhat less than that of the aircraft battery because of the immobilised electrolyte. The low utilisation factor of the negative active material is due to the oversized negative electrode for improved oxygen reduction and as 'charge reserve' (cf. Sections 7.5.3.2 and 8.1.1).

The rather poor values for the button cell in Table 5.4 are partly caused by the small size of this cell type, but also by the electrode type (pressed-powder electrodes).

Note: *The values in Table 5.4 indicate that the energy content of a battery is largely determined by the field of application the battery is designed for. The sole comparison of specific energy and energy density data, does not consider the required performance characteristics and may provide only limited information.*

Comparison between Tables 5.4 and 4.4 shows that the values for energy content in Wh/kg as well as energy content in Wh/dm^3 are of the same order of magnitude for nickel/cadmium batteries as those reported for lead-acid batteries. The utilisation of the active material is better for nickel/cadmium batteries in nearly all cases. But this advantage is compensated for by a larger share of passive parts to support the active material or conduct the current.

5.1.6 Heat Capacity (Specific Heat Content)

The temperature response of the battery to heat generation or heat loss is determined by the heat capacity of the battery, which can be calculated according to Eq. (2.81), p.36.

In Table 5.5, values are compiled for the substances included in the cell reaction, Eq. (5.5). The difference between the heat capacities of the solid substances and those of the electrolyte are similar to those of the lead-acid battery (cf. Table 4.5, p.110). But in most nickel/cadmium batteries, especially in sealed ones, the weight share of the electrolyte does not play such an important role because of the smaller electrolyte content. Polyamide is mainly used as separator material; to some extent also polypropylene or mixtures of both. Because of their rather high heat capacity, these substances can considerably influence the heat capacity of a battery, especially in small ones.

Substance	Molar weight g/mole	Heat capacity cal/(K·mole)	Heat capacity J/(K·Kg)
Ni	58.7	6.28	107
Fe	55.85	6.31	113
Cd	112.14	6.21	230
NiOOH	91.71	19 *	880 *
KOH, 1.20 g/cm^3			3300 **
KOH, 1.28 g/cm^3			3100 **
Polypropylene			2100
Polystyrene			1200
Polyamide			1900

Table 5.5 Heat capacity of the different components in nickel/cadmium batteries.
* estimated values ([13]);
** from ([14]);
remaining values from ([15]).

Table 5.6 gives the heat capacities of the batteries presented as examples in Table 5.4.

Type of Battery	Heat capacity kJ/(K·kg)	Heat capacity Wh/K·100 Ah	Percentage of electrolyte
Stationary, 315 Ah, pocket Plates	1.3	1.9	78%
Aircraft batt., 40 Ah	0.98	0.97	48%
Sealed round cell, 4 Ah	0.81	0.72	47%
Button cell, 0.06 Ah	0.52	0.72	41%

Table 5.6 Typical values of the heat capacity and share assigned to the electrolyte (1 J = 1 Ws; 1 kJ = 0.2778 Wh).

Table 5.6 shows that the heat capacities of the nickel/cadmium batteries are in the same order of magnitude as those for lead-acid batteries which have been compiled in Table 4.6, p.110. The small share of the electrolyte in the aircraft battery is due to the very limited amount of electrolyte in this battery type. The same situation applies for the sealed round cell. The heat capacity of the small button cell is mainly determined by the metallic housing.

5.2 Kinetic Effects

In nickel/cadmium batteries, positive and negative electrodes are based on different chemical elements, and the charge/discharge mechanism also is different in both electrodes.

5.2.1 The Reactions at the Negative Electrode

For the negative electrode, the situation is similar to that of the negative electrode of the lead-acid battery, because the reaction includes the dissolved state, as illustrated in Fig. 4.5, p.112, for the lead-acid system. Fig. 5.2 shows the corresponding reaction mechanism of the cadmium electrode in a simplified manner. Actually, slightly diverging structures of cadmium hydroxide can be formed, depending on the environmental conditions, and to some extent solid state reactions seem to be involved in the oxidation/reduction process (9).

The metallic cadmium (Cd) represents the charged condition, divalent Cd^{2+} ions represent the discharged material. When the cadmium electrode is discharged, cadmium metal is dissolved to cadmium ions (Cd^{2+}), but these are immediately precipitated as cadmium hydroxide ($Cd(OH)_2$) on account of its low solubility in alkaline solutions of about 10^{-5} mole/dm^3 (1.5 mg $Cd(OH)_2$ per dm^3). This solubility is about the same as that of lead sulphate in sulphuric acid (cf. Section 4.6.).

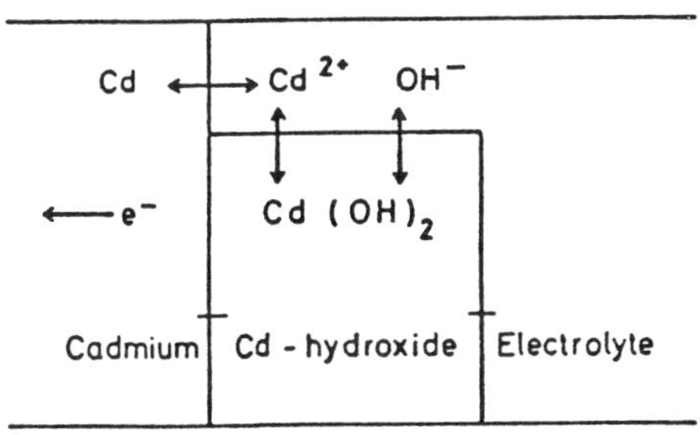

Fig. 5.2 Reaction mechanism of the cadmium/cadmium hydroxide electrode. Charge and discharge occur by dissolution of the reacting substances.

5.2.2 The Reactions at the Positive Electrode

The nickel ions are charged and discharged in the crystal lattice; they are not dissolved. To preserve electrical neutrality, a corresponding number of H^+ ions (protons) must migrate into the crystal lattice during the reduction (the discharge) of the Ni^{3+} or Ni^{4+} ions. When the nickel electrode is charged (oxidised), these protons have to leave the crystal lattice. This exchange of protons is connected with the formation of a reduced or oxidised crystallographic phase (16).

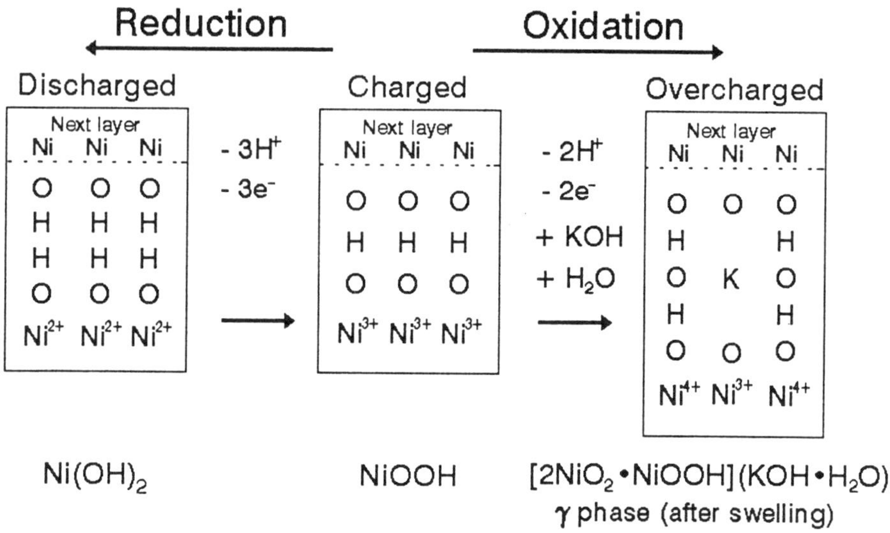

Fig. 5.3 Charge and discharge of the nickel hydroxide electrode with simultaneous release and absorption of protons (H^+ ions) and incorporation of a small amount of potassium ions (K^+).

The mechanism shown in Fig. 5.3 shows the electrochemical reactions of oxidation and reduction of nickel hydroxide in an idealised form. The various compounds involved in this scheme have been prepared by electrochemical and chemical methods and were identified by X-ray analysis. The active material in actual battery electrodes, however, usually shows poor crystallinity and varies in the stoichiometric compositions. These materials are characterised by their content of water and additives like potassium, and their more or less disordered crystalline structure.

Oxidation and reduction are solid state reactions, that are not performed completely in battery electrodes. Thus these reactions do not form definite chemical compounds, but gradually increased or decreased oxidation states of the nickel ions (Ni^{2+}, Ni^{3+} and Ni^{4+}). Deviations of the state of oxidation are the consequence, depending on the charging period and on the electrode potential ([17]). This reaction mechanism is absolutely different from the solution mechanism described for both electrodes of the lead-acid battery (Fig. 4.5, p.112) or the cadmium electrode (Fig. 5.2). In these electrodes, the oxidised and the reduced state are formed by definite chemical compounds; (deviations are very small, cf. the remark below Eq. (4.4), p.99). These states only can exist; there is nothing between them. In the nickel hydroxide electrode, varying oxidation states between the lowest and the highest value are possible, since several crystallographic phases are formed and also mixtures between them are possible. In view of the thermodynamic data, these peculiarities cause uncertainties in the derived data that have to be taken into account.

5.2.2.1 Peculiarities of the Nickel/Nickel Hydroxide Electrode

As already mentioned, Eqs. (5.4), (5.5), and (5.6) represent only approximations, because neither the charged state nor the discharged state is precisely described by these equations. The oxidised state, (the charged material) not only consists of Ni^{3+} ions, but may also contain Ni^{4+} ions, as indicated in Fig. 5.3. This means that the state of oxidation is not unambiguously determined, but may exceed the value corresponding to the trivalent nickel, because at increased electrode potential further oxidation of the nickel ion may occur.

As a consequence, 2·NiOOH in Eq. (5.5), should more precisely be written as

$$\boxed{u \cdot NiO_2 \; v \cdot NiOOH \; w \cdot H_2O} \quad (5.24)$$

u, v and w are factors describing the share of the three components.

The actual structures in battery electrodes are mixtures of non-stoichiometric compounds. Two different couples of charged/discharged species have been specified ([16]):

The β/β couple

β-phase of the reduced nickel hydroxide, $\beta\text{-}Ni(OH)_2$
⇔ β-phase of the oxidised nickel hydroxide, β-NiOOH (5.25)

The α/γ couple

α-phase of the reduced nickel hydroxide, $\alpha\text{-}Ni(OH)_2$
⇔ γ-phase of the oxidised nickel hydroxide, γ-NiOOH (5.26)

Both couples are characterised by different maximum oxidation state and different contents of KOH. The α/γ-couple absorbs about four times as much KOH than the β/β-couple ([3]). Both couples are not independent of each other. The main reaction in the nickel hydroxide electrode involves the charge and discharge of the couple β-NiOOH / $\beta\text{-}Ni(OH)_2$. Its oxidised form, the β-NiOOH, can be converted into γ-NiOOH by overcharging. Discharge of the γ-NiOOH forms $\alpha\text{-}Ni(OH)_2$ which gradually restructures back to the $\beta\text{-}Ni(OH)_2$. This leads to the relations illustrated in Fig. 5.4.

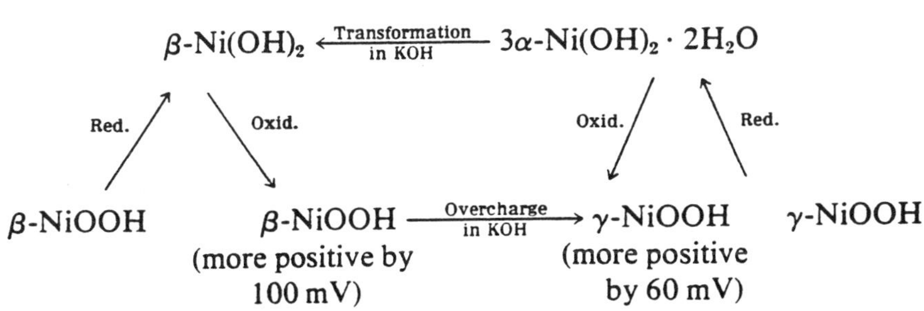

Fig. 5.4 The different modifications of nickel hydroxide and their relations according to ([18]). On the left, the ß/ß couple according to Eq. (5.25). On the right, the α/γ couple according to Eq. (5.26).

In actual electrodes, the situation is additionally complicated, because the phase diagram in Fig. 5.4 represents only the crystallised modifications of a whole spectrum of intermediate structures (precursor structures). The active material of an actual electrode, however, is characterised by poor crystallisation and a mixture of phases. Furthermore, ageing effects of transformation via the solution are reported ([19]).

Bode and co-workers ([20]) determined for the maximum oxidised state of the γ-nickel hydroxide the approximate composition

$$[4 \cdot NiO_2\ 2 \cdot NiOOH] \cdot (2 \cdot K\ 2 \cdot OH\ 2 \cdot H_2O) \tag{5.27}$$

This formula corresponds to the stoichiometric coefficient between O and Ni (the oxidation state) 1.83, compared to 1.5 for NiOOH and 2.0 for NiO_2. This formulation reflects the crystallographic situation indicated in Fig. 5.3. Experimentally, substances can only be realised with slightly diverging compositions, which correspond fairly closely to Eq. (5.27). Barnard and co-workers ([21]) confirmed these results and found the general formulation $M_{0.32} \cdot NiO_2 \cdot 0.7H_2O$ where M stands for Li^+, Na^+, K^+, and Rb^+.

Evidence of quadrivalent nickel ions was proved by spectroscopic experiments that indicated a nickel valence of 3.67 ([22]), which has also been stated for cobalt-containing nickel hydroxide ([23]). This value corresponds with the already mentioned stoichiometric coefficient at the maximum oxidation state of 1.83, which is about 60% above 1.5, that of the trivalent nickel hydroxide. This seems to be the maximum oxidation state that can be reached in aqueous electrolyte, because in this case the formula of γ-nickel hydroxide can be written in such a way that all the protons (H^+) remaining in the hydroxide form hydrogen bridges. Further oxidation would require much more energy. When the charging process is continued in aqueous solution, the system switches over to water decomposition without further increase of the oxidation state ([24]).

A remarkable fact in Eq. (5.27) is the small amount of potassium (K), required to stabilise this oxidation state. Improved oxidation, synonymous with increased charge acceptance, is observed when lithium ions are incorporated in the nickel lattice ([25]). The combination of co-deposited cobalt with lithium-containing electrolyte seems to enhance the effect, especially at elevated temperature ([26]). Cobalt addition above 11.1% forms a super-lattice within the hydroxide that obviously facilitates proton conduction ([27]). A characteristic weakness of the nickel hydroxide electrode, namely reduced charge acceptance at elevated temperature (cf. Fig. 3.8), can consequently be eliminated, at least partly, by cobalt addition.

The structures of the nickel hydroxide are dependent not only on the history of the electrode material but also on additives like cobalt ([28]), cadmium and zinc ([29]). Cobalt seems to promote a more complete oxidation of the $Ni(OH)_2$ ([30]). Furthermore, cobalt stabilises the nickel hydroxide and reduces self-discharge at elevated temperature, while the incorporation of cadmium increases the oxygen overvoltage. The latter is also achieved by addition of calcium compounds, especially CaF_2 ([31]). The addition of cadmium and zinc reduces the swelling of the active material caused by the volume increase connected with the introduction of protons (H^+ ions) and other components of the electrolyte, like H_2O, OH^- and K^+, between the layers of the crystal structure. This applies especially for the highly oxidisable γ-NiOOH.

One consequence of the complex composition of the active material in the positive electrode is that an unambiguous equilibrium is not established between the electrode potential and the oxidation state. Hysteresis is always observed when the nickel electrode is charged and discharged. The oxidation process (the charging process) cannot be reversed at the same electrode potential, but a considerable voltage drop indicates that the current reversal causes changes in the energy content of the active material. Fig. 5.5 shows stationary values for the oxidation and the subsequent reduction of the two couples shown in Fig. 5.4, the β-$Ni(OH)_2$/β-NiOOH and the α-$Ni(OH)_2$/γ-NiOOH couple. Each of the measuring points was established by polarising the electrode at the determined potential until the current attained a constant value. The filled circles in Fig. 5.5 mark the values observed during the oxidation (charging) of the electrode; the filled triangles represent the values determined during reduction (discharging).

The horizontal axis in Fig. 5.5 shows the potential of the electrode referred to the hydrogen electrode (in the alkaline electrolyte). The vertical axis shows the state of oxidation.

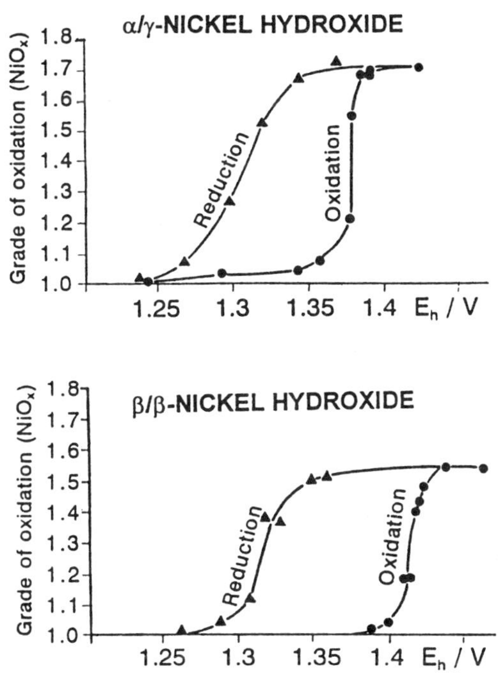

Fig. 5.5 Potentiostatic oxidation of $Ni(OH)_2$ and subsequent reduction of the formed NiOOH ([32]). In the top part: the α-$Ni(OH)_2$/γ-NiOOH couple. In the bottom part: the β-$Ni(OH)_2$/β-NiOOH couple.

Both charge/discharge couples show pronounced hysteresis, and their oxidation/reduction potentials also diverge.

5.2.2.2 Residual Capacity, Second Discharge Plateau

The terms 'residual capacity' or 'second discharge plateau' describe a phenomenon, observed with sintered nickel hydroxide electrodes (cf. the 'memory effect' in Section 11.1.3). Mainly after prolonged charging periods, but also after continuous cycling, an appreciable amount of charged active material (10 to 50%) is not discharged at high rates or is discharged at a voltage of about 300 mV below the normal voltage (the second discharge plateau). Fig. 5.6 shows an example.

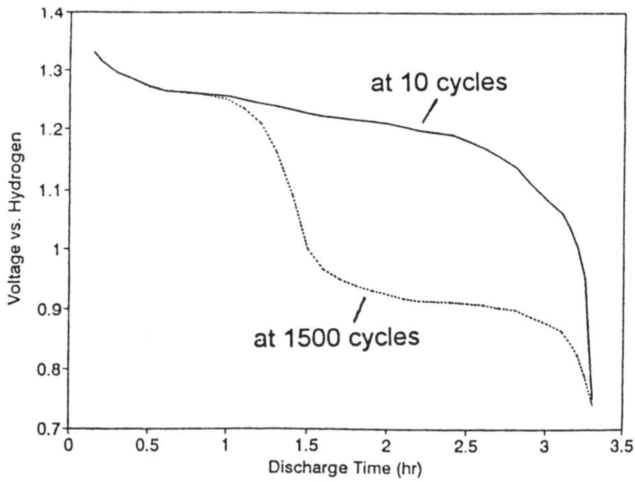

Fig. 5.6 Second discharge plateau of a sintered nickel electrode (Fig. 2 in ([33])).

The potential of the lower plateau is highly dependent on discharge rate, which indicates that it is caused by an additional resistance (cf. Fig. 3 in ([34]). It might be caused by semiconductor features but also by loss of electronic contact between the current collectors and the active material. A model based on semiconductor attributes and the parameters 'cross sectional area' and 'conducting length' predicts the general feature of the secondary discharge step (cf. ([5]), p. 135 f.), and indicates that the interface area between the substrate and the active material is an important parameter. On the other hand, experiments with thin film electrodes indicate that the formation of a low energy form of β-NiOOH can also cause a 'secondary discharge step' ([35]).

5.3 Self-discharge

The factors that cause self-discharge of nickel/cadmium batteries are:

1. The **decomposition** of the **charged nickel hydroxide** by separation of oxygen according to:

$$Ni_6O_6(OH)_6 + H_2O \Rightarrow Ni_6O_4(OH)_8 + 1/2 \cdot O_2 \quad (5.28)$$

since: $6 \cdot NiOOH \approx Ni_6O_6(OH)_6$

or

$$\boxed{2 \cdot NiOOH + H_2O \Rightarrow 2 \cdot Ni(OH)_2 + \tfrac{1}{2}\, O_2} \qquad (5.29)$$

This disintegration process decreases when the oxide is reduced, and comes to a standstill when a stable state of oxidation is reached. Theoretically a final oxidation state of 1.33 (Ni_3O_4) can be expected instead of 1.5 (NiOOH), which means that, at most, about half the charge would be lost. In practice, however, only 10 to 20% of capacity loss are caused by oxide decomposition. Then, the process is slowed down so far that more or less it can be neglected.

It is markedly increased at elevated temperature. At 45 °C, a steep decrease by nearly 20% is observed within two days. But even at that temperature further self-discharge does not proceed to more than 22% during 30 days, and does not depend on oxygen pressure ([36]).

In sealed nickel/cadmium batteries, oxygen can easily reach the negative electrode where it is consumed according to Eq. (5.32). Thus, oxygen separation at the positive electrode also causes self-discharge of the cadmium electrode.

2. The **'shuttle' of a redox system**, also discharges both electrodes. It is usually caused by nitrate ions (NO_3^-) ('**nitrate shuttle**)' that are dissolved in the electrolyte. When they reach the negative electrode by diffusion, they are reduced to NH_4^+:

$$\boxed{NO_3^- + 10 \cdot H^+ + 8 \cdot e^- \Rightarrow NH_4^+ + 3 \cdot H_2O} \qquad (5.30)$$

The NH_4^+ ions, however, do not stay at the negative electrode, but migrate back into the electrolyte and are oxidised to NO_2^- (nitrite) as soon as they reach the positive electrode again (the NO_3^- oxidation potential is not reached in open-circuit conditions). The cycle is repeated when the NO_2^- ions reach the negative electrode and again are reduced to NH_4^+ ions. Thus, the alternating oxidation to NO_2^- and reduction to NH_4^+ ions forms a 'shuttle' for the transport of electric charges between the two electrodes, and causes discharge of both electrodes.

Current transport by nitrate shuttle is effective, since six electrons are exchanged in each round when N^{3+} is reduced to N^{3-}:

$$\boxed{NO_2^- + 6 \cdot H_2O + 6 \cdot e^- \Rightarrow NH_4^+ + 8 \cdot OH^-} \qquad (5.31)$$

Furthermore, the nitrate shuttle runs continuously and causes permanent self-discharge of both electrodes.

Self-discharge caused by the nitrate shuttle is mainly observed in batteries with sintered electrodes, because nickel- and cadmium nitrates are used in the manufacturing process to precipitate the active material into the sintered substrate (cf. Section 11.2.4.1). Another source of this shuttle is degradation of the separator according to Eq. (5.34), p. 209.

3. The occurrence of **short-circuits** also affects both electrodes. It can be caused by penetration of the separator by small particles washed out of the active material, or by the growth of crystal bridges between the electrodes.
4. Self-discharge of the positive electrode can be caused by hydrogen that has been evolved during overcharging of nickel/cadmium batteries at too-high rates. But the effect is small, since the available amount of hydrogen is limited.

   In nickel/hydrogen and nickel/metal hydride batteries, however, hydrogen always is available within the cell and can cause considerable self-discharge of the positive electrode, especially in nickel/hydrogen batteries because of the high pressure (cf. Section 6.1.1.1).
5. Self-discharge of the negative electrode occurs when oxygen penetrates into a sealed cell through a leakage (or is released at the positive electrode according to Item 1). On account of the immobilised electrolyte, oxygen has access to the negative electrode surface and oxidises the cadmium according to

$$\boxed{Cd + 1/2 \cdot O_2 + H_2O \Rightarrow Cd(OH)_2} \quad (5.32)$$

Hydrogen evolution at the negative electrode, which mainly causes self-discharge in lead-acid batteries, does not occur at open-circuit voltage in the nickel/cadmium battery on account of the thermodynamic situation shown in Fig. 5.1.

5.3.1 Quantities of Self-Discharge

Self-discharge in nickel/cadmium batteries depends largely on the design, because different reactions cause self-discharge with the various electrode designs. In general, large surface areas of the active material favour self-discharge, as do several other parameters (e.g. narrow electrode spacing) that reduce the internal resistance. So, heavy-duty cells are more susceptible to self-discharge (cf., e.g., Fig. 26.11 in ([1])).

In Fig. 5.7 typical self-discharge curves are shown for vented types with pocket plates. Fig. 5.8 shows curves for button cells (with pressed powder electrodes and comparatively high internal resistance) and cylindrical cells with sintered electrodes. Both Figures illustrate the enormous influence of battery design on the discharge rate.

The vented battery loses about 25% of its capacity when stored at room temperature for one year. Button cells lose about 60% of their capacity during one year under such conditions, while the capacity loss of cylindrical cells in similar conditions is about 40% after only 3 months of storage time. This means there is a factor of more than 2 between the vented battery and the button cell. In the battery with sintered electrodes the large surface area of the plates and the nitrate shuttle are additional factors that increase the self-discharge rate.

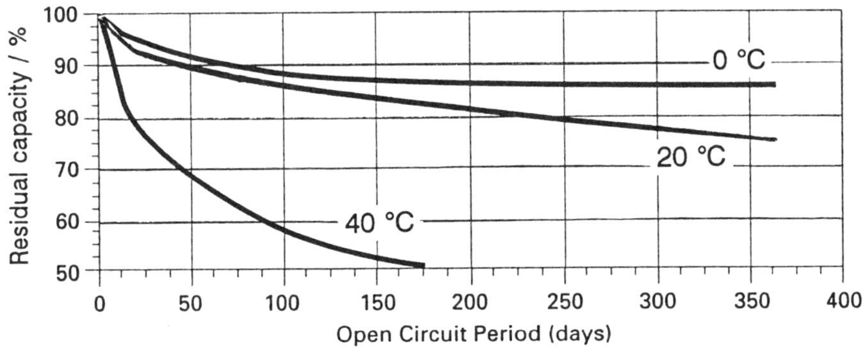

Fig. 5.7 Charge retention of a vented nickel/cadmium battery with pocket plates at different temperatures. (VANTAGE Battery of ALCAD Ltd ([37]).)

Fig. 5.7 also shows the strong increase of this self-discharge mechanism at 40 °C. According to the above Section, the initial steep part of the curves is mainly caused by decomposition of the nickel hydroxide.

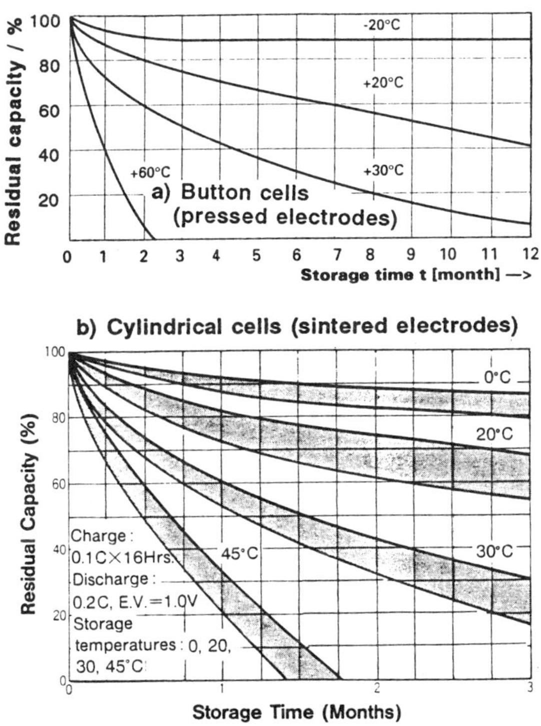

Fig. 5.8 Capacity loss during storage of sealed nickel/cadmium batteries: a) Button cells, VARTA DKZ ([38]); b) Cylindrical cells, SANYO N-Type, Fig. 4.1 in ([39]).

The strong influence of temperature on the self-discharge rate in both examples of Fig. 5.8 is mainly caused by the usual acceleration of electrochemical reactions (cf. Section 2.3.6), but may be increased by separator hydrolysis.

The self-discharge rate can also be influenced by additives. Cobalt addition to the nickel hydroxide, for example, reduces self-discharge at elevated temperatures.

5.4 Heat Effects

Heat effects in nickel/cadmium batteries are also determined by the reversible heat effect and the Joule effect (cf. Sections 2.2.4 and 2.4.1, respectively). Compared to lead-acid batteries, the reversible heat effect, according to Eq. (5.17), p.182, is larger and has the opposite sign i.e. it acts as a cooling effect during charging and contributes additional heat during discharge. As a consequence, vented nickel/cadmium batteries are more in danger of being overheated during discharge than during charging. *This is different for sealed nickel/cadmium batteries where the internal oxygen cycle is a most effective heat source when the battery is overcharged (cf. Figures 5.9 and 5.12).*

Fig. 5.9 shows heat generation in a vented nickel/cadmium battery when charged and discharged with a constant current (5-hour rate) and the charging voltage is limited to 1.65 V/cell. The calculation is based on the equilibrium voltage $E^0 = 1.3$ V (Eq. (5.10)) and the calorific voltage $E_{cal} = 1.44$ V (Eq. (5.19)). Due to the uncertain thermodynamic data, these calculations are only rough approximations, but correspond with practical experience.

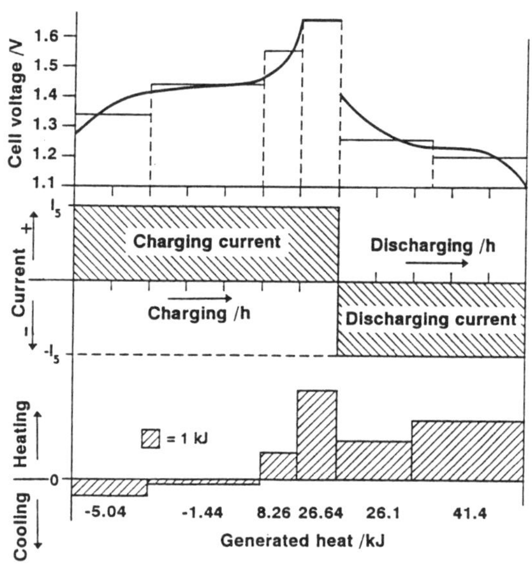

Fig. 5.9 Heat generation during charge and discharge of a vented nickel/cadmium battery. Charging with constant current I_5 (5-hour rate) until 1.65 V/cell is reached. Discharge also with I_5. In the top part, sections are shown that were used to calculate the average heat generation, shown in the bottom part. (Calculation based on $E^0 = 1.3$ V; $E_{cal} = 1.44$ V.) VARTA TS-type, values referred to 100 Ah of nominal capacity.

Fig. 5.9 shows that during the initial two sections of the charging period, slight cooling is observed on account of the reversible heat effect that consumes heat at a constant rate proportional to the current (cf. Eq. (2.42), p.22). With increasing cell voltage, the Joule effect is increased, and when the charging voltage exceeds 1.48 V/cell, water decomposition contributes an increasing amount of heat (cf. Section 2.6.1.4). Thus, during the final sections of the charging period, a growing amount of heat is generated.

In total 67.5 kJ were generated during discharging, while heat generation during charging only amounted to 28.4 kJ. This is mainly caused by the reversible heat effect that generates additional heat during discharge, while it compensates for heat generation during charging.

5.4.1 Heat Generation During Discharge

Fig. 5.10 shows discharge curves for a sealed nickel/cadmium battery with inserted curves for equilibrium and calorific voltage, E^0 and E_{cal}, respectively. Heat effects and the resulting total heat generation are derived from this Figure by graphical integration, and are listed in Table 5.7. On account of the less defined equilibrium voltage, these values are less accurate than those of lead-acid batteries in Section 4.5.

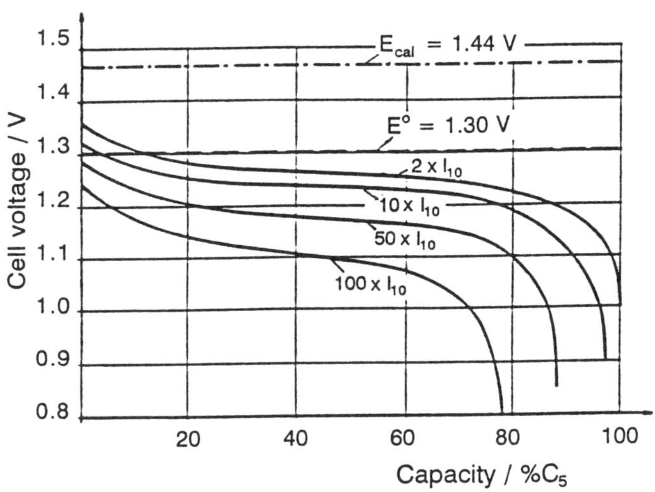

Fig. 5.10 Discharge curves for a sealed nickel/cadmium battery (cylindrical cell). Current ratings in multiples of the 10-hour discharge rate; end-of-discharge voltage between 1.0 and 0.8 V (cf. Eq. (3.4) in Section 3.2.1).
Dashed line: equilibrium voltage $E^0 = 1.3$ V (Eq. (5.10), p.179).
Dash-point line: calorific voltage $E_{cal} = E^0 + 0.14$ V (Eq. (5.19), p.182).
VARTA, RSP type, nominal capacity (5-hour rate) 1 Ah.

Fig. 5.10 shows clearly the excellent high-rate capability of nickel/cadmium batteries.

Even at a discharge current $100 \cdot I_{10}$, the discharge voltage is fairly constant and nearly 80% of the nominal capacity (5-hour rate) is available. The Figure shows furthermore, that E_{cal} is considerably higher than the equilibrium voltage E^o and always above the discharge curve. Thus the reversible heat effect during discharge always contributes a corresponding amount of additional heat.

1	Discharge rate Discharge period	$2 \cdot I_{10}$ 5 h	$10 \cdot I_{10}$ 0.97 h	$50 \cdot I_{10}$ 10.6 min	$100 \cdot I_{10}$ 4.7 min
2	Discharge current / A	2	1	5	10
3	Discharged energy / Wh	1.25	1.19	1.04	0.869
4	Q_{Joule} / Wh = % of discharged energy	-0.054 4.3%	-0.075 6.3%	-0.11 11%	-0.15 17%
5	Q_{rev} / Wh = % of discharged energy	-0.14 11.1%	-0.13 11.3%	-0.12 11.8%	-0.1 12.5%
6	Q_{total} / Wh = % of discharged energy	-0.19 15%	-0.21 18%	-0.23 23%	-0.26 30%
7	$Q_{total}/\Delta t$ / W	-0.039	-0.22	-1.33	-3.33

Table 5.7 Discharge parameters and values for Q_{Joule}, Q_{rev} and Q_{total} taken from Fig.5.10 by graphical integration. $Q_{total}/\Delta t$: average value for the whole discharge duration Δt. Sealed nickel/cadmium battery, cylindrical design, nominal capacity 1 Ah (5-hour rate).

The values in Table 5.7 show that the reversible heat effect represents the main heat source at low discharge rates. Q_{Joule} contributes less than half the heat generated in total (Q_{total}) for discharge rates up to 10 min (approximately $50 \cdot I_{10}$). The slight increase of the percentage of Q_{rev} compared to the discharged amount of energy (Line 5 in Table 5.7) at higher discharge rates is caused by the increasing voltage drop that reduces the Wh output compared to Ah output (Q_{rev} is strictly linked to the drawn amount of Ah).

Fig. 5.11 illustrates the situation. The nickel/cadmium battery produces considerably more heat at low or moderate discharge rates than a comparable lead-acid battery, for example the one that was the basis of the data shown in Fig. 4.12, p.124. For the nickel/cadmium battery, Line 6 in Table 5.7 shows that, even at the 5-hour discharge rate, the energy loss due to heat production is 15%, while the lead-acid battery under comparable discharge conditions generates almost no heat. This difference in heat generation diminishes with increasing discharge rates, because the Joule effect gains in importance, and the Joule effect is smaller in nickel/cadmium batteries on account of their lower internal resistance. But, in general, during discharge the heat generation of lead-acid batteries is smaller than that of nickel/cadmium batteries.

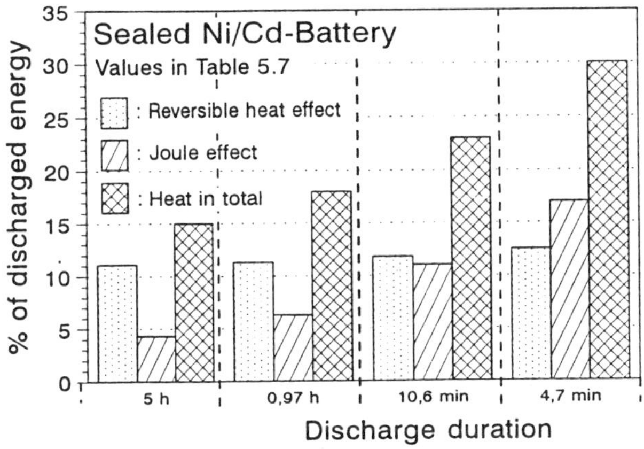

Fig. 5.11 Heat generation in the sealed nickel/cadmium battery, used as an example in Fig. 5.10, referred to the amount of discharged energy. (Lines 4, 5 and 6 in Table 5.7.) *Compared to Table 5.7, the signs of Q_{rev}, Q_{joule} and Q_{total} are reversed, to correspond with the usual practice that in heat calculations generated heat is positive which is expressed by $Q_{gen} = -Q_{total}$ in Equations 2.82 and 2.83, p.36.*

5.4.2 Heat Generation During Charging

When the battery is charged or overcharged, the secondary reactions gain importance: water decomposition in the vented design, and the internal oxygen cycle in sealed batteries. They consume the share of current that corresponds to overcharging. Fig. 5.12 illustrates heat generation by the various reactions when the battery is charged or overcharged at a constant voltage of 1.45 V/cell, the usual voltage for float charging of vented nickel/cadmium batteries, cf. Section 13.2.2.2.

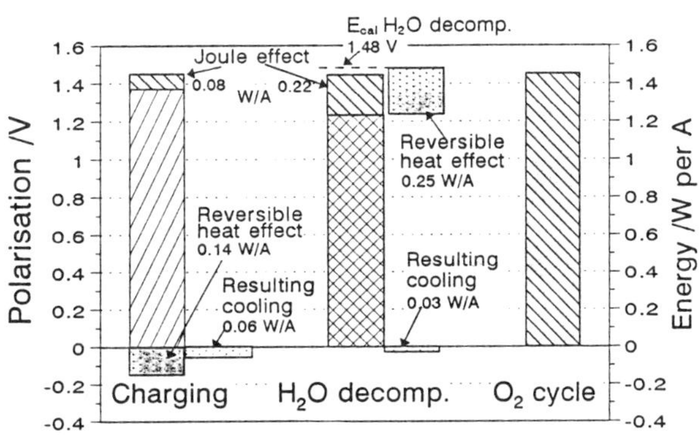

Fig. 5.12 Heat generation in nickel/cadmium batteries during charging or overcharging at a constant voltage of 1.45 V/cell. For the charging reaction 1.37 V/cell is assumed as 'equilibrium voltage' according to charging of the α/γ couple in Fig. 5.5.

The scale on the left in Fig. 5.12 gives the polarisation of the electrode. According to Eq. (2.73), p.34, $(E - E°)\cdot i = Q$, polarisation can be expressed as energy per current (W/A), as indicated by the scale on the right.

For the charging reaction an 'equilibrium voltage' of 1.37 V is assumed according to the 'charging' voltage of the α/γ couple in Fig. 5.5. For the reversible heat effect the equilibirum voltage of 1.31V is assumed. This is a discrepancy and illustrates the limited reversibility of the nickel hyrdoxide electrode. As a consequence the following calculation can only give rough approximations. The left column in the Figure shows that the charging reaction at this voltage is accompanied by small cooling effects. The Joule effect amounts to 0.08V (1.45V - 1.37V), while the reversible heat effect corresponds to 0.14V pr 0.14 W/A (1.45V - 1.31V). Thus a small cooling effect of 0.06 W/A results, which is uncertain, but shows that the heat effects are quite small.

Water decomposition (the middle column of Fig. 5.12) has the equilibrium voltage of 1.23 V/cell, so the Joule effect corresponds to 1.45 - 1.23 = 0.22 W/A. The calorific voltage E_{cal} is 1.48 V/cell (Section 2.6.1.3), thus the reversible heat effect amounts to 0.25 W/A. This means that the escaping gases hydrogen and oxygen take more energy away from the cell than corresponds to 1.45 V times current. So a small cooling effect of 0.03 W/A results. Heating by water decomposition only occurs when the cell voltage is higher than 1.48 V and the Joule effect exceeds the reversible heat effect. This corresponds also to the situation shown in Fig. 5.9.

The situation is different for sealed cells: the right column in Fig. 5.12 indicates that all the electric energy employed for the internal oxygen cycle causes heating. As mentioned in connection with Fig. 4.14 (p.126) the calorific voltage E_{cal} for this reaction is zero, so the whole column corresponds to the Joule effect.

Fig. 5.13 illustrates the heat evolution of a sealed nickel/cadmium battery during constant-current charging with a charge factor of 1.4 (such an amount of overcharge is usual for conventional charging methods but can only be applied to comparably small batteries < 10 Ah).

- The voltage curve at the top shows the gradual increase of charging voltage with charging time. The generated heat is calculated as an average value for different sections of this curve. The numbers beside the charging curve are the average voltages (V per cell) for the corresponding section.
- The middle Figure shows the (constant) current and its distribution between charging process and internal oxygen cycle.
- The bottom Figure shows the heat generation as average value for the different sections. The numbers are the heat in kJ (for comparison, converted to 100 Ah of nom. cap.).

  During the first two hours, the reversible heat effect exceeds the Joule effect and cooling is observed. So the number for this section is written below the zero line. (A corresponding slight cooling effect can be observed with the temperature curve in Fig. 13.18, p.431).

  When the charging process approaches completion, nearly all the current is used for the internal oxygen cycle, which causes much heat generation.

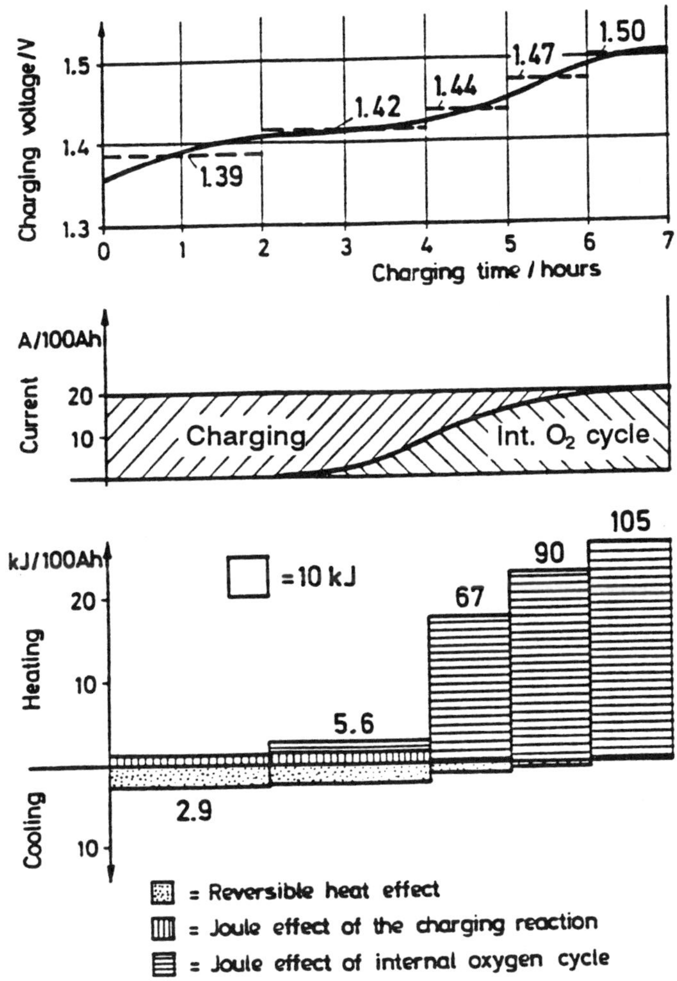

Fig. 5.13 Charging of a sealed nickel/cadmium battery with constant current 0.2 C(A). During 7 hours 140% of the nominal capacity are recharged which corresponds with a charge factor 1.4. For comparison, all values are converted to 100 Ah of nom. capacity. Actually, batteries of this type are in the size < 10 Ah. Middle: current distribution between charging and internal oxygen cycle. Bottom: heat generation as an average of the different sections (slight cooling during the first 2 hours).

Altogether 264.7 kJ = 73.53 Wh of heat are generated, referred to a nominal capacity of 100 Ah. These figures are larger than the 145 kJ/100 Ah of the valve-regulated lead-acid battery, shown in Table 4.11 (p.129). Actually, the difference is even larger when referred to the stored energy expressed in Wh. Then these Ah based values must be multiplied by the factor 1.5, since the 100 Ah lead-acid battery stores 1.5 Wh more due to its higher cell voltage (2.0 = 1.53 x 1.3). Thus the energy generation in a nickel/cadmium battery of comparable storage capacity corresponds to 397 kJ and is larger by nearly a factor of 3. The main reason for the increased heat generation of the sealed nickel/cadmium battery in this example is the high charge factor of 1.4 in Fig. 5.13. The charging factor for the lead-acid battery in Fig. 4.15 and Table 4.11 is about 1.10. This example clearly shows the strong influence of overcharging on heat generation caused by the internal oxygen cycle.

Fig. 5.13 shows that this heat is generated practically during the last 3 hours of the charging process. This means an average heat generation of 24.51 W/100 Ah for these three hours. The conclusion can be drawn that sealed nickel/cadmium batteries can be charged at a high rate so long as the current is actually used for charging and not for the internal oxygen cycle. Rapid charging methods, as described in Section 13, are always based on this principle.

Usually, battery manufacturers strongly advise the customer not to operate sealed nickel/cadmium batteries at constant-voltage float charge without monitoring (cf. Section 14), because of this heat generation during continuous overcharge on account of the internal oxygen cycle. The fact that this cycle can attain extremely fast rates, makes the situation really dangerous in regard to thermal runaway (cf. Section 4.5.3.6).

5.4.3 Heating of Batteries

As in Section 4.5.3, heating effects will be shown for the above examples to demonstrate typical values. Two borderline cases will be considered:

1. The maximum temperature difference ΔT between the battery and its surroundings that will be reached under adiabatic conditions, i.e. without any heat dissipation, according to Eq. (2.83), p.36.

2. The temperature difference ΔT_{bal} that is required between the container wall and the surrounding atmosphere to establish balance between generated and dissipated heat (Q_{gen} and Q_{diss}, respectively) according to Eq. (2.84), p.38.

In both cases, the temperature of the battery is assumed as uniform; temperature gradients within the battery are neglected. On account of the approximate character of the results, the specific surface areas are used for small, medium and large cells, which are compiled for lead-acid batteries in Table 4.13.

5.4.3.1 Heating Effects in Vented Batteries

Table 5.8 shows the maximum temperature rise that would result from a charge and discharge process of vented nickel/cadmium batteries, as shown in Fig. 5.9, when heat is not dissipated.

		1	2	3
		Generated heat Q_{gen}	Heat capacity C_{Batt}	ΔT
		kJ/100 Ah	kJ/(k·100 Ah)	°C
1	Charging	34.66	7	≈5
2	Discharging	67.5	"	≈10

Table 5.8 Maximum temperature difference ΔT that would be reached when charging or discharging occurred without heat dissipation. Heat generation data taken from Fig. 5.9. Heat capacity from Table 5.6 (p.186), 1 Wh = 3.6 kJ.

After charging, a temperature increase by 5 °C results; after discharge, the battery temperature is raised by 10 °C. These values are reached under adiabatic conditions, when heat dissipation virtually is excluded. In practice, some heat dissipation is always possible that reduces the temperature increase and finally leads to a balance between heat generation and heat dissipation. Table 5.9 shows the temperature differences ΔT_{bal} between the battery and its surroundings that achieves this balance.

		1	2	3	4
		Generated heat Q_{gen}		Heat dissipation Q_{diss}	ΔT_{bal}
		Wh/100 Ah	W/100 Ah	W/m^2	°C
1	Charging	7.89	1.12	11	≈3
1a	last 2 hours *	9.63	4.8 *	48	≈8
2	Discharging	18.75	3.75	37	≈6

Table 5.9 Temperature difference that balances heat generation and heat dissipation. Heat generation data from Fig. 5.9 (1 kJ = 0.2778 Wh). Heat dissipation in column 3 based on a cell surface area of 0.1 m^2/100 Ah according to Table 4.13 (p.131). * Data concern the last 2 hours of the charging process. ΔT_{bal} values from Fig. 2.12 (p.41).

In Table 5.9 two lines represent the charging period. Line 1 represents the average of the whole 7 hours of charging time, while Line 1a concerns only the last two hours, when heat actually is generated. The last Column shows that only 8 °C of temperature rise would achieve the desired balance.

Comparison between the Tables 5.8 and 5.9 emphasises the importance of heat dissipation. If there is no heat dissipation, each charge/discharge cycle raises the temperature of the battery by 15 °C. If, on the other hand, heat dissipation occurs as assumed in Fig. 2.12, p.41, the temperature of the battery is raised by about 5 °C, even when the battery is cycled continuously.

5.4.3.2 Heating Effects in Sealed Batteries During Discharge

Heating effects typical for small sealed Ni/Cd batteries shall be estimated in the following, based on discharge curves in Fig. 5.10 and derived discharge data in Table 5.7. Table 5.10 shows the maximum rise of temperature which occurs when heat cannot dissipate.

		1	2	3
	Discharge duration	Generated heat Q_{gen}		ΔT
		Wh/Ah	Wh/100 Ah	°C
1	5 h	0.19	19	26
2	0.97 h	0.21	21	29
3	10.6 min	0.23	23	32
4	4.7 min	0.26	26	36

Table 5.10 Maximum temperature difference ΔT that would be reached when discharging occurred without heat dissipation. Heat generation data taken from Table 5.7, Line 6 with $Q_{gen} = - Q_{total}$. Heat capacity 0.72 kJ/(k·100 Ah) (Table 5.6 (p.186)), 1 kJ = 0.2778 Wh.

The values in Table 5.10 show only a moderate increase with load that is increased by a factor of 50 between the 5 h and 4.7 min. discharge period. This has two reasons that are clearly shown in Fig. 5.11:

1. The reversible heat effect contributes a large share of heat generation, even at the highest load, because it amounts to about 11% of the converted energy. But the reversible heat effect is independent of the discharge rate (cf. Section 2.2.4).
2. Due to the low internal resistance of this battery, the Joule effect is comparatively small, and this also is true for its increase with load.

For both reasons, the generated heat does not differ too much, and the temperature of the battery is raised between 26 and 36 °C, despite the wide variation of the load.

The situation is completely different for the balancing temperature ΔT_{bal} that would be required if heat generation occured continuously by these currents. Table 5.11 shows the corresponding data.

		1	2	3	4
	Discharge duration	Generated heat Q_{gen}		Heat dissipation Q_{diss}	ΔT_{bal}
		Wh/100 Ah	W/100 Ah	W/m^2	°C
1	5 h	19	3.8	12.6	2
2	0.97 h	21	21.6	72	11
3	10.6 min	23	131	436	>20
4	4.7 min	26	333	1110	>>20

Table 5.11 Temperature difference that balances heat generation and heat dissipation. Heat generation data from Table 5.7 (1 kJ = 0.2778 Wh). Heat dissipation in column 3 based on a cell surface area of 0.3 m^2/100 Ah according to Table 4.13 (p.131) (small cells). ΔT_{bal} values from Fig. 2.12 (p.41).

Balance between heat generation and dissipation according to Fig. 2.12 is only reached for the 5 and 0.97 h discharge. In the other two Lines of Table 5.11, heat generation exceeds the values shown in Fig. 2.12. Actually, in these two cases heat generation is determined by the short discharge duration a long time before ΔT_{bal} is reached. So the finally reached temperature is limited to the values shown in Table 5.10.

5.4.3.3 Heating Effects in Sealed Batteries During Charging

Charging of sealed nickel/cadmium batteries causes heat generation mainly during overcharging as shown in Fig. 5.13.

- During the last 3 hours of the charging process, 262 kJ/100 Ah = 73 Wh/100 Ah are generated. When heat is not dissipated, a temperature rise of $\Delta T = 101$ °C results (under the assumption $C_{batt} = 0.72$ Wh/(k·100 Ah).
- The average heat generation during these 3 hours is 24.3 W/100 Ah. When the surface area amounts to 0.3 m^2/100 Ah (small cells in Table 4.13, p.131), heat dissipation of 81 W/m^2 is required for the balance $Q_{gen} = Q_{diss}$. According to Fig. 2.12, p.41 this causes a temperature difference of 14 °C.

This result emphasises the importance of heat dissipation. Without heat dissipation, a thermal runaway would be caused within the last 3 hours of the charging period, while a temperature rise of only 14 °C occurs under the usual conditions of heat dissipation, shown in Fig. 2.12.

5.5 The Electrolyte in Ni/Cd Batteries

Potassium hydroxide (KOH) solution is the normal electrolyte in nickel/cadmium batteries. Usually concentrations between 1.24 and 1.30 g/cm^3 are employed, occasionally up to 1.35 g/cm^3 (densities at 20 °C). Some characteristic data are listed in Table 5.12 .

KOH wt.%	Density (g/cm^3) at 20 °C	Molarity (mole /dm^3) at 20 °C	Freezing point °C	Specific Conductance ($\Omega^{-1}\cdot cm^{-1}$) at 25°C	at -20 °C
20	1.1818	4.212	-23.5	0.60	0.190
22	1.2014	4.710	-28	0.63	0.194
24	1.2210	5.223	-34	0.65	0.195
26	1.2408	5.750	-40.5	0.67	0.192
28	1.2609	6.292	-48.8	0.67	0.188
30	1.2813	6.851	-59	0.66	0.182
32	1.3020	6.851	-60*	0.65	0.176
34	1.3230	8.017	-52	0.63	0.167
36	1.344	8.626	-44	0.61	0.156

Table 5.12 Some characteristic data of potassium hydroxide. (First three columns from ([40]), others from ([41]). * Minimum: -65.2 °C at 31 wt.% KOH.

The conductivity shows a slightly marked maximum at about 27 wt.%. The position of this maximum depends on temperature. The decay of the conductance with temperature is significant: at minus 20 °C, it is less than one-third of the value at 25 °C, as the last two columns of Table 5.12 show.

Fig. 5.14 shows the dependence of the conductivity on temperature and KOH concentration. The reduction with decreasing temperature is considerable, and the maximum is slightly shifted to lower KOH concentrations. Comparison to Fig. 4.20 shows that the situation is fairly similar to that of the electrolyte in lead-acid batteries.

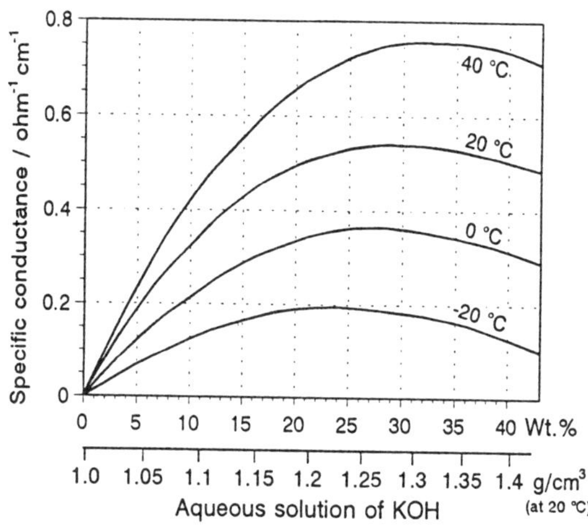

Fig. 5.14 Specific electric conductance of aqueous solution of potassium hydroxide and its dependence on temperature. Measured data (42).

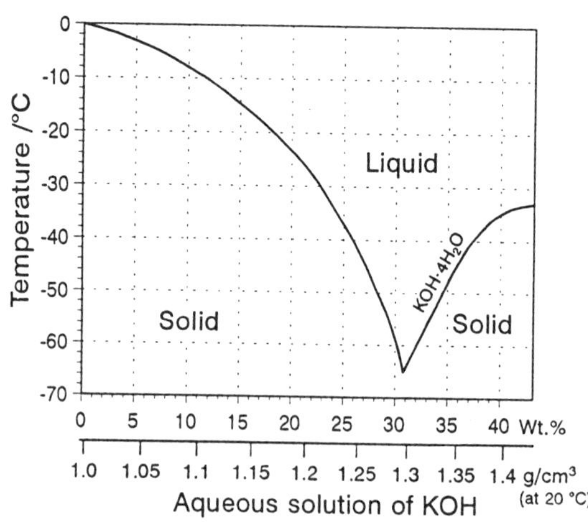

Fig. 5.15 Freezing temperature of aqueous solution of potassium hydroxide versus KOH concentration.

The freezing temperature of the electrolyte is shown in Fig. 5.1. The freezing curve of KOH solution shows that in the usual concentration range, applied in nickel/cadmium batteries (1.2 - 1.3 g/cm^3), the freezing point is below -30 °C. Due to the small concentration changes during charging and discharging, freezing problems compared to those in lead-acid batteries (cf. Section 4.6.2) are not observed in nickel/cadmium batteries. Nevertheless, freezing problems occasionally are observed during the recharge of nickel/cadmium batteries at high rates and very low temperature (e.g. ([10]), p.65). This may be caused by dilution of the electrolyte in the pore system during the recharge of the positive electrode, given by reversal of Eq. (5.4), p.177.

$$\boxed{Ni(OH)_2 + OH^- \Rightarrow NiOOH + H_2O + e^-} \qquad (5.33)$$

This reaction may locally cause such a dilution of the electrolyte that ice is formed. Increased resistance and reduced charge acceptance is the consequence.

5.5.1 Influence of Electrolyte Composition

The role of the electrolyte is complex. Mainly, it represents the medium for the ionic current. But it also influences the electrode composition, especially in the nickel hydroxide electrode. Although the electrolyte is not a component of the active material, like the sulphate ion in lead-acid batteries, comparatively small amounts of the electrolyte components are also incorporated, as has been mentioned with Fig. 5.3, p.188.

As an example of the influence of such incorporations on the capacity, Fig. 5.16 shows the capacity of the nickel hydroxide electrode achieved in various electrolytes at various temperatures.

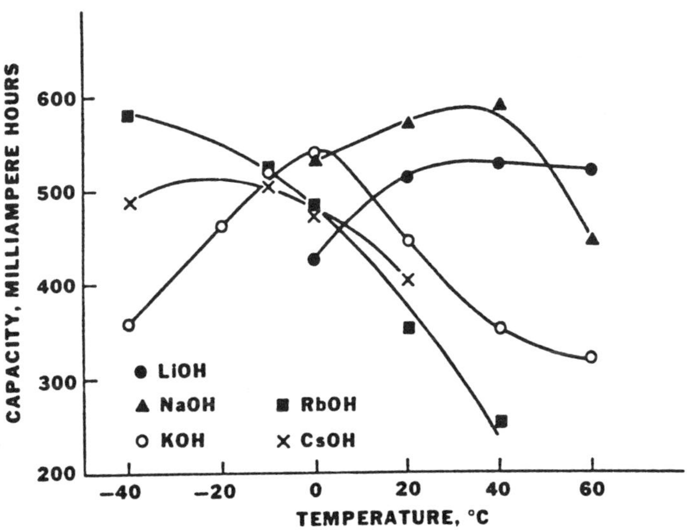

Fig. 5.16 Capacity of nickel hydroxide test electrodes in various electrolytes. Charge at the 5-hour rate with a charge factor 1.5. Discharge at the 2-hour rate. (Fig. 10 in ([43]).)

At elevated temperatures, higher capacities are achieved in LiOH and NaOH electrolytes. Measurements of the gassing rates during charging confirmed that the electrolyte influences charge acceptance.

Lithium addition is well known in flooded nickel/cadmium batteries to improve cycle stability of the nickel hydroxide electrode (cf., e.g., ([6]), p. 82).

5.5.2 Sodium Hydroxide (NaOH) Electrolyte

Nearly 100% of the nickel/cadmium batteries are equipped with diluted potassium hydroxide (KOH) as electrolyte. Sodium hydroxide (NaOH) is used only in special types of sealed nickel/cadmium batteries. The disadvantage of sodium hydroxide electrolyte is its reduced conductance compared to potassium hydroxide. Fig. 5.17 shows conductance curves for both electrolytes at 25 °C.

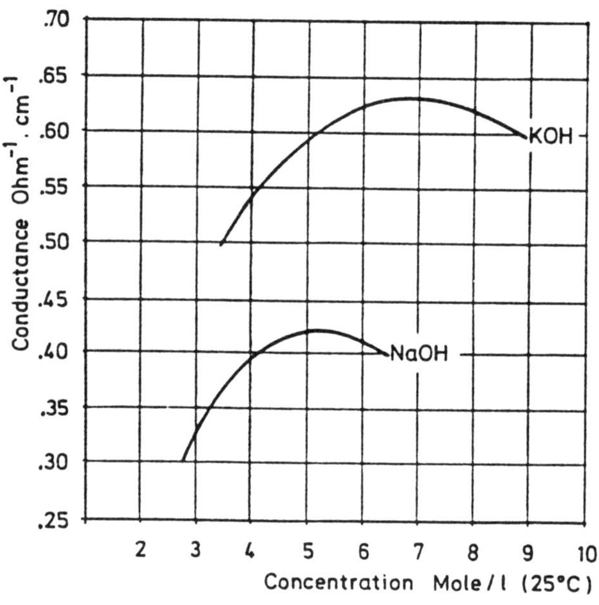

Fig. 5.17 Conductance of potassium and sodium hydroxide electrolyte at 25 °C (from ([44])).

The advantage of sodium hydroxide electrolyte is the reduced tendency to seep out of the seal. This is of special interest for batteries that are directly soldered to printed circuit boards and therefore have to be leak-proof for a long service life.

As a consequence, sodium hydroxide electrolyte is only used in special batteries, where low temperature operations and high currents drains are not required, but any leakage must be prevented, even after long service life. This applies to button cells for memory backup ([45]).

5.5.3 Immobilised Electrolyte

Immobilisation of the electrolyte in sealed nickel/cadmium batteries is always achieved with absorbent fibre felts, whose features are described in Section 4.6.4.1 for lead-acid batteries. Gelled electrolyte is not used for nickel/cadmium batteries.

The critical height of the cell, mentioned for lead-acid batteries, is not relevant, because the size of sealed nickel/cadmium batteries is generally limited by thermal problems.

The preferred separator material for sealed nickel/cadmium batteries is polyamide (trade name Nylon) because of its excellent wettability in alkaline electrolyte. The disadvantage of polyamide is hydrolysis, which gradually decomposes the separator, especially at high temperatures. Hydrolysis means that the bond C-N is destroyed according to

$$\boxed{R\text{-}C\text{-}NH_2 + H_2O \Rightarrow R\text{-}C\text{-}OH + NH_3} \qquad (5.34)$$

This reaction is not only critical in regard to the disintegration of the separator, but the formed NH_3 , since the resulting NH_3 can cause a 'nitrate shuttle', as described in Section 5.3, and can so increase the self-discharge.

A more stable fibre material is polypropylene, especially at high temperatures. However, polypropylene has poor wettability, and the electrolyte does not soak sufficiently. For this reason, most manufacturers employ a mixture of polyamide and polypropylene. If the polyamide suffers hydrolysis in such a blend, the polypropylene remains as a supporting framework and preserves the mechanical stability of the separator. The polyamide, although mechanically weakened, absorbs enough.

Pre-treated fibres of polypropylene are used occasionally. But such a pre-treatment is not stable enough for extended service life. Grafted polypropylene, which is equipped with hydrophilic groups by radiation-aided polymerisation, is claimed to be superior in this respect (46). So, hydrophilic groups of acrylic acid can be crosslinked to polyethylene or poypropylene fibres by UV radiation (47). Such separator material is used to an increasing extent in nickel/metal hydride, but also in nickel/cadmium batteries.

5.6 Corrosion

Compared to the lead-acid battery (cf. Section 3.6), the corrosion effects with nickel are negligibly small, and the general opinion that they can be neglected is right. But, in certain circumstances, minor effects of corrosion may also be observed with nickel/cadmium batteries.

Metallic nickel is not a stable phase at the equilibrium potential of the nickel/nickel hydroxide electrode. For thermodynamic reasons, $Ni(OH)_2$ is formed at electrode potentials 0.2 V higher than that of the $Cd/Cd(OH)_2$ electrode, which is far below the positive electrode potential.

However, a compact metallic nickel surface is hardly oxidised even at elevated potentials during overcharge of the positive electrode. Only on extremely large surface areas, as in sintered substrates or foam substrates, may some conversion of the substrate material into nickel hydroxide be discovered after prolonged overcharge or a large number of charge/discharge cycles. A further exception is the impregnation process of sintered electrodes, where nitric acid increases the corrosion rate (cf. Section 11.2.4.1) and can cause the 'memory effect' by formation of intermetallic Ni/Cd compounds (cf. Section 11.1.3).

More dangerous than the corrosion of nickel is that of iron, since iron in the positive active material reduces the oxygen overvoltage, and so increases oxygen and heat generation during charging and overcharging. Such a contamination of the active material can occur, when nickel-plated iron is used as substrate in the positive electrode, and the nickel layer is not dense. A capacity loss of the battery up to 30% can result (cf. ([10]), p. 44).

References

([1]) D.Linden, (Editor) 'Handbook of Batteries', 2nd Edition, McGraw-Hill Inc., New York, 1995, Chapters 23

([2]) W.Jungner, Swed.Pat. 10,177 (1899)

([3])R.Barnard, C.F.Randell and F.L.Tye, J. Appl. Electrochem. **10** (1980) p. 109

([4]) D.D.Macdonald and M.L.Challingsworth, J. Electrochem. Soc., **140** (1993), p. 606, Table 1

([5]) R.Barnard, C.F.Randell and F.L.Tye, J. Appl. Electrochem., **10** (1980) p.127

([6]) S.U.Falk and A.J.Salkind, 'Alkaline Storage Batteries', John Wiley & Sons, New York, 1969

([7]) W.M.Latimer, 'Oxidation Potentials', 2nd Edition, Englewood Cliffs, Prentice-Hall Inc. (1959), p. 199

([8]) P.Ness, unpublished internal reports (VARTA), 1968

([9]) R.Barnard, 'Cadmium in Alkaline Solution', J. Appl. Electrochem., **11** (1981), 217, (a survey on literature, 128 references)

([10]) VARTA Batterie AG (Editor) 'Sealed Nickel Cadmium Batteries', © VDI Verlag GmbH, Düsseldorf, 1982, Fig. 5/1. p. 59

([11]) C.A.Hampel, 'The Encyclopedia of Electrochemistry', © Reinhold Publishing Corporation, New York, 1964, p.423

([12]) P.Ness, unpublished internal reports (VARTA), 1968

([13]) K.Kleinsorgen and D.Dennstedt, internal VARTA report,1986

([14]) Intern. Critical Tables, McGraw Hill Book Co., New York, 1929, Vol.V, p. 115

([15]) 'CRC Handbook of Chemistry and Physic', 69th Edition, CRC Press, Inc, Boca Raton, Florida, 1989, p. D51

([16]) H.Bode, K.Dehmelt and J.Witte, Z.anorg. allg. Chemie, 1969 (**366**), 8, and esp. p. 9

([17]) R.Barnard and C.F.Randell, J. Appl. Electrochem., **13** (1983), p. 97

(18) VARTA Batterie AG (Editor), 'Sealed Nickel Cadmium Batteries', VDI-Verlag 1982, p.37; see also S.U.Falk and A.J.Salkind, 'Alkaline Storage Batteries', John Wiley & Sons, New York, 1969, Fig. 2.2, p.50.
(19) A.Delahay-Vidal, F.Portemer, B.-Beaudoin, K.Tekaia-Elhsissen, P.Genin and M.Figlarz *in* D.A.Corrigan, 'Nickel Hydroxide Electrodes', Proc. Vol. 90 - 4, The Electrochem. Soc.Inc., Pennington, U.S.A., 1990, p. 44
(20) H.Bode, K.Dehmelt und J.Witte, Z.anorg. allg. Chemie, 1969 (**366**), p. 15
(21) R.Barnard, C.F.Randell, F.L.Tye, Journal of Applied Electrochemistry **11**, (1995), p.517
(22) D.A.Corrigan and S.L.Knight, J.Electrochem.Soc., **136** (1989), p. 613
(23) B.C.Cornilsen, X.Shan and P.L.Loyselle *in* D.A.Corrigan, 'Nickel Hydroxide Electrodes', Proc. Vol. 90 - 4, © The Electrochem. Soc.Inc., Pennington, U.S.A., 1990, p. 89
(24) W.Dennstedt, VARTA internal report (1986), unpublished
(25) W.Dennstedt, unpublished internal VARTA report (1981)
(26) A.K.Sood *in* D.A.Corrigan, 'Nickel Hydroxide Electrodes', Proc. Vol. 90 - 4, The Electrochem. Soc.Inc., Pennington, U.S.A. 1990, p. 163
(27) A.H.Zimmermann and A.H.Phan, in Proc. Symp. 'Hydrogen and Metal Hydride Batteries', Editors D.Bennett and T.Sakai, © The Electrochemical Society, Inc, Proceedings Vol. 94-27, 1994, p.341
(28) B.C.Cornilsen, X.Shan and P.L.Loyselle *in* D.A.Corrigan, 'Nickel Hydroxide Electrodes', Proc. Vol. 90 - 4, The Electrochem. Soc.Inc., Pennington, U.S.A., 1990, p. 82
(29) M.E.Unates, M.E.Folquer, J.R.Vilche and A.J.Arvia *in* D.A.Corrigan, 'Nickel Hydroxide Electrodes', Proc. Vol. 90 - 4, The Electrochem. Soc.Inc. Pennington, U.S.A., 1990, p. 134
(30) J.McBreen *in* D.A.Corrigan, 'Nickel Hydroxide Electrodes', Proc. Vol. 90 - 4, © The Electrochem. Soc.,Inc., Pennington, U.S.A., 1990, p. 61
(31) K.Ohta, K Hayashi, H.Matsuda, J.Toyoguchi and M.Ikoma in 'Hydrogen and Metal Hydride Batteries' Proc. Vol. 94-27, © The Electrochem. Soc., Inc. Pennington, NJ, 1994, p. 296
(32) H.Bode, K.Dehmelt und J.Witte, Z.anorg. allg. Chemie, 1969 (**366**), p. 8
(33) A.H. Zimmerman in D.Bennett and T.Sakai (Editors), Hydrogen and Metal Hydride Batteries, Proc. Vol. 94-, © The Electrochemical Society, Pennington, 1994, p.268 (a review with 80 references)
(34) R.Barnard, G.T.Crickmore, J.A.Lee, F.L.Tye, Journal of Applied Electrochemistry **10**, (1980), p.61
(35) R.D.Armstrong and HongWang, Electrochimica Acta, **36** 1991 p.759
(36) M.Ikoma, Y.Hoshina, and I.Matsumoto J., Electrochem. Soc., **143** (1996), p. 1904
(37) J.N.Taylor(ALCAD Ltd), Proc. 17th Intelec Meeting, The Hague 1995, © IEEE Communications Soc., p.778
(38) VARTA Batterie AG, 'Sealed Ni-Cd Cells - Sales Program and Technical Handbook', © VARTA Batterie AG, Daimlerstr.1, D-73749 Ellwangen/Jagst, Germany, Figs. 2.14 and 3.10
(39) SANYO, Cadnica, Sealed Type Nickel-Cadmium Batteries, Engineering Handbook, © 1995 by SANYO Electric Co., Ltd, Osaka, Japan
(40) 'CRC Handbook of Chemistry and Physics', CRC Press, Boca Raton, 1988, p. D-245

(41) S.U.Falk and A.J.Salkind, 'Alkaline Storage Batteries', John Wiley & Sons, New York, 1969, Figs. 8.7 and 8.15

(42) D.Spahrbier, internal VARTA report, April 1963

(43) E.J.Rubin and R.Baboian, J. Electrochem. Soc., **118** (1971), p. 428

(44) VARTA Batterie AG (Editor), 'Sealed Nickel Cadmium Batterie', © VDI-Verlag GmbH, Düsseldorf, 1982, p. 33

(45) S.Tateishi, Y Takahashi and K.Fuchida, Proc. 5th Intelec Meeting, Tokyo 1983, © IEEE

(46) A.Kita, S.Yoshida and S.Tanso,Proc.32nd Intern. Power Sources Symp., The Electrochem. Soc., Pennington, N.J., 1986

(47) R.Singleton and D.Barker, Batteries International, Issue **28**, July 1996, p. 35

6. The Nickel/Hydrogen and Nickel/Metal Hydride Battery Systems

Hydrogen is used instead of cadmium as active material in the negative electrode in nickel/hydrogen and nickel/metal hydride battery systems, while the positive electrode and the electrolyte are the same as in nickel/cadmium batteries. Nickel/hydrogen and nickel/metal hydride batteries in principle represent the same battery system. They are distinguished from each other by the mode of storing the 'active material' of the negative electrode, the hydrogen, and by the negative electrode material that in both systems does not directly participate in the electrode reaction, but mediates as a catalyst for hydrogen oxidation and reduction.

- In **nickel/hydrogen batteries**, noble metals like platinum are used as catalytic surface in the negative electrode. Hydrogen (H_2) is stored as a gas under high pressure.
- In **nickel/metal hydride batteries** special alloys act as the catalytic electrode surface and simultaneously absorb the hydrogen (H_2), forming metal hydrides under low pressure. Thus the negative electrode material simultaneously catalyses the reaction and stores the hydrogen that is formed during charging.

Nickel/hydrogen batteries have been widely applied for many years in aerospace applications. For widespread commercial use, these batteries are too expensive on account of the required noble metal for catalysts and the stable cell container that has to withstand high pressures of the stored gaseous hydrogen.

Competition with the sealed nickel/cadmium battery became possible only after special alloys had been developed that store the hydrogen as metal hydride and replace the noble metal. Absorption of hydrogen in metals as a hydride was a well known fact, but the number of hydride forming elements was limited and their storage parameters not adequate for battery technology. Since the seventies a new group of hydride forming metals, the intermetallic compounds, have been discovered and special preparation methods for them developed. These are alloys of two or more metals that have a certain stoichiometry and structure. The number of possible combinations is nearly unlimited, and combinations are possible that absorb and desorb hydrogen at a desired pressure and temperature. At the surface of these alloys, hydrogen oxidation and reduction is so fast that noble metals are not required. With such alloys, the nickel/metal hydride battery, also called the 'metal hydride battery' or 'hydride/nickel oxide battery', could be developed which has gained a considerable market share within a few years and replaces nickel/cadmium batteries to an increasing extent, and such a replacement causes no problems due to the negligible difference of the cell voltage. Two reasons promoted this fast development:

- Cadmium is a toxic metal. This does not cause problems as long as the battery is in service. But it causes disposal problems, because it is not possible to prevent rundown batteries from being thrown into general rubbish tips. When such waste is treated in incineration plants, cadmium is vaporised and requires expensive filtering of the gas, and the resulting ash is so contaminated with cadmium that it must be classified hazardous. But also when the battery waste is dumped in rubbish pits, cadmium will be released gradually and can spoil the ground-water. For this reason, a substitute for cadmium has urgently been sought.
- The nickel/metal hydride battery stores about up to nearly 100% more energy, compared to a nickel/cadmium battery of comparable design and size. This is of great importance for portable devices with an appreciable need of energy, like cam-corders, telephones, and tools.

Because of the second item, the nickel/metal hydride battery system not only replaces the unwanted cadmium but also represents an improved storage system.

6.1 Nickel/Hydrogen Batteries

Nickel/hydrogen batteries, known also as the hydrogen/nickel oxide batteries, have been developed for aerospace applications and are still the number one energy storage system in many of these applications ([1]). This system employs the same positive electrode and also the same electrolyte as described for the nickel/cadmium battery in Chapter 5. The negative electrode, however, is different from a conventional battery electrode and corresponds to a fuel-cell electrode that oxidises hydrogen (H_2) to H^+ ions during discharge, or effects the reversed reaction, the reduction $H^+ \Rightarrow H_2$ at a potential below the hydrogen equilibrium when the battery is charged. In contrast to a conventional storage battery, the electrode material is not transformed, but acts only as a catalytic surface that provides or absorbs the electrons. The 'active material', the hydrogen, is not stored in the electrode. In its charged state, it remains as gaseous hydrogen within the cell that has to withstand a corresponding high pressure. During discharge, water is formed that is consumed by the discharge reaction of the positive electrode.

Note: *The catalytic effect of the negative electrode is not a pure surface effect but includes absorption and the formation of metal hydrides. But, the amount of platinum or palladium, used in such batteries, is so small that the storage capacity of the electrode has no relevance (cf. Section 6.3).*

The generation of gaseous hydrogen during charging and its consumption during discharging means that the internal gas pressure is proportional to the state of charge. This offers the ability to check the state of charge by measurement of the internal cell pressure.

The discharge reaction of the negative electrode is:

$$\boxed{H_2 \Rightarrow 2{\cdot}H^+ + 2{\cdot}e^-} \qquad (6.1)$$

which has to be completed as a chemical reaction

$$\boxed{H_2 + 2{\cdot}OH^- \Rightarrow 2{\cdot}H_2O + 2{\cdot}e^-} \qquad (6.2)$$

Its reversal (from right to left) represents the charge.

For catalytic reasons, the electrode surface has to contain noble metals to make Eq. (6.1) as well as its reversal, sufficiently fast. Usually platinum is applied on account of its outstanding catalytic capabilities and its chemical stability. It is usually Teflon bonded or covered with microporous Teflon membranes to make the surface hydrophobic and so prevent the formation of thicker layers of water that would hinder gas access.

To ensure that the gaseous hydrogen (H_2) reaches the negative electrode fast enough, the electrolyte has to be immobilised to provide direct access for the gas to the surface. Fibre felts and knitted textiles of mineral fibres are used (cf. Section 32.3.3 in ([1])). These separators are based on asbestos fibres (fuel-cell-grade asbestos paper) or zircon fibres stabilised with yttrium (Zircar cloth). They provide high temperature resistance and chemical resistance for long-life batteries.

The positive electrode is the same as in nickel/cadmium batteries. Its (simplified) reaction equation for discharge (Eq.(5.4)) is :

$$\boxed{2{\cdot}NiOOH + 2{\cdot}H_2O + 2{\cdot}e^- \Rightarrow 2{\cdot}Ni(OH)_2 + 2{\cdot}OH^-} \qquad (6.3)$$

(For details of the nickel hydroxide electrode cf. Section 5.2.2.)

Combination of Eq. (6.3) with Eq.(6.2) leads to the cell reaction:

$$\boxed{2{\cdot}NiOOH + H_2 \Rightarrow 2{\cdot}Ni(OH)_2} \qquad (6.4)$$

Eq. (6.4) describes the discharge. NiOOH is reduced to $Ni(OH)_2$ and hydrogen is consumed. During charge, this equation is reversed. The hydrogen developed during charge is stored in each cell under high pressure. The cells therefore have to withstand pressures up to about $5{\cdot}10^6$ Pa (50 bar).

Advantages of the nickel/hydrogen battery system are:

- A fairly high specific energy of 50 to 60 Wh/kg depending on design and size.

- The ability to withstand a great number of deep discharges (40,000 cycles at 40% DOD).
- A calendar life of 15 years.
- The battery can be overcharged and single cells reversed without damage.
- The state of charge can easily be controlled by measurement of the H_2 pressure.
- Each cell is hermetically sealed. No maintenance is required.

Disadvantages are:

- Its high price which limits the application more or less to aerospace projects. (*More recently, efforts have been made (2) to reduce the price and develop a nickel/hydrogen battery also for remote terrestrial applications like photo-voltaic stations*).
- High self-discharge rate of about 5% per day at room temperature, due to direct reaction between hydrogen and nickel hydroxide (cf. Section 6.1.1.1).
- Energy density is comparatively low and amounts to 20 to 40 Wh/dm^3, because of the volume required for gas storage and the shape of the pressure vessel.

The above listed advantages give reasons for the widespread use of this battery system in the aerospace field. On the other hand, more recent developments indicate that even in space applications the nickel/hydrogen system might be displaced by the less expensive nickel/metal hydride system (3).

6.1.1 Secondary Reactions

The thermodynamic situation is very similar to that in the nickel/cadmium battery described in Section 5.1 and shown in Fig. 5.1. The positive electrode is identical, and the equilibrium potential of the hydrogen electrode is only 0.02 V lower than that of the cadmium electrode.

1. Oxygen evolution occurs at an electrode potential above 1.23 V, which is slightly below the open-circuit potential of the positive electrode according to

$$\boxed{4\cdot OH^- \Rightarrow O_2 + 2\cdot H_2O + 4\cdot e^-} \qquad (6.5)$$

This oxygen evolution cannot be avoided, and is the main reaction at the positive electrode during overcharging.

2. Oxygen reduction according to

$$\boxed{O_2 + 2\cdot H_2O + 4\cdot e^- \Rightarrow 4\cdot OH^-} \qquad (6.6)$$

is the reversal of oxygen evolution and is possible below 1.23 V. Oxygen is reduced as soon as it reaches the negative-electrode surface, and the sum of Eqs. (6.5) and (6.6) forms the internal oxygen cycle during overcharging (cf. Chapter 8).

If no current flows through the cell, any oxygen that had been generated within the cell, is removed by direct catalytic reaction (recombination of oxygen and hydrogen), because the negative electrode acts like a recombination device (cf. Section 7.5.2).

3. Hydrogen oxidation is the discharge reaction at the negative electrode according to Eq. (6.2). But hydrogen oxidation is also possible at the positive electrode as an unwanted secondary reaction. At open-circuit, reduction of the nickel hydroxide is the counter reaction; and Eq. (6.4)

$$2{\cdot}NiOOH + H_2 \Rightarrow 2{\cdot}Ni(OH)_2$$

is the result. This direct reaction between H_2 and Ni(OOH) causes the high self-discharge rate of nickel/hydrogen batteries (cf. Section 6.1.1.1).

4. An **internal hydrogen cycle** is established when the battery is discharged so deeply that the positive electrodes are completely discharged. Then, hydrogen is evolved according to the reversal of Eq. (6.2) at the reversed nickel hydroxide electrode:

$$2{\cdot}H_2O + 2{\cdot}e^- \Rightarrow H_2 + 2{\cdot}OH^- \qquad (6.7)$$

which is oxidised at the at the negative electrode according to Eq. (6.2), the normal discharge reaction:

$$H_2 + 2{\cdot}OH^- \Rightarrow 2{\cdot}H_2O + 2{\cdot}e^-$$

So, the negative electrode is not reversed, but remains at its normal discharge potential, since hydrogen (H_2) is now supplied by the reversed positive electrode. The sum of both reactions forms an internal hydrogen cycle that is fast enough to carry the discharge current due to the high rate of hydrogen oxidation at the negative electrode.

This internal hydrogen cycle is an important secondary reaction, since it allows deep discharges that include the reversal of individual cells. This is one of the features that make nickel/hydrogen batteries so suitable for space applications where such deep discharges occasionally cannot be avoided.

Note: *Prolonged reversal of the positive electrode can harm the capacity of the positive electrode. For example, it can reduce oxides, like cobalt oxide, which are added to stabilise the electronic conductivity of the active material.*

5. Corrosion of the nickel or nickel-plated substrate is very slow as mentioned in Section 5.6.

6.1.1.1 Self-Discharge

The relative high self-discharge is a disadvantage of nickel/hydrogen batteries. Within ten days, it amounts to about 50% and 80% of capacity at 20 and 30 °C, respectively (cf. Fig. 32.15 in ([1])). It is caused by the direct reaction of the charged active material with hydrogen according to Eq. (6.4)

$$2{\cdot}NiOOH + H_2 \Rightarrow 2{\cdot}Ni(OH)_2$$

The rate of this reaction can be determined by measuring the heat evolution that corresponds to ΔH given by Eq. (6.9). Fig. 6.1 shows the results of such measurements by microcalorimetric technique in comparison with the observed capacity loss during 24 hours. The results of both methods agree quite well.

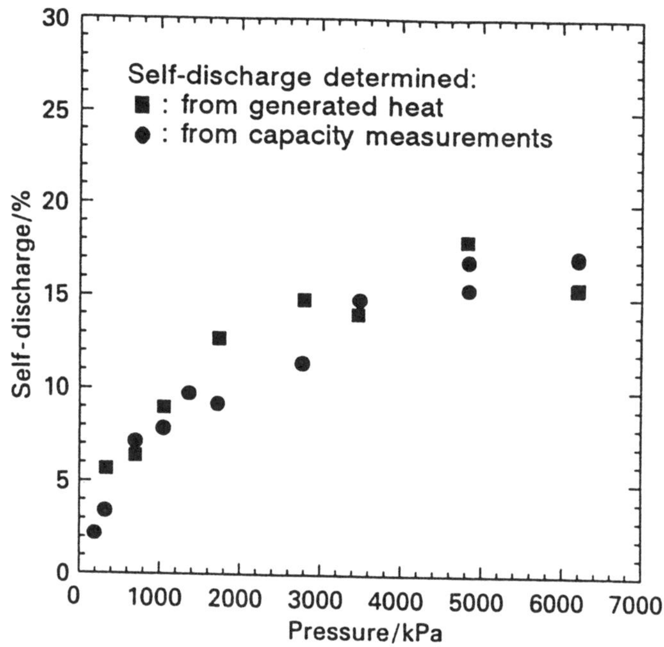

Fig. 6.1 Nickel/hydrogen battery. Percentage of self-discharge at 25 °C after 24 h as a function of hydrogen pressures. NiOOH electrode containing 10% Cobalt. Fig. 5 in (4).

Reduced self-discharge at low pressures can be observed in Fig. 6.1. It indicates that the diffusion rate through the wetting layer at the electrode surface is rate limiting (cf. Section 2.6.3.2). This was demonstrated by measurements with electrodes in flooded, fully wet, and starved states (Fig. 3 in (4)). Furthermore, these microcalorimetric measurements showed that H_2 oxidation at the nickel substrate surface is a comparatively slow reaction.

6.2 Thermodynamic Parameters

Nickel/hydrogen and Nickel/metal hydride batteries are based on the same cell reaction (Eq.(6.4)). For this reason, the following derivation of thermodynamic data applies to both systems, *except the heat effects caused by hydrogen absorption and desorption in metal hydride batteries, described in Section 6.5.1.* On account of the uncertainties connected with the nickel hydroxide electrode, the derived data can only be approximate, as has been explained in the Sections 5.1.1 and 5.2.

When for the nickel electrode the same data are used as in Section 5.1.1, i.e. the doubly enframed data of Table 5.1, the reaction equation must be written as:

$$\boxed{2\cdot(\beta\text{-NiOOH}\cdot H_2O) + H_2 \Leftrightarrow 2\cdot Ni(OH)_2 + 2\cdot H_2O} \tag{6.8}$$

and the set of data is:

	Substance	Enthalpy ($H^{o,s}$) kJ/mole	Free Enthalpy ($G^{o,s}$) kJ/mole
1	H_2	0	0
2	β-NiOOH·H_2O	-676	-561
3	β-$Ni(OH)_2$	-537.8	-453.5
4	H_2O	-285.8	-237.2

Table 6.1 Thermodynamic data: standard values of enthalpy and free enthalpy (Gibbs free energy) of formation concerning the chemical compounds used in Eq (6.8). Values in Lines 2, 3, and 4 are the doubly enframed ones in Table 5.1, p.178.

When these values are inserted for the 'reactants' and 'products' into Eq. (6.8), according to Eq. (2.7), (p.11), ΔH and ΔG result.

$$\Delta H = 2 \text{ x } (-537.8) + 2 \text{ x } (-285.8) + 2 \text{ x } 676$$

$$\boxed{\Delta H = -1647 + 1352 = -295 \text{ kJ}} \tag{6.9}$$

$$\Delta G = 2 \text{ x } (-453.5) + 2 \text{ x } (-237.2) + 2 \text{ x } 561$$

$$\boxed{\Delta G = -1381 + 1122 = -259 \text{ kJ}} \tag{6.10}$$

and the reversible heat effect according to Eq. (2.4)

$$\boxed{Q_{rev} = T\cdot\Delta S = \Delta H - \Delta G = -36 \text{ kJ}} \tag{6.11}$$

The negative sign of Q_{rev} means that the reversible heat effect contributes additional heat during discharge, but reduces heat generation during charging (as in nickel/cadmium batteries, cf. Eq. (5.9), p.179).

The equilibrium voltage E^o that results when ΔG (259 kJ = 71.9 Wh) is inserted in Eq. (2.6) is:

$$\boxed{E^o \approx 1.34 \text{ V}} \tag{6.12}$$

The temperature coefficient is derived when ΔS (=T·ΔS/298.2 = -120.7 J = -0.0335 Wh) is inserted in Eq. (2.16), p.13:

$$\boxed{dE^o/dT \approx -0.6 \text{ mV/K}} \tag{6.13}$$

The values are similar to those of the nickel/cadmium battery system. As far as the approximate character of the equilibrium voltage is concerned, the same arguments are true as were mentioned in connection with Eq. (5.10), p.179.

Note: *For the temperature behaviour of nickel/metal hydride batteries which are based on the same cell reaction, the heat effects of the hydrogen absorption and desorption reactions are important. Hydrogen absorption is always an exothermic reaction and contributes additional heat during charging. These heat effects depend on the alloy applied for hydrogen storage (cf. Section 6.5.1).They can compensate or even over-compensate the reversible heat effect* Q_{rev} *given by Eq. (6.10). For this reason, the characteristic feature of nickel/cadmium and nickel/hydrogen batteries, that the charging reaction causes cooling of the battery, may not apply for nickel/metal hydride batteries.*

6.2.1 Weight-Dependent Quantities

The weight-depending data of nickel/hydrogen batteries are advantageous, due to the low specific weight of hydrogen. According to Eq. (6.4) and the molar weights of NiOOH (91.72 g, Table 5.3) and H_2 (2.02 g), only 185.5 g of active material are required to produce $2 \cdot F = 53.6$ Ah, which corresponds to 71.9 Wh when multiplied by the equilibrium voltage 1.34 V. This results in the fairly high value for the specific energy:

$$\frac{71.8}{0.1855} = 387 \quad \text{Wh / kg} \tag{6.14}$$

which is more than twice that of the lead-acid battery (Eq.(4.22)) and much better than the 214 Wh/kg of the nickel/cadmium battery (Eq.(5.23)).

Even if, for a more realistic evaluation, the water-containing ß-NiOOH (Line 2 in Table 6.1) with a molar weight of 109.7 is inserted instead of NiOOH, the required weight to produce 71.98 Wh would be 0.222 kg and the result would be 324 Wh/kg for the maximum in theory. This is also much better than the comparable values for the lead-acid and nickel/cadmium system. However, the reductions between this theoretical figure and the actual values for nickel/hydrogen batteries are even larger than those described in Sections 4.2.2 and 5.1.5, because of the high pressure resistance of the cell container and the required volume for gas storage. About 50 Wh/kg and 90 Wh/dm^3 are reached in practice.

Nickel/metal hydride batteries do not afford such a stable container, but additional weight is contributed by the metal that stores hydrogen. In practice, a specific energy of about 60 Wh/kg is reached for smaller batteries. Larger types reach 80 Wh/kg. For the energy density, at maximum, about 200 Wh/dm^3 are attained (cf. Table 6.6).

6.3 Hydrogen-Absorbing Alloys

A serious disadvantage of the nickel/hydrogen battery system is the high pressure, always connected with the storage of gaseous hydrogen. But hydrogen can be stored at fairly low pressure when it is absorbed by metals which form hydrides. Palladium, for example, absorbs between 0.4 and 2 wt.% of hydrogen, depending on the grade of its dispersion (between bulk metal and platinum black). This means that 1 cm^3 palladium (≈ 12 g) is able to store up to 0.1 gram of hydrogen, which corresponds to 1200 cm^3 of gaseous H_2 at 25 °C and 1 atm (cf. Table 2.6, p.48). The apparent density of the hydrogen within the metal even exceeds the density of 0.07 g/cm^3 of liquid hydrogen at 20 K, which demonstrates the enormous hydrogen-storage capability of such metals.

Fig. 6.2 shows a hydrogen absorption/desorption curve. Initially hydrogen is absorbed with increasing pressure, until the formation of the MH_n phase starts. Then the pressure remains approximately constant, until the metallic phase is converted into hydride. Desorption of hydrogen occurs at lower pressure and the hysteresis between absorption and desorption pressure indicates the energy loss between absorption and desorption.

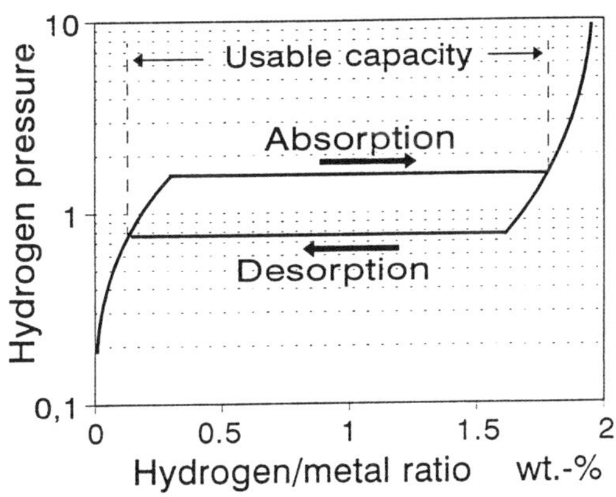

Fig. 6.2 Characteristic relation between pressure and hydrogen content in a metal hydride system. ('Pressure/composition isotherm', also called 'PCI curve' or 'pressure-temperature-composition curve' .) With increasing temperature the absorption/ desorption plateau is shifted to higher H_2 pressures (cf. Fig. 6.4).

***Note**: The hydrogen content is often expressed by the ratio between the number of hydrogen atoms per metal atom or per formula unit. In the case of a $LaNi_5$ alloy, 1 per metal atom is equivalent to 6 per formula unit (cf. Figs. 6.4 and 6.5).*

At maximum about 1.5 percent by weight of hydrogen is absorbed (hydrogen/metal relation plotted on the horizontal axis, cf. Fig. 6.4). This means that about 0.015 grams of hydrogen can be stored per gram of metal.

According to Eq. (6.1) one mole H_2 means a charge transfer of 2·F, which leads to the electrochemical equivalent

$$\boxed{\text{2.02 grams of } H_2 \text{ correspond with 53.6 Ah}} \quad (6.15)$$

or

$$\boxed{1\text{ g } H_2 \sim 26.5\text{ Ah}} \quad (6.16)$$

Consequently the capacity of a metal hydride electrode which contains 1.5 percent hydrogen by weight amounts to utilisation figures of about 400 mAh/g at maximum, and can easily compete with that of cadmium electrodes, as comparison with the values in Table 5.3 (p.184) shows.

Less noble metals are also able to store considerable amounts of hydrogen. This has, for example, been shown when Raney-nickel electrodes, were applied for battery electrodes in connection with fuel cell development ([5]).

The equilibrium hydrogen pressure for an hydride is given by the Van't Hoff equation

$$\boxed{\ln(p) = \frac{\Delta H}{R \cdot T} - \frac{\Delta S}{R}} \tag{6.17}$$

with: p : equilibrium H_2 pressure

ΔH and ΔS enthalpy and entropy respectively of hydride formation

R : molar gas constant

T : temperature

When p is plotted on a logarithmic scale versus 1/T, straight lines result with the slope $\Delta H/R$ as shown in Fig. 6.3 . $\Delta S/R$ is given by the intercept $1/T = 0$.

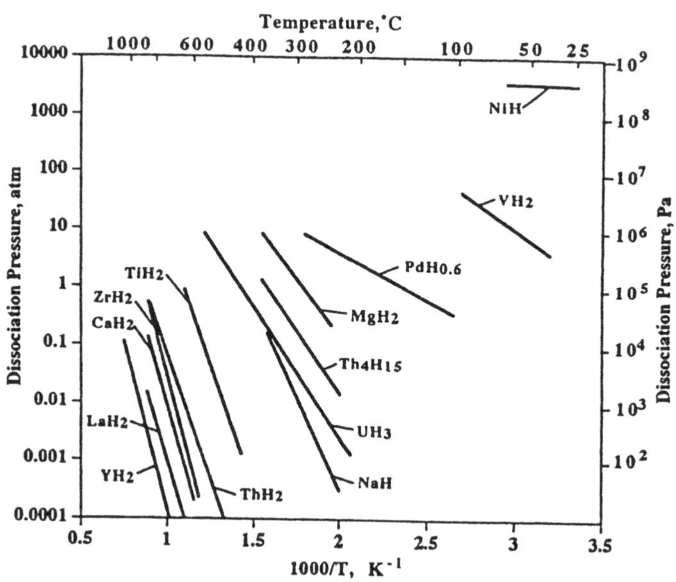

Fig. 6.3 Van't Hoff lines of elemental hydrides. (Fig. 2 in ([6]).)

A great variety of elemental hydrides exists. For stable compounds like YH_2, the hydrogen equilibrium pressure is even low at high temperature. Weak hydrides like NiH, on the other end of the scale, exhibit high H_2 pressures already at room temperature.

6.3.1 Intermetallic Compound Hydrides

Despite the great number of elemental hydrides, little practical use was possible, since only a few offer reasonable equilibrium pressure at room temperature. This situation changed when intermetallic compounds were developed that combine strong-hydride forming elements with those that form weak hydrides. An appropriate ratio between the two components can be used to tailor metal hydrides of the desired decomposition pressure.

Intermetallic compounds are alloys of two or more metallic elements, having narrow integer stoichiometries. The number of possible intermetallic compounds is nearly unlimited. The growing demand for hydrogen storage inspired intensified research, and a great number of alloys have been discovered or developed that form metal hydrides ([7]), providing hydrogen storage in a suitable range of pressure and temperature (surveys are given in ([8])). They are divided into groups A_uB_v, based on composition and crystal structure. A represents a strong hydride forming element and B the weak hydride forming component. For practical purposes A and B are at least partially substituted by different elements or combinations of them. Due to their complex structure and due to the ability to substitute A or B with a number of elements, they offer an enormous variety of parameters for hydride formation.

Combinations that proved to be suited for hydrogen storage in batteries are known as AB, AB_2, A_2B and AB_5 (cf., e.g., ([6])). Table 6.2 shows their characteristic composition and some features.

	A_uB_v class (Basis)	Actual Components	Storage capability	Remarks
			Ah/kg	
1	AB_5 ($LaNi_5$)	**A**: Mischmetal*, La, Ce, Ti **B**: Ni, Co, Mn, Al	≈ 300 at max. (e.g. ([9]))	At present, mostly used alloy group
2	AB_2 ($TiNi_2$)	**A**: V, Ti **B**: Zr, Ni (+ Cr, Co, Fe, Mn)	≈ 400 at max. (e.g. ([10])	Basis of 'multicomponent alloys' ([11])
3	AB (ZrNi)	**A:** Zr, Ti **B:** Ni, Fe, Cr, V		Early used for hydrogen storage in cars and as electrodes ([12])
4	A_2B (Ti_2Ni)	**A:** Mg, Ti **B:** Ni		

Table 6.2 Classes of intermetallic compounds used for negative electrodes of nickel hydride batteries. Alloys corresponding to Lines 1 and 2 are nowadays mainly applied. * Mischmetal (Mm) is an unrefined rare earth mixture (mainly Ce, La, Nd, and Pr). Its composition depends on the ore ([6]), but 'synthetic mischmetals' are also known ([13]).

Fig. 6.4 shows pressure/composition isotherms of an AB_5 alloy. For each temperature the upper and lower curves denote absorption and desorption, respectively. Compared to the schematic drawing of Fig. 6.2, a slight slope of the plateau is observed, as is usual in practice. The horizontal axis is given in mole hydrogen per mole metal. So the ratio 1 means 6·H per alloy formula, or with the mole weights of La, Ni and Al = 138, 59 and 27 respectively, 6 g hydrogen per 423 g of alloy. This means a storage capability of about 1.4 wt.%.

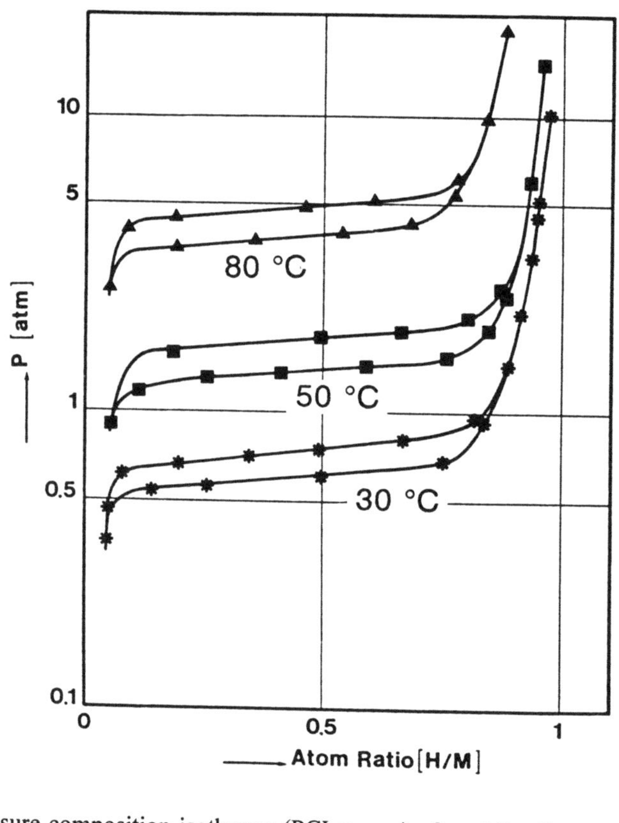

Fig. 6.4 Pressure composition isotherms (PCI curves) of an AB_5 alloy with nickel partly substituted by aluminium ($LaNi_{4.7}Al_{0.3}$) (Fig. 2.18 in (14))

AB_5 alloys are now mainly used for batteries (15). Mischmetals, which are unrefined mixtures of rare earths, are mostly substituted instead of lanthanum, mainly because they are less expensive than pure La (6) although such substitutions also are required to stabilise the alloy and prevent its premature degradation (16). Fig. 6.5 shows that a combination of mischmetal with nickel and several substitutes for nickel can approach fairly well the 'ideal' absorption-desorption isotherms of $LaNi_5$.

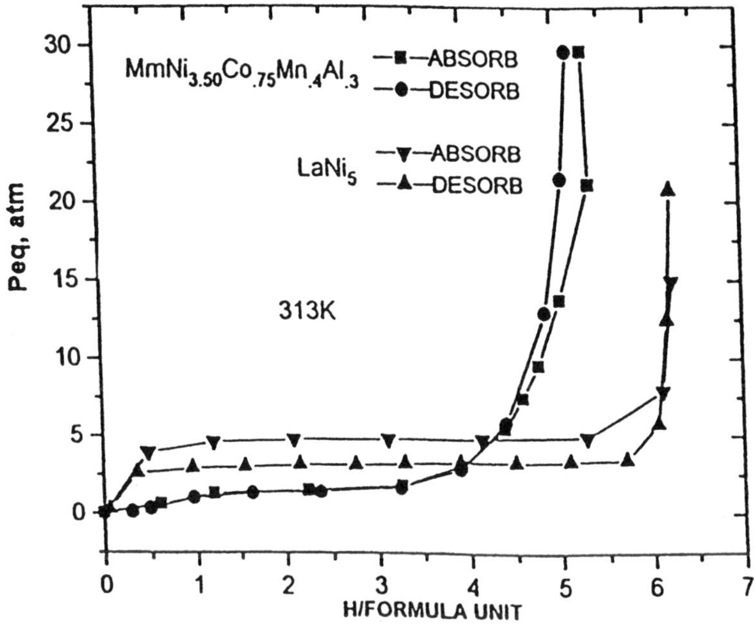

Fig. 6.5 Absorption-desorption isotherms of $LaNi_5$ and an AB_5-group alloy with mischmetal in place of A and a combination of Ni, Co, Mn and Al in place of B. (Fig. 2 in ([17]).)

The example in Fig. 6.5 shows the general situation that not only lanthanum is substituted by the mischmetal, but also nickel is partially substituted by other metals.

The AB_2 type in Line 2 of Table 6.2 is based on vanadium, titanium, zirconium and nickel. Especially for these intermetallic compounds, the stoichiometry is not well defined. So, some alloys ascribed to this group may be named A_2B or AB as well. They often have the crystallographic structure of Laves phases and are called 'Zr-Ni-V-Cr multicomponent alloys' (e.g. the 'Ovonic alloys' ([11])).

Titanium/nickel intermetallic phases (A_2B and AB type) were developed in the seventies with great assiduity when the aim was hydrogen storage for road vehicles with hydrogen combustion engines ([18]). They were also used as battery electrodes ([12]). The results showed that the electricity (Ah) which can be drawn from such an electrode depends largely on the structure of the metal. From Ti_2-Ni-$H_{2.5}$ alloys, a maximum of 170 Ah/kg could be drawn, although the hydrogen content corresponds to 430 Ah/kg. Sintered Ni-Ti-H_x mixtures, by comparison, delivered 300 to 320 Ah/kg.

6.4 Low Pressure Nickel/Hydrogen Batteries

Metal hydrides were first used in 'low pressure Ni/H_2 cells' only for storing hydrogen ([19]). The main reasons for such a development were:

1. Reduction of the cost for the pressure resistant container.
2. Reduction of the self-discharge rate due to decreased internal H_2 pressure.

The design of these batteries is similar to that of nickel/hydrogen batteries. Platinum or palladium is still used to generate hydrogen (H_2) or to catalyse the reverse reaction, the oxidation to H^+ ions. But the catalyst is only contained in a surface layer, while the bulk of the negative electrode consists of an intermetallic compound, e.g. $LaNi_5$, which absorbs the hydrogen and forms a hydride. This main part of the negative electrode only stores the hydrogen and is not in direct contact with the electrolyte.

Oxygen, which is generated during charging and especially during overcharging, might react with the hydrogen-storing alloy and cause degradation by oxidation. To prevent this, an auxiliary carbon electrode is connected to the negative electrode. This auxiliary electrode forms an outer surface layer which does not only reduce the oxygen, but protects the storing part of the electrode against degradation by oxygen.

Batteries of this design proved to be as effective as conventional nickel/hydrogen batteries, but the internal pressure is kept below 6 atm.

6.5 Negative Electrodes for Nickel/Metal Hydride Batteries

The direct use of metal hydrides as negative electrodes was not satisfactory as long as pure titanium/nickel or lanthanum/nickel compounds were employed. The degradation in the corrosive alkaline environment proved to be one major problem ([20]). This degradation is mainly caused by lattice expansion, along with hydrogen absorption and oxidation. A continuous decay of the capacity resulted to about 50% after 100 cycles.

In the battery, the hydrogen-storing alloy is in direct contact with the electrolyte and acts as the electrode as well. So an interface layer is formed that has to resist corrosion and oxidation attacks and may undergo significant structural and compositional changes ([21]). This layer plays an important role: while the bulk of the electrode material determines the capability for hydrogen absorption and so the capacity, that interface layer is decisive for most performance data. Load capacity, given by the exchange-current density, and cycle life are such parameters. Improvements were tried by a double phase technique ([22]). The bulk of the electrode is covered in this case with a surface layer of similar material but different crystallographic phases.

AB_5 alloys had to be modified for their direct use as an electrode by substituting parts of the lanthanum (La) and the nickel (Ni) with other elements. The introduction of mischmetal instead of lanthanum has been mentioned in Section 6.3.1. But also other combinations of elements are applied, and with such 'multicomponent alloys' like $La_{0.9}Ti_{0.1}Ni_{2.5}Co_{2.0}Al_{0.5}$ and $La_{0.8}Nd_{0.2}Ni_{2.5}Co_{2.4}Si_{0.1}$ (Nd : neodymium, one of the 'rare earth' elements like lanthanum) more than 1000 full cycles were reached with less than

30% capacity loss of the negative electrode ([23]). Partial substitution of lanthanum by cerium improves the corrosion resistance of the electrode material and significantly reduces capacity decay during cycling ([17]). (*Presumably this is the main reason that the use of mischmetal instead of lanthanum improves electrode performance, because of its high content of cerium.*) Obviously, this is not solely caused by the reduction of the volume change per cycle due to cerium. Cerium moreover seems to form a protective oxide layer on the metal surfaces. This is confirmed by experiments on electrodes coated with tetravalent cerium (CeO_2) ([24]). A reduction of the capacity decay per cycle has been observed, independently of whether the electrode was coated by wet chemical methods or the alloy powder was coated by ball milling.

Alloys of the AB_2 type, based on vanadium, titanium, zirconium and nickel (V-Ti-Zr-Ni alloys) have been developed especially for battery applications under the name 'Ovonic alloys' ([25]). Batteries with these alloys are claimed to reach 1400 charge/discharge cycles ([11]).

6.5.1 Heat Effects with Hydride Formation

The formation of a stable hydride is always connected with heat generation on account of the necessarily negative value of the enthalpy of formation ΔH. For this reason, hydrogen absorption during charging produces heat and is opposed to the reversible heat effect Q_{rev}, given by Eq. (6.11), p.219.

ΔH is a measure for the strength of the chemical bond between metal and hydrogen, and varies widely from metal to metal. The varying slopes of the Van´t Hoff lines in Fig. 6.3 indicate this variation. Table 6.3 shows some examples.

	Hydride	Enthalpy of formation ΔH
		kJ/mole H
1	YH_2	-114
2	TH_2	-68
3	MgH_2	-37
4	$PdH_{0.6}$	-20
5	$NiH_{0.5}$	-3

Table 6.3 Enthalpy of formation for some elemental hydrides. Values from Table 6.6.1 in ([26]). These values correspond to the shape of the concerned Van't Hoff Lines in Fig. 6.3.

Data of some representative alloys, used in battery electrodes, are shown in Table 6.4.

All the values in Table 6.4 are rather similar, which follows from the condition that the equilibrium pressures of suitable hydrides have to be within a certain range.

	Hydride basis	Enthalpy of formation ΔH
		kJ/mole H_2
1	TiFe	-28.1 to - 33.8 ◊
2	$LaNi_5$	-29.4 to -31.6 *; 28.5 **
3	LaCeNiSn	-32.5 **
4	LaCeNiCoSn	-38.2 **
5	LaNiCoSn	-42.5 **
6	LaNiSn	-29.4 **

Table 6.4 Enthalpy of hydride formation for some intermetallic compounds hydrides.
◊ Values from Table 3.2 in ([27]).
* Values from Table 3.1 in ([27]).
** Values from Table 1 in ([28]).

Comparison with the reversible heat effect (Eq. (6.11)) shows that the heat effect by hydride formation or decomposition is in the same order of magnitude. The negative sign of ΔH in Table 6.4 means that heat is generated when the hydride is formed, i.e. when the battery is charged. As a consequence, during charging, heat generation by hydride formation more or less compensates for cooling, caused by the reversible heat effect. The considerable cooling effect, observed with nickel/cadmium batteries (cf. Fig. 5.11, p.199, and Fig. 5.13, p.201) is not to be expected with nickel/metal hydride batteries. This is shown by the charging curve in Fig. 13.22. On the other hand, heat generation during discharging is reduced in nickel/metal hydride batteries by the cooling effect of hydride decomposition (+ΔH).

6.5.2 Manufacture of the Negative-Electrode Material

The hydrogen storing alloys are usually not manufactured in battery plants, but are provided by specialised suppliers (some references are given in Section 2.5 of ([14])). Selection of the raw materials and preparation of the hydrogen-absorbing alloy are most important, because not only its composition but also its structure and the distribution of the alloying elements are decisive for capacity and service life ([29]).

To achieve an intimate mixture, melting of the components is the usual starting process. The raw materials have to be checked thoroughly, because small additions of foreign metals can change the properties of the alloys markedly. The melting as well as the solidification process have to be controlled and an important point is to ensure that the melt becomes homogeneous. Special melting and cooling techniques have to be applied to avoid segregation. Normally, annealing is applied to eliminate microsegregation by solid state diffusion and to allow the system to establish a homogeneous composition and full structural order.

At the end of the process, the alloy is grinded and pulverised material delivered to the battery manufacturer.

6.6 General Features of the Nickel/Metal Hydride System

The characteristics of nickel/metal hydride batteries are very similar to those of sealed nickel/cadmium batteries. The cell voltage differs by only 20 mV, and charging as well as discharging performance are so alike that both battery systems can be replaced by each other in all normal applications. The discharge curves in Fig. 6.6 confirm this: they differ from each other only by the discharge duration.

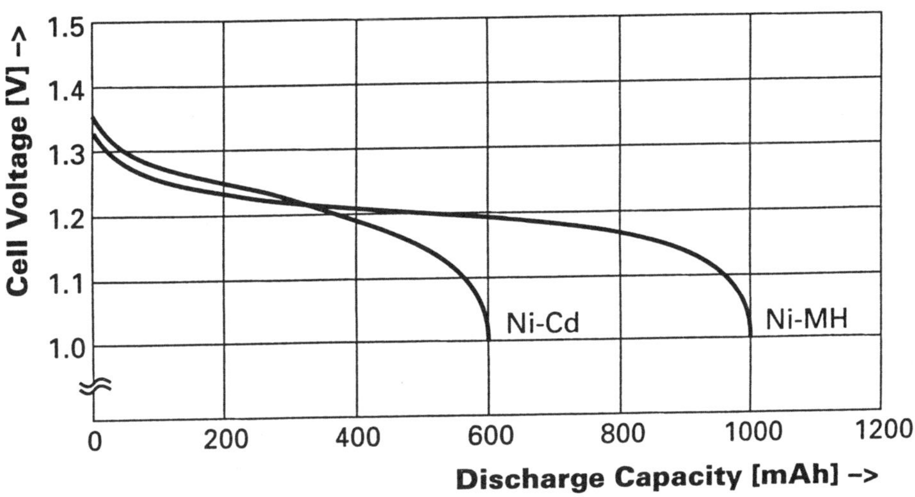

Fig. 6.6 Comparison of discharge voltage and capacity between nickel/cadmium and nickel/metal hydride batteries of the same size ([30]). Cylindrical cells, AA type, discharging current 0.2 CA (≈ 5 hours).

In general, metal hydride electrodes are inferior to cadmium electrodes at extremely high rates and low temperatures, because of the limited rate of hydrogen desorption. In practice, however, discharge performance under these conditions is mainly a question of electrode thickness. Thus, such a comparison depends largely on the special battery design, and nickel/metal hydride batteries with thin electrodes can give a better performance at low temperatures than nickel/cadmium batteries with thicker electrodes.

The required weight and the theoretical utilisation values of the substances employed in Eq. (6.4), can be evaluated from the 71.9 Wh (ΔG in Eq. (6.10) that are converted per multiple of the cell reaction. Table 6.5 shows the resulting figures. These values are based on equilibrium and thus represent maximum values.

	Substance	Molar weight g	Required quantities based on Eq. (6.4) (2F = 53.6 Ah) g/Ah	Utilisation values Ah/kg	Ah/lb.
1	NiOOH	91.72	3.421	292.3	132.6
2	$Ni(OH)_2$	92.71	3.458	289.2	131.2
3	H_2	2.016	0.0376	26,587	12,040
4	H_2 storing alloy	---	≈1.25	≈800	≈400

Table 6.5 The components in Eq. (6.4) and the required hydride forming alloy. Molar weights, required quantities and (maximum) utilisation values derived from thermodynamic data. Values in Line 4 based on the hydrogen storage capability of 1.5 wt.%.

The values of the nickel electrode are the same as those in Table 5.3, p.184, for nickel/cadmium batteries. The higher cell capacity of nickel/metal hydride batteries results from a larger volume of the positive electrode since the negative electrode requires less space. The enormous storage capability of hydrogen per weight unit is demonstrated by the utilisation values in Line 4.

The utilisation values of the active materials shown in Table 6.5 are never reached in actual batteries. In the positive electrode, values between 230 and 280 Ah/kg are reached for nickel hydroxide (cf. Table 5.4, p.184). The utilisation of hydrogen is limited to the absorption/desorption plateau of the pressure/composition isotherm (Fig. 6.2).

Table 6.6 shows data of various types of nickel/metal hydride batteries and utilisation values of the active material reached in practice.

The values in Table 6.6 accentuate the dependence of such data on battery design and size. The two cylindrical batteries in Lines 1 and 2 are of the same design and differ only in size. The small prismatic battery is based on similar electrodes, but they are folded (cf. Fig. 12.3, p.401).

The moderate values of storage capability for the button cells in the Lines 4 to 6 are mainly caused by the small size of this cell type, and to a certain extent by the 2 mm thick pressed-powder electrodes, positive as well as negative (as in Fig. 11.13, p.381). (This thickness does not count too much at the 5-hour rate). The weight-related values are largely reduced by the heavy metallic container of these small batteries.

Lines 7 and 8 concern batteries, especially designed for electric vehicle propulsion. The high-power version is capable of 200 W/kg and 550 W/dm^3. For the high-power cell in Line 7, specific energy and energy density are slightly reduced on account of the reinforced conducting elements (cf. Fig. 12.4, p.401).

Comparison between Tables 6.6 and 5.4, p.184, shows that the values for energy content in Wh/kg as well as energy content in Wh/dm^3 are considerably higher for nickel/metal hydride batteries.

	Type of battery (always single cells)	Nominal capacity (Ah)	Discharge rate (hours)	Storage capability Wh/kg	Ah/kg	Wh/dm^3
1	Cylindrical (4/4A) *	2.6	5	60	52	200
2	Cylindrical (AA) *	0.6	5	55	46	190
3	Small prismatic cell *	0.55	5	42	35	140
4	Button cell ◊	0.32	5	32	27	90
5	Button cell ◊	0.20	5	24	20	72
6	Button cell ◊	0.012	5	12	10	44
	Prismatic prototypes for electric vehicles					
7	High power cells **	60	3	60	52	155
8	High energy cell **	150	3	80	67	190

Table 6.6 Nickel/metal hydride batteries, typical values of energy content per cell weight and volume in various cell designs.
* Positive foam electrode, negative pressed (0.7 mm), data from(30, 31).
◊ Positive and negative electrode 2 mm thick, pressed, data from (32).
** Positive foam electrode, negative pressed (0.7 mm), data from (33).

6.6.1 Self-Discharge

Self-discharge in nickel/metal hydride batteries is based on the same reaction that has been described for nickel/hydrogen batteries in Section 6.1.1.1. But, due to the low hydrogen equilibrium pressure of the alloys used, the rate of self-discharge is reduced so far that it only attains rates known for nickel/cadmium batteries (cf. Fig. 5.8, p.195).

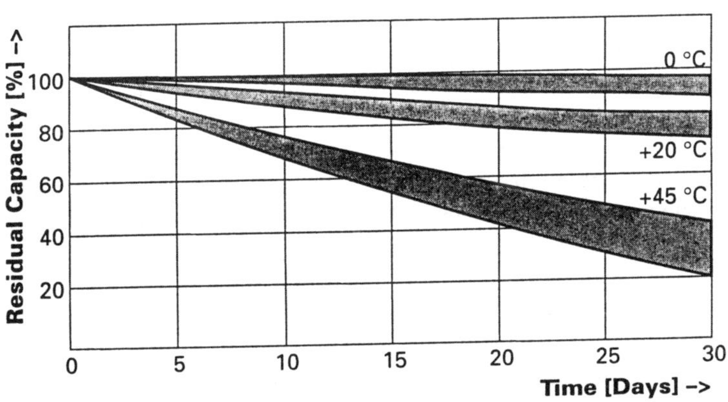

Fig. 6.7 Self-discharge of cylindrical nickel/metal hydride batteries (Lines 1 and 2 in Table 6.6) at various temperatures. (Fig. 2.11 in (30).)

Self-discharge is strongly influenced by the thickness of the electrodes as comparison between Fig. 6.7 and 6.8 shows.

The self-discharge rate is in the same range as that of nickel/cadmium batteries (cf. Fig. 5.8).

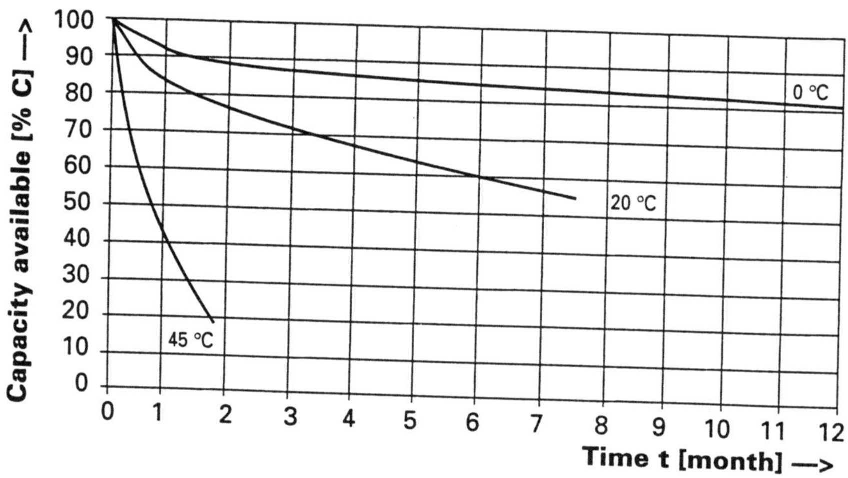

Fig. 6.8 Self-discharge of nickel/metal hydride button cells (Lines 4 to 6 in Table 6.6) at various temperatures. (Fig. 2.12 in ([32]).)

The rate of self-discharge is reduced with progressing discharge of the positive electrode.

6.6.2 Heat Effects

The heat capacity of nickel/metal hydride batteries is similar to that of nickel/cadmium batteries, as shown in the two bottom lines of Table 5.6.

As mentioned in Section 6.5.1, hydride formation and decomposition largely compensates for the reversible heat effect. This means that for nickel/metal hydride batteries $E_{cal} \approx E^o$, and heat generation is more or less determined by the Joule effect. Thus heat generation during discharge is much smaller than in nickel/cadmium batteries (cf., e.g., the heat contributions, shown in Fig. 5.11, p.199). On the other hand, heat generation during charging is higher in nickel/metal hydride batteries, since the cooling effect of the reversible heat effect is compensated by the exothermic formation of the metal hydride. For this reason, fast charging methods have to be applied more carefully for nickel/metal hydride batteries.

On the whole, a slightly increased heat generation is to be expected in nickel/metal hydride batteries, since hydrogen absorption/desorption is not quite reversible, as indicated by the hysteresis of the corresponding curves (Figs. 6.2, 6.4, and 6.5). So the absorption/desorption reaction generates a small amount of additional heat.

References

([1]) J.M.Dunlop, 'Nickel Hydrogen Batteries' *in* D.Linden, 'Handbook of Batteries and Fuel Cells', © McGraw-Hill Book Co., New York 1995, Chapter 32

([2]) D.M.Bush and J.F.Sindorf *in* 'Stationary Energy Storage: Load Leveling and Remote Applications', The Electrochem. Soc. (1988) Proc. Vol 88-11 p.318

([3]) D.Coates and R.D.Wright, *in* Proc. Symp. 'Hydrogen Storage Materials, Batteries and Electrochemistry', The Electrochemical Society, Inc, Proceedings Vol. 92-5, 1992, p.199

([4]) A.Visitin, A.Anani, S.Srinivasan, and A.J.Appleby, Journal of Applied Electrochemistry **25**, (1995), p.833

([5]) E.W.Justi and A.W.Winsel, 'Kalte Verbrennung (Fuel Cells)', Franz Steiner Verlag, Wiesbaden ,1962, p.69 (in German)

([6]) G.Sandrock in Proc. Symp. 'Hydrogen and Metal Hydride Batteries, Editors D.Bennett and T.Sakai, © The Electrochemical Society, Inc, Proceedings Vol. 94-27, 1994, p.1 (a survey with 60 references).

([7]) W.M.Mueller, J.P.Blackledge and G.G.Libowitz, 'Metal Hydrides', Academic Press, New York, 1988

([8]) L.Schlapbach (Editor), Hydrogen *in* Intermetallic Compounds I and II, Vols. 63 and 67 *in* Topics in Applied Physics, © Springer Verlag, Berlin - Heidelberg 1988 and 1992
G.G.Libowitz in Proc. Symp. 'Hydrogen Storage Materials, Batteries and Electrochemistry', The Electrochemical Society, Inc, Proceedings Vol. 92-5, 1992, p.3

([9]) J.S.Eager and V.J.Puglisi, Nickel-Metal Hydride Performance and Environmental Advantages, *in* The Fourth International Rechargeable Battery Seminar, Deerfield Beach, Florida, March 1992, Shawmco, Inc. P.O.Box 9495, Tulsa, OK 74137 USA

([10]) K.Kajita and Y.Uetani, Ni/MH Batteries using Muliphase-Multicomponent Alloys, *in* The Tenth International Seminar on Primary and Secondary Battery Technology an Application, Deerfield Beach, Florida, 1993, © Florida Educational Seminars, Inc.

([11]) M.A.Fetcenco, S.Venkatesan and S.R.Ovshinsky *in* Proc. Symp. 'Hydrogen Storage Materials, Batteries and Electrochemistry 1992', The Electrochemical Soc.,Proc. Vol.92-5, p. 141

([12]) E.W.Justi, H.H.Ewe, A.W.Kalberlah, N.M.Saridakis and M.H.Schaeffer, Energy Conversion, **10**, 1970, p. 183
M.A.Gutjahr, H.Buchner, K.D.Beccu, and H.Säufferer, in Power Sources, Vol.IV, Edited by D.H.Collins, Oriel Press, Newcastle upon Tyne, 1973, p. 79

([13]) S.Mukerjee, J.McBreen, J.J.Reilly, J.R.Johnson, G.Adzic, M.P.S.Kumar, W.Zhang and Srinivasan, in Proc. Symp. 'Hydrogen and Metal Hydride Batteries', Editors D.Bennett and T.Sakai, © The Electrochemical Society, Inc, Proceedings Vol. 94-27, 1994, p.33

([14]) A.Percheron-Guégan and J.-M.Welter, Preparation of Intermetallics and Hydrides, *in* L.Schlapbach (Editor), Hydrogen in Intermetallic Compounds I, Vol.63 of Topics in Applied Physics, © Springer Verlag, Berlin-Heidelberg 1988, p.11

([15])T.Sakai, K.Muta, H.Miyamura, N.Kuriyama and H.Ishikawa, *in* Proc. Symp.'Hydrogen Storage Materials, Batteries and Electrochemistry', The Electrochemical Society, Inc, Proceedings Vol. 92-5, 1992, p.59., esp. p.62

(16) A Percheron-Guégan, M.Latroche, J.C. Achard, Y.Chabre and J.Bouet, in Proc. Symp. 'Hydrogen and Metal Hydride Batteries', Editors D.Bennett and T.Sakai, © The Electrochemical Society, Inc, Proceedings Vol. 94-27, 1994, p.196 (a review with 37 ref.)

(17) G.D.Adzic, J.R.Jonson, J.J.Reilly, J.McBreen, and S.Mukerjee, in Proc. Symp. 'Hydrogen and Metal Hydride Batteries', Editors D.Bennett and T.Sakai, © The Electrochemical Society, Inc, Proceedings Vol. 94-27, 1994, p.24

(18) H.Buchner, M.A.Gutjahr, K.D.Beccu and H.Säufferer, Z.Metallkde.,**63** (1972), 497

(19) T.L.Markin, R.Bennett, N.J.Bridger and R.M.Dell, 'Power Sources 8', Proc. of the 12th Internat.Power Sources Symp., Brighton 1980, p.445

(20) H.F.Bittner and C.C.Badcock, J. Chem. Soc., **130** (1983), 193 C

(21) L.Schlapbach, F.Meli, and A.Zuettel, in Proc. Symp. 'Hydrogen and Metal Hydride Batteries', Editors D.Bennett and T.Sakai, © The Electrochemical Society, Inc, Proceedings Vol. 94-27, 1994, p.102

(22) P.H.L.Notten and P. Hokkeling, J. Electrochem. Soc., **138** (1991), p. 1877

(23) J.R.van Beek, H.C.Donkersloot and J.J.G.Willems, 'Power Sources 10', Proc. 14th Internat.Power Sources Symposium, Brighton, 1984, p.317

(24) M.P.S. Kumar, W.Zhang, K.Petrov, G.D.Adzic, J.R.Johnson, J.J Reeilly, and R.B.Schwarz, in Proc. Symp. 'Hydrogen and Metal Hydride Batteries', Editors D.Bennett and T.Sakai, © The Electrochemical Society, Inc, Proceedings Vol. 94-27, 1994, p.111

(25) M.A.Fetcenko, S.Venkatesan, K.C.Hong and B.Reichman, 'Power Sources 12', Proc. 16th Internat. Power Sources Symposium, Bournemouth, 1988

(26) R.Griessen and T.Riesterer, Heat of Formation Models, *in* L.Schlapbach (Editor), Hydrogen in Intermetallic Compounds I, Vol.63 of Topics in Applied Physics, © Springer Verlag, Berlin-Heidelberg 1988, p.219

(27) T.B.Flanagan and W.A. Oates, Thermodynamics of Intermetallic Compound-Hydrogen Systems, *in* L.Schlapbach (Editor), Hydrogen in Intermetallic Compounds I, Vol.63 of Topics in Applied Physics, © Springer Verlag, Berlin-Heidelberg 1988, p.49

(28) W.Zhang, MB.S.Kumar, a. Visitin, an S.Srinivasan, in Proc. Symp. 'Hydrogen and Metal Hydride Batteries', Editors D.Bennett and T.Sakai, © The Electrochemical Society, Inc, Proceedings Vol. 94-27, 1994, p.249

(29) M.Tadokoro and coworkers, Proc. Symp. 'Hydrogen Storage Materials, Batteries and Electrochemistry 1992' The Electrochemical Soc.,Proc. Vol.92-5, p.92

(30) VARTA, 'Ni/MH Cylindrical and Prismatic Cells, Sales Program and Technical Handbook', 43221e, © by VARTA Batterie AG, Daimlerstr.1, D-73479 Ellwangen/Jagst, Germany

(31) SANYO 'Nickel-Metal Hydride Rechargeable Batteries *Twicell*', SANYO Electric Co, Ltd, Sumoto-city, Hyogo, Japan

(32) VARTA, 'Ni/MH Button Cells, Sales Program and Technical Handbook', 43321e, © by VARTA Batterie AG, Daimlerstr.1, D-73479 Ellwangen/Jagst, Germany

(33) K.Brandt, U.Köhler, E.Niggemann, F.Böttcher, and F:J.Kruger, Proc. 13th Intern. Seminar on Primary and Secondary Battery Technology and Application, March 1996, Boca Raton, Florida

7. Loss of Electrolyte and Water

The electrolyte is an essential component of the battery. It provides the ionic conductance and has to be present in the correct concentration and at a sufficient volume. Otherwise, the internal resistance of the battery is increased and its capacity reduced, especially when parts of the electrodes are no longer in contact with the electrolyte. In lead-acid batteries, in addition, the electrolyte contributes components of the electrode reactions. This increases the dependence of the capacity on the available amount of electrolyte.

From a flooded battery, electrolyte can be lost by spilling, if a leakage is formed in the container, or when the battery is tilted too much. Loss of electrolyte can also be caused by the formation of fumes that are carried out of the battery by a gas stream.

The volume of the electrolyte is also reduced by water loss. The three battery systems treated in this book are based on aqueous electrolytes: diluted sulphuric acid or diluted potassium hydroxide, and the diluting component of the electrolyte, the water, can be lost by evaporation as described in Section 2.6.5, or by electrolytic decomposition that occurs as soon as the cell voltage is higher than 1.23 V, as has been pointed out in Section 2.6.1. In lead-acid batteries, grid corrosion is an additional water consuming reaction (cf. Eq. (4.27), p.118).

Thus electrolyte or water loss can happen in four different ways:

1. Direct loss of electrolyte by spilling, container leakages, or as fumes (Section 7.1).
2. Loss as water vapour by evaporation. This can be direct loss by permeation of vapour through the container walls (Section 2.6.5.2). Another kind of vapour loss is the saturation of hydrogen and oxygen gases that escape from the cell.
3. Loss of water by electrolytic decomposition and escape of gaseous hydrogen or oxygen.
4. Loss of oxygen from the aqueous component by absorption of oxygen when PbO_2 is formed from Pb by grid corrosion (Section 7.2.2.3).

Direct loss of electrolyte mainly concerns flooded batteries. The danger of spillage was one of the motives to immobilise the electrolyte and make the batteries spill-proof. Fumes of electrolyte are formed at the surface of the liquid electrolyte, and fast transport of these out of the cell requires heavy gassing rates. For both reasons, loss of electrolyte by fumes is also mainly associated with flooded batteries and intense gassing rates.

Water loss according to the items 2 to 4 is of special interest for valve-regulated lead-acid batteries, because it cannot be compensated for by refilling, and the initial amount of electrolyte includes only little reserve, due to its immobilisation. On the other hand, water loss is unavoidable. Hydrogen evolution at the negative electrode can never be completely suppressed, and oxygen is consumed at the positive electrode by grid corrosion. Furthermore, these batteries are usually encapsulated in plastic containers that are permeable to a certain extent to water vapour and gases. For all these reasons, water loss is one of the major causes of capacity decrease with valve-regulated lead-acid batteries ([1]).

For nickel/hydrogen and nickel/cadmium batteries of virtually sealed design and encapsulated in metallic housings, all four possibilities of electrolyte or water loss are not relevant. Even, when these batteries are equipped with a resealing valve which opens only in an emergency, (cf. the note in Section 1.1), water loss can be neglected as long as the battery is operated more or less properly.

7.1 Direct Loss of Electrolyte, Formation of Fumes

Direct loss of electrolyte by spilling or by leakage of the cell container can be neglected in valve-regulated lead-acid, sealed nickel/cadmium, and nickel/metal hydride batteries since they do not contain liquid electrolyte, on account of the immobilisation described in Sections 4.6.4 and 5.5.2. For this reason, electrolyte cannot be spilled from intact batteries and, even if the container shows cracks, only a few droplets of electrolyte can leave the cell. To be spill-proof is one of the advantages of this battery design.

Electrolyte fumes are formed in batteries with liquid electrolyte when rising gas bubbles burst at the surface of the electrolyte. Fine droplets of sulphuric acid or potassium hydroxide are sprayed as an aerosol. If the particles are fine enough, they can float in the gas and may be carried away with the gas stream leaving the cell. For this reason, electrolyte fumes are mainly observed when the battery is heavily overcharged.

The quantity of electrolyte that is lost from batteries as fumes can be neglected. Nevertheless, even traces of sprayed electrolyte can be a nuisance, because the droplets are deposited in the area surrounding the battery opening.

- Sulphuric acid remains there for a time because the acid itself is not vaporised; rather its concentration is governed by the relative humidity, as described in Section 2.6.4.5. Corrosion occurs if the acid fumes reach metallic surfaces; furthermore, the wet surface of the cells gets dirty by absorbing dust, and so the insulation between cells and against the ground can be reduced.

- Potassium hydroxide, when in contact with the atmosphere, is gradually converted into carbonate according to

$$\boxed{KOH + CO_2 \Rightarrow KHCO_3} \quad (7.1)$$

  The generated carbonate forms a white crust of crystals on the cell which may affect the insulation resistance within the battery or between the battery and the ground.

From nickel/cadmium and nickel/metal hydride batteries, gas does not escape under normal operational conditions, thus fumes of electrolyte also cannot leave the cell. In valve-regulated lead-acid batteries, liquid electrolyte is not present; consequently, gas bubbles cannot burst at the surface. Furthermore, there is no continual gas stream leaving the battery to carry the fumes. Thus, also these batteries normally do not produce electrolyte fumes.

Note: *The formation of carbonate crystals is sometimes observed with sealed nickel/cadmium batteries. However, these precipitations are not caused by fumes, but by electrolyte creeping through the seal.*

7.2 Water Decomposition in Lead-Acid Batteries

Fig. 7.1 illustrates the situation for the lead-acid battery. It corresponds to Fig. 4.1 (p.97), but here the secondary reactions are inserted as curves that indicate their rates. The electrode potential again is plotted at the x-axis; the negative electrode range on the left and the positive electrode range on the right. The two hatched vertical columns represent the equilibrium potentials of the negative and positive electrodes. The width of these columns indicates that these potentials are influenced by the concentration of sulphuric acid (cf. Fig. 2.3, p.19). The potential difference of 1.23 V between the beginning of hydrogen and oxygen evolution marks the water decomposition voltage, according to Eq. (2.101), p.45.

The broken curves correspond to current/potential curves for gas evolution and corrosion rates, cf. Fig. 7.9, p.253. They indicate the rate of these reactions as a function of the electrode potential. The gradual increase of these curves at the beginning indicates that these reactions occur slowly, as long as the potential difference relative to their origin, the overvoltage (cf. Section 2.3.3), remains fairly small. When, however, the overvoltage exceeds a certain value, the two curves for gas evolution show a steep increase. This means that hydrogen as well as oxygen generation can gain in volume enormously when the electrode potentials are altered accordingly.

Fig. 7.1 shows that hydrogen and oxygen evolution already occur at the equilibrium potentials of the negative and positive electrodes, respectively, although at a comparably low rate. This causes self-discharge of both electrodes according to Eq. (4.23), p.116, and Eq. (4.26), p.118, and results in self-discharge rates as shown in Table 4.8, p.119.

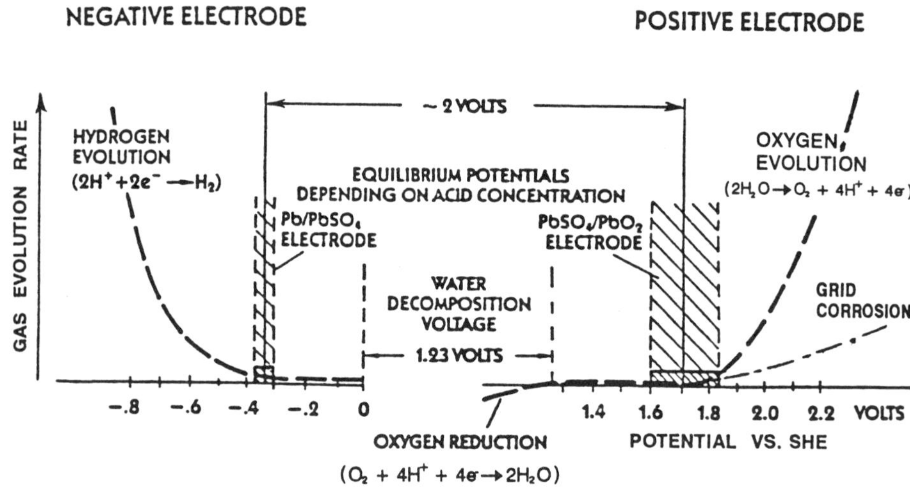

Fig. 7.1 Equilibrium potentials of the negative and positive electrodes in a lead-acid battery, and rates of hydrogen and oxygen evolution and grid corrosion indicated by reaction rate/potential curves. *(Reaction rates can be expressed as current densities, then the hydrogen-evolution rate is equivalent to a **negative** current, as shown in Fig. 7.9.)*

Corrosion of the positive grid that leads to the protecting PbO_2 layer, described in Fig. 4.26, p.153, also is an electrochemical reaction. The corrosion rate is at its minimum at about 40 to 80 mV above the $PbSO_4/PbO_2$ potential, as mentioned in Section 4.7.4.1. At an electrode potential below this minimum, corrosion increase is caused by destabilisation of the protecting layer. At overvoltages larger than this minimum, the corrosion rate also increases exponentially with increasing electrode potential, but this increase is significantly smaller than that of the oxygen-evolution rate, as indicated in Fig. 7.1 (cf. also Fig. 4.28, p.161 and Fig. 7.5, p.242)

Oxygen reduction is possible at an electrode potential below the equilibrium potential of the oxygen electrode of about 1.23 V vs. SHE (Standard Hydrogen Electrode) in acid solutions (dependence on concentration of the electrolyte cf. Section 2.6.1.3). Thus it is always possible at the negative electrode. Together with oxygen generation at the positive electrode, it forms the internal oxygen cycle, which is the main reaction when valve-regulated lead-acid batteries are overcharged. The rate of this reaction is largely determined by the rate of oxygen transport. In flooded batteries, it is greatly hindered due to the slow diffusion rate of oxygen dissolved in the electrolyte. Thus, in stationary batteries of the vented type with usual plate spacing, the internal oxygen cycle is limited to a few mA per 100 Ah and in general is not noticed under normal float conditions (cf. Section 2.6.3.1). Only when batteries are float charged at quite low voltages with resulting low float currents, can varying efficiency of the internal oxygen cycle in the individual cells split the battery into groups of deviating float voltages ([2]). An increased rate of oxygen reduction is furthermore observed in flooded batteries that have lost so much water that the connecting elements or even parts of the electrodes are no longer immersed in the electrolyte and can easily be reached by the oxygen gas.

7.2.1 TAFEL Lines

As shown in Section 2.3.3, for many electrochemical reactions the current is an exponential function of the electrode potential. This causes a very strong increase of the current/potential curve, which suggests plotting the current in a logarithmic scale to cover a wide range. Such a semilogarithmic plot has the further advantage that the current/voltage curve forms a straight line, provided that the polarisation is so large that the reverse reaction can be neglected (cf. Section 2.3.3.1). In the lead-acid battery, this is true for the hydrogen and oxygen gas-evolving reactions, since they are polarised by more than 0.3 V already at the open-circuit voltage of the battery. Consequently, the voltage/log(current) relationship is a TAFEL equation according to Eq. (2.62), p.28:

$$\eta = E - E^{o} = a + b \cdot \log(|i|) \qquad (7.2)$$

Such a linearisation more easily allows quantitative considerations.

Fig. 7.2 shows such curves, calculated according to Eq. (2.61), p.28 for hydrogen and oxygen evolution at various temperatures.

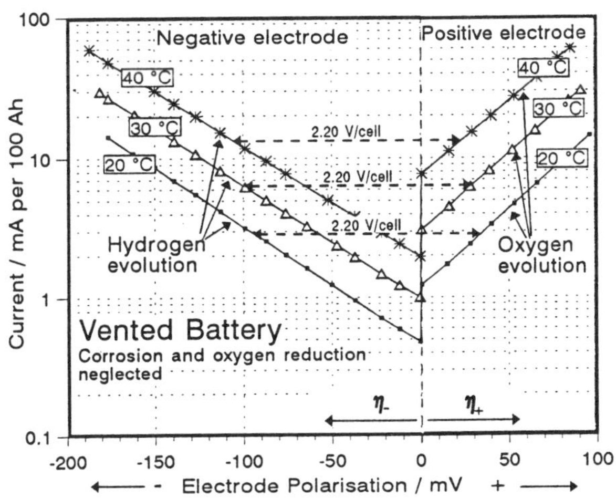

Fig. 7.2 Hydrogen- and oxygen-evolution rates versus electrode polarisation, calculated for a 'standard vented battery' (TAFEL lines). The zero point of the horizontal axis includes the open-circuit voltage of the battery, (*2.08 V/cell at the assumed acid density of 1.24 g/cm³*). (Fig. 2 in (3).)

***Note 1**: The relationship between gas-evolution rates and nominal capacity in Fig. 7.2 is, strictly speaking, not correct. It presupposes a fixed ratio between the 'electrochemical effective surface area' and the nominal capacity or rated capacity of the electrode, because current density can only be referred to a surface area. Several parameters like porosity of the active material and plate thickness will influence this ratio. However, the effective surface area is not known, and the above relationship is a workable approximation. In practice, mainly the vertical position of the TAFEL lines varies for different battery types.*

Fig. 7.2 shows the characteristic slopes of hydrogen- and oxygen-evolution curves versus electrode polarisation: the equivalent current is increased by one order of magnitude, when the electrode polarisation is increased by about 120 and 90 mV for the negative and positive electrodes, respectively.

Note 2: *Such data depend, of course, to some extent on the kinetic data of the battery concerned. The 'standard vented battery' used here represents an average that fairly well approximates modern stationary batteries. Tafel lines for hydrogen and oxygen evolution have been confirmed by many experiments. A slope 'b' (the b factor), of about 0.12 V per current decade is nearly always found for hydrogen evolution; while the corresponding slope for oxygen evolution is between 0.07 and 0.14 V per current decade (cf., e.g., ([4])).*

The general rule is confirmed that currents are approximately doubled, when the temperature is increased by 10 °C (Section 2.3.6). Temperature effects on O_2 evolution are stronger than on H_2 evolution (which corresponds to experimental results).

The inserted arrows represent the situation when the battery is on float charge at 2.20 V/cell. With the data, used for this calculation, polarisation values of positive and negative electrode amount to 30 and -90 mV, respectively, at 20 °C. At a constant float voltage, η_+ is reduced by about -6 mV per 10 °C with increasing temperature, while η_- is correspondingly increased (cf. Fig. 8.19, p.296). This effect is important in regard to temperature compensation of float voltage (cf. Section 13.1.1.2).

An experimental example for TAFEL lines and the balance between float current and cell voltage is shown in Fig. 7.3. To gain these values, the fully charged battery had been overcharged with stepwise increased currents (filled circles). Constant voltage was attained before the current was switched to the next step. The corresponding procedure was applied subsequently with decreasing current steps (open circles). Little difference is observed between the two series of measurements. This confirms that an approximately stable state was reached at each step.

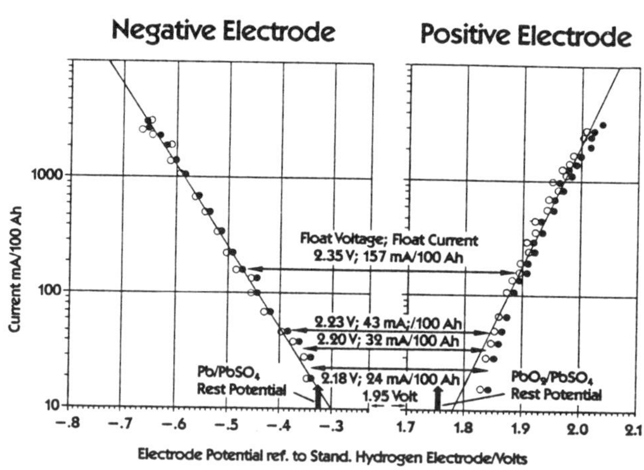

Fig. 7.3 TAFEL lines during overcharge of a flooded lead-acid battery (VARTA bloc 614) (N.C. 50Ah) after 3 yrs of float-charge service. Acid density 1.24 g/cm^3; Room temperature.

The two lines in Fig. 7.3 represent slopes of -0.12 V per current decade for hydrogen evolution and 0.08 V/decade for oxygen evolution. The experimental data fit both these curves well. Some deterioration from the TAFEL lines can be observed at low currents. This is mainly caused by the fact that at low float voltages the current shares consumed by grid corrosion and the internal oxygen cycle gain importance (cf. Fig. 7.5). Thus the simple consideration of only hydrogen and oxygen generation is increasingly disturbed.

The mutual dependence between overcharge current and overcharge voltage is illustrated by the double arrows in Fig. 7.3. When the voltage is prefixed, a certain current is the consequence; when the current is prefixed, a certain voltage results. In both cases, the polarisation of the electrodes attains values that fulfil the general rule that equal currents flow through both electrodes.

Addition of the slopes of the two TAFEL lines (0.12 + 0.08 V/current decade) leads to the well known rule for flooded lead-acid batteries that an increase of the overcharge voltage by 0.2 V increases the current tenfold which can be expressed by:

$$\boxed{\frac{E_1}{E_2} = 0.2 \cdot \log\left(\frac{i_1}{i_2}\right)} \tag{7.3}$$

Fig. 7.4 shows such current/voltage curves. The lines are drawn with a slope of 0.2 V/decade. The influence of the temperature is considered according to the general rule that the reaction rates are doubled per 10 K of temperature increase (Eq. (2.68), p.33). Although the slope of the lines and their temperature dependence are rough approximations, the inserted experimental data fit fairly well to the lines.

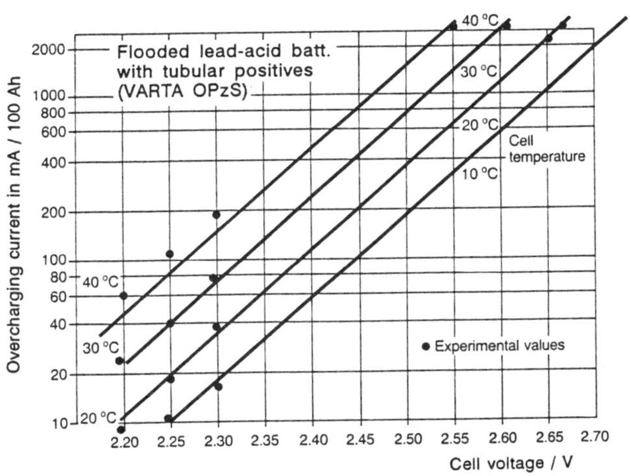

Fig. 7.4 Current voltage curves for a flooded lead-acid battery in semilogarithmic scale. The lines are drawn according to the rules:

1. A voltage increase of 0.2 V causes a current increase by one order of magnitude,
2. Temperature increase by 10 K doubles the current (cf. Fig. 7.2).

The filled circles are experimental values.

Current/voltage curves as shown in Fig. 7.4 are more or less general for flooded lead-acid batteries. For different battery types these curves are shifted upwards or downwards, since the current value at a certain cell voltage is determined by cell design (cf. above Note). Batteries with narrowly spaced thin plates have a higher float voltage. The grid alloy is also of influence, and in batteries with antimonial grid alloys the curves are gradually shifted to higher current values on account of the increased hydrogen-evolution rate (cf. Section 7.2.3.2).

Note 3: *The straight lines in Fig. 7.4 are often called TAFEL lines. Strictly speaking, however, the TAFEL relation can only be applied to the $I = f(U)$ relation of a single-electrode reaction. (It has been recognised by J.TAFEL for hydrogen evolution at noble-metal electrodes in 1905.)*

In the Figures 7.2 to 7.4, only hydrogen and oxygen evolution have been considered. In Fig. 7.1, however, two further secondary reactions are mentioned: grid corrosion and oxygen reduction. Fig. 7.5 shows a plot in semilogarithmic scale that also considers these reactions. Furthermore, the charging characteristics of both electrodes are indicated by broken curves. These charging reactions have very steep current/potential curves, since these reactions are very fast, and small polarisations already cause high current rates. Their curves are not linearised by the semilogarithmic plotting, and with progressing charge, their slope decreases since less and less discharged material is available. In the Figure, this is indicated by the curved arrows.

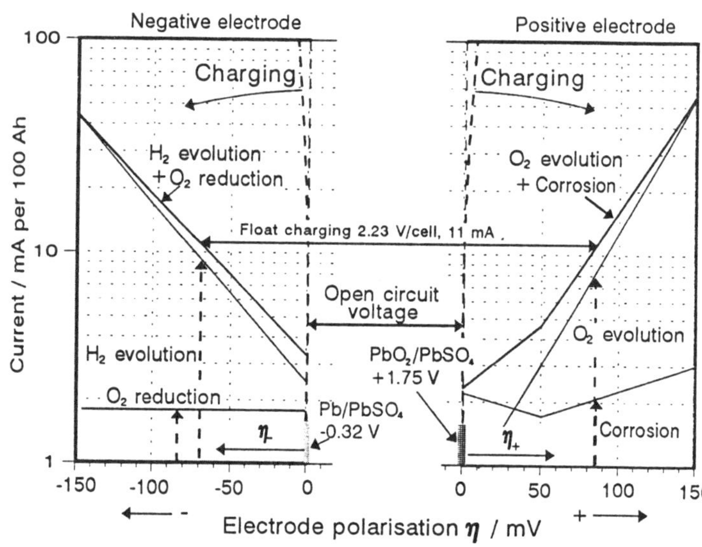

Fig. 7.5 Charging curves, TAFEL lines for hydrogen and oxygen generation and corrosion, and total-current curves for a flooded lead-acid battery in a semilogarithmic plot. Polarisation is referred to the equilibrium potentials of the $PbSO_4/PbO_2$ and the $Pb/PbSO_4$ electrodes i.e. cell voltage $E = E^o + \eta_+ - \eta_-$. The open-circuit voltage E^o = 2.07 V/cell corresponds to an acid density of 1.22 g/cm^3. For oxygen reduction a limiting-current density of 1.7 mA/100 Ah is assumed (cf. Note 3 below).

When the battery is fully charged, only the secondary reactions remain. Fig. 7.5 illustrates this situation: the current that flows through the negative electrode is the sum of hydrogen evolution and oxygen reduction (H_2 evolution + O_2 reduction), while at the positive electrode, the current mainly causes oxygen evolution, but corrosion always contributes. At increased polarisations, the contributions of O_2 reduction and corrosion are negligibly small, but at small float voltages they disturb the simple TAFEL lines as has been mentioned in connection with Fig. 7.3.

Hydrogen evolution, oxygen evolution and corrosion occur independently according to their individual current/potential curves. The slopes of the hydrogen- and oxygen-evolution curves are 0.12 and 0.08 V/decade, respectively. (*As already mentioned, these data depend on the kinetic parameters of the considered battery.*)

Corrosion depends on electrode potential in a more complex way. As described in Section 4.7.4.1, it is at its minimum at 40 to 80 mV above the $PbSO_4/PbO_2$ equilibrium potential. At a more positive potential it follows more or less a TAFEL relation as indicated in Fig. 4.28, p.161, with a slope of about 0.25 V/decade of current. In Fig. 7.5, this behaviour is approximated by two TAFEL lines: a falling one at small polarisation values and an increasing one at higher polarisations. This is, of course, an arbitrary approximation but reflects the general experience fairly well. Corrosion depends on quite a number of parameters, like grid alloy composition and microstructure, and corrosion can furthermore be a function of time. The values used here were selected for model calculations ([5]) (cf. Section 8.3.3). They are based on experimental data and a larger number of data in the literature. Although they are of a more or less general character, larger deviations are possible for individual batteries.

Oxygen reduction is assumed as a constant current, because of the limitation by oxygen diffusion. The value of 1.7 mA/100 Ah used here corresponds to the estimation given in Section 2.6.3.1 (cf. Note 4 below).

The resulting polarisation of both electrodes is determined by the general rule that in total, the current flow through the positive electrode equals the current flow through the negative electrode. This is shown for float charging at 2.23 V/cell by the corresponding vertical arrows: both electrodes are polarised so that the current equivalent of corrosion + oxygen evolution equals that of hydrogen evolution + oxygen reduction.

Note 4: *A simple plot of TAFEL lines, as shown in Fig. 7.5, does not consider the fact that O_2 reduction at the negative electrode is linked to O_2 evolution at the positive. At very small polarisation values, oxygen reduction may no longer be limited by oxygen transport, but by the rate of oxygen generation at the positive electrode. The O_2 reduction current is then smaller than the limiting current and additional polarisation of the negative electrode is required to generate a corresponding increased amount of hydrogen (cf. Section 8.3).*

Fig. 7.6 illustrates the situation for a flooded battery: on the left during charging when most of the current is used for recharge, and on the right during overcharging or float charging when the battery is fully charged and only the secondary reactions remain. During charging, hydrogen evolution and (to a small extent) oxygen reduction occur as secondary reactions at the negative electrode, while oxygen evolution and grid corrosion are secondary reactions at the positive electrode. As long as enough discharged material is available, the charging reaction dominates and the secondary reactions are small in comparison. With progressing recharge, the efficiency of the charging reaction gradually decreases, as indicated by the downwards turned arrows, corresponding to the curved arrows in Fig. 7.5. When the cell voltage remains constant, simultaneously, the current is reduced, and the shares of the various reactions are changed. But the current in total always must be the same for both electrodes.

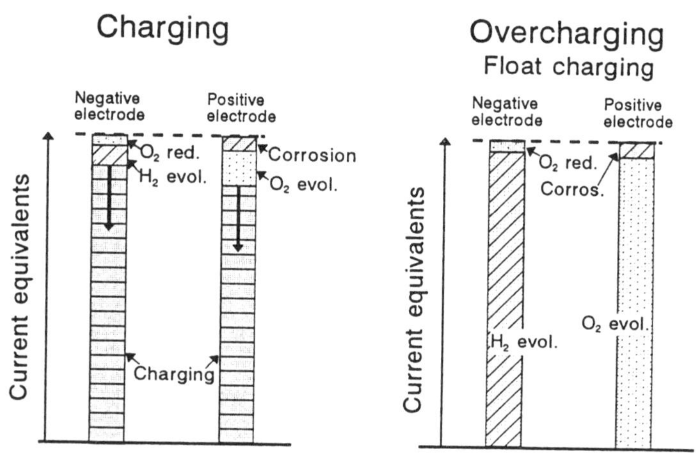

Fig. 7.6 Current contributions of the various reactions during charging and overcharging of flooded lead-acid batteries. Each reaction follows its individual current/voltage relation. Polarisation of positive and negative electrodes automatically amounts to values where the same current flows through both electrodes. On the left, the situation during charging; on the right, overcharging or float charging.

7.2.2 The Rate of Water Loss, Self-Discharge

At open-circuit voltage, the rates of the secondary reactions are determined by the intersections between the rest or equilibrium potential of the negative and positive electrodes and the corresponding gas-evolution and corrosion curves. The secondary reactions form mixed potentials with the corresponding discharge reaction that cause self-discharge according to Eqs.(4.23) to (4.27) in Section 4.4.1.

In Fig. 7.5, the hydrogen-evolution rate at zero overvoltage is equivalent to 2.5 mA per 100 Ah of nominal capacity. Such a rate is equivalent to a self-discharge rate of about 2% per month, which is typical for modern stationary batteries (cf. Table 4.8 in Section 4.4.3).

An oxygen-reduction rate equivalent to about 2 mA/100 Ah, as shown in Fig. 7.5, is only reached during float charge, when oxygen is generated at the positive electrode at a sufficient rate and has to pass only the small distance between the electrodes in the battery. Oxygen intake through the vent of a flooded battery does not cause a significant self-discharge rate, since the transport of dissolved oxygen from the head space to the electrodes is too slow on account of the small diffusion coefficient. In valve-regulated batteries, however, oxygen intake can cause a considerable self-discharge rate because of the rapid gas transport within the immobilised electrolyte.

At the positive electrode, two self-discharge processes occur simultaneously: oxygen evolution and grid corrosion. Both are balanced by discharge of the active material according to Eqs. (4.26) and (4.27), p.118. Extrapolation of the TAFEL line for oxygen evolution in Fig. 7.5 results to an equivalent of about 0.6 mA/100 Ah, and this characterises the usual situation at open-circuit i.e. that the rate of oxygen evolution amounts to only a fraction of that of hydrogen evolution, which equals 2.3 mA/100 Ah for this example. Corrosion usually is more important for self-discharge of the positive electrode and can cause considerable damage during prolonged storage periods (cf. Section 4.7.1). In Fig. 7.5 it amounts to about 2 mA, and normally it can be expected in such an order of magnitude.

7.2.2.1 Water Loss by Hydrogen Evolution

The rate of self-discharge of the negative electrode determines the minimum hydrogen evolution of the fully charged battery. Only during discharge does the potential of the negative electrode attain positive values compared to its equilibrium, and H_2 evolution is further reduced. During any kind of charging, the electrode potential is more negative and the hydrogen evolution increased. Also during float charging, a certain polarisation is required to preserve the fully charged state (cf. Section 13.1.1.1), and this always means increased hydrogen evolution. Consequently,

- the self-discharge rate describes the minimum hydrogen evolution that can be achieved for the fully charged lead-acid battery.

Nearly all the hydrogen gas that is generated escapes from the cell. Only negligible traces of hydrogen are oxidised ($H_2 \Rightarrow 2{\cdot}H^+ + 2{\cdot}e^-$) at the positive electrode and remain within the cell, forming an 'internal hydrogen cycle'.

Hydrogen evolution means water loss, since the hydrogen is taken from the water content of the electrolyte and the reaction

$$2{\cdot}H^+ + 2{\cdot}e^- \Rightarrow H_2 \qquad (7.4a)$$

must always be completed by a corresponding oxidation reaction

at the positive electrode, e.g. $$H_2O \Rightarrow ½O_2 + 2{\cdot}H^+ + 2{\cdot}e^- \qquad (7.4b)$$

to the reaction $$\boxed{H_2O \Rightarrow H_2 + ½O_2} \qquad (7.4)$$

For this reason, the hydrogen-evolution rate does not only mean self-discharge of the negative electrode, but also water loss from the battery.

Reaction (7.4b) describes the direct oxygen evolution at the positive electrode, but this complementary reaction can also be self-discharge of the positive electrode or grid corrosion according to Eqs. (4.26) or (4.27), p.118. The result is the same.

Such a water loss also occurs when the negative electrode firstly had suffered self-discharge according to Eq. (4.23):

$$Pb + H_2SO_4 \Rightarrow H_2 + PbSO_4 \qquad (7.5a)$$

and then is recharged (Eq. (4.3)) $$PbSO_4 + 2 \cdot H^+ + 2 \cdot e^- \Rightarrow Pb + H_2SO_4 \qquad (7.5b)$$

with the result $$\boxed{2 \cdot H^+ + 2 \cdot e^- \Rightarrow H_2} \qquad (7.5)$$

that has to be completed corresponding to Eq. (7.4).

Some values for the equivalents of hydrogen-evolution reaction are listed in Table 7.1.

H_2-evol. rate	Capacity loss		H_2 evolution*		Water loss**	
	per month	per year	cm^3/hour	cm^3/day	g/month	g/year
1 mA/100 Ah	0.72%	8.76%	0.46	11.0	0.24	2.9
2 mA/100 Ah	1.44%	17.5%	0.91	21.8	0.48	5.9
4 mA/100 Ah	2.88%	35.0%	1.83	43.9	0.97	11.8
8 mA/100 Ah	5.76%	70.1%	3.66	87.8	1.93	23.6

Table 7.1 Self-discharge rate of the Pb electrode by hydrogen evolution, and related equivalent values of capacity loss, hydrogen-evolution rate and water loss.
All values related to 25 °C and one cell with 100 Ah of nom. capacity.
* 53.6 Ah per mole H_2 ($2 \cdot e^-$ in Eq. (7.4a)); equivalent to 24.5 dm^3 per mole H_2 at 25 °C (cf. Table 2.6, p.48).
** 1 mole $H_2 \approx$ 18.02 g H_2O.

The initial water content of a compact designed battery is about 10 g per Ah of nominal capacity (≈16 g of acid of the density 1.28) *in starved-electrolyte designs it can be only 8 g of water per Ah.* Thus, 10 g H_2O loss per 100 Ah ≈ 1% of the initial water content. The self-discharge rate equivalent to a hydrogen-evolution rate of 2.5 mA/100 Ah, as shown in Fig. 7.5, corresponds to a water loss of 7.4 g/100 Ah per year. This means that 74 g of water would be lost by this reaction in 10 years, which are about 7.4% of the initial water content.

7.2.2.2 Water Loss by Oxygen Evolution

Self-discharge caused by oxygen evolution corresponds to the above described hydrogen evolution. It represents the minimum value of the oxygen-evolution rate of a fully charged positive electrode. Only at a potential lower than the equilibrium, is the oxygen-evolution rate reduced, but then the electrode is discharged. Oxygen evolution also means water loss, since

the reaction $$H_2O \Rightarrow \tfrac{1}{2}O_2 + 2 \cdot H^+ + 2 \cdot e^- \qquad (7.6a)$$

has to be completed by $$2 \cdot H^+ + 2 \cdot e^- \Rightarrow H_2 \qquad (7.6b)$$

to the reaction $\boxed{H_2O \Rightarrow ½O_2 + H_2}$ (7.6)

This is also true when the oxygen primarily was absorbed by self-discharge according to (Eq. (4.26), p.118). When the electrode subsequently is recharged,

self-discharge: $PbO_2 + H_2SO_4 \Rightarrow ½O_2 + PbSO_4 + H_2O$

is compensated by recharge: $PbSO_4 + 2{\cdot}H_2O \Rightarrow PbO_2 + H_2SO_4 + 2{\cdot}H^+ + 2{\cdot}e^-$

with the result $\boxed{H_2O \Rightarrow ½O_2 + 2{\cdot}H^+ + 2{\cdot}e^-}$ (7.7)

Simultaneously, hydrogen ions (H^+) must be reduced at the negative electrode according to Eq. (7.6b) to compensate for the amount of H^+ ions generated by Eq. (7.7). This can be compensated by self-discharge of the negative electrode.

A table corresponding to Table 7.1 could be established with the only difference being that the evolved gas volume would be halved, since O_2 is equivalent to $4{\cdot}F$ = 107.2 Ah. Capacity loss and water loss would be identical to the values Table 7.1.

Self-discharge of the positive electrode by oxygen evolution, is usually insignificant on account of the low oxygen-evolution rate, as indicated by Fig. 7.5.

7.2.2.3 Water Loss by Corrosion

Self-discharge by corrosion is not at its minimum at the open-circuit potential, as shown in Fig. 7.5 . Corrosion acts in two ways:

1. The lead (Pb) that forms grids and conducting elements in the positive electrode group is converted into lead dioxide (PbO_2), and the conducting cross section of these elements is gradually reduced (cf. Section 4.7, especially Fig. 4.27, p.155).
2. By this reaction, oxygen is absorbed that is taken from the water content of the electrolyte. So, corrosion also causes water loss.

When the corrosion occurs directly by an anodic current, the reaction

$$\boxed{Pb + 2{\cdot}H_2O \Rightarrow PbO_2 + 4{\cdot}H^+\ 4{\cdot}e^-} \quad (7.8)$$

generates protons (H^+) that have to be converted into H_2 by reduction at the negative electrode, either by a corresponding current or by self-discharge of the negative electrode.

When primarily oxygen was absorbed by self-discharge, according to Eq. (4.28)

$$PbO_2 + Pb + 2{\cdot}H_2SO_4 \Rightarrow 2{\cdot}PbSO_4 + 2{\cdot}H_2O \quad (7.9a)$$

and subsequent recharge according to the reversal of Eq. (4.4), p.99, follows

$$2\ PbSO_4 + 4{\cdot}H_2O \Rightarrow 2{\cdot}PbO_2 + 2{\cdot}H_2SO_4 + 4{\cdot}H^+ + 4{\cdot}e^- \quad (7.9b)$$

the result is also $\boxed{Pb + 2{\cdot}H_2O \Rightarrow PbO_2 + 4{\cdot}H^+ + 4{\cdot}e^-}$ (7.9)

Table 7.2 shows some values for lead conversion and water loss. The water loss corresponds to that in Table 7.1. This applies also to capacity loss, not shown in Table 7.2.

Corrosion rate	Lead conversion into PbO_2			Water loss	
	g/year	cm^3/year	% of grid *	g/year	**
1 mA/100 Ah	17	1.5	1.7	2.9	0.3%
2 mA/100 Ah	34	3	3.4	5.9	0.6%
4 mA/100 Ah	68	6	6.8	11.8	1.2%
8 mA/100 Ah	135	12	13.5	23.6	2.4%

Table 7.2 Lead conversion into lead dioxide and water loss equivalent to the corrosion rate related to 100 Ah of nom. capacity. Equivalent for lead corrosion: 1.9325 g Pb/Ah (Eq. (4.34), p.154). Density of lead: 11.34 g/cm^3 (Table 4.18, p.150). * based on 1000 g of grid weight per 100 Ah of nom. capacity. ** based on an initial water content of 10 g per 100 Ah of nom. capacity.

7.2.2.4 Water Loss of the Cell

Self-discharge and water loss of the complete cell is determined by the process with the highest rate.

- Self-discharge occurs independently at each electrode. The normally higher self-discharge of the negative electrode might not be noticed in a subsequent discharge, since in practice the negative electrode often is oversized.
- When the battery is recharged, the electrode with the lower discharge rate is overcharged and produces a corresponding amount of gas.

7.2.3 Parameters Influencing Water Loss and Self-Discharge

The rate of self-discharge caused by hydrogen and oxygen evolution or by corrosion is closely connected with the rate of these secondary reactions, since these reactions are the slow components of the corresponding mixed potentials (cf. Section 2.3.5). For this reason, parameters that influence these secondary reactions also affect the rates of self-discharge and water loss.

Temperature is a parameter that strongly influences electrochemical reaction rates. The general approximation that temperature increase by 10 K doubles the reaction rates by a factor of 2, as described in Section 2.3.6, applies also to hydrogen and oxygen evolution, as is shown in Fig. 7.2.

The rate of corrosion is also approximately doubled per 10 °C of temperature increase. This means that self-discharge and water loss are approximately doubled. At 40 °C, for example, these reactions occur four times faster than at 20 °C.

The state of charge affects the gassing reactions, because hydrogen as well as oxygen evolution cannot occur at a surface covered by lead sulphate ($PbSO_4$). As a consequence, hydrogen evolution, synonymous with the self-discharge rate of the negative electrode, decreases when the negative electrode is partly discharged. The vertical position of the TAFEL line is then shifted towards the bottom. For this reason, self-discharge caused by hydrogen evolution is gradually reduced with progressing discharge. This is a very important fact for valve-regulated batteries, where balance is required between self-discharge of the negative electrode and grid corrosion in the positive. A too-high initial hydrogen-evolution rate is thus gradually adjusted to the required value by discharge of the negative electrode (cf. Section 8.3.2).

7.2.3.1 Influences on the Oxygen-Evolution Rate

The special situation of the grid material in the positive electrode of lead-acid batteries has been described in Section 4.7. The high electrode potential means that all metals, usually applied as current conductors, are converted into their oxidised form and dissolved as ions, unless they are completely covered by a passivating layer. For this reason, most metals - released from the corroding grid alloy or dissolved as contamination in the electrolyte - lose their electrochemical effectiveness as soon as they come into contact with the positive electrode.

Organic compounds are also destroyed by oxidation by the high potential. To some extent, this oxidation may cause self-discharge of the positive electrode when the battery is resting on open-circuit.

The beneficial influence of **Antimony** on the cycle stability of PbO_2 electrodes has been mentioned in Section 4.7.3. Increased oxygen overvoltage at lead dioxide electrodes by antimony addition has been reported occasionally but was not confirmed in general.

Phosphoric acid has long been used as an additive to stabilise the cycle performance of lead dioxide electrodes (cf. Section 10.1.1.2). One of its influences is the increase of oxygen overvoltage at PbO_2 electrodes ([6]).

Cobalt has some influence on oxygen overvoltage at PbO_2 electrodes. It can be introduced into the battery as a metallic additive to the positive grid alloy, but it may also be added to the electrolyte as a solution of cobalt ions (e.g. $CoSO_4$). It is well known that the service life of lead dioxide counter-electrodes in the electrolytic zinc process can be extended significantly when cobalt is added to the electrolyte ([7]), because cobalt reduces the oxygen overvoltage and, at a given current density, the electrode becomes less polarised. Less polarisation is synonymous with a lower electrode potential and means reduced corrosion attack, according to Fig. 4.28. When this practice was transferred to lead-acid batteries, improved corrosion resistance was observed when grids were overcharged ([8]). But already in 1958 J.J.Lander ([9]) pointed out that such an improved overcharge resistance, observed in special test conditions, might not play a major role in the service life of the battery. Actually, the influence of cobalt on the oxygen overvoltage is mainly in a potential range far above that of the positive electrode in the lead-acid battery in normal service conditions, and negative effects like oxidation of the separator or increased hydrogen evolution usually outweigh the advantages of cobalt addition ([10]).

7.2.3.2 Influences on the Hydrogen-Evolution Rate

The situation of the negative electrode is absolutely different in relation to contamination by foreign metals, due to its negative potential. Only the outstanding high overvoltage of hydrogen evolution at the lead surface reduces hydrogen evolution to a tolerable small rate. But this overvoltage is reduced as soon as the lead surface is contaminated by metals of lower hydrogen overvoltage. High overvoltage means that the reaction rate is correspondingly hindered, and such a kinetic barrier is always in danger of being removed. Such a contamination occurs when metals with an equilibrium potential above that of the lead electrode are precipitated at the electrode surface and remain there as a stable metal. This concerns not only impurities of the electrolyte, but also additives that are leached out from the corrosion products of the positive grid alloy (cf. Table 4.21, p.158).

The outstanding high hydrogen overvoltage of lead in comparison with some other metals is illustrated by Fig. 7.7 (cf. also ([11])). It shows rates of hydrogen evolution at the various metals. The horizontal axis represents the overvoltage ($E - E^o$) (cf. Eq. (2.60), p.28). The zero point reflects the equilibrium potential for hydrogen evolution.

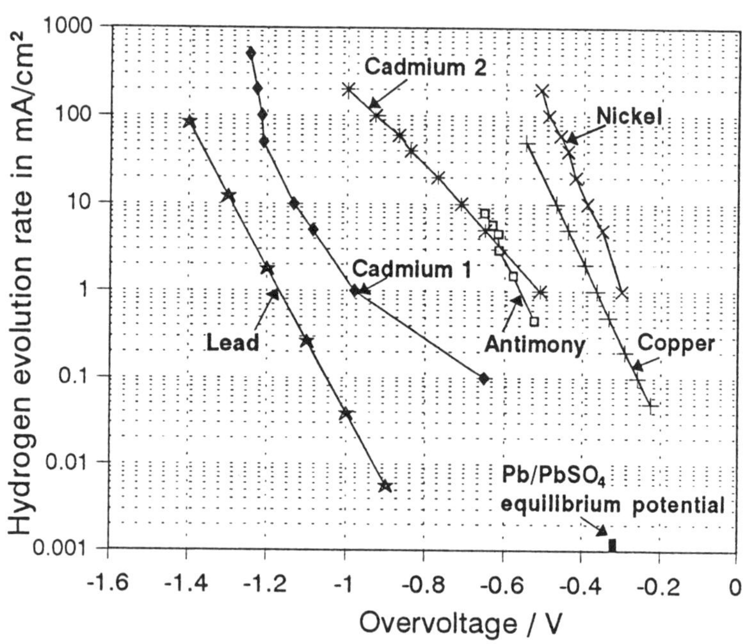

Fig. 7.7 Hydrogen-evolution rates vs. overvoltage at different electrode materials in sulphuric acid. Lead ([12]); Cadmium 1 and Cadmium 2 ([13]); Antimony ([14]); Copper ([15]); Nickel ([16]).

Fig. 7.7 shows that most of the hydrogen-evolution curves are fairly well approximated by TAFEL lines, and the slopes of these lines are rather similar for the different materials, except for cadmium. For this metal, two different series of results are shown in Fig. 7.7. These diverging results make the limitations obvious when overvoltage figures are compared. A general problem is caused by the surface area to which the current density is referred. The 'true surface area' may be different to the geometric surface area, and this difference depends largely on the previous treatment of the electrode surface. Furthermore, hydrogen evolution is impeded by many organic substances resulting in increased overvoltage. The medium values of the 'Cadmium 1' curve may indicate such a contamination, because a straight line between the lowest and highest value gives a line parallel to the 'Cadmium 2' line.

Fig. 7.7 clearly shows the outstanding situation of lead in relation to hydrogen overvoltage. Hydrogen is evolved much faster at all the other metals in this Figure. If the negative electrode of the lead-acid battery with its rest potential of - 0.35 V were short-circuited with a nickel electrode, self-discharge would occur at the high rate of about 1 mA per cm^2 of the nickel electrode. Balance of such a high hydrogen-evolution rate would cause spontaneous discharge of the lead electrode.

The equilibrium potentials E^o of these metals in acid solution are:

Reaction	E^o	
Cd/Cd^{2+}	-0.42 V	cf. Section 2.2.3.2
Ni/Ni^{2+}	-0.250 V	
Sb/Sb_2O_3	+0.152 V	
Cu/Cu^{2+}	+0.337 V	

Table 7.3 Equilibrium potentials of the metals in Fig. 7.7 vs. standard hydrogen electrode.

Copper, antimony, and nickel are precipitated at the negative electrode surface even at the open-circuit potential. Cadmium is only precipitated when the electrode is sufficiently polarised during charging. The hydrogen-evolution rates in Fig. 7.7 are referred to the pure metals. The increase caused by precipitation would only be proportional to the fraction of the surface area that is covered by the foreign metal.

The same situation is caused if these metals are dissolved, e.g. as a result of grid corrosion. When they reach the surface of the negative electrode, they are precipitated at the surface and reduce the hydrogen overvoltage. Such an effect is well known for antimony and is often called 'antimony poisoning' of the negative electrode. The reason that the word 'poisoning' is used only in connection with antimony is simply that only antimony is added in the range up to 10 wt.% to the grid alloy (cf. Table 4.21, p.158). Fig. 7.7 shows that nickel and copper are more effective for hydrogen evolution, but in a battery normally they never reach comparably high concentrations. As soon as the antimony content of the grid alloy is reduced to values comparable to these other additives, 'antimony poisoning' largely diminishes.

The effect of such a contamination on float current balance in a vented lead-acid battery is illustrated in Fig. 7.8: when the hydrogen overvoltage is decreased, the hydrogen evolution curve is shifted upwards, and the negative electrode is correspondingly depolarised. If the float voltage is maintained, a corresponding additional polarisation of the positive electrode occurs, so that current balance is re-established between negative and positive electrodes, but at an increased current.

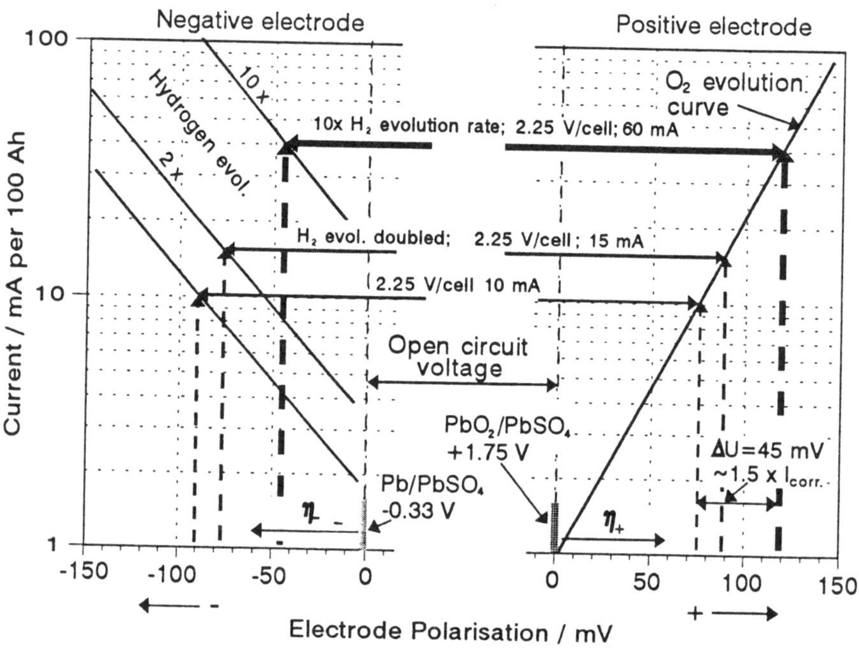

Fig. 7.8 Effect of increased hydrogen evolution on the float current balance in a vented lead-acid battery.

For the vented battery in Fig. 7.8 a 'normal' float current of 10 mA/100 Ah is assumed at 2.25 V/cell. The slopes of the TAFEL lines correspond to those in the preceding Figures 7.3 and 7.5. When the hydrogen-evolution rate is doubled (2x line in the Figure), the polarisation of both electrodes is shifted to more positive values by about 12 mV, and the float current is increased to 15 mA/100 Ah. At a ten times greater hydrogen-evolution rate, a float current of 60 mA results, and the potential of both electrodes is shifted by about 45 mV. Thus the increased hydrogen generation causes a considerable larger float current, with a correspondingly increased water loss and gassing rate which again may give rise to problems with acid fumes (cf. Section 7.1). Furthermore, corrosion of the positive grid is increased due to the increased potential. As indicated in Fig. 7.8, an increase by the factor 1.5 is to be expected when the 'normal' polarisation of about 75 mV is close to the minimum of corrosion (cf. Fig. 7.5).

In valve-regulated batteries, such a contamination does not markedly influence the polarisation of the negative electrode, since it already is quite small. But the increased hydrogen evolution causes a correspondingly increased water loss and may result in partly discharged negative electrodes, as described in Section 8.3, cf. esp. Figures 8.15 and 8.21. Furthermore, the increased polarisation of the positive electrode contributes additional ageing factors due to grid corrosion and oxygen consumption.

Organic substances are often not destroyed by contact with the negative electrode, although they might be changed by reducing processes, especially in the presence of hydrogen. They can increase the hydrogen overvoltage, when they cover parts of the surface and so reduce the active surface area. Such substances are added to the negative active material in small quantities as 'expanders' (cf. Section 10.1.2).

7.3 Water Decomposition in Ni/Cd Batteries

Fig. 7.9 illustrates the situation for nickel/cadmium batteries.

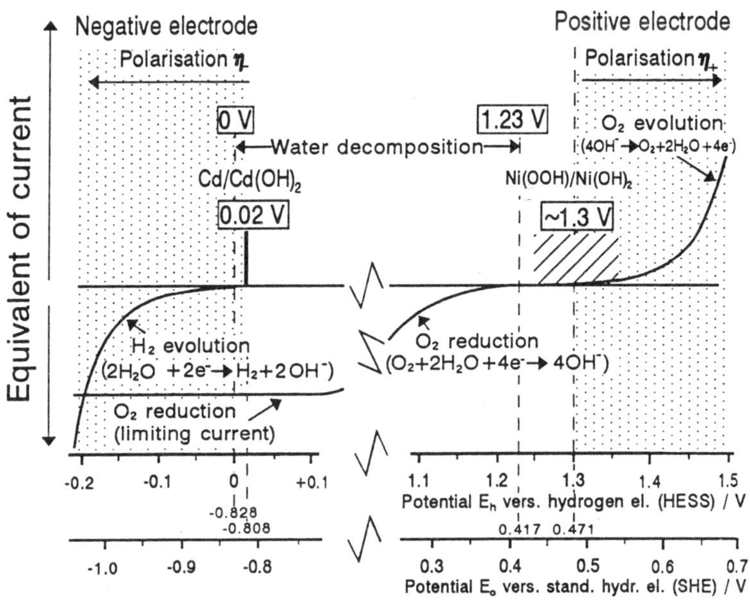

Fig. 7.9 Equilibrium potentials of the negative and positive electrodes in a nickel/cadmium battery and current/voltage curves of hydrogen evolution, oxygen evolution, and oxygen reduction. (The latter forms a limiting current).

It corresponds to Fig. 5.1, p.175, but the secondary reactions are inserted as current/voltage curves which indicate their rates referred to the electrode potential. The x-axis shows the electrode potential. Two scales are employed: the upper one represents the potential E_h referred to the hydrogen electrode in the same solution (HESS). As pointed out in Section 5.1.2.1, the equilibrium potentials for all the reactions in Fig. 7.9 are shifted in the same manner with changing pH value (cf. Eq. (5.16), p.181). Consequently, the E_h scale is independent of the electrolyte concentration E_h.

The lower scale represents the electrode potential E_o referred to the standard hydrogen electrode (SHE) for 1.250 g/cm^3 density of the KOH solution at 25 °C. *The electrolyte density has to be considered in this case, because, referred to this scale, all potentials in the Figure, are shifted in like manner with KOH concentration (cf. Section 5.1.2.1).*

The most remarkable difference from the lead-acid battery is that the equilibrium potential of the negative electrode, the cadmium electrode, is more positive than the potential when hydrogen starts to be evolved. As a consequence, self-discharge of the cadmium electrode caused by hydrogen evolution does not occur. An important further consequence is that hydrogen evolution can completely be avoided, if the (negative) polarisation of the cadmium electrode during charging is kept so small that the hydrogen equilibrium potential is not reached. Use of this fact is made in sealed nickel/cadmium batteries where the internal oxygen cycle depolarises the negative electrode (cf. Section 8.1). The avoidance of hydrogen evolution is one of the presuppositions that the nickel/cadmium battery can be sealed hermetically.

Oxygen evolution cannot be avoided at the positive electrode, because already the equilibrium potential of this electrode exceeds the equilibrium value of oxygen evolution, and a situation results similar to that of the positive electrode in the lead-acid battery described above.

Oxygen reduction occurs at an electrode potential below +1.23 V. Thus it is always possible at the negative electrode. In Fig. 7.9 it is inserted as a constant current, a limiting current, because its rate is limited by the oxygen-transport rate. In flooded batteries this limiting current is as low as that in vented lead-acid batteries, as indicated in Fig. 7.5, p.242. In sealed nickel/cadmium batteries with immobilised electrolyte, oxygen-transport rates can be fairly high, due to void space within the separator and narrow spacing of the electrodes. Overcharging at the 5-hour rate is possible for most designs without hydrogen evolution.

Corrosion of the nickel substrate in the positive electrode is negligible. For this reason, water loss by grid corrosion is negligible in nickel/cadmium batteries.

When a vented nickel/cadmium battery is overcharged, both electrodes are polarised, the positive to a more positive potential, the negative to a more negative potential, until the evolution rates of oxygen and hydrogen equal each other. Both gases are then produced in a stoichiometric ratio, and the result is water decomposition according to Eq. (2.93), p.44, as in vented lead-acid batteries (cf. Fig. 7.3, p.240) . In the sealed design, hydrogen evolution can be avoided and grid corrosion is negligible. As a result, oxygen evolution and oxygen reduction are the only secondary reactions caused by overcharge. The consequence is the 'undisturbed internal oxygen cycle' as shown in Fig. 8.3, p.273.

7.3.1 TAFEL Lines in Nickel/Cadmium Batteries

TAFEL lines as described above for hydrogen and oxygen evolution in the lead-acid battery cannot be applied for nickel/cadmium batteries in the region of the open-circuit voltage. The important prerequisite does not apply that the potential is shifted far enough from the equilibrium value so that the reverse reaction can be neglected (cf. Fig. 2.8, p.29). For hydrogen evolution, the equilibrium potential is even below the open-circuit potential of the cadmium electrode; and also the equilibrium potential of oxygen evolution is close to the open-circuit potential of the positive electrode. Thus, for small polarisation values, straight lines cannot be expected for the relation between gas-evolution rate and overvoltage.

But at increased overvoltage the above prerequisite is fulfilled and TAFEL lines represent also the relation between electrode potential and current or cell voltage and current.

Fig. 7.10 shows TAFEL lines for sealed nickel/cadmium batteries during overcharge. These curves correspond to those for the valve-regulated lead-acid battery in Fig. 8.20, p.296.

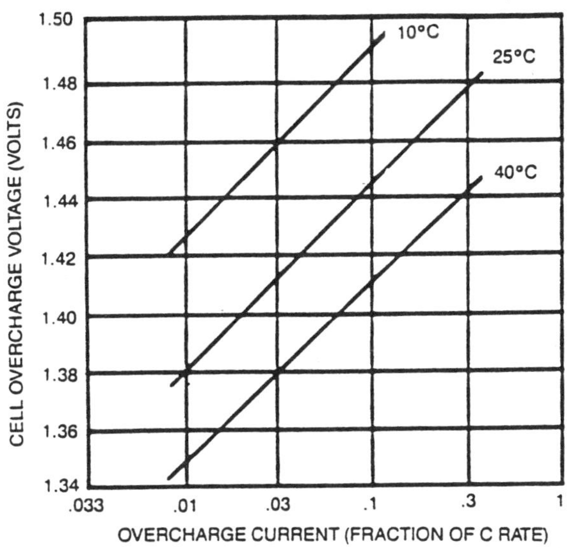

Fig. 7.10 Overcharge current versus cell voltage for sealed nickel/cadmium batteries. (From ([17]).)

The b factor of the TAFEL lines is 60 mV per current decade in this example, compared to about 100 to 140 mV in the case of valve-regulated lead-acid battery according to Figs. 8.27 and 8.20.

The temperature dependence of these TAFEL lines also approximates to the rule in Eq.(2.68), p.33, that the reaction rate, i.e. the current, doubles per 10 °C of temperature increase. At 40 °C, the current increases the factor by 3 when compared to 25 °C. This results in an increase factor of 2.1 per 10 °C. A greater increase, namely by a factor 4.6 per 15 °C, is observed when the temperature is raised from 10 to 25 °C. The corresponding 'ARRHENIUS factor' would be 2.75. These values underline the approximate character of such rules.

7.4 Water Decomposition in Ni/MeH Batteries

The thermodynamic situation of nickel/metal hydride batteries is quite similar to that of nickel/cadmium batteries. It is shown in Fig. 8.9, p.281, which corresponds to Fig. 7.9. At the positive electrode, oxygen evolution as a secondary reaction is unavoidable as described in Section 7.3. When the negative electrode is polarised to values below its equilibrium potential, hydrogen is evolved. But in this battery system, hydrogen evolution is not a secondary reaction that occurs at a comparatively low rate. At the special alloys used for hydrogen storage, hydrogen evolution is fast and water loss would be correspondingly fast, if the cell were not sealed.

Hydrogen would also be lost without polarisation from the negative electrode of a vented cell since the hydrogen concentration in the surrounding atmosphere is close to zero, and the H_2 storing alloy would gradually give off its hydrogen.

For both reasons, nickel/metal hydride batteries can only be used as sealed ones, and water loss has no relevance for them.

7.5 Methods to Reduce Water Decomposition

Water loss is a serious demerit of conventional lead-acid and nickel/cadmium batteries because maintenance at regular intervals is required for its compensation, to avoid premature failure of the battery. Furthermore, water loss causes concentration changes of the electrolyte. Last but not least, water decomposition means the generation of hydrogen and oxygen, and hydrogen forms an explosive mixture with oxygen for all hydrogen contents between 4 and 76 vol.%, and the gas leaving the battery carries acid or caustic-solution fumes (cf. Section 2.6.6).

The main reason for water loss is the electrolytic decomposition. So, many efforts have been made to reduce the water decomposition, such as the introduction of alloys with low antimony content or antimony-free alloys as described in Section 4.7.3. However, in the lead-acid battery, generation of hydrogen and oxygen can never be completely suppressed, as pointed out above.

7.5.1 Minimised Overcharge

The generation of hydrogen and oxygen depends largely on the electrode potential. Consequently, the cell voltage should be kept as low as possible to minimise the polarisation of both electrodes. However, there are serious obstacles that limit this simple method to reduce water loss:

- The nickel hydroxide electrode requires charging at a high potential because of the complex relation between electrode potential and state of charge (Section 5.2.2.1). Otherwise, the electrode does not reach full capacity.
- In the lead-acid battery, the charging voltage can be much reduced, when the charging periods are long. This is done for stationary batteries during float charging. However, acid stratification occurs, as described in Section 4.6.3 when lead-acid batteries are repeatedly discharged and recharged without intermediate periods of overcharge at increased cell voltage, or without forced acid agitation (Section 7.5.1.2).

For both reasons, minimum overcharge is not possible for nickel/cadmium and nickel/metal hydride batteries, and for lead-acid batteries it is only possible in special applications.

7.5.1.1 Maintenance-Free Starter Batteries

As starter batteries for motor cars, 'maintenance-free lead-acid batteries' that are not based on the internal oxygen cycle have been on the market for many years. They are flooded lead-acid batteries that do not require refilling during their entire service life.

This is due to the special operating conditions of the car battery. The usual private car runs for only about 500 hours per year. During this period the battery is charged or overcharged, and the charging voltage is limited to about 2.35 V per cell. For the rest of the time, when the car is parked, the starter battery is standing on open-circuit voltage or discharged with an extremely low current by some accessories like electric clocks. The amount of current drawn from the battery during these rest periods is not worth mentioning.

Furthermore, normally, starter batteries are subject to only shallow discharges, since the normal engine-cranking process requires only a small fraction of their capacity.

These conditions are favourable for low water loss:

- The overcharge periods are short compared to those of a stationary battery, which is overcharged 8760 hours per year.
- In normal operating conditions, discharging/charging cycles do not occur.

On the other hand, the service life of the starter battery is limited to two to five years, depending on the environmental conditions. This limited service life is the consequence of the special design, to achieve extremely high-load performance at a low weight of the battery.

On account of the operating conditions and the limited service life, starter batteries with liquid electrolyte can be designed in such a way that they do not require refilling during service life. Low antimony content (< 2% Sb by weight) in the grid of the positive electrode and antimony-free grid materials in the negative electrode or antimony-free grids for both electrodes are employed to reach this goal.

Starter batteries with immobilised electrolyte and internal oxygen cycle are also on the market, but their volume is limited (cf. Section 10.8.4).

7.5.1.2 Forced Acid Agitation in Lead-Acid Batteries

To avoid acid stratification, conventional lead-acid batteries in traction applications are overcharged by 10 to 20% per cycle to produce gas bubbles for mixing the electrolyte, as mentioned in Section 4.6.3. However, this method is not very effective and causes not only increased water loss, but also further disadvantages like increased corrosion attack and heat generation, besides the fact that energy is wasted. A superior method is to mix the acid using air-lift pumps. This method has been common for many years in submarine batteries ([18]) and is also used for solar applications. More recently, it has been adopted for specially designed batteries for load-levelling applications ([19]) for electric vehicles ([20]), and also for the normal traction application ([21]).

Fig. 7.11 shows the arrangement of an air-lift pump. Gas bubbles are generated in a vertical tube by feeding small amounts of air from a central supply. These bubbles reduce the specific weight of the liquid inside the tube, which results in an upward movement of this mixture of acid and gas bubbles. With this convection stream, acid is lifted from the bottom and poured above the top of the plates. Thus, complete mixing of the electrolyte can be achieved with a small amount of air and within a short period of time.

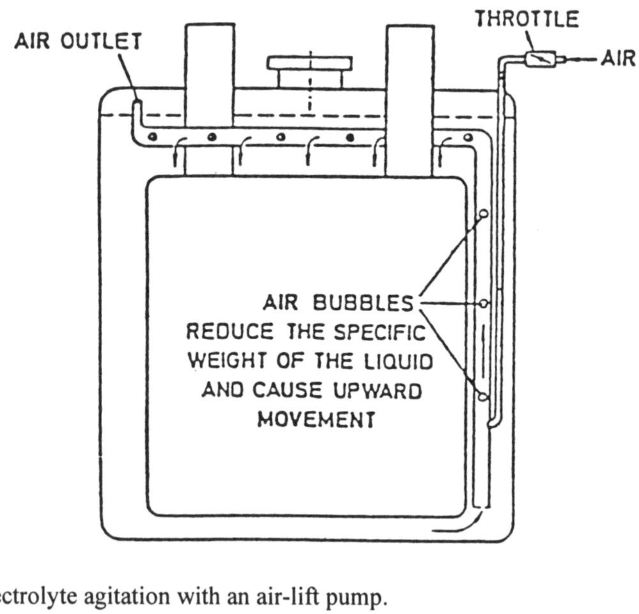

Fig. 7.11 Electrolyte agitation with an air-lift pump.

The throttle shown in Fig. 7.11 reduces the air stream and thereby the air pressure for the individual cell. This inlet for the individual cell has to be small compared to the main supply, to secure uniform gas flow for each cell.

7.5.2 Catalytic Gas Recombination

Another way to reduce water loss in conventional lead-acid batteries is to use catalytic devices to recombine the gaseous hydrogen and oxygen to water (cf. Eq. (6.5)):

$$\boxed{H_2 + \tfrac{1}{2}O_2 \Rightarrow H_2O} \qquad (7.10)$$

This idea is very old; as early as 1912 a patent was granted to Edinson ([22]) to remove gases from storage batteries by using an electrically heated platinum wire in the gas phase for the recombination of hydrogen and oxygen. More modern devices are based on catalysts that can operate or at least start operation at room temperature. In general, finely distributed palladium, sometimes also platinum, are applied as a catalyst which is part of a 'recombinator' that is used instead of the vent plug (Fig. 7.12).

Hydrogen and oxygen escaping from the cell are recombined to yield water vapour with the aid of the catalyst. The recombination reaction (Eq. (7.10)) is the reversal of Eq. (2.93), p.44, and is heat-intensive (ΔH = - 285.8 kJ/mole, according to Eq. (2.99), p.44). Consequently, catalytic recombination devices in general are faced with the problem of heat dissipation.

Fig. 7.12 shows sections of such a devices. Both examples are based on the same principle: traces of the catalytic active material are spread on a large surface that is passed by the gas stream. A protective layer of porous carbon, often mixed with oxides ([23]), absorbs substances that would deactivate ('poison') the catalyst. Antimony, sulphur and arsenic, for example, are such substances.

In the design on the left in Fig. 7.12, the gases have to flow through a tube of porous ceramic that is filled with the protecting material before they reach the central cylindrical body that contains the catalyst. The produced water vapour is condensed at the container wall. Gases that have not been recombined, leave through the opening after passing a porous plug that reduces the flow rate. With increased overcharging current, the temperature of the container wall increases and may reach critical values that destroy the plastic material. Maximum load of such a recombinator therefore, depends on its size. In general, these devices are mainly used for stationary batteries with limited overcharging currents.

A different design is shown on the right, where the catalyst is connected to the metallic cover of the device. The gas does not pass the catalyst, but flows along its surface. Also in this design, a protective layer has to be passed by the gas before it reaches the catalyst.

In the device on the right hand side, the heat produced by the recombination itself is transferred across the cover. This 'recombination heat' amounts to about 70% of the heat generated in total. About 30% of the total heat is generated by the condensation of the formed water vapour at the side walls. The dashed arrows indicate both processes of heat dissipation. This design is superior in regard to heat dissipation. Furthermore, it is safe from the point of view of overcharging, because the recombination process is self-limiting. When the recombination rate exceeds a certain value, the stream of water vapour flowing back from the catalyst stops further access of the hydrogen/oxygen mixture to the catalyst. Thus the recombination rate and the heat generation are reduced, as is the efficiency, because a larger share of gas must leave the device without being recombined. On the other hand, this design is inferior at low gas flow rates, since there is no forced flow through the substrate that contains the catalyst.

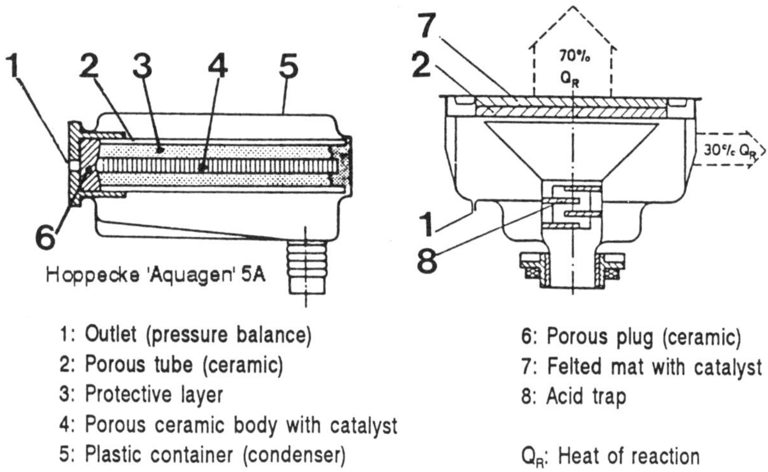

Fig. 7.12 Sectional view of catalytic recombination plugs. On the left, a design that uses a porous ceramic tube as substrate for the catalyst ([23]). In the design on the right (VARTA), a felted mat forms the substrate ([24]).

The efficiency of a catalytic-recombination device is always limited to less than 100%, because a part of the gas leaves the device without being recombined, together with some water vapour. In stationary applications, 85% efficiency is reached at maximum. When the battery is cycled, the situation is aggravated by the fact that hydrogen and oxygen are not evolved at stoichiometric rates during the charging process (cf. Fig. 4.16, p.104).

Occasionally, the heat released by the recombination process was used to monitor the charging process, since maximum heat generation in the recombinator is reached when both electrodes are charged completely, and hydrogen and oxygen are evolved at stoichiometric rates ([25]). A drawback of this method is that increased hydrogen evolution in traction batteries with progressing service life causes a too-early end of the charging process (antimony contamination of the negative electrode).

7.5.3 Reversal of Gas Evolution, Internal Gas Cycles

Another possibility to reduce gas escape are internal gas cycles: gas that is evolved at one electrode is consumed at the counter electrode by the reversed reaction. Since hydrogen and oxygen are evolved, such internal cycles are possible for both gases based on the reversal of the water decomposition reaction (cf. Section 2.6.1)

The **internal hydrogen cycle** is established in acidic medium by:

Hydrogen evolution at the negative electrode $$2 \cdot H^+ + 2 \cdot e^- \Rightarrow H_2 \quad (7.11a)$$

that is balanced by the reversed reaction, the oxidation of hydrogen,

at the positive electrode $$H_2 \Rightarrow 2 \cdot H^+ + 2 \cdot e^- \quad (7.11b)$$

or in alkaline medium:

Hydrogen evolution at the negative electrode $$2 \cdot H_2O + 2 \cdot e^- \Rightarrow 2 \cdot OH^- + H_2 \quad (7.12a)$$

that is balanced by the reversed reaction, the oxidation of hydrogen,

at the positive electrode $$2 \cdot OH^- + H_2 \Rightarrow 2 \cdot H_2O + 2 \cdot e^- \quad (7.12b)$$

For thermodynamic reasons, these reactions are possible, as indicated in Figs. 4.1 and 5.1. The results would be:

- Hydrogen, evolved at the negative electrode, does not escape from the battery.
- A slight concentration gradient for H^+ or OH^- ions would be observed, since these ions are consumed at the negative and generated at the positive electrode. But the resulting concentration gradients are kept small on account of equalisation by diffusion (cf. Fig. 8.2, p.271).
- All the electrical energy used for this reaction, produces heat corresponding to the Joule effect (Eq. (2.73)). Heat generation, since Eqs. (7.11a) and (7.11b) or Eqs. (7.12a) and (7.12b) form an electrochemical couple with the equilibrium voltage $E^o = 0$, and thus the actual cell voltage is overvoltage η, according to $\eta = E - E^o$ (Section 2.3.3).

The general problem of the internal hydrogen cycle is the low reaction rate of hydrogen oxidation (Eq. (7.11b)) at the positive electrodes of the battery systems treated in this book (cf. the following Section). For this reason, the internal hydrogen cycle has no relevance for the overcharge of lead-acid, nickel/cadmium, and nickel/metal hydride batteries. But it is effective during the overdischarge of nickel/metal hydride batteries as shown in Section 8.2.1.

The corresponding **internal oxygen cycle** is in acidic medium:

Oxygen evolution at the positive electrode $$2 \cdot H_2O \Rightarrow O_2 + 4 \cdot H^+ + 4 \cdot e^- \quad (7.13a)$$

is balanced by the reversed reaction, the reduction of oxygen,

at the negative electrode $$O_2 + 4 \cdot H^+ + 4 \cdot e^- \Rightarrow 2 \cdot H_2O \quad (7.13b)$$

or in alkaline medium:

Oxygen evolution at the positive electrode $$4 \cdot OH^- \Rightarrow O_2 + 2 \cdot H_2O + 4 \cdot e^- \quad (7.14a)$$

is balanced at the negative electrode by $$O_2 + 2 \cdot H_2O + 4 \cdot e^- \Rightarrow 4 \cdot OH^- \quad (7.14b)$$

The thermodynamic situation is similar to that mentioned above for the internal hydrogen cycle. But oxygen reduction according to Eqs. (7.13b) and (7.14b) is fast enough to establish such an internal cycle, provided that the oxygen can reach the surface of the negative electrode correspondingly fast. This can be achieved by immobilised electrolyte. The internal oxygen cycle is of great importance and forms the basis of the maintenance-free batteries discussed in this book. It prevents the escape of oxygen during charging and overcharging, prevents H_2 generation in nickel/cadmium batteries, and widely reduces H_2 escape from lead-acid batteries. So the batteries can be overcharged with the only result that all the electrical energy used for this cycle is converted into heat and negligible concentration profiles are established within the cell (cf. Fig. 8.2, p.271). This internal oxygen cycle is treated in more detail in Chapter 8.

7.5.3.1 Hydrogen Oxidation

From a theoretical standpoint, hydrogen oxidation according to Eqs. (7.11b) or (7.12b) should occur as soon as gaseous (or dissolved) hydrogen molecules reach an electrode more positive than the equilibrium potential of the hydrogen electrode ($E_h > 0$). But hydrogen oxidation occurs only slowly at nickel hydroxide electrodes and is negligibly slow at lead dioxide electrodes.

In nickel/cadmium batteries, traces of hydrogen are gradually consumed.. For this reason, occasional overcharge of sealed nickel/cadmium batteries can be tolerated if they are equipped with a resealing vent that closes again when the overpressure producing gas is released. The rest of the hydrogen gas is gradually removed by oxidation according to Eq. (7.12b).

The high self-discharge rate of nickel/hydrogen batteries was mentioned in Section 6.1 (cf. Fig. 6.1, p.218). It is caused by hydrogen oxidation that is balanced by discharge of the nickel hydroxide according to Eq. (6.4). The same mechanism is effective in nickel/metal hydride batteries, but due to the low pressure of hydrogen gas, self-discharge rates are in the range of 25% per month at 20 °C (Fig. 6.7, p.231). This means that 25 Ah are lost per 100 Ah within one month, which is equivalent to a current of 35 mA, a current that corresponds to a charging or discharging rate of about 3000 hours. This rate of hydrogen oxidation is by orders of magnitude smaller than the usual charging or discharging rates, and it is far below the usually tolerated overcharging currents in the range of the 5-hour rate. Under these premises, an effective internal hydrogen cycle cannot be established, even in batteries with nickel hydroxide electrodes.

The general experience for lead-acid batteries is that hydrogen oxidation can be neglected, and hydrogen gas evolved at the negative electrode has to leave the battery ([26]).

Results of pressure-decay measurements were published that indicate hydrogen-oxidation rates (internal hydrogen cycle) above the self-discharge rate of about 2% per month ([27]). But these results are not confirmed by general experience, and in normal conditions obviously only a considerably smaller portion of the hydrogen-evolution rate can be oxidised at the lead dioxide surface ([28]). Consequently, the minimum amount of hydrogen, leaving the valve-regulated lead-acid battery, will be only slightly below the self-discharge rate, if hydrogen is not otherwise lost.

Note: *The precise measurement of small amounts of hydrogen is difficult, because of the tendency of hydrogen to permeate many materials (cf. Table 2.13, p.61). Not only the container material but also the tubing material, used for the gas measurement, has to be scrutinised, as well as the sealing substances. In general, only metallic tubes should be used for precise gas measurements. But the hydrogen-collecting vessels and the liquid in them may also cause serious hydrogen loss.*

A further parameter markedly influencing hydrogen evolution is the state of charge of the negative electrode (cf. Section 7.2.3). Partially discharged negative electrodes can give the wrong impression of extremely low hydrogen evolution.

7.5.3.2 Oxygen Reduction

The reversal of oxygen evolution, namely the reduction of oxygen, occurs when the potential of the concerned electrode attains values below the oxygen equilibrium potential. It occurs in acid medium according to:

$$\boxed{O_2 + 4 \cdot H^+ + 4 \cdot e^- \Rightarrow 2 \cdot H_2O} \tag{7.15}$$

and in alkaline medium according to:

$$\boxed{O_2 + 2 \cdot H_2O + 4 \cdot e^- \Rightarrow 4 \cdot OH^-} \tag{7.16}$$

Both reactions normally are not possible at the positive electrodes, since their potential always remains above 1.23 V compared to the hydrogen electrode. But oxygen reduction occurs at the negative electrodes as soon as oxygen reaches their surfaces. It is also called **recombination** as mentioned in the Note on p.272.

In the literature, the oxygen-reduction reaction is often assumed to include the discharge of the negative electrode. So, in lead-acid batteries, oxygen is presumed to react with the negative active material under formation of the oxide:

$$\boxed{2 \cdot Pb + O_2 \Rightarrow 2 \cdot PbO} \tag{7.17}$$

that reacts with sulfuric acid

$$\boxed{2 \cdot PbO + 2 \cdot H_2SO_4 \Rightarrow 2 \cdot PbSO_4 + 2 \cdot H_2O} \tag{7.18}$$

and finally is recharged, according to

$$\boxed{2 \cdot PbSO_4 + 4 \cdot H^+ + 2 \cdot e^- \Rightarrow 2 \cdot Pb + 2 \cdot H_2SO_4} \tag{7.19}$$

When Eqs. (7.17) to (7.19) are added, the result is:

$$\boxed{O_2 + 4 \cdot H^+ + 4 \cdot e^- \Rightarrow 2 \cdot H_2O}$$

which is identical with Eq. (7.15).

This sequence of reactions often is called 'chemical recombination'. It includes self-discharge according to Eq. (4.24) (p.116), conversion of PbO into $PbSO_4$ according to Eq. (4.25) and subsequent recharge. Oxidation of lead according to Eq. (7.17) can, however, only occur above the equilibrium potential of the negative electrode. This means that the potential of the lead electrode must be equal to or more positive than the open-circuit potential.

In proper float conditions, the negative-electrode potential should always be kept a little more negative than its equilibrium value, to secure the state of full charge for the negative electrode. Consequently, in normal float conditions, only Eq. (7.15) is possible. Only when the electrode potential equals or is more positive than its equilibrium value, can lead oxidation according to Eqs. (7.17) and (7.18) occur, with the result that the negative electrode is partly discharged, since then recharge according to Eq. (7.19) is not possible.

A corresponding situation applies to the cadmium electrode in nickel/cadmium batteries.

As a consequence, oxygen reduction during float charging must occur directly without intermediate oxidation of the negative electrode according to Eq. (7.17). This does, however, not exclude the formation of surface layers as described below (Item 2).

Eqs. (7.15) and (7.16) describe overall reactions. Both give no information in regard to the true reaction path. This is much more complicated and not yet known in full detail. In general, two electrons are not exchanged in one reaction step, but require a sequence of reactions. Two mechanisms are discussed ([29]):

1. A reduction mechanism that includes radical groups like OH and HO_2, and is in an acid medium:

$$O_2 + e^- \Rightarrow O_2^- \qquad (7.20a)$$

$$O_2^- + H^+ \Rightarrow HO_2 \qquad (7.20b)$$

$$HO_2 + e^- \Rightarrow HO_2^- \qquad (7.20c)$$

$$HO_2^- + H^+ \Rightarrow H_2O_2 \qquad (7.20d)$$

$$H_2O_2 + e^- \Rightarrow OH + OH^- \qquad (7.20e)$$

$$OH + e^- \Rightarrow OH^- \qquad (7.20f)$$

$$2{\cdot}OH^- + 2{\cdot}H^+ \Rightarrow 2{\cdot}H_2O \qquad (7.20g)$$

The result of this sequence is identical to Eq. (7.15).

For alkaline medium, Eqs (7.20b) and (7.20d) have to be changed into

$$O_2^- + H_2O \Rightarrow HO_2 + OH^- \qquad (7.20b')$$

$$HO_2^- + H_2O \Rightarrow H_2O_2 + OH^- \qquad (7.20d')$$

Then the sequence of Eqs. (7.20a) to (7.20f) results in Eq. (7.16).

In both versions, the sum of Eqs. (7.20a) to (7.20d) is the generation of H_2O_2:

$$\boxed{O_2 + 2{\cdot}H^+ + 2{\cdot}e^- \Rightarrow H_2O_2} \qquad (7.21)$$

Hydrogen peroxide (H_2O_2) can actually be measured when oxygen is reduced in alkaline or neutral solutions (cf., e.g., [30]). In acid solutions its stability is too low to observe this intermediate compound.

2. The other mechanism of oxygen reduction assumes the formation of metal/oxygen layers at the surface of the electrode (adsorption layers, surface oxides, or hydroxides that allow the conversion of O_2 molecules into O^{2-} ions):

$$2\cdot Me + O_2 \Rightarrow 2\cdot Me\text{-}O \tag{7.22a}$$

$$2\cdot Me\text{-}O + 4\cdot H^+ + 4\cdot e^- \Rightarrow 2\cdot Me + 2\cdot H_2O \tag{7.22b}$$

The sum of both reactions also is Eq. (7.15).

In alkaline medium, Eq. (7.20b) has to be modified into

$$2\cdot Me\text{-}O + 2\cdot H_2O + 4\cdot e^- \Rightarrow 2\cdot Me + 4\cdot OH^- \tag{7.22b'}$$

and Eq. (7.16) results.

7.5.3.3 Auxiliary Electrodes in Lead-Acid Batteries

One possible way to achieve hydrogen as well as oxygen consumption in the battery is to use auxiliary electrodes of suitable materials. To overcome the problem of too low a transport rate of the dissolved gas, these auxiliary electrodes must be in direct contact with the gaseous phase above the electrodes

However, this solution causes new problems, because the 'suitable substances' for hydrogen and oxygen consumption influence other electrochemical processes. As soon as a noble-metal electrode is connected to the negative electrode of a lead-acid battery a dramatic increase of hydrogen evolution occurs, on account of reduced hydrogen overvoltage (cf. Fig. 7.7). As a consequence, a diode has to be inserted between the lead electrode and the auxiliary electrode, as shown in Fig. 7.13, which allows the current to flow only in the required direction.

An even more complicated situation exists for auxiliary electrodes that consume hydrogen. These electrodes have to be connected to the positive electrode of the battery. Such active electrode materials, however, increase unwanted oxygen evolution unless this current is blocked by a diode. Also, serious corrosion problems are to be expected at the auxiliary electrode on account of the high potential of the PbO_2 electrode.

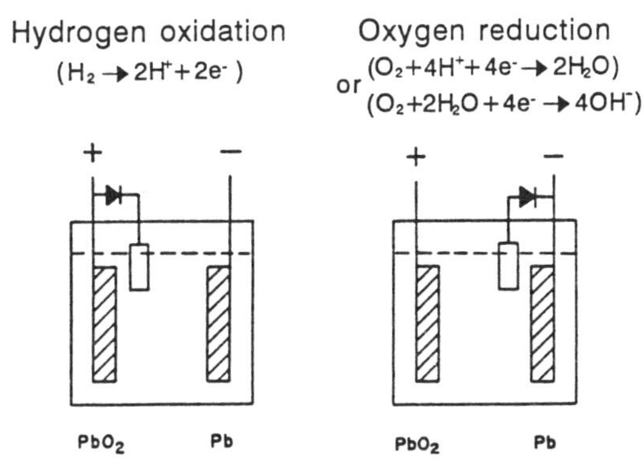

Fig. 7.13 Arrangement of auxiliary electrodes to achieve oxygen and hydrogen consumption (from (31)).

Ruetschi and Ockerman presented a flooded lead-acid battery that was virtually sealed with the aid of auxiliary electrodes ([31]). Oxygen reduction is achieved with a porous carbon electrode impregnated with silver, while hydrogen oxidation is achieved with a palladium-impregnated carbon electrode. Furthermore, in this paper, all questions concerning auxiliary electrodes are described in detail.

More recently, tungsten-carbide electrodes have been proposed to achieve flooded, but sealed, lead-acid batteries ([32]).

But even if gas escape is completely suppressed by these auxiliary electrodes, grid corrosion remains as a secondary reaction, and gradual discharge of the positive electrode would not be compensated for. For this reason, the above mentioned virtually sealed lead-acid battery would gradually lose its capacity, because of the self-discharge of its positive electrode according to Eq. (4.27), p.118.

7.5.3.4 Auxiliary Electrodes in Sealed Nickel/Cadmium Batteries

In sealed nickel/cadmium batteries, auxiliary electrodes are applied in practice, at least in the form of carbon layers on the surface of the negative electrode (cf. Sections 11.5.1.1 and 11.5.3.2). They increase the oxygen-reduction rate and thereby the maximum rate of the internal oxygen cycle (see next Chapter). As a consequence, increased overcharge currents can be permitted, which means shorter charging periods.

Such auxiliary electrodes are effective in two respects:

- The surface area of the negative electrode is enlarged by substances of high porosity.
- The materials of the auxiliary electrodes are usually mixed with hydrophobic substances (e.g. fluorinated copolymers like TEFLON) that reduce the thickness of the wetting electrolyte film, that forms the diffusion layer, and thereby improve the access of the gas to the electrode surface.

The composition of such auxiliary electrodes is similar to that of air-consuming electrodes in primary batteries. Active carbon mixed with lampblack mostly form the basis.

Direct contact can be established between the auxiliary electrode and the negative electrode in the nickel/cadmium battery. Diodes, as shown in Fig. 7.13 for the lead-acid battery, are not required, because of the more positive potential of the cadmium electrode (cf. Fig. 7.9). Hydrogen cannot be evolved as long as the potential of the negative electrode remains above this value. Consequently, the problem that the auxiliary electrode might enhance the hydrogen-evolution rate is not relevant. On the other hand, increased overcharge rates are of special interest for nickel/cadmium batteries, used as battery packs for tools (cf. Section 13.2.2). In such a battery, individual cells have to withstand relatively high overcharges rates.

References

(1) F.J.Vaccaro and P.Casson, Proc. 9th Intelec Meeting, Stockholm, © IEEE Comm. Soc., 1987, p. 128

(2) W.B.Brecht, Proc. 9th Intelec Meeting, Stockholm, 1987, © IEEE Communications Soc., p.99

(3) D.Berndt, R.Bräutigam and U.Teutsch, Proc. 17th Intelec Meeting, The Hague, 1995, © IEEE Communications Soc., p.1

(4) H.Bode, 'Lead-Acid Batteries', John Wiley & Sons, New York, 1977, p. 315

(5) U.Teutsch, Proc. 1st Telescon Meeting, Berlin 1994, ©IEEE Communications Soc., p.89

(6) H.Döring, K.Wiesener, J.Garche, and W.Fischer, Journal Power Sources, **38** (1992), p. 261

(7) M.Rey, P.Coheur and H.Herbiet, Trans. Electrochem. Soc., **73** (1938), p. 315

(8) R.K.Galgali, P.V.Vasudeva Rao and H.Udupa, J. Electrochem. Soc, **126** (1979), p. 725.

(9) J.J.Lander, J. Electrochem. Soc., **105** (1958), p. 289

(10) D.Berndt, Metall, **35** (1981), p. 45 (in German)

(11) K.J.Vetter, 'Electrochemical Kinetics', Academic Press, New York, London 1967, Fig. 197, p. 539 (cf. C.A.Hampel, 'The Encyclopedia of Electrochemistry' Reinhold Publishing Corporation, New York,1964, p. 871)

(12) 'Gmelins Handbuch d.anorg. Chemie', Syst.Nr.47, Blei, Teil B 2, Verlag Chemie, 1972, p.79

(13) 'Gmelins Handbuch d.anorg. Chemie', Syst.Nr.33, Cadmium, Verlag Chemie, 1959, p.254
Cadmium 1: M.Knobel, P.Caplan and M Eiseman, Trans.Am.Electrochem. Soc.,**43** (1923), pp. 55 and 64.
Cadmium 2: A.G.Percerskaja and V.V.Stender, Zhurn. fiz. Chim., **24** (1950), p. 856

(14) 'Gmelins Handbuch d.anorg. Chemie', Syst.Nr.18, Antimon, Teil B 2, Springer-Verlag, Berlin, 1974, p 212

(15) 'Gmelins Handbuch d.anorg. Chemie', Syst.Nr.60, Kupfer, Teil A 2, Verlag Chemie, 1955, p.1106

(16 'Gmelins Handbuch d.anorg. Chemie', Syst.Nr.57, Nickel, Teil A II, 2, Verlag Chemie, 1968, p.568

(17) Gates Energy Products (Editor), 'Sealed Rechargeable Batterie', Application Manual, Gates Energy Products Inc., Gainesville, Florida, 1989, p. 58 (now HAWKER Batteries)

(18) N.E.Bagshaw, 'Batteries on Ships', Research Studies Press (John Wiley & Sons), 1982, p.106

(19) G.D.Rodriguez W.C.Spindler and D.S.Carr, J Power Sources, **31** (1990), p. 311

(20) K.R.Bullock, B.K.Mahato, G.H.Brilmyer and G.L.Wierschem *in* 'Advances in Lead-Acid Batteries', The Electrochem. Soc. Proceedings, Vol.84-14, p. 451

(21) W.Huber, Batteries International, Issue 4, July 1990, p.48

(22) T.A.Edinson U.S.Pat. 1,016,874 (1912)

(23) German Patent DE 20 08 218 (Sept.30, 1976, Accumulatorenwerk Hoppecke Carl Zoellner & Sohn, 5798 Hoppecke)

(24) K.Ledjeff and A.Winsel, J.Power Sources, **8** (1982), p. 211 (in German)

(25) German Patent 26 38 899, Accumulatorenwerke Hoppecke Carl Zoellner & Sohn, 5000 Köln, 28.8.76

(26) J.S.Symanski, B.K.Mahato and K.R.Bullock, J. Electrochem. Soc., **135** (1988), 548

(27) B.K.Mahato, E.Y.Weissman and E.C.Laird, J. Electrochem. Soc., **121** (1974), p. 13
M.Maja and N.Penazi, J. Power Sources, **25** (1989), p. 229

(28) H.Dietz, M.Radwan, J.Garche, H.Döring and K.Wiesener, J. Appl. Electrochem., **21** (1991), p. 221.

H.Dietz, J.Garche and K.Wiesener, 'Proc.Unesco Expert Workshop, Contribution of Electrochemistry to Energy Conservation and Saving and Environmental Protection', Gaussig Castle, Oct.30 - Nov.3 1989, p.117

(29) K.J.Vetter, 'Electrochemical Kinetics', Academic Press, New York, London, 1967
Original German Edition: Springer-Verlag, Berlin, 1961, p.159

(30) A.J.Bard and L.R.Faulkner, 'Electrochemical Methods - Fundamentals and Applications, © John Wiley & Sons, Inc., New York, 1980, p. 175

(31) P.Ruetschi and J.B.Ockerman, Electrochem. Techn., **4** (1966), p. 383

(32) I.Nikolov, G.Papazov, D.Pavlov, T.Vitanov and V.Naidenov, J.Power Sources, **31** (1990), p.69

G.Papazov, I.Nikolov, D.Pavlov, T.Vitanov, P.Andreev and M.Bojinov, ibid., p.79

8. The Internal Oxygen Cycle

The three battery systems treated in this book are characterised by the internal oxygen cycle i.e. oxygen that is generated during charging or overcharging at the positive electrode is subsequently reduced at the negative electrode. This cycle is formed by the two reactions:

1. Oxygen evolution at the positive electrode at the rate $i(O_2 \text{ evol.})$;
2. Oxygen reduction at the negative electrode at the rate $i(O_2 \text{ reduct.})$;

where $i(O_2 \text{ evol.})$ and $i(O_2 \text{ reduct.})$ describe the reaction rates as current equivalents.

When all the oxygen evolved at the positive electrode is subsequently reduced at the negative electrode the efficiency of this cycle is 100%, and the relation holds:

$$\boxed{i(O_2 \text{ evol.}) = i(O_2 \text{ reduct.})} \tag{8.1}$$

Oxygen generation at the positive electrode is thus counterbalanced at the negative electrode, and the created oxygen cycle absorbs the overcharging current that otherwise would decompose water into hydrogen and oxygen. Actually, 100% of efficiency can only be reached in nickel/cadmium and nickel/metal hydride batteries that are virtually sealed, while in valve-regulated lead-acid batteries 100% efficiency only can be approached (cf. Section 8.3.5).

The internal oxygen cycle can be written for acid solutions

$i(O_2 \text{ evol.})$ at the positive electrode $$2\cdot H_2O \Rightarrow O_2 + 4\cdot H^+ + 4\cdot e^- \tag{8.2a}$$

$i(O_2 \text{ reduct.})$ at the negative electrode $$O_2 + 4\cdot H^+ + 4\cdot e^- \Rightarrow 2\cdot H_2O \tag{8.2b}$$

and for alkaline solutions

$i(O_2 \text{ evol.})$ at the positive electrode $$4\cdot OH^- \Rightarrow O_2 + 2\cdot H_2O + 4\cdot e^- \tag{8.3a}$$

$i(O_2 \text{ reduct.})$ at the negative electrode $$O_2 + 2\cdot H_2O + 4\cdot e^- \Rightarrow 4\cdot OH^- \tag{8.3b}$$

In both cases, the reaction at the positive electrode is balanced by the reaction at the negative electrode, and the current consumed by this cycle does not change the chemical state within the cell.

The principle of the internal oxygen cycle is shown in Fig. 8.1.

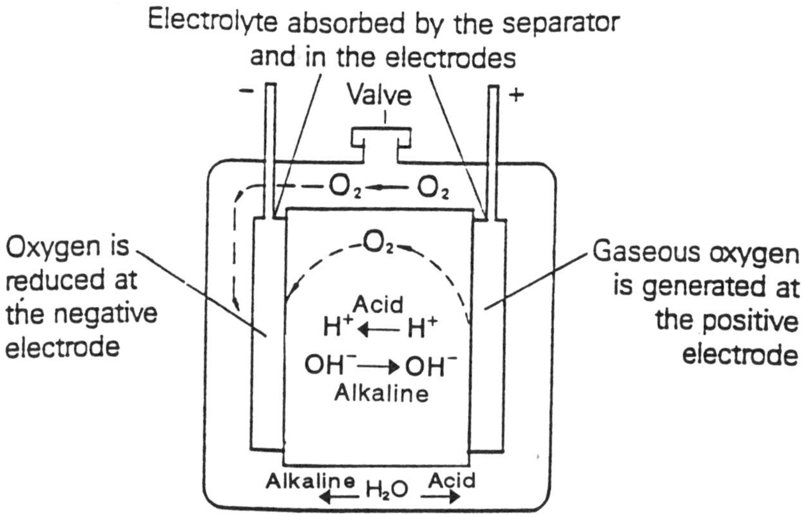

At the negative electrode		At the positive electrode
$O_2 + 4H^+ + 4e^- \rightarrow 2H_2O$	acid	$2H_2O \rightarrow O_2 + 4H^+ + 4e^-$
$O_2 + 2H_2O + 4e^- \rightarrow 4OH^-$	alkaline	$4OH^- \rightarrow O_2 + 2H_2O + 4e^-$

Fig. 8.1 The internal oxygen cycle.

Gaseous oxygen (O_2) generated at the positive electrode passes to the negative electrode, and is reduced to O^{2-} ions. These ions do not exist in the electrolyte, but are converted into water or OH^- ions as indicated below the Figure. To complete the cycle, protons (H^+) migrate from the positive to the negative electrode in acid solution, while in alkaline medium OH^- ions migrate from the negative to the positive electrode. The oxygen generated at the positive electrode thus remains in the cell (cf. Note on p.272).

Prerequisites for the internal oxygen cycle are:

- The cell has to be encapsulated by a sealed container to prevent the escape of the formed gaseous oxygen.
- Oxygen transport from the positive to the negative electrode has to be fast enough to carry the overcharging current. This is achieved by immobilised electrolyte, where most of the distance is surmounted by diffusion in the gaseous phase, and only a thin wetting film has to be passed by dissolved oxygen (cf. Sections 4.6.4 and 5.5.3).

The maximum rate of the internal oxygen cycle is determined by the transport rate of oxygen to the surface of the negative electrode. This is expressed by the limiting oxygen-reduction current in Fig. 7.9, p.253.

The internal oxygen cycle does not cause chemical changes in the electrodes. Only a small concentration gradient results in the electrolyte, as illustrated in Fig. 8.2 for a lead-acid battery.

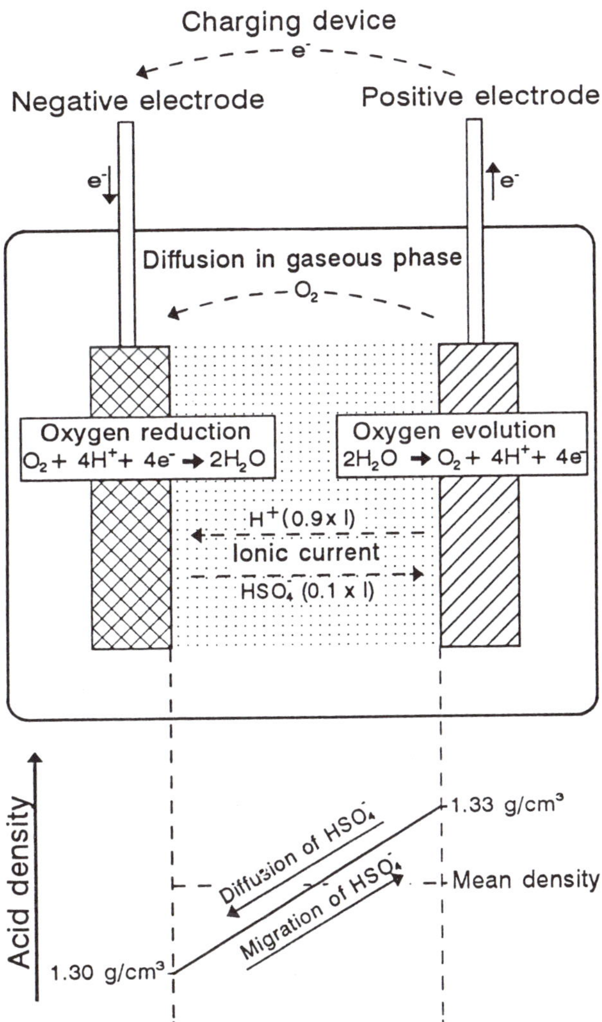

Fig. 8.2 Ion transport and resulting concentration gradient caused by the internal oxygen cycle in a lead-acid battery. (The secondary reactions hydrogen evolution and grid corrosion are not considered, cf. Fig. 8.11, p.284.) The shown concentration figures were determined after 6 months of continuous overcharging at 2.3 V/cell. (They depend on cell design.)

The ionic current is carried by H^+ and HSO_4^- ions according to their transference numbers (Table 2.3, p.25), which means that about 90% of the ionic current is carried by H^+ ions and 10% by HSO_4^- ions, as shown in Fig. 8.2. As a result, the acid concentration is increased at the positive electrode in continuous overcharge conditions (float charging).

If only the current were effective, a continuously increasing concentration gradient between the positive and negative electrodes would be generated. Diffusion, however, acts as a counter-reaction that tends to level out the imbalance of acid concentration throughout the cell. Consequently, a stationary gradient of acid concentration is established, when the cell is overcharged for sufficient time.

In normal service conditions, these concentration gradients can be neglected. But after prolonged overcharge at a fairly high voltage, acid depletion may slightly reduce the capacity of the negative electrode at high loads.

The situation in nickel/cadmium and nickel/metal hydride batteries is similar. The transport numbers are 0.22 and 0.78 for the K^+ and the OH^- ions, respectively (Table 2.3, p.25). This means that about 22% of the ionic current is carried by the K^+ ion, while the more mobile OH^- ion carries about 78%. As a consequence, a slight increase of the electrolyte concentration is observed at the negative electrode during continuous overcharge. Also here, the concentration gradient is moderated and finally limited by the counteracting diffusion.

The internal oxygen cycle as shown in Fig. 8.1 can be realised in nickel/cadmium and nickel/metal hydride batteries, since the overcharging reaction can be confined to this internal cycle. In lead-acid batteries, this situation can only be approached, since two secondary reactions disturb this cycle: a certain generation and escape of hydrogen is unavoidable and grid corrosion consumes oxygen. As a consequence, lead-acid batteries cannot be sealed hermetically but require a valve for gas escape, and a certain gradual water loss cannot be avoided (cf. Section 8.3).

Note: *The internal oxygen cycle is often called 'recombination' and valve-regulated lead-acid batteries that are based on this cycle are called 'recombinant' batteries. This is true in so far as the reduced oxygen (O^{2-}) is reunited (recombined) with protons (H^+) to water (H_2O). But the internal oxygen cycle cannot consume gaseous hydrogen (H_2) and therefore must not be mistaken with the catalytic recombination described in Section 7.5.2.*

8.1 Sealed Nickel/Cadmium Batteries

The internal oxygen cycle was first realised in sealed nickel/cadmium batteries. The thermodynamic situation shown in Figs. 5.1 and 7.9, p.175 and p.253, repsectively, favours this principle, since the equilibrium potential of the cadmium electrode is more positive than the equilibrium potential of the hydrogen electrode. As long as the negative electrode is not polarised by more than 20 mV to more negative potentials, hydrogen is not evolved. Such a situation is automatically achieved by the internal oxygen cycle, since oxygen reduction occurs at a more positive potential. When this reaction is fast enough to carry the total overcharging current, the negative electrode requires no polarisation for hydrogen evolution. *In this situation, oxygen is sometimes called a 'depolariser', since it makes polarisation of the negative electrode unnecessary; oxygen 'depolarises' the negative electrode.*

When the positive electrode is fully charged, oxygen evolution (i(O_2 evol.)) is caused by the current. When this oxygen subsequently is reduced at the negative electrode, an equivalent current (i(O_2 reduct.)) is required. Thus, oxygen evolution and reduction are the only two reactions maintained by the overcharging current (float charging). When i(float) denotes the overcharging or float current, this means:

$$\boxed{i(\text{float}) = i(O_2\ \text{evol.}) = i(O_2\ \text{reduct.})} \tag{8.4}$$

Then the overcharging current forms the internal oxygen cycle, as shown in Fig. 8.1, and simultaneously satisfies the condition that the same current flows through both electrodes.

Fig. 8.3 illustrates this by a simplified pattern.

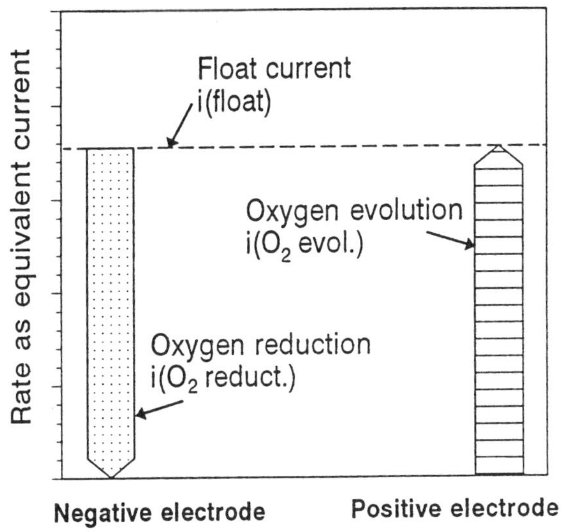

Fig. 8.3 Internal oxygen cycle during overcharge of a sealed nickel/cadmium battery. Oxygen evolution and oxygen reduction balance each other.

The situation in Fig. 8.3 is true only as long as 100% of the oxygen evolved at the positive electrode is reduced at the negative electrode. The limit of this reduction rate is determined by the maximum rate of oxygen transport from the positive to the negative electrode (cf. Section 2.6.3). It is represented by the broken line in Fig. 7.9, p.253, (the limiting-current density, which means the maximum of i(O_2 reduct.)). It depends on a number of parameters, like void space in the cell for gas diffusion, surface area of the negative electrode, thickness of the wetting layer, and internal gas pressure. Fig. 8.4 shows as an example the relation between overcharging current and internal pressure for two batteries of the same capacity but different design. The prismatic battery (the lower curve in Fig. 8.4) has electrodes comparatively spaced. The microporous separator is not pressed much and provides much void space. The cylindrical cell, on the other hand, is a compact construction with a highly pressed separation.

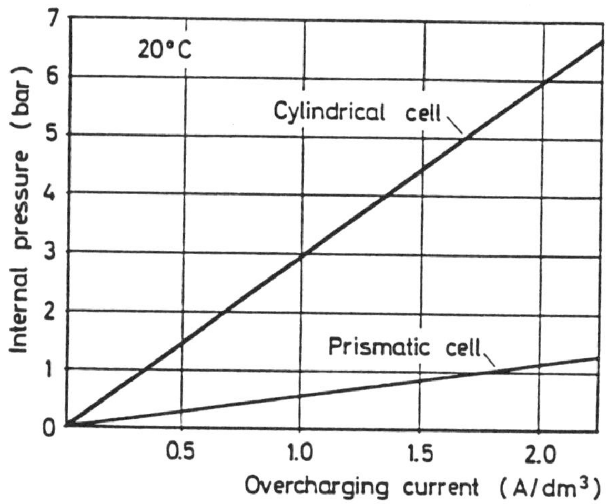

Fig. 8.4 Internal pressure versus overcharge current for two sealed nickel/cadmium batteries of different design but same capacity ([1]). Both measurements after 30% overcharging (overcharge factor of 1.3).

Both batteries were overcharged with constant currents and the resulting internal pressure indicates the partial O_2 pressure that is required for this current. The strict proportionality between gas pressure and overcharge current indicates that diffusion of dissolved oxygen through the wetting layer at the negative electrode is of important influence in both batteries, since this diffusion rate depends directly on oxygen pressure on account of the O_2 solubility (cf. Section 2.6.3.2). Obviously, this wetting layer is thinner (or the surface area larger) in the prismatic cell, and so a higher float current is reached at a lower internal pressure.

If the overcharging current is so high that the internal pressure exceeds the opening value of the valve, a certain portion of the evolved oxygen cannot reach the negative electrode fast enough but escapes from the cell. As a consequence, the remaining oxygen-reduction rate does not cover the overcharging current, and the potential of the negative electrode is automatically shifted to more negative values until hydrogen is generated at an adequate rate (i(H_2 evol.)), to compensate for the missing oxygen reduction. (The electrode is correspondingly polarised.) As a consequence, simultaneously with the oxygen, an equivalent amount of hydrogen has to leave the cell.

Fig. 8.5 illustrates this.

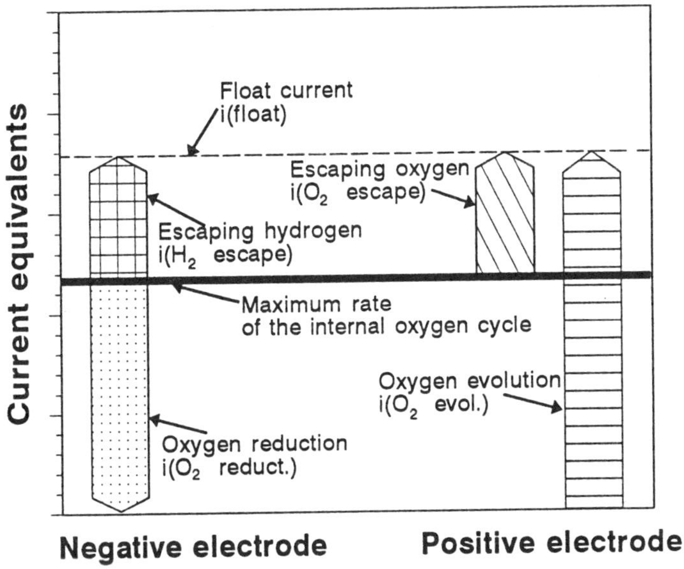

Fig. 8.5 Current distribution in a sealed nickel/cadmium battery, overcharged at a rate that exceeds the oxygen-reduction rate that can be reached at maximum (*at the highest internal pressure that is possible before the valve opens or the cell is destroyed*). The oxygen that cannot be reduced has to leave the cell together with an equivalent amount of hydrogen.

The situation is described by two equations. The first one means that the same current must flow through both electrodes:

$$\text{i(positive electrode)} = \text{i(negative electrode)} = \text{float current} \quad (8.5a)$$

which can be written as:

$$\text{i}(O_2 \text{ evol.}) = \text{i}(O_2 \text{ reduct.}) + \text{i}(H_2 \text{ evol.}) = \text{i(float)} \quad (8.5b)$$

As a second relation, the equivalent of oxygen that is reduced is given by the difference between evolved and escaped oxygen where i(O_2 escape) represents the equivalent of the oxygen which has to leave the cell:

$$\text{i}(O_2 \text{ reduct.}) = \text{i}(O_2 \text{ evol.}) - \text{i}(O_2 \text{ escape}) \quad (8.6)$$

When Eq. (8.6) is inserted into Eq. (8.5b), the result is

$$\text{i}(O_2 \text{ escape}) = \text{i}(H_2 \text{ evol.}) \quad (8.7)$$

During such inadmissible overcharges, the sealed nickel/cadmium battery operates to a certain extent like a vented type, losing hydrogen and oxygen in a stoichiometric ratio (as current equivalents) and losing water at an equivalent rate (cf. Table 7.1, p.246).
Note: *Actually, in nickel/cadmium batteries the negative electrode contains a surplus of active material that is still uncharged when the positive electrode is fully charged. This*

'charge reserve' is described in the following Section. Thus, when the overcharging current exceeds the maximum rate of the internal oxygen cycle, for a short initial period of time, the missing oxygen at the negative electrode is not compensated by hydrogen evolution, rather the remaining $Cd(OH)_2$ is charged. Strictly speaking, the situation described by Eq. (8.5b) applies only when also the negative electrode is fully charged, including the 'charge reserve'.

8.1.1 Charge Reserve

Hydrogen evolution must be avoided, since hydrogen is not consumed within the battery as has been described in Section 7.5.3.1. Hydrogen gas, if formed at the negative electrode, increases the internal pressure until the valve opens or the cell bursts. Hydrogen is not generated as long as the negative electrode retains its potential above the hydrogen-equilibrium potential, which means that the potential of the negative electrode has to be kept quite close to its equilibrium value (cf. Fig. 7.9, p.253). Two conditions must be fulfilled to achieve this:

- The potential of the negative electrode must be stable; i.e. charged material (Cd) as well as discharged material ($Cd(OH)_2$) must be present to maintain the equilibrium potential.
- The overcharge current must not exceed the maximum oxygen-reduction rate.

To secure a stable electrode potential, care has to be taken that some active material of the negative electrode remains uncharged when the battery is fully charged. This is achieved by oversizing the cadmium electrode, as shown in Fig. 8.6. The dotted blocks indicate the shares of capacity that actually can be utilised. The block on the left represents the electricity that can be drawn from the positive electrode and recharged again. This usable capacity of the positive electrode is synonymous with the capacity of the cell.

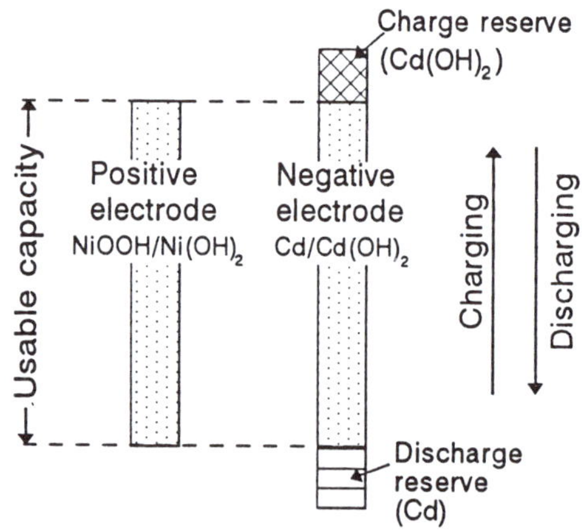

Fig. 8.6 Cell balance of a sealed nickel/cadmium battery.

The capacity of the negative electrode is represented by the block on the right. Its capacity is oversized, compared to the positive electrode. To match the design conditions shown in Fig. 8.6, the two electrodes must be assembled in a definite state of charge. Some of the active material in the negative electrode must remain uncharged when the positive electrode is fully charged and the cell is sealed. This share of capacity is called **'charge reserve'**. To achieve this layout, the negative electrode is discharged to a definite extent before it is assembled with a fully charged positive electrode. After the battery is sealed, this layout remains largely unchanged in normal operating conditions.

The charge reserve in the negative electrode complies with the first of the two above conditions; it stabilises the potential of the negative electrode. When the battery is charged, the two dotted blocks in Fig. 8.6 are converted from the discharged into the charged state at a uniform rate. When the positive electrode reaches the fully charged state, oxygen evolution occurs instead of the charging reaction, and this oxygen is transported to the negative electrode and reduced. When the overcharging current only generates oxygen at the positive electrode, and subsequently this oxygen is reduced at the negative electrode, no current is left for further charging of the negative electrode. So, the charge reserve remains in its discharged state for an unlimited period of overcharge. The charge reserve acts as a 'potentiostat' to prevent the negative electrode from becoming polarised to a more negative potential which would cause hydrogen evolution. Such a protection also applies during short periods of too heavy overcharges as mentioned in the Note at the end of the preceding Section.

Fig. 8.7 illustrates the consequences in battery practice. The broken curve represents the charging curve of a vented cell. The increase of the cell voltage during the initial stage of the charging process is mainly caused by polarisation of the positive electrode. The negative electrode, the cadmium electrode, is hardly polarised at that stage on account of the high charge acceptance.

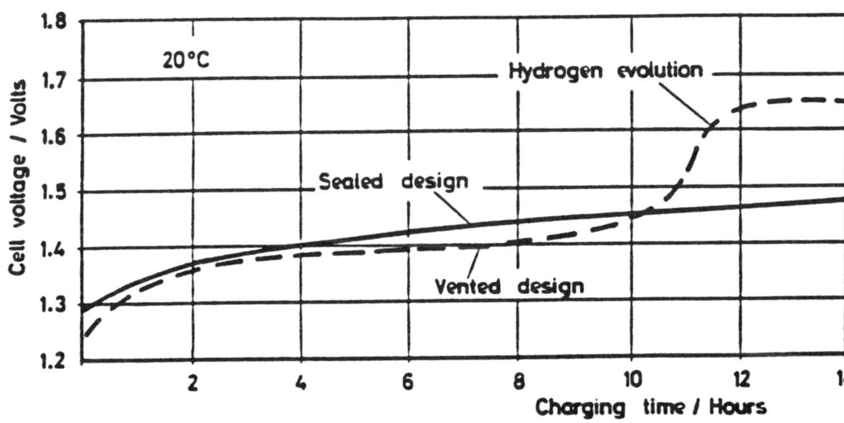

Fig. 8.7 Charging curves of nickel/cadmium batteries of similar size, but vented and sealed design. Constant-charging current I_{10} (10-hour rate).

After about ten hours, the negative electrode approaches the state of full charge. In the flooded battery, polarisation sets in and both electrode potentials are shifted to more positive and negative values, sufficient for the required rate of oxygen and hydrogen evolution, respectively. The increase of the cell voltage shown by the broken curve in Fig. 8.7 is mainly caused by polarisation of the negative electrode.

The charging curve for the sealed design shows also the gradual increase of the cell voltage caused by the increasing potential of the positive electrode. After about ten hours, the positive electrode reaches the state of full charge, as for the vented type, and an increasing share of the current causes oxygen generation. In this cell, however, the potential of the negative electrode is stabilised by the charge reserve until oxygen comes from the positive electrode. So the negative electrode is not polarised to generate hydrogen. Consequently, the corresponding increase of the cell voltage is not observed, as shown by the solid curve in Fig. 8.7.

Note: *When the battery is charged by larger currents, a maximum of the voltage is observed when full charge is approached (cf. Fig. 13.15, p.429 and Fig. 13.18, p.432). This voltage maximum is the result of the increased heat generation by the internal oxygen cycle (cf. Fig. 5.13, p.201). The resulting temperature increase reduces the overvoltage required for oxygen evolution and so reduces the cell voltage. This voltage decay often is used as a signal to end the charging process.*

8.1.2 Discharge Reserve

Fig. 8.6 shows that the negative electrode is also oversized from the point of the view of discharge. A share of active material of the negative electrode remains charged when the positive electrode is discharged completely. This part of the capacity is called 'discharge reserve ' or 'overdischarge reserve'. Reasons for this overcapacity are:

- The active material of the negative electrode, cadmium (Cd), is subject to ageing factors more than the positive active material. The discharge reserve compensates to some extent for these ageing effects.
- With increasing discharge rates, the utilisation of the active material in the negative electrode decreases more than that of the positive active material. Thus the capacity of the negative electrode is more affected at high loads and may then determine the capacity of the cell. When a number of cells are switched in series in a battery, the negative electrodes in individual cells are in danger of being reversed in the case of a deep discharge with high load. The discharge reserve delays such a reversal and may prevent virtual reversal of the concerned cells before the discharge of the whole battery is terminated.

8.1.3 Reversal Protection, (Antipolar Material)

When the nickel/cadmium battery is sealed, the capacity layout (Fig. 8.6) remains unaltered in normal service conditions. Hydrogen (H_2) is not generated as long as the charging current does not exceed the maximum rate of the internal oxygen cycle. Further secondary reactions, such as corrosion, can be neglected.

A danger of hydrogen generation may, however, arise in batteries consisting of a number of cells switched in series. When such a battery suffers a deep discharge, on account of capacity deviations some of the cells are overdischarged and reversed while the rest are still delivering energy. In a cell which gets reversed, the dischargeable part of the active material of the positive electrode is converted into $Ni(OH)_2$ and cannot be discharged further. If the discharge of the battery is continued, the potential of this electrode is shifted in the negative direction until, finally, the electrode is reversed and hydrogen evolution occurs. In a completely sealed cell, this hydrogen evolution increases the internal pressure, which finally may destroy the seal or at least cause water loss in cells with re-sealing valves.

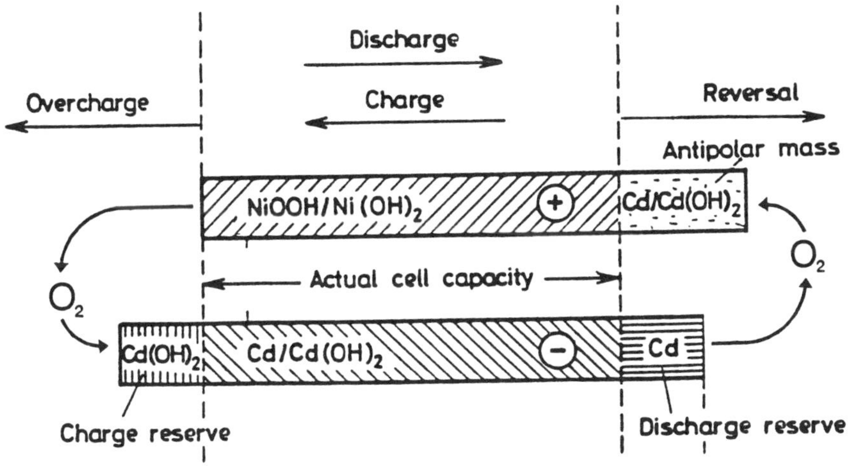

Fig. 8.8 Balance of electrode capacities (cell balance) in a sealed nickel/cadmium battery with reversal protection by 'antipolar mass' (antipolar material).

To prevent this, 'antipolar material' is applied, which results in the capacity layout shown in Fig. 8.8. 'Antipolar material' means that active material, normally used in the negative electrode, is added to a definite extent to the positive electrode. As long as the battery is working in normal conditions, this material remains oxidised to $Cd(OH)_2$ i.e. in its discharged state. On reversal, the antipolar material must be charged, which means the $Cd(OH)_2$ must be reduced to cadmium (Cd), and this process stabilises the potential of the reversed electrode at the potential of the cadmium electrode and prevents the generation of hydrogen until all the $Cd(OH)_2$ is converted.

If the capacity layout of Fig. 8.8 applies, the discharge reserve of the negative electrode is exhausted before the antipolar material in the positive is charged completely. Consequently, the negative electrode is converted completely into $Cd(OH)_2$ and the electrode is reversed to the potential of oxygen generation. When this stage is reached, the reversed cell operates in the same way as the normal cell during overcharge. Oxygen is generated at the now positive cadmium hydroxide electrode, and oxygen is consumed at the antipolar material, which now acts as a $Cd/Cd(OH)_2$ electrode. Thus the reversed cell attains a stable state without hydrogen evolution.

Nowadays, antipolar material is used only for button cells or coin cells (cf. Section 11.5.1) because this cell design is absolutely sealed and would be destroyed by the internal overpressure caused by hydrogen evolution. In sealed nickel/cadmium batteries of cylindrical or prismatic design (cf. Sections 11.5.2 and 11.5.3), antipolar material usually is no longer applied. These cells are generally equipped with valves that can release overpressure but are closed again afterwards (cf. Section 11.4). With these resealing valves, occasional overcharges and reversals can be accepted. The cell only loses some electrolyte, which may not influence the performance data too much, if the level of abuse does not exceed certain limits. Traces of hydrogen are furthermore gradually oxidised at the positive electrode, as mentioned in Section 7.5.3.1.

The main reason to abandon antipolar material is that neither space nor weight need be sacrificed. The general tendency to increase capacity more and more supported such a development. Beyond that, reversal protection by antipolar mass cannot be achieved perfectly, but is limited to certain discharge conditions.

8.1.4 Time-Dependent Changes

The cell balance for sealed nickel/cadmium batteries in Figs. 8.6 and 8.8 is usually assumed to remain unchanged during the whole service life. This is true only as an approximation, because irreversible chemical changes are connected with the charging and discharging processes.

One of these changes is the recrystallisation of the active material of the cadmium electrode, mentioned in Section 11.1.2. This recrystallisation leads to a slight reduction of the available capacity of the negative electrode.

Separator oxidation or hydrolysis (cf. Section 5.5.3) also causes changes of the cell balance, because oxygen or water is consumed thereby.

Another effect that changes the capacity of the positive electrode is the oxidation of nickel substrates described in Section 5.6 (nickel corrosion). But, compared to the corrosion rates observed for lead in the lead-acid battery (cf. Section 4.7), the corrosion rate of metallic nickel is extremely low and can usually be neglected.

8.2 Nickel/Metal Hydride Batteries

Nickel/metal hydride batteries are based on the same positive electrode as nickel/cadmium batteries. For this reason, the part of Fig. 8.9 that concerns the positive electrode is identical with this part of Fig. 7.9, p.253. For thermodynamic reasons, oxygen evolution cannot be avoided at the positive electrode. But this overcharging reaction is compensated by oxygen reduction at the negative electrode, and so an internal oxygen cycle is formed. As in Fig. 7.9, oxygen reduction is inserted as a limiting current, because its rate is limited by the oxygen-transport rate through or around the immobilised electrolyte.

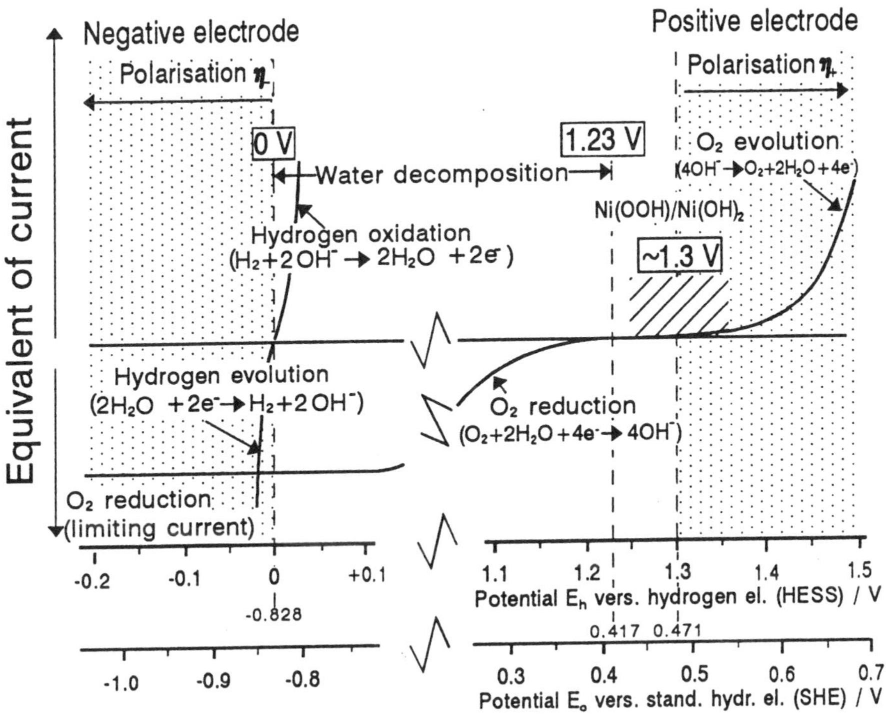

Fig. 8.9 Equilibrium potentials of the negative and positive electrodes in a nickel/metal hydride battery, and current/voltage curves for O_2 and H_2 evolution and H_2 oxidation. For the x-axis, the electrode potential, two scales are employed. The upper one represents the potential E_h referred to the hydrogen electrode in the same solution (HESS); the lower scale E_o referred to the standard hydrogen electrode (SHE) for 1.250 g/cm^3 density of the KOH solution at 25 °C (cf. Fig. 7.9).

At the negative electrode, oxidation of hydrogen ($H_2 \Rightarrow 2{\cdot}H^+ + 2{\cdot}e^-$) is the discharging reaction. In the alkaline electrolyte this corresponds to Eq. (6.2), p.215.

$$H_2 + 2{\cdot}OH^- \Rightarrow 2{\cdot}H_2O + 2{\cdot}e^- \quad (8.8)$$

Due to the catalytic activity of the storage alloy, even at very small polarisations this reaction occurs at a high rate, as indicated by the steep current/voltage curve in Fig. 8.9. The reversal of Eq. (8.8), the reduction of hydrogen, i.e. the charging reaction of this electrode

$$2{\cdot}H_2O + 2{\cdot}e^- \Rightarrow H_2 + 2{\cdot}OH^- \quad (8.9)$$

is also fast and is represented by a correspondingly steep current/voltage curve in Fig. 8.9.

8.2.1 Charge Reserve - Discharge Reserve

Different from nickel/cadmium batteries, hydrogen evolution cannot be suppressed in nickel/metal hydride batteries, since it is the charging reaction of the negative electrode. But hydrogen does not cause an intolerable increase of the internal pressure as long as absorption capacity is available. To secure this situation, a '**charge reserve**' is required as shown in Fig. 8.10. This means that absorption capacity in the negative electrode is still available when the positive electrode reaches the state of complete charge i.e. the negative electrode is still partly discharged, and the internal pressure is kept within the usual range. The current that flows through the positive electrode is then used for oxygen evolution, and this oxygen is subsequently reduced at the negative electrode. As in nickel/cadmium batteries, an internal oxygen cycle is formed, described by

$$\boxed{i(\text{float}) = i(O_2\ \text{evol.}) = i(O_2\ \text{reduct.})} \qquad (8.10)$$

The overcharging current in total is consumed by this internal oxygen cycle. No current is left for hydrogen evolution, and so the 'charge reserve' remains unchanged.

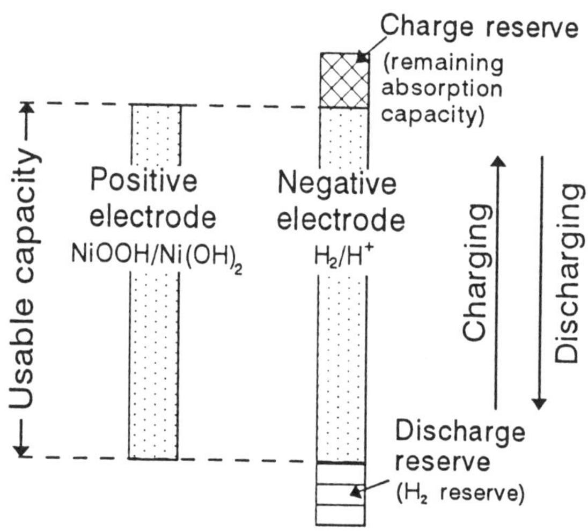

Fig. 8.10 Balance of electrode capacities (cell balance) in a nickel/metal hydride battery with 'charge reserve' and 'discharge reserve'.

The '**discharge reserve**' in Fig. 8.10 ensures that the negative electrode still is partly charged when the positive electrode is completely discharged. When discharge of the cell continues, the remaining 'reduced active material', the still available H_2, stabilises the potential of the negative electrode, while the positive electrode is reversed. At the reversed positive electrode, hydrogen is evolved according to:

$$\boxed{2\cdot H_2O + 2\cdot e^- \Rightarrow H_2 + 2\cdot OH^-} \qquad (8.11)$$

which is oxidised at the negative electrode according to its normal discharge reaction (Eq. (8.8):

$$\boxed{H_2 + 2{\cdot}OH^- \Rightarrow 2{\cdot}H_2O + 2{\cdot}e^-} \qquad (8.12)$$

As a result, the negative electrode is not further discharged. It remains at its normal discharge potential, and the combination of Eqs. (8.11) and (8.12) form an internal hydrogen cycle, as mentioned for nickel/hydrogen batteries in Section 6.1.1. This cycle causes a stable state of the overdischarged cell according to

$$\boxed{i(\text{overdischarge}) = i(H_2\ \text{evolution}) = i(H_2\ \text{reduction})} \qquad (8.13)$$

Due to this internal hydrogen cycle, nickel/metal hydride batteries are protected against reversal and 'antipolar material', as described in Section 8.1.3 for nickel / cadmium batteries, is not required.

8.3 Valve-Regulated Lead-Acid Batteries

The valve-regulated lead-acid battery is also based on the internal oxygen cycle. But the situation is different from that of sealed nickel/cadmium and nickel/metal hydride batteries because of two secondary reactions that occur even at open-circuit voltage, and at a rate that cannot be neglected. These are:

1. **Hydrogen evolution at the negative electrode**

and

2. **Corrosion of the grid and conducting elements connected with the positive electrode**

Both can be expressed as current equivalents

$$\boxed{i(H_2\ \text{evol.})}$$

where 1 mA is equivalent to the evolution of 0.448 cm^3 of H_2 per hour at 20 °C (Eq. (2.114) and Table 2.7, p.49),

and

$$\boxed{i(\text{corrosion})}$$

where 1 mA is the equivalent to the corrosion of 1.926 mg Pb per hour (Eq. (4.34), p.154).

The minimum hydrogen-evolution rate is the self-discharge rate at the open-circuit potential of the lead electrode: about 1 to 4 mA/100 Ah (cf. Section 7.2.2.1 and Table 4.8, p.119). This gas volume has to leave the cell, since the internal hydrogen cycle is negligible, as mentioned in Section 7.5.3. Thus, according to Table 4.8, hydrogen-evolution rates between 0.5 and 2 ml/100 Ah/hour are to be expected, that escape periodically through the valve if hydrogen is not otherwise lost (cf. note at the end of Section 7.5.3.1, p.263).

A corrosion rate of 1 mA/100 Ah was mentioned for the positive grid in Table 4.8, p.119. This is in the same order of magnitude as the above hydrogen evolution rate. *It is, of course, only a rough figure, because the relationship between grid surface and nominal capacity varies for different battery designs and the corrosion rate depends on quite a number of parameters.*

The consequence of these secondary reactions is that the valve-regulated lead-acid battery is not a virtually closed system, since hydrogen escapes from the battery and oxygen is consumed by corrosion. Both reactions mean water loss, although to only a small amount that normally only causes a slight capacity loss during service life. But, under unfavourable conditions (elevated temperature), water loss can be a serious ageing factor (cf. Section 15.2.3).

On account of these secondary reactions, the 'ideal internal oxygen cycle' can only be approximated, and Fig. 8.1, p.270, has to be modified for valve-regulated lead-acid batteries as shown in Fig. 8.11.

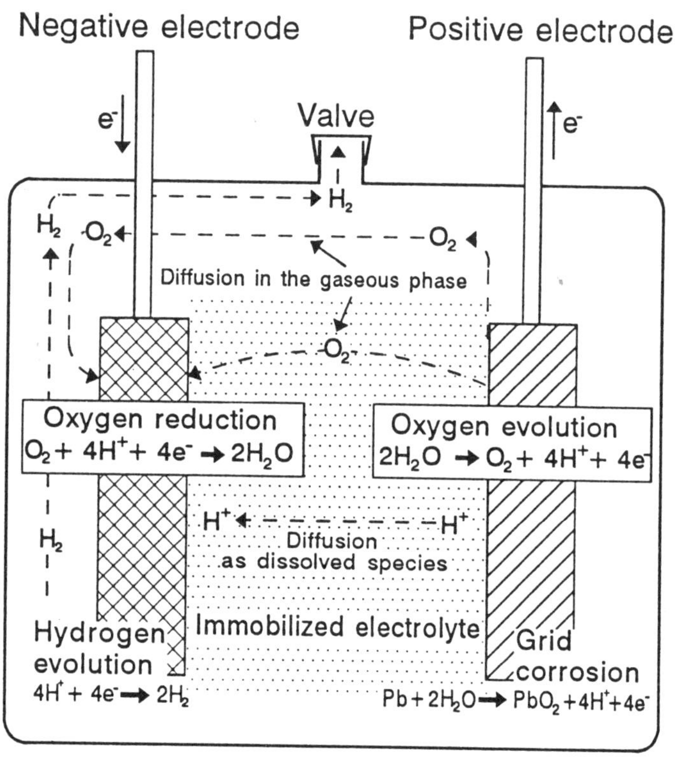

Fig. 8.11 The internal oxygen cycle in valve-regulated lead-acid batteries that unavoidably is accompanied by evolution and escape of hydrogen, and by corrosion of the positive grid.

Fig. 8.12 shows a simple scheme of the situation during overcharging or float charging, i.e. when the battery is fully charged and only the secondary reactions remain.

It corresponds to the right hand part of Fig. 7.6, p.244, and comparison of both Figures indicates the difference between the flooded and the valve-regulated designs. The situation of the positive electrode is the same in vented and in valve-regulated batteries: oxygen evolution is the main reaction during overcharging. At the negative electrode, however, in the valve-regulated design, oxygen reduction is the main reaction with the result that hydrogen evolution is largely reduced.

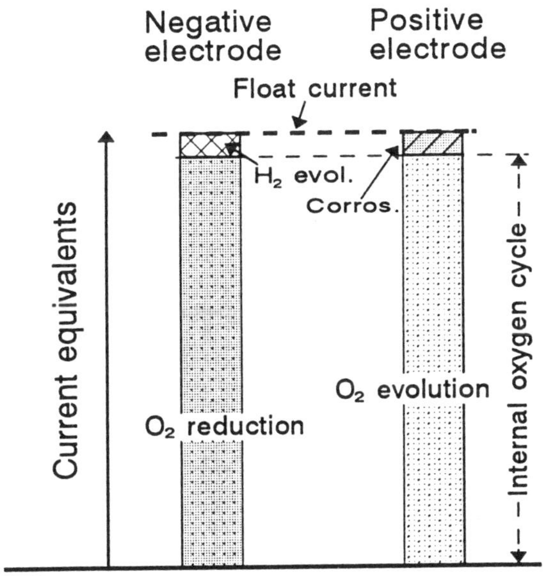

Fig. 8.12 Current contributions of the various reactions during overcharging (or float charging) of valve-regulated lead-acid batteries. (The columns are not to scale, since the internal oxygen cycle normally amounts to more than 99% of the float current.)

The fact that equal currents flow through both electrodes means:

$$i(\text{float}) = i(O_2 \text{ evol.}) + i(\text{corrosion}) = i(O_2 \text{ reduct.}) + i(H_2 \text{ evol.}) \tag{8.14}$$

That is, oxygen evolution plus grid corrosion must equal oxygen reduction plus hydrogen evolution, as shown in Fig. 8.12.

The effect of the internal oxygen cycle on the current/potential relation is illustrated by Fig. 8.13. It corresponds to Fig. 7.5, p.242, and shows the different float-charge behaviour of the flooded and the vented design, provided that the basic parameters for hydrogen and oxygen evolution are the same in both battery designs.

Oxygen evolution, hydrogen evolution, and corrosion follow their individual current/voltage relations. In the vented battery, the rates of O_2 reduction (O_2 red. (fl)) and grid corrosion (corrosion (fl)) are small compared to hydrogen and oxygen evolution, and the polarisation of the positive and negative electrodes is largely determined by hydrogen and oxygen evolution (H_2(fl) and O_2(fl) in the Figure). Both approximately balance each other and are nearly equivalent to the float current of 11 mA/100 Ah at the float voltage of 2.23 V/cell in this example. The resulting polarisation is similar for positive and negative electrodes, it depends on the kinetic parameters, expressed by the factors a and b of the corresponding TAFEL lines. In battery practice, corrosion (corrosion (fl)) and O_2 reduction (O_2 red. (fl)) usually are not noticed as current-consuming reactions, since water loss is fairly well approximated by the float current equivalent.

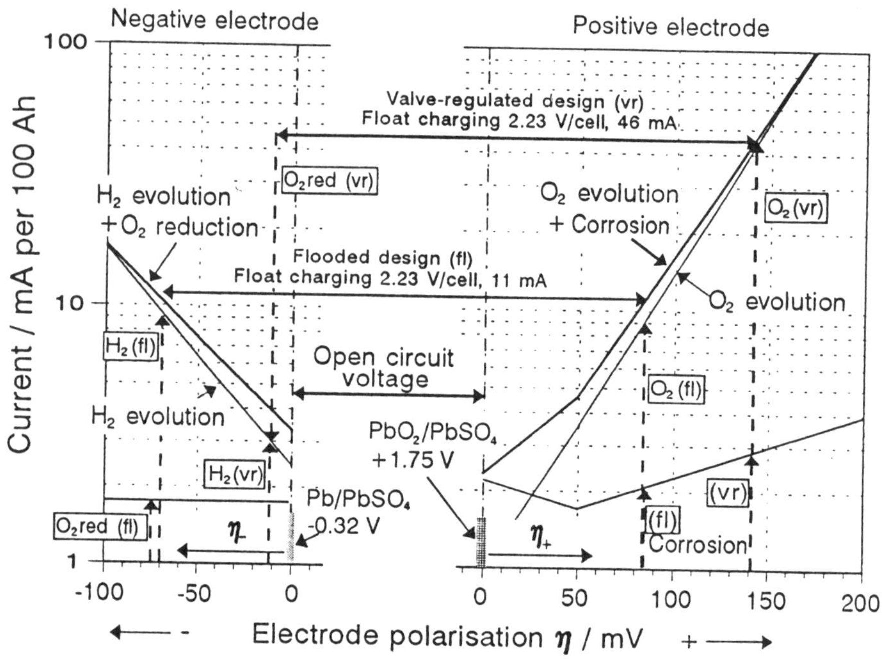

Fig. 8.13 Float charging of a flooded (vented) and a valve-regulated lead-acid battery, when the kinetic parameters of oxygen evolution (O_2), hydrogen evolution (H_2), and corrosion are considered as unchanged in both battery designs. (fl) and (vr) denote the parameters of the flooded and the valve-regulated design, respectively. The vertical columns denote the equilibrium potentials (assumed acid density 1.24 g/cm^3, E^o = 2.08 V). Cell voltage is $E = E^o + \eta_+ - \eta_-$.

In the valve-regulated design, oxygen reduction (O_2red (vr)) is the main reaction at the negative electrode that (nearly) compensates oxygen evolution (O_2(vr)) at the positive electrode. Consequently, the negative electrode is not polarised, but remains close to its equilibrium potential, and hydrogen evolution (H_2(vr)) is reduced to about the self-discharge rate. With the same float voltage (2.23 V/cell), the potential of the positive electrode is shifted to a more positive value, and the oxygen-evolution rate is correspondingly increased. The result is an increased float current in the valve-regulated design (46 mA/100 Ah in this example), but the hydrogen-evolution rate is reduced to about 3 mA/100 Ah.

An important characteristic of the internal oxygen cycle is that oxygen reduction at the negative electrode as the main overcharging reaction does not depend on electrode polarisation, but is determined by oxygen evolution at the positive electrode and its subsequent transport to the negative electrode. Polarisation of the negative electrode is only required to complement the gap between O_2 evolution and O_2 reduction that may be caused by a low efficiency of the internal oxygen cycle, or by a too-high corrosion rate as will be shown in Section 8.3.2.

8.3.1 Limited Recombination Efficiency

When the recombination efficiency is below 100%, a corresponding amount of the evolved oxygen is not reduced at the negative electrode, i.e.

$$\boxed{i(O_2 \text{ reduct.}) < i(O_2 \text{ evol.})} \qquad (8.15)$$

This means that a certain portion of the generated oxygen escapes from the cell, since it cannot reach the surface of the negative electrode fast enough under the given maximum pressure. Eq. (8.15) can be written:

$$\boxed{i(O_2 \text{ reduct.}) = i(O_2 \text{ evol.}) - i(O_2 \text{ escape})} \qquad (8.15a)$$

when $i(O_2$ escape) denotes the escaped amount of oxygen. (*The main reason for such a reduced efficiency is a too low transport rate of oxygen when too much electrolyte fills the pore system required for gas diffusion. But a leakage that allows oxygen escape would also cause this effect.*)

To achieve current balance, a corresponding additional amount of hydrogen is evolved at the negative electrode, as shown in Fig. 8.14. As a consequence, the evolved hydrogen together with the non-reduced portion of evolved oxygen escapes from the cell.

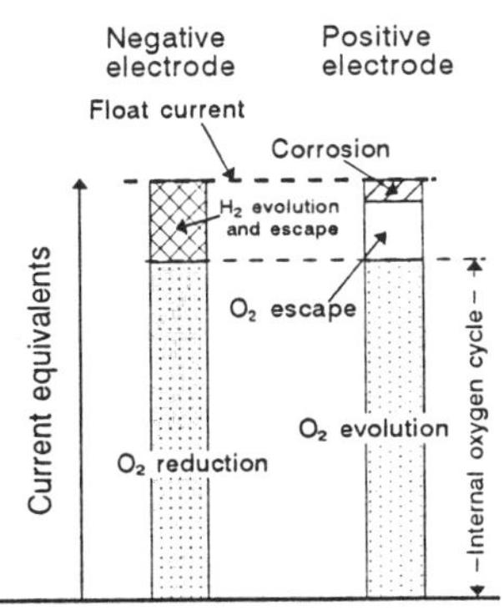

Fig. 8.14 Current contributions of the various reactions during overcharging when the efficiency rate of the internal oxygen cycle is low.

Fig. 8.14 corresponds to Fig. 8.5, p.275. The battery operates to a certain degree like a vented one. Positive and negative electrodes are polarised to potentials above and below their equilibrium values; a mixture of oxygen and hydrogen escapes from the battery, and water loss is correspondingly increased.

8.3.2 Current Balance at 100% Recombination Efficiency

At 100% recombination efficiency, all the oxygen evolved at the positive electrode is subsequently reduced, i.e.

$$\boxed{i(O_2 \text{ reduct.}) = i(O_2 \text{ evol.})} \tag{8.16}$$

From Eq. (8.14) it then follows that grid corrosion and hydrogen evolution must also be balanced:

$$\boxed{i(H_2 \text{ evol.}) = i(\text{corrosion})} \tag{8.17}$$

as shown in Fig. 8.12, p.285. If this condition cannot be fulfilled, current balance is disturbed and discharge reactions occur for compensation, as shown in Fig. 8.15.

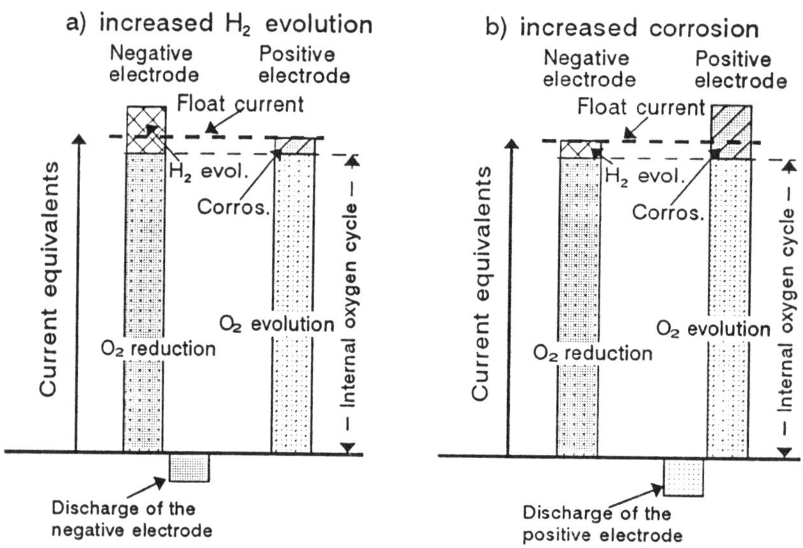

Fig. 8.15 Current contributions of the various reactions during overcharging, when the efficiency of the internal oxygen cycle is 100%, but hydrogen-evolution and corrosion rates are not balanced.

Part a) in Fig. 8.15 shows the situation when the hydrogen-evolution rate exceeds the corrosion rate. The reason may be a too high self-discharge rate of the negative active material or oversized negative electrodes. To satisfy the rule that the same current must flow through both electrodes, the too large cathodic current at the negative electrode (i(O_2 reduct.) + i(H_2 evol.)) has to be compensated for by an anodic process, as is the discharge of the active material. Under such circumstances, the potential of the negative electrode is shifted to a value slightly above its equilibrium, so that the discharge reaction becomes possible (cf. Fig. 8.21, p.297).

The demand that equal currents flow through both electrodes means:

$$\begin{aligned} \text{i(float)} &= \text{i(O}_2\text{ evol.)} + \text{i(corr.)} \\ &= \text{i(O}_2\text{ reduct.)} + \text{i(H}_2\text{ evol.)} + \text{i(Pb discharge)} \end{aligned} \tag{8.18}$$

where i(Pb discharge) has the opposite sign than i(O_2 reduct.) and i(H_2 evol.). This compensation occurs automatically by discharge according to Eqs. (4.23) or (4.24), p.116, because of the O_2 that is left in the cell. It depolarises the negative electrode so far that it reaches its equilibrium potential or values slightly above. *(If a portion of the left O_2 escapes through the vent, a corresponding reduction of the efficiency of the internal oxygen cycle according to Section 8.3.1 would be the consequence.)*

As has been mentioned in Sections 4.4.1 and 7.2.3, the hydrogen-evolution rate gradually decreases with progressing discharge of the negative electrode. Thus the excess gassing of the negative electrode is gradually decreased until balance between hydrogen evolution and corrosion is reached. Then, current balance is stable, but the negative electrode is partly discharged.

Note: *In practice, partly discharged negative electrodes are often observed in valve-regulated batteries after prolonged float service when 100% recombination efficiency or values close to it are reached on account of water loss (cf. Section 8.3.4). The usual slight oversizing of the negative electrodes or loss of capacity of the positive electrode favour this situation. In principle, this is not surprising, since at low float voltages it is difficult to achieve and maintain the balance between hydrogen evolution and grid corrosion as demanded by Eq. (8.17). But this situation is not really dangerous since a stable state of the battery is automatically reached, although at a reduced capacity, on account of the partly discharged negative electrodes (cf. also Sections 8.3.3 and 8.3.4).*

Part b) in Fig. 8.15 shows the corresponding situation with a too-high corrosion rate. In this case the anodic current through the positive electrode (i(O_2 evol.) + i(corrosion)) must be compensated for by a cathodic process, discharge of the positive electrode, and the current balance is:

$$\begin{aligned} \text{i(float)} &= \text{i(O}_2\text{ evol.)} + \text{i(corr.)} + \text{i(PbO}_2\text{ discharge)} \\ &= \text{i(O}_2\text{ reduct.)} + \text{i(H}_2\text{ evol.)} \end{aligned} \tag{8.19}$$

where i(PbO_2 discharge) has the opposite sign than i(O_2 evol.) and i(corr.). Then the positive electrode is no longer polarised but remains at its equilibrium potential or slightly below, and is gradually discharged (cf. Fig. 8.24, p.300). In practice, this situation would be much more critical then the above example (a) in Fig. 8.15, since corrosion is not reduced with progressing discharge, but is even increased when the equilibrium potential of the positive electrode is approached. For this reason, the grid alloy of the positive electrode has to be as corrosion resistant as possible.

The following model simulations show that such situations with zero polarisation of one of the electrodes mainly occur at low float voltages (cf. Figs. 8.21 and 8.24, p.297 and p.300). At a sufficient high float voltage, Eq. (8.17) is automatically fulfilled, also at 100% efficiency of the internal oxygen cycle. Then both electrodes are polarised so that discharge is not possible. But, simultaneously, hydrogen evolution and grid corrosion are increased.

8.3.3 Model Simulations

The above Sections illustrate the complex balance between the various reactions that characterises float charging of valve-regulated lead-acid batteries. The float voltage (E) is composed of the open-circuit voltage and the polarisation of positive and negative electrodes:

$$E = E^o + \eta_+ - \eta_- \tag{8.20}$$

with E^o: equilibrium or open-circuit voltage

η_+; η_-: polarisation of the positive and negative electrode, respectively

Polarisation is the decisive parameter that determines the rates of the various electrochemical reactions, like oxygen and hydrogen evolution and grid corrosion. Oxygen reduction, however, the main reaction at the negative electrode, does not depend on the polarisation of this electrode but is closely linked to oxygen evolution at the positive electrode. The polarisation values actually reached depend on the balance between all reaction rates under observance of the rule that the same current flows through both electrodes:

$$\text{i(positive electrode)} = \text{i(negative electrode)} \tag{8.21}$$

as already has been described in the preceding Sections.

A further condition that should be fulfilled is:

$$\eta_+ > 0 \quad \text{and} \quad \eta_- < 0 \tag{8.22}$$

Otherwise, the concerned electrode is not kept fully charged as described above.

The details of such a complex balance cannot be derived from simple drawings of TAFEL lines or similar deliberations. To elucidate this situation, mathematical treatment is required that considers the rates of the various reactions and their mutual interdependence. Such a mathematical model for the float charge situation has been developed ([2]) that considers the gassing reactions, oxygen reduction, grid corrosion, and their dependence on temperature, but also discharge reactions that might occur as described in Section 8.3.2. The model is based on the current/polarisation relationship for the different reactions that considers the kinetic parameters of the reactions and the influence of temperature (cf. Eq. (2.58), p.28). The resulting system of equations is solved numerically. The reliability of this model has been confirmed by comparison between model simulations and a larger number of experiments ([3]). The following examples demonstrate that such a model provides a better understanding of the complex balance between the various reactions.

These examples show calculated results for a battery with an efficient internal oxygen cycle (99.5%). They are based on a hypothetical battery that closely corresponds to modern valve-regulated batteries, suited for high load applications (electrode area ≅ 3000 cm^2 per 100 Ah, plate distance ≅ 0.3 cm). (*Kinetic parameters of this 'Standard Battery' are compiled in Table 1 in* ([4])). Most of the used values were derived from thoroughly carried out experiments, and their compatibility with published data has been checked. The kinetic parameters for corrosion are based on literature. Due to these data, the model simulations provide a well based general insight for the valve-regulated battery design. Nevertheless, for individual battery designs, smaller or larger deviations are to be expected.

In this context, it should be noted that the model uses an equivalent current of grid corrosion, irrespective of whether the actual corrosion reaction is $Pb \Rightarrow Pb^{4+}$ or $Pb \Rightarrow Pb^{2+}$, and irrespective of the kind of corrosion attack (homogeneous, intergranular etc.). To estimate the actual grid disintegration, such calculated corrosion-current densities have to be combined with experimental results for corrosion damage at the specific electrode potential and temperature.

8.3.3.1 Example 1, the 'Standard Battery'

In Fig. 8.16, electrode reaction rates are plotted as current equivalents vs. electrode polarisation for the above mentioned 'Standard Battery'. Hydrogen evolution, oxygen evolution and grid corrosion are represented by TAFEL lines. As in Fig. 7.5, p.242, corrosion is represented by the combination of two TAFEL lines.

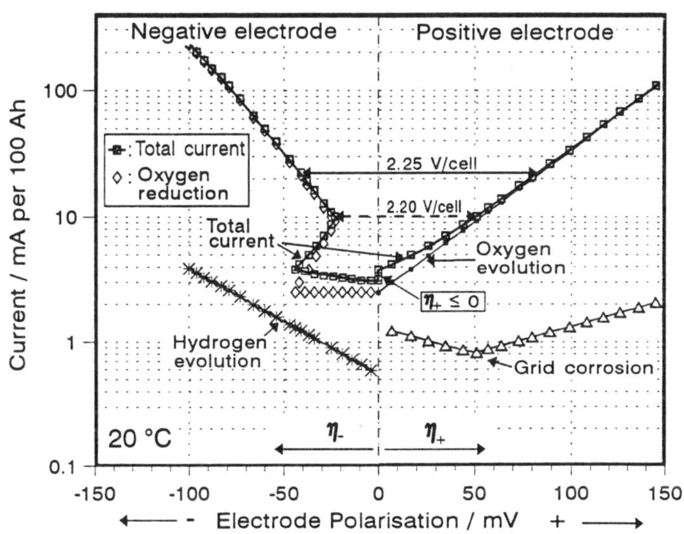

Fig. 8.16 Calculated current densities as a function of electrode polarisation for a valve-regulated lead-acid battery at 20 °C. Acid density 1.28 g/cm^3; open-circuit voltage 2.13 V/cell. Efficiency of the internal-oxygen cycle assumed as 99.5%. (Fig. 2 in ([4]).)

In Fig. 8.16, at very low float voltages, causing currents below 4 mA/100 Ah, grid corrosion is not balanced by hydrogen evolution. As a consequence, polarisation of the positive electrode is reduced to zero or slightly below ($\eta_+ \leq 0$), and discharge of the positive active material occurs for compensation according to Eq. (8.19), i.e. the too-high anodic current (i(O_2 evol.) + i(corros.)) is reduced by the cathodic i(PbO_2 discharge). This is the section $\eta_+ \leq 0$ in the total-current curve, where polarisation of the positive electrode remains very close to zero, since a few millivolts of negative polarisation are enough for sufficiently high discharge currents. Thus, in the section $\eta_+ \leq 0$ the potential of the positive electrode is approximately constant, and oxygen evolution remains constant. As a consequence of the close connection between oxygen generation and oxygen reduction at the assumed high efficiency of the internal oxygen cycle, this also applies to oxygen reduction. This causes the horizontal section of the oxygen-reduction curve that leads to a local minimum of η_- at the value of -0.045 V in Fig. 8.16.

At higher float voltages (currents above 4 mA/100 Ah in Fig. 8.16), hydrogen evolution balances grid corrosion and the positive electrode is also polarised. The decreasing η_-, observed for currents just above the $\eta_+ \leq 0$ section, is caused by the negative slope of the corrosion curve, since the decrease of corrosion causes a corresponding increase of oxygen evolution. When the corrosion has reached its minimum and starts to increase with polarisation, a turning point is also observed in the oxygen-reduction curve and in the total-current curve at the negative electrode. So, the shape of the corrosion curve is projected onto the negative-current curve.

The double arrow that marks a float voltage of 2.20 V/cell is very close to this point. Float charging at this voltage means minimum corrosion of the positive grid, and hydrogen evolution also is at its minimum due to the local minimum of η_-. In this example, both electrodes are sufficiently polarised to keep them fully charged at a cell voltage of 2.20 V.

At float voltages above 2.20 V/cell, a slightly increasing polarisation of the negative electrode is observed, since increasing grid corrosion causes a growing oxygen deficit that has to be compensated for by a correspondingly increased hydrogen-evolution rate. As a result, corrosion and hydrogen-evolution rates are slightly increased. This applies, e.g., to the float voltage of 2.25 V/cell, also marked by a double arrow. Simultaneously, the float current is increased from 10 to 22 mA/100 Ah.

Comparison of Fig. 8.16 with Fig. 7.5, p.242, clearly shows the more complex situation compared to vented batteries. In valve-regulated batteries, the current/polarisation relations can be expressed as TAFEL lines only for increased cell voltages. At low float voltages, interdependencies between the reactions disturb the simple picture (cf. Fig. 8.20).

Fig. 8.17 shows results of the same model simulation. Here polarisation of positive and negative electrodes are plotted against cell voltage. These curves reflect the characteristic shape of the current/polarisation curves in Fig. 8.16. The 'voltage of minimum corrosion' E_m corresponds to the local polarisation minimum of the negative electrode in Fig. 8.16 that is reached at about 2.20 V/cell.

Some characteristic features for valve-regulated lead-acid batteries can be derived:

- Effects of cell voltage are stronger on positive- electrode polarisation. η_- is less affected by cell voltage changes.
- Positive electrode polarisation always decreases with decreasing float voltage. At float voltages below E_m, the polarisation of the positive electrode decreases rapidly with decreasing cell voltage, because of the increased corrosion rate.
- Negative-electrode polarisation, and hence hydrogen evolution, reaches a local minimum at E_m, the voltage of minimum corrosion.

In the example of Fig. 8.17, the positive electrode is not polarised at float voltages below 2.17 V/cell. At such low float voltages the positive electrode would gradually be discharged.

The broken horizontal lines indicate polarisation values that are desired for an 'ideal' float charging. $\eta_+ = 60$ mV coincides more or less with the minimum of corrosion that can be expected between 40 and 80 mV above the $PbO_2/PbSO_4$ equilibrium potential. $\eta_- = -30$ mV ensures that full charge of the negative electrode is maintained and includes a certain reserve for float charging at lower temperatures, since η_- is reduced with decreasing temperature, as can be seen in Fig. 8.19.

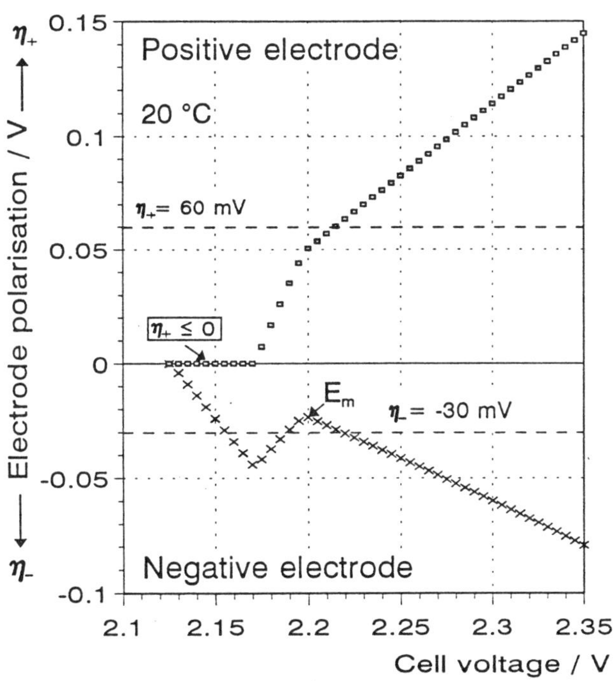

Fig. 8.17 Electrode polarisation versus float voltage. Values as in Fig. 8.16. E_m = voltage of minimum corrosion.

Influence of temperature: since most electrochemical reactions depend in a similar way on temperature, the shape of the i/η curves is basically the same for different battery temperatures, provided that oxygen-reduction efficiency is sufficiently high. (*Oxygen transport by diffusion is less accelerated by temperature increase than the electrochemical reactions. For this reason, a slight reduction of this efficiency with increasing temperature can be expected*). Fig. 8.18 shows voltage/polarisation curves at different temperatures when the efficiency of the internal oxygen cycle remains at 99.5%.

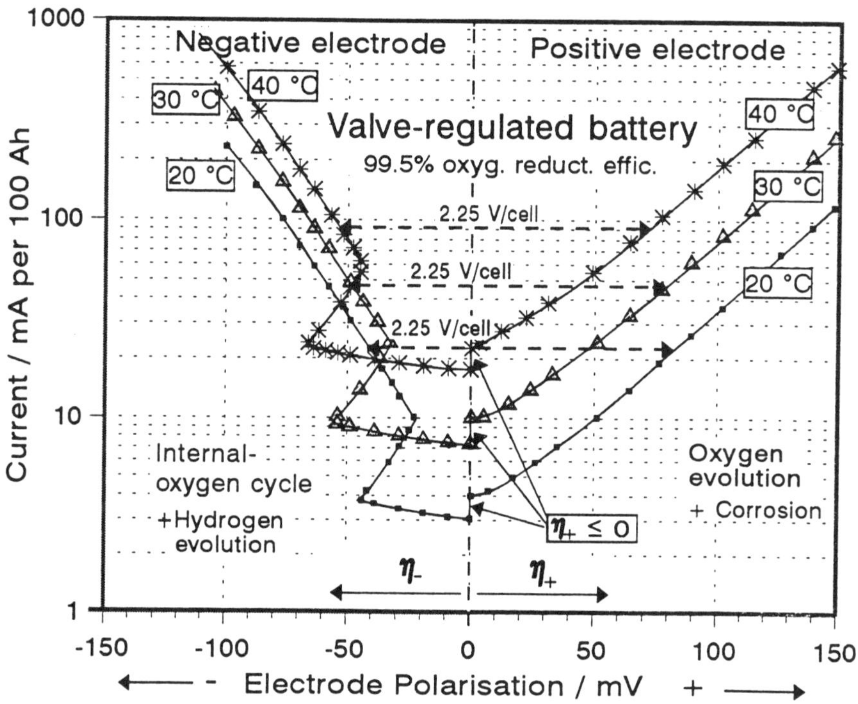

Fig. 8.18 Calculated current/polarisation curves at various temperatures, calculated for the valve-regulated battery according to Fig. 8.16. The 20 °C curve corresponds to the total current curve in Fig. 8.16. (Fig. 10 in ([3]).)

A slight shift of the 2.25 V arrows to the left is observed with increasing temperature. It means that polarisation of the positive electrode is reduced while that of the negative electrode is increased with temperature. Because of the complex shape of the i/η_- curve, the extent of this shift depends on cell voltage. This is more clearly shown in Fig. 8.19 where polarisation values are plotted versus cell voltage.

Characteristic features are:

- At constant float voltage, polarisation of the positive electrode (η_+) decreases with increasing temperature, while polarisation of the negative electrode (η_-) increases.

The voltage of minimum corrosion E_m is an important parameter:

- At float voltages below E_m, electrode polarisation is quite sensitive to temperature. As a consequence, E_m should be the lower limit for float voltage. At E_m, grid corrosion is at its minimum and water loss is at its local minimum.
- With increasing temperature, the voltage of minimum corrosion (E_m) is shifted to more positive values.

These features are of greatest interest with regard to the question of whether temperature influence can be compensated for by changing the float voltage. This is discussed in more detail in Section 13.1.1.2.

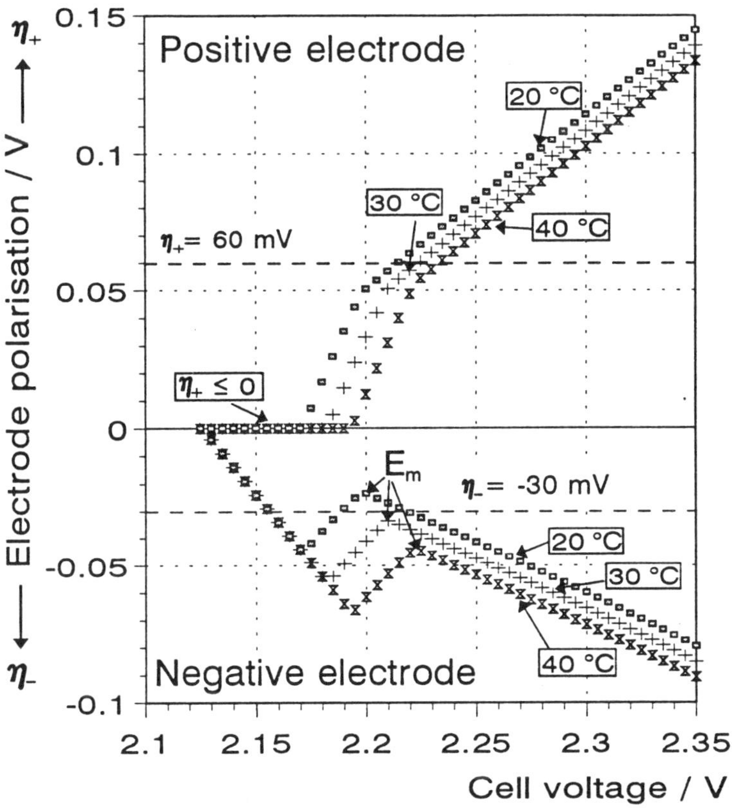

Fig. 8.19 Electrode polarisation versus float voltage at various temperatures. Values as in Fig. 8.18. E_m = voltage of minimum corrosion.

Current versus cell voltage curves are shown in Fig. 8.20. They reflect the characteristic features of the curves in Fig. 8.19. Due to the complex interdependence of the reaction rates, especially at low cell voltages, they are different from simple straight lines, as shown in Fig. 7.4, p.241, for vented batteries.

The first very flat section of these curves corresponds to the '$\eta_+ \leq 0$' sections in Fig. 8.19. In this float-voltage range, the positive electrode suffers gradual discharge, and the Figure shows a slight expansion of this section to higher float voltages with increasing temperature. The contiguous section with a steep slope (one decade of current increase per 53 mV) starts with anodic polarisation of the positive electrode, and the steep slope is caused by decreasing η_- values with current increase, observed for all three curves in Fig 8.19. With increasing temperature, this part of the curve is shifted to higher float voltages.

The slope of the current/voltage curves changes again when η_+ reaches the value that corresponds to the corrosion minimum. Above that value, straight lines are observed, and the current increases by one decade per 143 mV. This is a stronger dependence compared to the same increase per 200 mV, usually observed for vented batteries (cf. Fig. 7.4, p.241). The reason is the smaller polarisation of the negative electrode in valve-regulated batteries.

Note: *The slopes of such curves and the positions of the changes of inclination depend on kinetic parameters of all the involved reactions, i.e. they depend on specific parameters of the concerned battery (cf. Figs. 8.26 and 8.27).*

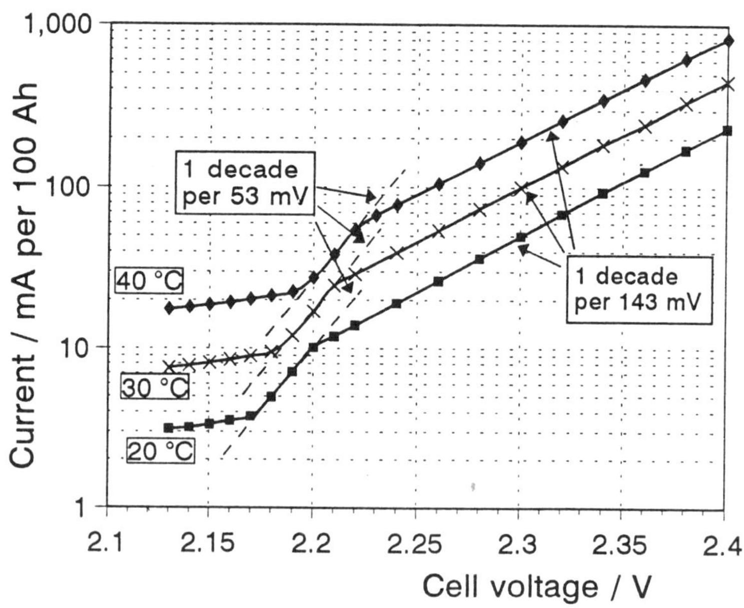

Fig. 8.20 Current versus cell voltage at different temperatures. Calculated for values as in Fig. 8.18. (Fig. 11 in (3).)

8.3.3.2 Example 2, Hydrogen Evolution Exceeds Corrosion

The influence of increased hydrogen evolution is demonstrated in Fig. 8.21 where the exchange-current density for hydrogen evolution is assumed to be twice as high as in the 'Standard Battery'. That is, the TAFEL line for hydrogen evolution is shifted so that the current at a corresponding polarisation is doubled. Such an increase can result from catalytic effects of foreign metal ions (alloy additives), but also from an increased electrochemically active surface area of the negative electrode (oversized negative electrodes).

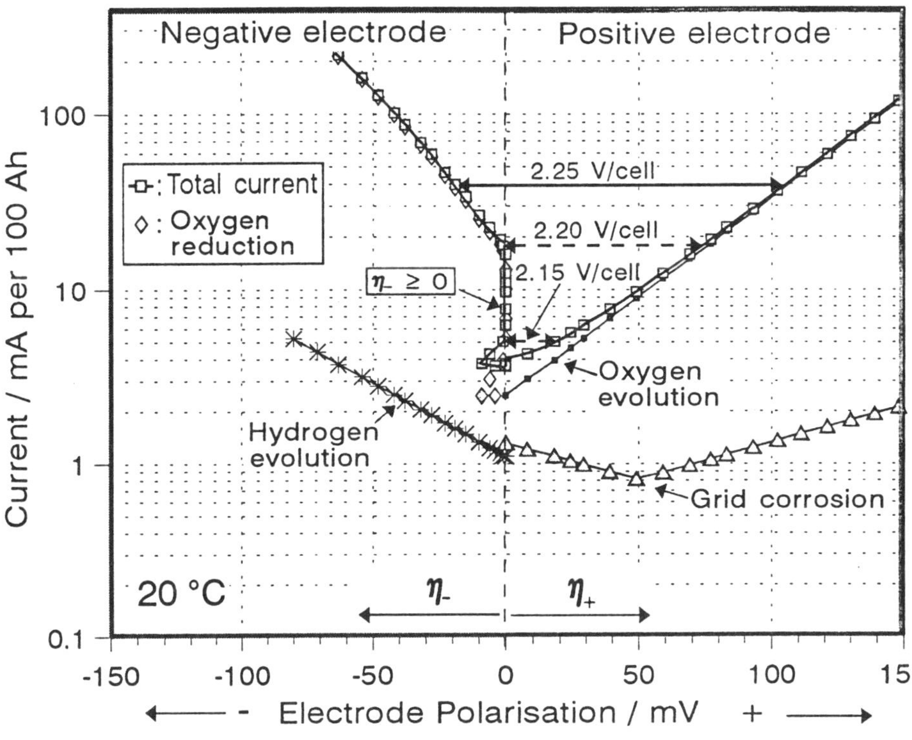

Fig. 8.21 Valve-regulated battery with increased hydrogen evolution at 20 °C. Currents vs. electrode polarisation. (Fig. 4a in ([4]).)

As a consequence of the increased hydrogen-evolution rate, the negative electrode requires less polarisation to evolve an sufficient equivalent of hydrogen. Thus, the total-current curve of the negative electrode is shifted to lower polarisation values, as comparison of Fig. 8.21 with Fig. 8.16 shows. At float voltages below 2.20 V/cell (indicated by the arrows in the Figure), the polarisation of the negative electrode is zero or even slightly positive, since H_2 evolution exceeds grid corrosion and current balance between both electrodes is only possible by discharge of the negative electrode according to Eq. (8.18), p.289. This occurs in a range of positive-electrode polarisations close to the corrosion minimum (23 mV < η_+ <77 mV), where the corrosion-current density is lower than the hydrogen-evolution current density at $\eta_- \approx 0$ mV and balance between both cannot be established. As a consequence, the negative electrode is gradually discharged at float voltages between 2.15 and 2.20 V/cell, as indicated by $\eta_- \geq 0$.

On the other hand the reduced polarisation of the negative electrode increases that of the positive electrode. For this reason, increased hydrogen evolution stabilises the charging conditions for the positive electrode.

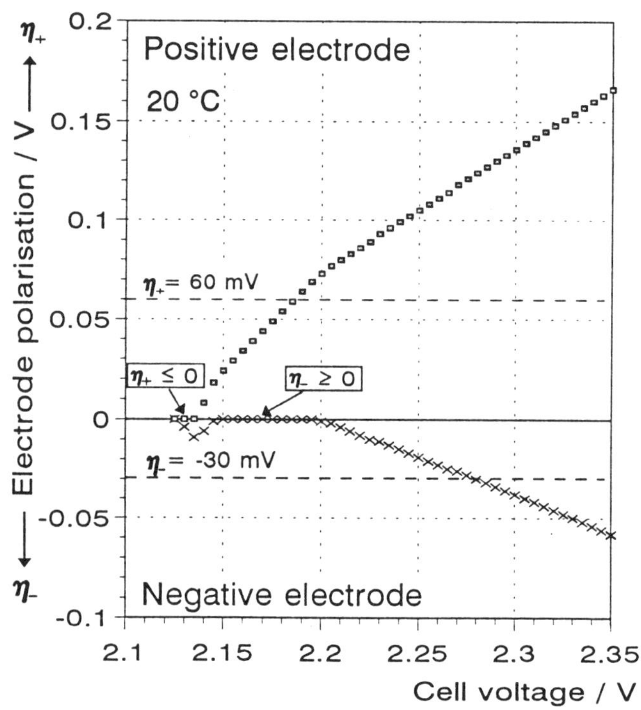

Fig. 8.22 Polarisation of the positive and negative electrodes vs. float voltage. Values as in Fig. 8.21.

Fig. 8.22 emphasises the message given in Fig. 8.21. It shows the same values as Fig. 8.21, but drawn in a different relation. As in Fig. 8.17, the horizontal axis represents the float voltage and the corresponding polarisation values of positive and negative electrodes are drawn vertically. Both polarisation curves reflect more or less the total-current curves in Fig. 8.21.

Compared with Fig. 8.17, now the critical range of float voltage for the positive electrode ($\eta_+ \leq 0$) is reduced to a very small range at extremely low float voltages (*this effect is hardly to be seen in Fig. 8.21*). At a cell voltage of 2.15 V, the positive electrode reaches nearly 30 mV of polarisation, and the line $\eta_+ = 60$ mV is reached at a float voltage of 2.18 V/cell. So the positive electrode is always on the safe side, as far as sufficient polarisation is concerned. At higher float voltages, increased grid corrosion will even be a disadvantageous effect of such high polarisation values. For the negative electrode the range of zero polarisation or even slight positive polarisation values ($\eta_- \geq 0$) covers a relatively wide float voltage range. This shows that in the case of increased hydrogen evolution the negative electrode is in danger of being discharged even at float voltages of about 2.2 V/cell.

Fortunately, in practice such a discharge would not continue endlessly, since the hydrogen-evolution rate is reduced with progressing discharge of the negative electrode, as has been mentioned in Section 7.2.3 (cf. also the Note in Section 8.3.2, p.289).

As has been shown in Fig. 8.19, temperature has a marked effect on electrode polarisation. Increased temperature increases the polarisation of the negative electrode. Thus, for the negative electrode, the situation would be more critical at low temperatures but less critical at elevated temperatures.

The voltage of minimum corrosion E_m is not to be seen in Fig. 8.22, since it is located within the $\eta_- \geq 0$ range.

Fig. 8.23 shows the dependence of the current float voltage under the above described conditions.

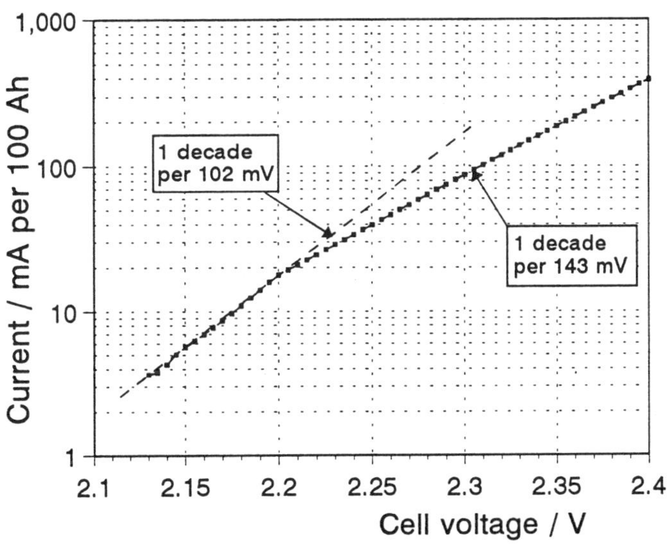

Fig. 8.23 Current vs. float voltage. Values as in Fig. 8.21.

In the voltage range where the negative electrode is not polarised ($\eta_- \geq 0$), a current increase of one decade is observed per 102 mV. This corresponds, of course, to the slope of the oxygen-evolution curve (cf. Fig. 8.27, p.303). At higher float voltages, the slope of the curve corresponds to that in Fig. 8.20, since the same data have been used for both examples.

Note: *Since the balance between hydrogen evolution and grid corrosion is the decisive parameter, a correspondingly reduced corrosion rate compared to hydrogen evolution would have similar results as described in Figs. 8.21 to 8.23.*

8.3.3.3 Example 3, Corrosion Exceeds Hydrogen Evolution

As an example demonstrating the opposite, calculated results for a battery with a quarter of the hydrogen-evolution rate compared to the 'Standard Battery' are shown in Figures 8.24 and 8.25. (*According to the above Note, a comparable increase of the corrosion rate would have the same results.*) This means that now the balance between hydrogen evolution and grid corrosion is disturbed by a too-small hydrogen-evolution rate. *(A too-high corrosion rate would cause the corresponding result*). The reduced hydrogen evolution requires increased polarisation of the negative electrode to generate the amount of hydrogen that is needed to balance grid corrosion. As a consequence, the total-current curve for the negative electrode is shifted to increased polarisation (a more negative electrode potential).

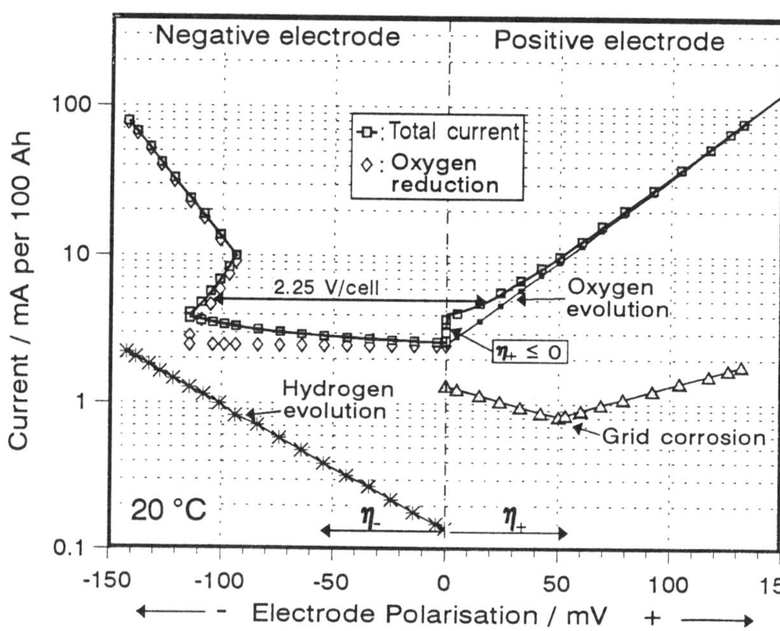

Fig. 8.24 Valve-regulated battery with reduced hydrogen evolution at 20 °C. Currents vs. electrode polarisation.

According to the previous discussions, it is expected that under these conditions the critical float-voltage range is extended where the positive electrode is discharged.

The polarisation curves versus float in Fig. 8.25 underline this situation. There is a wide range of float voltage where the polarisation of the positive electrode is zero or even slightly negative ($\eta_+ \leq 0$) and as a consequence it might not be kept fully charged, while the negative electrode always is polarised sufficiently.

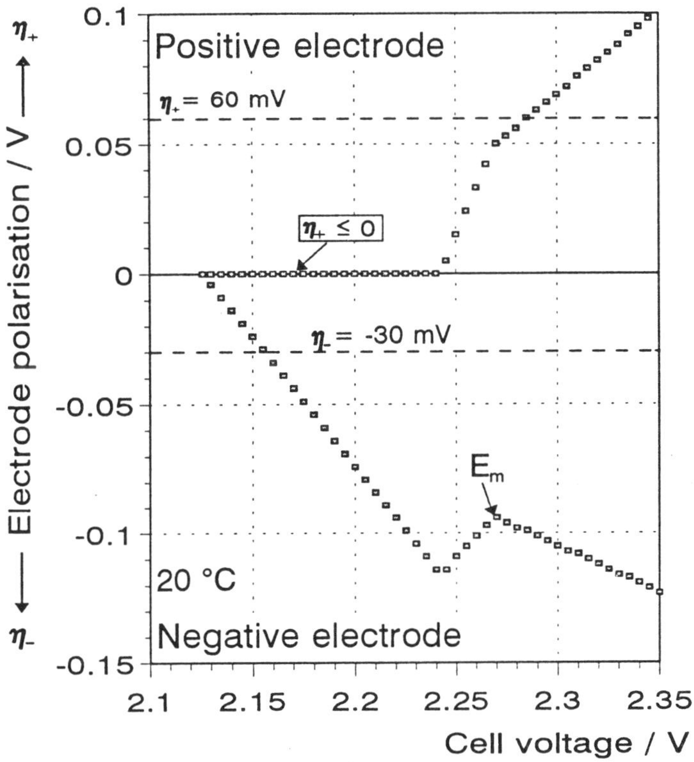

Fig. 8.25 Polarisation vs. float voltage. Values as in Fig. 8.24.

Therefore, batteries with extremely low hydrogen-evolution rates are prone to gradual discharge of the positive electrode, when floated at comparatively low voltages. With proceeding discharge, oxygen evolution is reduced and balance re-established as in the above described case of a too-high hydrogen evolution. The situation can, however, be aggravated by increased corrosion, when after an extended period of such an unfavourable float operation, the protecting PbO_2 layer at the grid surface is disintegrated.

Note: *Too strong a hindrance of hydrogen evolution can also have unfavourable effects in vented lead-acid batteries that are floated at low voltages. Too strong polarisation of* *the negative electrodes reduces the current too far at low float voltages and may result in gradual discharge of the positive electrodes. As a well known practice in the field of stationary lead-acid batteries, traces of palladium are added to catalyse hydrogen evolution. The negative electrodes are then depolarised and the positives shifted to potential values that ensure proper charging* (5).

8.3.4 Experimental Results

In Fig. 8.26 curves, calculated with the above model, are compared to experimental data. The Figure corresponds to Fig. 8.19 with slightly changed kinetic parameters due to the experimental batteries. (*Values at very low float voltages were not made in this case, partly since they are uncertain on account of the unstable capacity in that region of low electrode polarisation.*)

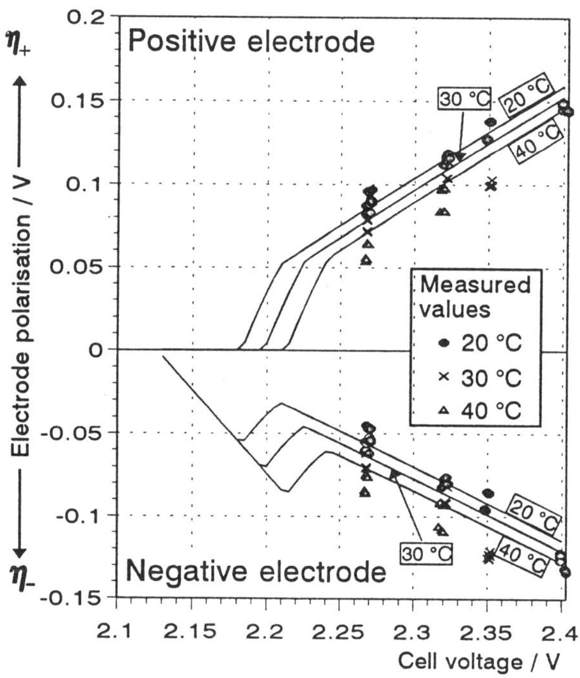

Fig. 8.26 Electrode polarisation versus float voltage, based on cells with high efficiency of the internal oxygen cycle. Experimental data from two cells. (Fig. 15 in (3).)

As in Fig. 8.19, polarisation of the positive electrodes is reduced with increasing temperature, while that of negative electrodes is increased. All experimental values correspond to this trend, and the temperature effect seems to be even stronger with these values. This is caused by a reduction of the efficiency of the internal oxygen cycle with increasing temperature. Such a reduction was observed in the experiments but has been neglected in the model simulations. A minimum efficiency of 98% was found at 2.35 V and 30 °C.

Measured oxygen concentration in the head space of the cells ranged from 2 vol.% at 2.27V/20 °C to 17 vol.% at 2.40 V/20 °C or 2.35 V/30 °C, a result that also indicates such a reduced recombination efficiency at elevated temperature.

In battery practice, the ideal balance between hydrogen evolution and grid corrosion is often not achieved. Frequently very small polarisation values are observed for the negative electrode ([6]). A too-high hydrogen-evolution rate can be the reason, partly due to the usual battery design that slightly oversizes the negative electrodes to compensate for capacity loss. The result corresponds to Figures 8.21 to 8.23 in Section 8.33.

Such experimental float-current/potential curves are shown in Fig. 8.27. They concern a valve-regulated battery design that was based on a special alloy in the positive electrode that contained 1% of antimony to stabilise cycling performance ([7]). Even this small antimony content contributed to a slightly increased hydrogen-evolution rate. The batteries were overcharged in a constant voltage regime at different temperatures and voltages. Each potential measurement was taken under stationary conditions, which means that at each voltage step an adequate period was spent to establish constant current.

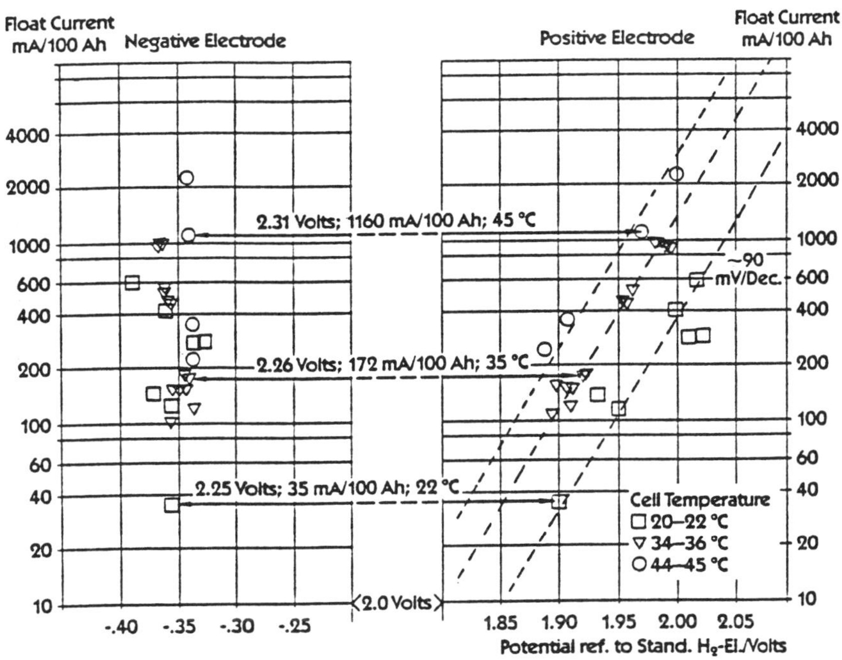

Fig. 8.27 Potentials of negative and positive electrodes in valve-regulated lead-acid batteries during overcharging with different currents at various temperatures. VARTA solid bloc (1986 design); 60 Ah nom. cap.

The behaviour of the positive electrode corresponds with that of vented lead-acid batteries: when the electrode potential is increased by approximately 90 mV, the current increased tenfold. The Figure also shows that the current is approximately doubled when the temperature rises by 10 °C, in accordance with the general rule (Section 2.3.6, Eq. (2.67)).

The negative electrode shows only slight polarisation. At all temperatures, no polarisation is observed up to a current of 400 mA/100 Ah, which indicates an extended region with $\eta_- \geq 0$ according to Fig. 8.21. At higher currents, slight polarisation of the negative electrode starts at low temperatures, while at high temperatures, the potential always remains near the open-circuit value. This indicates that in these batteries, hydrogen evolution outbalances the corrosion rate and that partly discharged negative electrodes are to be expected. The latter was confirmed by corresponding examinations.

The cell voltage, obtained by adding the potentials of the positive and negative electrodes, in this case amounts to about 90 mV and so corresponds to the region of the curve in Fig. 8.23, p.299, with a current increase of one decade per 102 mV of cell-voltage increase.

As mentioned in the Note on p.289, the discharge of the negative electrode is not continued to complete discharge of the negative electrode, but the discharge rate gradually slows down, since H_2 is not evolved at the $PbSO_4$ surface. As a consequence, with progressing discharge, hydrogen evolution gets smaller and smaller until a stable current balance is reached within the cell without further discharge of the negative.

When this current balance is reached, a stable situation exists between the current flow through both electrodes, and the state of the partly discharged negative electrode remains unchanged, since no current is left to fill up its capacity. Thus the negative electrode remains partly discharged for all the service life, unless the corrosion rate is increased or the efficiency of the internal oxygen cycle is reduced so far that the sum of oxygen reduction + hydrogen evolution is smaller than the current that flows through the positive electrode, i.e. oxygen evolution + grid corrosion. This is one of the problems in VLRA batteries that the negative electrode cannot be recharged separately. Overcharging of the cell only accelerates the internal oxygen cycle, unless it occurs so extensively that the cell partly acts like a vented one and oxygen and hydrogen escape. But this means that the efficiency of the internal oxygen cycle is reduced, and a corresponding amount of water is lost. Occasionally such a recovery of the negative electrodes is aimed by regular boost charging (cf. Section 13.1.5). But after this recharge the negative electrodes will be discharged again until balance between hydrogen-evolution and corrosion rates is re-established.

The discharged share of the active material in the negative electrode corresponds to the 'charge reserve' in sealed nickel/cadmium batteries (Section 8.1.1). In the latter, this charge reserve is intended to keep the negative electrode depolarised, and so prevent the generation of hydrogen. In lead-acid batteries such a 'charge reserve' is not necessary, since H_2 is also evolved at the open-circuit voltage and can never be suppressed completely. On the contrary, depolarisation of the negative electrode means higher polarisation of the positive and so increased corrosion of the positive grid (cf. Section 8.3.6).

8.3.5 Efficiency of the Internal Oxygen Cycle

The internal oxygen cycle is characterised by two parameters:

1. Its **maximum rate** which means the maximum current that can be carried by this cycle. Further increase of the current would cause a corresponding loss of oxygen (cf., e.g., Fig. 8.5, p.275). This maximum rate is represented by a limiting current (cf., e.g., Fig. 8.9, p.281). Influencing parameters are, e.g., void space between positive and negative electrodes, temperature and internal gas pressure (cf. Fig. 8.4, p.274).
2. The **efficiency of the internal oxygen cycle** that is given by:

$$\boxed{\eta = i(O_2 \text{ reduct.})/i(O_2 \text{ evol.})\cdot 100} \quad \% \qquad (8.23)$$

It is often also called 'recombination efficiency' ([8]), and describes the ratio between oxygen evolution and oxygen reduction within the cell.

In sealed nickel/cadmium batteries and in nickel/metal hydride batteries, the efficiency of the internal oxygen cycle is virtually 100%. In valve-regulated lead-acid batteries, this efficiency can be very close to 100% but a certain amount of oxygen is necessarily lost, together with the hydrogen that escapes through the valve. It can be derived from float current and gas measurements, provided that the following conditions apply:

- The battery is fully charged and charge or discharge reactions do not occur at positive and negative electrodes.
- Oxygen does not intrude into the cell by a leakage, and oxygen loss by other corrosion processes (separators, conducting elements of negative electrode) can be neglected.
- Unnoticed hydrogen loss is minimised.

The current balance is then given by Eq. (8.14), p.285:

$$\boxed{i(\text{float}) = i(O_2 \text{ evol.}) + i(\text{corrosion}) = i(O_2 \text{ reduct.}) + i(H_2 \text{ evol.})}$$

Oxygen evolution and reduction rates ($i(O_2$ evol.) and $i(O_2$ reduct.)) can be expressed by

$$\boxed{i(O_2 \text{ evol.}) = i(\text{float}) - i(\text{corrosion})} \qquad (8.24)$$

and

$$\boxed{i(O_2 \text{ reduct.}) = i(\text{float}) - i(H_2 \text{ evol.})} \qquad (8.25)$$

The corrosion rate (i(corrosion)) can be substituted by a further relation, since

$$\boxed{i(O_2 \text{ evol.}) - i(O_2 \text{ reduct.}) = i(O_2 \text{ escape})} \qquad (8.26)$$

when $i(O_2$ escape) is the amount of oxygen that escapes from the cell. *This is only true when no oxygen consumption occurs in the cell except i(corrosion).* When this relation is used to substitute $i(O_2$ evol.) or $i(O_2$ reduct.) in the above Eq. (8.14), the corrosion rate is expressed as the difference between hydrogen evolution and oxygen escape:

$$\boxed{i(\text{corrosion}) = i(H_2\ \text{evol.}) - i(O_2\ \text{escape})} \tag{8.27}$$

With these relations the efficiency of the internal oxygen cycle can be written as:

$$\boxed{\eta = \frac{i(\text{float}) - i(H_2\ \text{evol.})}{i(\text{float}) - i(\text{corrosion})} \cdot 100} \quad \% \tag{8.28}$$

or

$$\boxed{\eta = \frac{i(\text{float}) - i(H_2\ \text{evol.})}{i(\text{float}) - i(H_2\ \text{evol.}) + i(O_2\ \text{escape})} \cdot 100} \quad \% \tag{8.29}$$

η can thus be derived from current and gas measurements.

In stationary valve-regulated batteries floated in the range between 2.23 and 2.32 V per cell, the efficiency of the internal oxygen cycle is usually more than 99%, at least after a certain ageing period (cf. the following Section).

Note: *Gas measurements can be very erratic mainly for two reasons:*

- *Gases, escape very fast through leakages (cf. Table 2.8, p.51). Accurate sealing of the container including the wall openings for tubes and reference electrodes is required.*
- *Gases, especially hydrogen, permeate through plastic materials (cf. Table 2.13, p.61), and the permeation of hydrogen is favoured since the partial pressure in the surrounding atmosphere is zero. Increased thickness of the container walls and metallic tubing is recommended.*

Despite all precautions, the accuracy of gas measurements is limited, and a certain loss is unavoidable (cf., e.g., Section 4.2 in ([3])). For this reason, it seems questionable to include gas measurements in battery standards, unless it can be ensured that adequate measuring equipment is available everywhere.

Weight measurements are unsuited to determine the efficiency of the internal oxygen cycle, because water loss by corrosion (i(corrosion)) is not realised by weight measurements, because the oxygen remains in the battery.

8.3.6 Time-Dependent Changes, Cell Balance

On account of the secondary reactions, the valve-regulated lead-acid battery is an unstable system. Hydrogen evolution and grid corrosion represent unavoidable reactions that result in loss of water. At an adaquate low float voltage, their rates are small, provided the efficiency of the internal oxygen cycle is close to 100%. But a certain water loss is unavoidable that gradually changes the cell balance and the performance data (cf., e.g., Figs. 4.24 and 4.25, p.150 and 151). Under unfavourable conditions, water loss can even cause premature failure.

The efficiency of the internal oxygen cycle can be insufficient at the beginning of service life, if the battery contains too much electrolyte. But water loss increases the efficiency and also the maximum rate of the internal oxygen cycle because of the increased void space for oxygen transport. Thus, the valve-regulated lead-acid battery approaches the state of minimum water loss with proceeding service life (cf. the Note below). But the water loss means also an increase of its internal resistance.

Note: *A high maximum rate of the internal oxygen cycle is sometimes demanded, supposedly to avoid water loss. But, for valve-regulated lead-acid batteries in normal operational conditions, a high maximum rate of the internal oxygen cycle is not a substantial question, especially for stationary applications, because lead-acid batteries are charged at constant voltage and overcharging currents are low (cf. Chapter 13.1.1). The internal oxygen cycle must only be able to carry these low currents. This is different for sealed nickel/cadmium batteries, which usually are charged (and overcharged) with fairly high constant currents which require a correspondingly fast rate of the internal oxygen cycle.*

The efficiency of the internal oxygen cycle, however, must be high also in valve-regulated lead-acid batteries. Otherwise, water loss occurs as indicated in Fig. 8.14, p.287.

The gradual changes of the initial cell balance in the valve-regulated lead-acid battery are a fundamental difference from the sealed nickel/cadmium battery. As pointed out above, it would make no sense to oversize the negative electrodes in valve-regulated lead-acid batteries. The excessive charge of the negative electrode would be reduced by discharge until balance is achieved. For this reason, the terms 'charge reserve' and 'discharge reserve' have no relevance for valve-regulated lead-acid batteries.

8.3.7 Consequences for Grid alloys

Hydrogen evolution and grid corrosion determine water loss, one of the major ageing factors. Both must be small and remain so during service life. According to the preceding Sections, a balance between these side reactions is also desired.

Corrosion of the positive grid must be kept as low as possible, not only to preserve the conductivity of the grid but also to reduce water loss of the electrolyte. Increased grid corrosion reduces oxygen evolution and leads to increased polarisation of the negative electrode with additional hydrogen evolution (cf. Fig. 8.15b, p.288). Furthermore, alloying additives that are released by corrosion might increase hydrogen evolution. For these reasons, the grid alloy plays an important role in valve-regulated lead-acid batteries (cf. Section 4.7).

References

(1) VARTA Batterie AG (Editor), 'Sealed Nickel Cadmium Batteries', VDI Verlag GmbH., Düsseldorf 1982, p 71

(2) U.Teutsch, Proc. of the 1st Telescon Meeting, Berlin, 1994, © IEEE Comm. Soc., 1994, p.89

(3) D.Berndt, R.Bräutigam, and U.Teutsch, Proc. of the 17th Intelec Meeting, The Hague 1995, © IEEE Comm. Soc., 1995, p. 1

(4) D.Berndt and U.Teutsch, J. Electrochem. Soc., **143** (1996), p. 790

(5) W.B.Brecht Proc. 9th Intelec Meeting, Stockholm, 1987, IEEE Communications Soc., p.99

(6) W.E.M.Jones and D.O.Feder, Proc. 18th Intelec Meeting, Boston, 1996, © IEEE Communications Soc., p.358

(7) D.Berndt and H Franke, Proc. 9th Intelec Meeting, Stockholm, 1987, IEEE Communications Soc., p.115

(8) J.Szymborski and B.Burrows, 'Power Sources 9', Academic Press Inc., London ,1983, p.113

9. Battery Technology

The previous chapters concerned fundamental laws, electrochemical reactions and data important for the electrochemical system. The following chapters deal with the batteries and their performance. Consequently, the reacting substances are no longer described as chemical symbols in electrochemical equations but as 'active materials' in the form of powders and pastes that are supported by 'passive design elements', like the conducting grids and separators.

The design principle of a battery was shown in Fig. 2.2, p.8, namely two electrodes immersed in the electrolyte. In practice, often a manifold of positive and negative electrodes are switched in parallel and connected to the terminals of the battery by current-conducting elements ('bus-bars', 'pole bridges'); the electrodes are narrowly spaced, and a separator is arranged between them to avoid short-circuits. Often the separator simultaneously provides support. The realisation of this principle can be achieved in many ways, and in battery practice many such variations do actually exist for different reasons:

- Above all, a battery has to be adapted to its special purpose, and this adaptation may require a special construction.
- Another reason for varying battery design is manufacturing costs, which are influenced by production methods. It is a question of the number of batteries made, and which manufacturing method will be the cheapest one. As a consequence, different manufacturers may use different production methods and variations of battery technology to produce batteries for the same application.
- To some extent, the existing mechanical equipment of the plant is a deciding factor for battery design. High investment expenses may be prohibitive to switch over to new methods.

As a result, an enormous number of different battery designs made by various methods are available on the market, and an attempt to present a comprehensive description would far exceed the scope of this book. Only some sections of battery technology and design can be touched on in the following chapters. The goal of these chapters is to show how the technical product 'battery' depends on the basic features of the battery system concerned.

The development of battery technology goes back to the beginning of battery manufacture. Many methods originally were purely derived by experience. Today, most of the manufacturing processes are well based on theory. Nevertheless, a considerable portion of experience is still required for making batteries, since some of the included chemical processes are so complex that they cannot be treated quantitatively by theory.

9.1 Materials for Battery Components

The technology that can be applied depends largely on the materials that are suited for making the battery concerned.

In lead-acid batteries, lead and its components are used in positive and negative electrodes as the active material. But all the passive components are also made of lead or must be covered by lead as has been described in Sections 4.7 and 4.8.

- In the positive electrode, all metals that normally are used for current-conducting elements are not stable on account of the high potential. Lead can be used as current conductor, but a certain corrosion attack is unavoidable and grid design has to consider a corresponding corrosion reserve. The development of corrosion-resistant alloys and suited methods to manufacture grids were initiated by this problem.
- In the negative electrode, the outstanding high hydrogen overvoltage of lead reduces spontaneous self-discharge to a tolerable limit (Section 7.2.3.2). All other metals would spoil the battery by intolerable high self-discharge of the negative electrode. This also has consequences for positive grid alloys: critical additives, like antimony, have to be avoided when hydrogen evolution must be minimised.

A further consequence of this situation is that metallic containers cannot be used for lead-acid batteries. Contact with the positive electrode would cause corrosion, and traces of the metal that are dissolved would spoil the negative electrode. For this reason, lead-acid batteries originally were installed in wooden trays, coated with lead or rubber, or in glass vessels. Nowadays plastic material is mostly used for cell containers.

In nickel/cadmium and nickel/metal hydride batteries, nickel or nickel-plated steel can be used for supporting and conducting elements.

- The potential of the nickel hydroxide electrode is about 0.5 V lower than that of the lead dioxide electrode and in the alkaline electrolyte the corrosion of metallic nickel can be neglected.

- Nickel can also be used as conducting material in the negative electrode of Ni/Cd batteries, since the potential of the cadmium electrode is about 20 mV above the starting potential for hydrogen evolution, and so the lower hydrogen overvoltage of nickel has no relevance.

This situation allowed the development of a great variety of supporting elements. Nickel or nickel-plated steel can also be used for battery containers.

Until the second world war, the battery manufacturers made most of the materials and parts used themselves. Nowadays more and more of these components are supplied by specialised plants.

9.2 The Active Material

The active material comprises the substances that constitute the cell reaction. Consequently, the active material is the most essential part of the battery, and battery technology has to aim at optimum constitution and performance of the active material for the expected application. This does not only concern the chemical composition but also the physical structure and its stability. Highly specialised methods have been developed to fulfil these requirements

Electrochemical reactions occur at the phase boundary between the active material and the electrolyte. Sometimes these reactions include dissolution of the substances, but also solid-state reactions, as in the nickel hydroxide electrode, can be included. To make sure that a sufficient rate of reaction is achieved, the surface of the reacting materials has to be large. Otherwise, the kinetic parameters would reduce the reaction rate too much. Table 9.1 shows the surface areas of different active materials.

Substance	B.E.T. Surface (1) m^2/g
Lead dioxide, $Pb0_2$	4 - 6 (2)
Lead , Pb	0.3 - 0.6
Cadmium, Cd	2 - 3
Nickel hydroxide, NiOOH	20 - 60

Table 9.1 B.E.T. surface areas of the active materials in lead-acid and nickel/cadmium batteries.

References

([1]) B.E.T. means: determination of the true surface area by nitrogen absorption, S.Brunauer, P.H.Emmet and E.Teller, J. Amer. chem. Soc., **60** (1938), p. 309

([2]) H.Bode, 'Lead-Acid Batteries', John Wiley & Sons, New York, 1977, Table 3.14, p. 260

10. Lead-Acid Battery Technology

It was pointed out at the beginning of this book that the valve-regulated lead-acid battery primarily is a lead-acid battery. Thus, no special technology exists for valve-regulated lead-acid batteries; basically they are manufactured in the same way as conventional lead-acid batteries, except that the electrolyte is immobilised, and a valve that only allows the escape of gases, replaces the vent cap. For this reason, the following review of battery technology applies to conventional lead-acid batteries as well, but is limited to items that are of interest for the valve-regulated design.

10.1 The Active Material

Some general features of the active material, especially its large surface area, were already pointed out in Chapter 9. Figs. 10.1 to 10.3 show the typical microscopic appearance in charged and discharged state. Although certain features are characteristic, microscopic pictures of this kind may vary considerably, because of the different parameters that influence the formation of the crystals when a substance is precipitated. Furthermore, the charge/discharge conditions and the age of the battery influence the morphology of the active material (cf., e.g., ([1])). *(A detailed description of the components of the active material, in lead-acid batteries, including the sulphuric acid, and their chemical properties is given in ([2]).)*

10.1.1 The Active Material of the Positive Electrode

Fig. 10.1 shows the 'lump' structure typical of the charged active material of the positive electrode, at least for a fairly new electrode. These 'lumps' are porous agglomerates. About 50% of their volume is occupied by lead dioxide; the other 50% is pores. A large share of micropores produces the high surface area shown in Table 9.1.

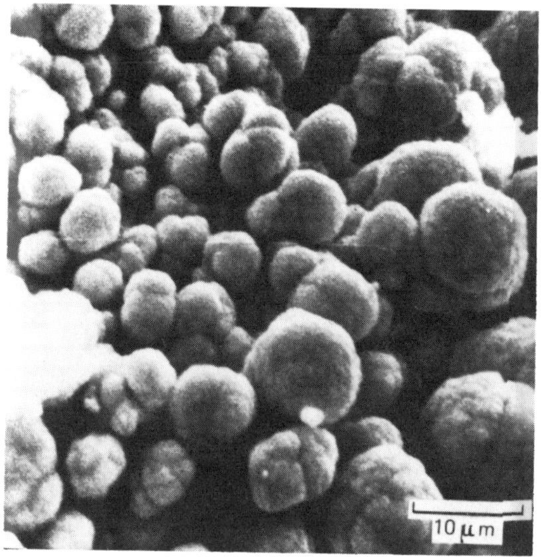

Fig. 10.1 Charged PbO_2 electrode. Surface appearance of the active material. (Fig. 3/5 in ([3]).)

Two polymorphs of lead dioxide occur in the active material: α-PbO_2 and β-PbO_2 ([4]). α-PbO_2 is generated in a more alkaline environment ([5]). This concerns parts of the plate where the active material suffers limited acid access during the formation process (cf. Figs. 10.18 and 10.20 in Section 10.7). When the battery has been discharged and is charged again, only β-PbO_2 is formed on account of the acidic environment. For this reason, the content of α-PbO_2 decreases with the number of discharge/charge cycles, especially for a high initial α-PbO_2 content (cf. ([2]), p.267, esp. Figs. 3.27 and 3.28). Plates containing much α-PbO_2 show reduced initial capacity, which increases gradually on account of the conversion of α-PbO_2 to ß-PbO_2. There was some indication that plates with high α-PbO_2 content would outperform other plates in standby applications ([5]). So, the α/β-PbO_2 ratio has been specified occasionally for stationary lead-acid batteries. But these observations have not been confirmed in general ([6]). Other experiments indicate that the structure of lead dioxide agglomerates, which can be influenced by the formation process, is important for cycle stability ([7]) (cf. also the item 'paste mixing' in Section 10.7).

The typical appearance of a discharged electrode, positive or negative, is shown in Fig. 10.2. The crystals consist of lead sulphate ($PbSO_4$). Comparison of Figs. 10.1 and 10.2 illustrates clearly how radically the active material is changed when the electrode is discharged or charged. This change is caused by the solution mechanism of the electrode reaction (cf. Fig. 4.5, p.112). The active material is dissolved and precipitated as a new compound. It is obvious that it is difficult to reverse this reaction completely and re-establish the initial structure of the active material, after a far-reaching conversion of this kind. Such a far-reaching conversion is connected with extended discharge at low rates, denoted as 'deep discharge' (cf. Section 3.4). For this reason, repeated deep discharging is detrimental to the active material and a strong relation exists between depth of discharge and the number of discharges that can be achieved with the battery ([8]) (cf. also Figs. 3.10 and 3.14 in Section 3.4).

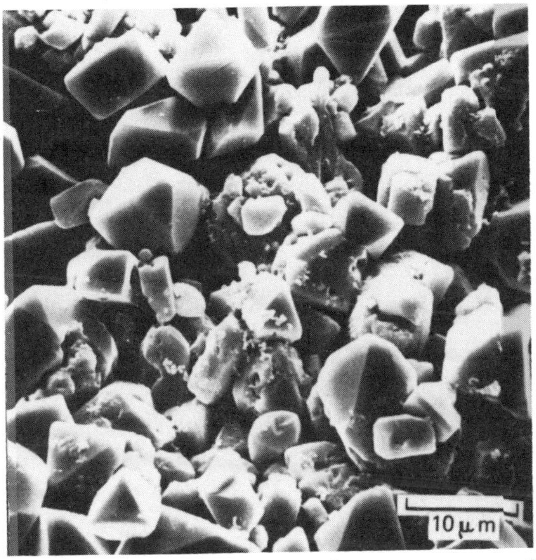

Fig. 10.2 Discharged PbO_2 or Pb electrode. Surface appearance of the active material Comparatively coarse $PbSO_4$ crystals are to be seen at the surface. (Fig. 3/7 in ([3]).)

10.1.1.1 Ageing Factors of the Positive Active Material

The active material of the positive electrode is prone to lose its mechanical strength when repeated discharge/charge cycles occur, because the alternating dissolution and precipitation processes convert the agglomerate structure shown in Fig. 10.1 into an accumulation of fine crystals ([9]). So, the active material suffers degradation and part of it may fall off the plate as fine particles. This process is called '**shedding**'. It is also described as '**soft positives**'. Shedding is only the outward appearance of a more general ageing process which means that the active material is prone to disintegration of its electronic conductivity and mechanical strength. This causes the so called '**premature capacity-loss**' ([10]) (a survey with references is given in ([11])). It becomes evident as a decreasing utilisation factor with increasing cycles.

Shedding of the positive active material is a characteristic ageing feature in conventional lead-acid batteries when they are charged and discharged frequently. In valve-regulated lead-acid batteries, the active material cannot fall off the electrode, because of the immobilised electrolyte. But it also often suffers premature capacity-loss.

In a model that considers the active material as an aggregate of spheres ('Kugelhaufen') it is explained by a gradual increase of the ohmic resistance, mainly in the connecting region of the individual particles of the active material ([12]). These connecting regions establish the electrical contact between the individual particles of the active material. They are decisive for the ohmic resistance because of the minimised cross sectional areas in these bridging zones. The structure of the connecting regions between the particles of active material is largely influenced by the conditions when these regions are re-established during the charging process. For this reason, it is understandable that the charging conditions are important for the stability of the active material and that in many cases, after a premature decay, full capacity can be regained with suitable charge/discharge procedures ([13]). For this reason, the premature capacity-loss is sometimes called '**reversible capacity-decay**'.

Another model assumes that gel zones are formed by hydrated lead dioxide ($PbO(OH)_2$) and act as bridging elements between the crystallite particles. Electrons can move along the polymer chains of this gel and so cause electronic conductivity between the crystalline zones ([14]).

Quite often, simultaneously with the capacity decay, the formation of a barrier layer of lead sulphate ($PbSO_4$) is observed between the grid and the active material ([15]). In view of the explanation given above, this layer may be the final stage of the process. When the ohmic resistance of the active material is increased, the charge/discharge reaction is restricted to the area close to the grid surface. Then, deep discharge must happen to this part of the active material, causing the high concentration of sulphate.

In general, a better cycle stability of the active materials is observed with lead-acid batteries with immobilised electrolyte compared to flooded batteries of corresponding design ([16]). The reasons are:

- The stronger mechanical support of the active material by the glass mat or the gelled electrolyte helps to maintain contact with the grid.
- The utilisation of the active material is reduced with immobilised electrolyte. As a consequence, the depth of discharge is reduced, which also stabilises the active material.

10.1.1.2 The Influence of Antimony, Tin and Phosphoric Acid

Antimony (Sb) and Tin (Sn) are usually not added to the active material, but both are alloying components of the grid (cf. Section 4.7). They are gradually released from the grid by corrosion, and permeate the active material by dissolution and diffusion.

The above described 'premature capacity-loss', i.e. a decay of the utilisation factor, became especially evident when antimony-free alloys were introduced, and such batteries operated in charge/discharge cycle regimes. For this reason, this effect is also called '**antimony-free effect**', although it is also observed with grids containing antimony. The mechanism of this effect has not yet been explained in detail, but antimony has a strong influence on the stability of the active material that cannot be compensated by special pre-treatment or design of the electrodes ([17]).

When specimens of pure lead and a 5% antimony alloy were periodically oxidised and reduced, lead oxide layers were observed with different structures:

- coarse and insulating with the antimony-free electrode,
- fine and of low resistance with the antimony alloy ([18]).

The origin of such insulating layers may explain the high resistance also established within the active material when antimony is not present.

The 'Kugelhaufen model', mentioned in the preceding Section, explains the beneficial influence of antimony by improved conductivity of the zones that connect the spheres. According to the gel model, antimony decreases the crystallinity of PbO_2 and so increases the conductivity by the gel zones ([19]), and especially influences the structure of the corrosion layer that intermediates between grid and active material ([20]).

Tin addition to the positive-grid alloy also has a capacity stabilising effect, but this apparently concerns only the boundary between grid and active material.

Phosphoric acid (H_3PO_4) is added in small amounts to the electrolyte. A beneficial effect on cycle stability has long been known for phosphoric acid, which has been used as an additive in conventional lead-acid batteries for many years to improve cycle stability, although the disadvantage of a slightly reduced capacity had to be accepted ([21]). Addition of phosphoric acid to the electrolyte improves long-term capacity and reduces the formation of sulphate layers around the grid ([22]). The addition of 20 to 35 g/l phosphoric acid was protected by patent for valve-regulated lead-acid batteries with gelled electrolyte ([23]).

Extensive experiments showed ([24]) that, at low H_3PO_4 concentrations, $Pb_3(PO_4)_2$ acts as an intermediary in the corrosion of Pb to PbO_2. Obviously, the phosphoric acid influences the formation of lead dioxide (PbO_2) on account of its strong adsorption and leads to a fine grain structure of the positive active material ([25]). However, in spite of the repeated use of phosphoric acid in lead-acid batteries, some questions on its interaction are still to be elucidated ([26]).

10.1.2 The Active Material of the Negative Electrode

The typical appearance of active material in the negative electrode is shown in Fig. 10.3.

Fig. 10.3 Charged Pb electrode. Surface appearance of the active material. (Fig. 3/6 in ([3]).)

Fairly fine crystals of metallic lead can be observed. A number of additives are required to preserve this fine structure. Otherwise the metallic crystals grow from recharge to recharge and cause a decrease of capacity because of the reduced true surface area.

The additives required for the negative active material are called expanders because they 'expand' the surface area of the lead electrode, or at least preserve its large true surface area. In general, expanders in total are added to less than 1 wt.%.

Usual expanders are:

Barium sulphate ($BaSO_4$), which has an orthorhombic crystal structure similar to that of lead sulphate ($PbSO_4$). It is assumed that barium sulphate crystals act as nucleants for lead sulphate crystals while the plate is discharged. The precipitation of a large number of lead sulphate crystals is promoted by these nucleants, and coarsening of the discharged active material is thus retarded.

Organic additives such as **lignin** or its derivatives are mostly used as expanders. The application of these expanders goes back to the time when thin wooden sheets were used as separators. Premature capacity-decay of the negative electrode was observed when these sheets were replaced by 'wood-free' separators. As a remedy, wood flour or sawdust or derivatives of wood hydrolysis were added to the negative active material. Humic acid also was in widespread use. Today, chemically pretreated derivatives of lignin like **ligno-sulphonic acid** or **ligno-sulphonate** are mostly applied. They are also based on wood, but are better specified than the untreated natural products.

Organic expanders prevent the growth of dendritic lead crystals. Obviously, parts of the crystal surface are blocked by the fairly large molecules of these compounds, and further regular growth of the lead crystals is suppressed. Small crystals with a large surface area are the consequence. The effect of organic expanders is enhanced by barium sulphate ([27]), presumably on account of the formation of finely dispersed lead sulphate promoted during the preceding discharge.

Usually, lignin derivatives were assumed to increase the hydrogen overvoltage (decreased hydrogen evolution at the negative electrode), but experiments, reported more recently, did not confirm this ([28]).

Expanding effects are known for some organic compounds added to the electrolyte, like **alizarine red** ([29]) or **methyl orange** ([30]). However, these substances are mainly of interest for flooded lead-acid batteries, where they can be used to refresh the expanding effect if required.

Lampblack (or **soot**) is also added as an 'expanding' substance to the active material for the negative electrode. It seems to have an important function as a storage medium for the organic expanding substances ([31]).

Oil is often added to the active material of negative electrodes. It is not generally referred to as an expander, although it has expanding effects. The main reason for its addition is that oil supports the preservation of charge when the negative electrode is dried in the charged state (cf. Section 10.7.1). Furthermore, oil inhibits hydrogen evolution to some extent, which means that the self-discharge of the negative electrode is slightly reduced.

10.2 The Plate Support (The Grid)

The plate support in lead-acid batteries is usually called 'grid'. In most batteries the grid has to provide both mechanical support for the active material and electronic conductivity for the collected current.

Fig. 10.4 shows the design of a grid and its horizontal and vertical cross sections. This Figure indicates the origin of the term 'grid' for this kind of plate support.

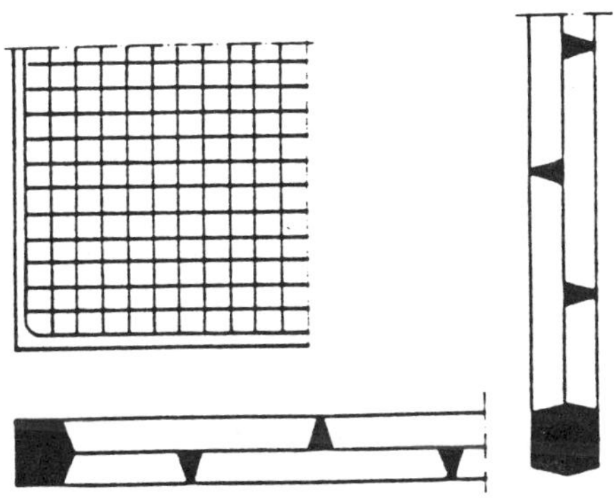

Fig. 10.4 Grid of a lead-acid battery with thicker walls and a rectangular mesh. One corner and cross-sections in vertical and horizontal directions. (From ([32]).)

The network of lead wires must be designed to provide optimum mechanical support to the pellets of active material that later will fill the void space (cf. pasting in Section 10.7). But sufficient conductivity has also to be provided by the grid. Furthermore, the internal resistance of a battery is influenced by the ratio between the grid surface and the amount of active material ([33]).

The layout of a grid can vary markedly, as Fig. 10.5 shows. The reason for such different designs is to improve the conductivity of the grid by running more wires towards the lug connecting the plate with the top bar and the terminal post (cf. Fig. 10.10, p.325). So the resistance can be reduced, but the grid weight maintained.

Grids for positive and negative electrodes are usually similar. In batteries designed for extended service life, the positive grid is made heavier to provide corrosion reserve (cf. Section 4.7).

The mechanical support of the active material is of minor importance when additional support is provided, e.g. by envelopes or tubes. This also applies to valve-regulated lead-acid batteries, since the plates are packed fairly tightly between the more or less compressed separators, and free electrolyte is not present to carry away active material from the plate. So, the grid is not designed primarily to firmly clasp the pellets of active material. Rather, current-conducting qualities and corrosion resistance are emphasised. Fig. 10.6 shows a positive grid designed with this premise. Cylindrical rods are applied for the horizontal and vertical wires.

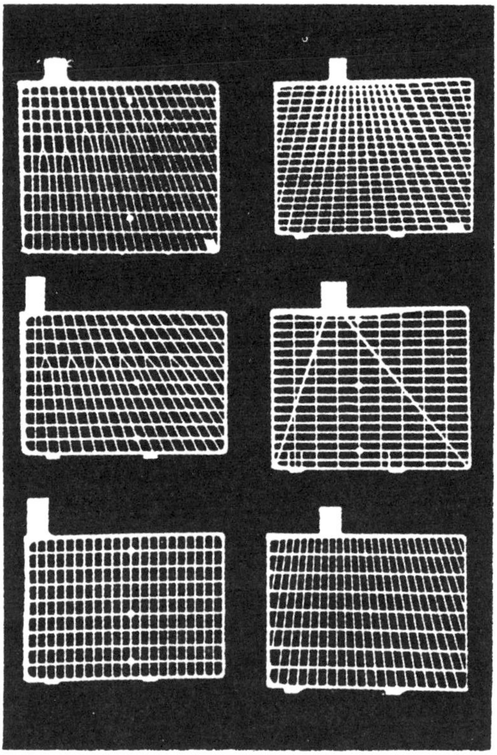

Fig. 10.5 Book-mould-cast thin grids of different design ([34]).

Fig. 10.6 Positive grid of a valve-regulated lead-acid battery with cross section of the pasted plate. (VARTA solid bloc, from ([35]).)

The cross section of the pasted plate in Fig. 10.6 shows that the grid wires are embedded within the active material. This central position of the current conductor effects a fairly uniform current distribution across the active material, which leads to improved utilisation. Beyond that, the cylindrical shape of the wires means minimum corrosion attack on account of the low surface area compared to the volume.

The tubular-plate design for positive electrodes, shown in Fig. 10.7, is common mainly in European countries for batteries with larger capacities. In this plate design, the conducting elements are separated from the components that contribute mechanical support. The grid consists of vertical lead rods in the centres of tubes that are formed by woven, braided or nonwoven fabrics (cf. also Fig. 10.12, p.327).

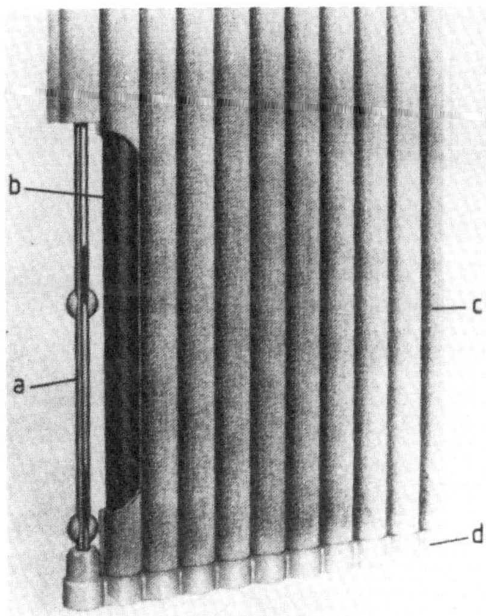

Fig. 10.7 Section of a tubular plate.
a: Lead-alloy spine (grid). b: Active material (PbO_2).
c: Tube (in this example, fabric of polyester fibres). d: Bottom seal of plastic caps.

The advantage of tubular plates is the comparatively high utilisation of the active material, which results in a rather low weight in relation to capacity.

These features of the tubular plate are caused by two facts, namely:

- The central current-collecting spine produces uniform current flow across the active material.
- The mechanical support by the tube allows the use of fairly light active material. This means high porosity and a high utilisation factor.

Disadvantageous for tubular plates is the fact that a minimum tube diameter between 6 and 8 mm is required for economic production. But the tube diameter corresponds with plate thickness, and lead-acid batteries with such thick plates are inferior for high-rate discharge. Valve-regulated lead-acid batteries with tubular plates are consequently used for stationary applications with bridging times in the range of hours ([36]), and for traction applications ([37]).

10.2.1 Grid-Manufacturing Methods

Different methods are used to manufacture grids for lead-acid batteries ([38]). The grid types in Figs. 10.4 and 10.5 are made by gravity casting in book moulds, while grids for tubular plates are usually pressure-cast.

New grid-manufacturing methods have been developed for batteries produced in large numbers, and with highly corrosion-resistant alloys that are too soft to be handled in the conventional way as single plates. These methods are increasingly used for vented types of starter batteries and also for valve-regulated batteries. The common feature of these methods is that they produce the grid continuously.

Expanded lead has been applied for grid production on a large scale since the 1970s (for a description of the method, cf., e.g., ([39])). Usually, the method starts from rolled or drumcast lead sheet that is slotted and expanded to form an endless ribbon of grids. (*Wrought basic material, produced by rolling a thick slab down to the desired thickness, is adjudged to have excellent corrosion resistance when produced under specified conditions* ([40]).) Sometimes the lead sheet is drumcast and expanded in one operation ([41]). The expanded ribbon is cut into single plates after being filled with active material ('pasted', cf. Section 10.7.1).

Fig. 10.8 shows at the top an example of such a single grid (*which is usually not available, because the material can only be handled as a band*).

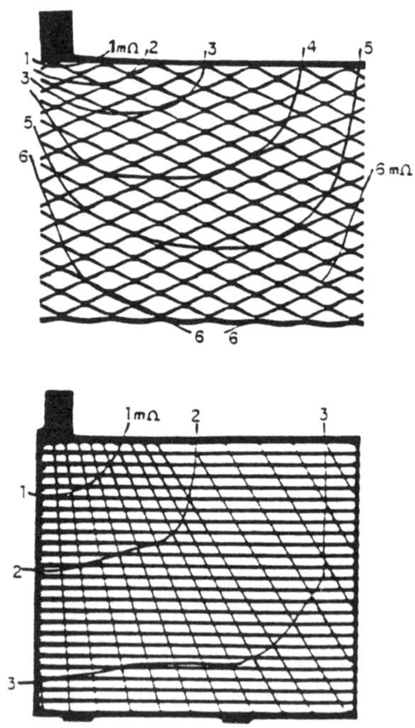

Fig. 10. 8 Top: Expanded lead grid. Bottom: Book-mould-cast grid. Both with ohmic-resistance map. (From ([42]).)

The horizontal diamonds are formed by the expansion process. The single wires are wrought when the sheet is expanded and, as a consequence, the grid takes a slightly three-dimensional structure. The top bar and the lug, are punched from the unslotted part of the strip. The method is quite economical, because the expanding process produces little scrap.

The bottom part of Fig. 10.8 shows a book-mould-cast grid. For comparison, the ohmic resistances measured from the lug are drawn in both pictures. This comparison indicates a disadvantage of the expanding method. The position of the diamonds is determined by the expanding process. For this reason, the current-conducting path cannot be optimised for the expanded grid. As a result its ohmic resistance is higher when about the same amount of lead is used for both grid designs.

However, the total ohmic grid resistance between the lug and the opposite edge of the grid does not apply in the battery. This would only be true if the current had to flow through the whole grid before entering the active material. Actually, the current that is supplied from the lug leaves the plate surface in part and flows to the opposite electrode. The current flow along the plate is thereby gradually reduced until it reaches zero at the bottom of the plate. The voltage drop along the plate is lowered on account of the diminishing current. As a consequence, the '**effective grid resistance**', the voltage drop along the grid, when the battery is discharged or charged, is only about 1/3 of the grid resistance measured between the lug and the opposite corner ([43]). This, to some extent, modifies the effect indicated by the two parts of Fig. 10.8.

Another way to produce battery grids continuously is **drum casting** or **continuous grid casting**. The principle is shown in Fig. 10.9.

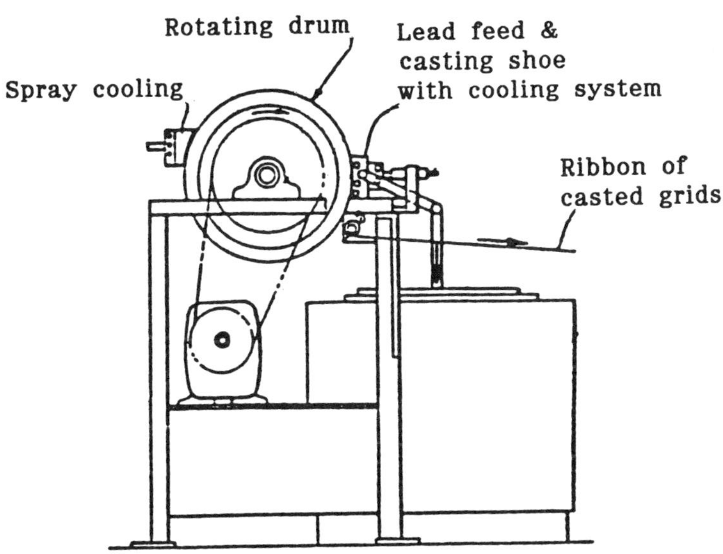

Fig. 10.9 Principle of continuous grid casting (from ([44])).

Drum casting is widely used to cast endless metal bands. The molten metal is fed to a casting shoe sliding on a rotating drum. The casting shoe is shaped to the contour of the drum surface and slightly pressed against the drum. With a plane surface of the drum, a corresponding plane metal strip is produced. For grid casting, cavities are cut in the drum surface that correspond with the grid mould. These cavities are filled with liquid lead, supplied by a slot in the casting shoe. Cooling devices in the casting shoe cause extremely fast solidification of the lead in the cavities, with the result that a continuous string of grids can be stripped off the drum at the end of casting as shown in Fig. 10.9.

The method is restricted to thin grids (< 1 mm); otherwise solidification within a fraction of a second would not be possible, because of the amount of material to be cooled down. Furthermore, the fine structure of the material due to the quenching effect (the extremely fast cooling), makes it prone to increased corrosion. Both effects initially limited the application of the method to the manufacture of grids for negative electrodes. Combination with subsequent rolling that reduces the thickness by a factor of between 2 and 3.5 seems to improve tensile strength and corrosion resistance. Connection with a suitable pasting machine that covers the very thin grid on both sides with active material ('overpasting'), allows also the production of positive plates ([45]).

Drum-cast grids are not symmetrical, because the cavities that form the mould are engraved only in the drum, while the casting shoe is smooth. But this is not relevant in view of the fairly thin grids produced by the method. But the design of the grid is arbitrary, and grid resistance can be minimised by suitable arrangement of the grid wires.

Punching is a another way to produce grids continuously (cf. Fig. 10.24, p.343). The disadvantage of this method is that a large proportion of the original lead sheet forms scrap, and has to be remelted and rolled again. But the method allows production of comparatively thin grids from pure lead or lead/tin alloys, which on account of their softness cannot be handled by other methods.

10.3 Connecting Elements (Top Lead)

Different from the design principle of the battery shown in Fig. 2.2, p.8, usually several electrodes of each polarity are connected in parallel per cell to enlarge the surface area, except in cells with wound electrodes (cf. Figures 10.24 and 10.25, p.343 and 344). Furthermore, a battery is mostly formed from a number of cells that have to be connected with one another. Special connections are required when several cells are housed in one multicell container as a so called 'block battery' (cf., e.g., Fig. 10.22, p.342).

Fig. 10.10 shows a plate group typical of prismatic (single) cell design.

Depending on the layout of the battery, the outside plates can be of positive or negative polarity. It is more customary to have negative plates at both outsides, and the group consists of n positive and (n+1) negative plates, as shown in Fig. 10.10. But the opposite combination is also used, as well as a symmetrical layout with n positive and n negative plates.

Fig. 10.10 Plate group of a prismatic valve-regulated lead-acid battery. 1: Negative plate. 2: Positive plate. 3: Separator. 4: Plate lug. 5: Top bar, bus bar, strap, or pole bridge. 6: Terminal.

The lug of each plate is welded to the connecting bar. This welding can be done by hand or, in more automated production lines, by machines. Then the top bar (with the terminal if required) is cast while the lugs are simultaneously dipped into the mould. This procedure, where the production of the top bar (the strap) and its welding to the lugs are combined, is called COS (cast-on straps) welding ([46]). The vulnerability of these connecting parts to corrosion was mentioned in Section 4.7.6.2 . *A survey of alloys for the cast-on process is given in ([47]).*

The connection between the individual cells within a block battery, the intercell connector, should have minimum resistance. A direct connection through the partition between adjacent cells is usual. Fig. 10.11 illustrates two methods used for lead-acid batteries.

The '**up and over connection**' is used when the intercell connector is manually welded. This connection is illustrated in the top part of Fig. 10.11. To permit the welding, the ends of both top bars to be connected are L-shaped, and a slot is cut from the partition wall at these ends, as shown in the Figure. Through this slot a bridge is established by hand-welding between the upright ends of both 'L's. The space between this bridge, the partition wall and the lid of the battery container can be used to connect the cells for common ventilation, if only one valve is applied for the whole battery (cf. Fig. 10.24, p.343). Accurate sealing between the bridge and the container elements is necessary for complete separation.

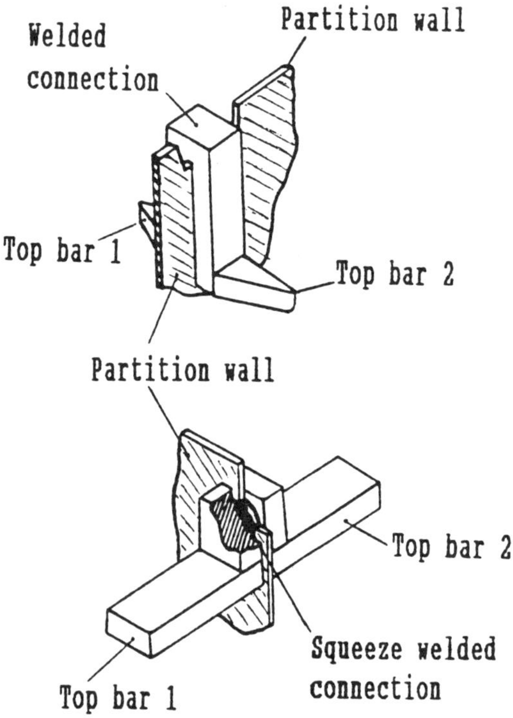

Fig. 10.11 Intercell-connector designs for valve-regulated lead-acid batteries. Top part: 'Up and over' connection between cell No. 1 (Top bar 1) and cell No. 2 (Top bar 2). Mainly used for manual welding. Bottom part: Squeeze-welded connection, used with welding machines. (From ([48]).)

The bottom part of Fig. 10.11 shows the **'through the partition connection'** usually produced by welding machines. A circular or elliptical opening is provided in the partition wall. The ends of the L-shaped top bars are pressed together by special pliers so that contact is established between them through the opening in the partition wall. Simultaneously, high-current pulses flow through the pliers, the top-bar ends and their area of contact. On account of the small cross section of the contacting area, the current resistance is high, and the material begins to melt and forms a welded connection between the two top bars, while the squeezed lug ends simultaneously are sealed to the partition wall. The quality of this welding and sealing depends on pressure, welding current, pulse duration and welding time (cf. ([49])).

10.4 Separators

Some aspects of separators for valve-regulated lead-acid batteries were described in Section 4.6.4.

The glass-mat separator is inserted between the plates, as in Fig. 10.10. For optimum function, a definite compression must be provided by the container. Another way to achieve definite and permanent compression, is the use of plastic ribbons (band pressed elements (50)). The trend is toward more fine fibres and fairly high compression factors (up to 30%) to reduce stratification problems, and the addition of hydrophobic material is claimed to ensure fast oxygen transport, even in a 'wet' separator (51).

When the electrolyte is immobilised as a gel, a microporous separator is required, because the gelled electrolyte itself does not prevent the penetration of lead dendrites from the negative to the positive electrode. Separators, as used in conventional lead-acid batteries, are applied for this kind of valve-regulated lead-acid battery, based on polyethylene (52), PVC or resorcinol resins. A survey including the specific data is given in (53). Fig. 10.12 shows how the separation usually is arranged for a tubular cell. A corresponding situation applies to pasted plates. The thickness of the separator, including ribs that can be on both sides, affects the distance between the plates. Larger ribs point to the positive electrode to provide an increased volume for acid to compensate for the higher acid dilution by water generation (cf. Eq. (4.4), p.99). When the separator does not have ribs, perforated distance pieces or glass mats are applied in addition to the separator to provide sufficient spacing of the electrodes.

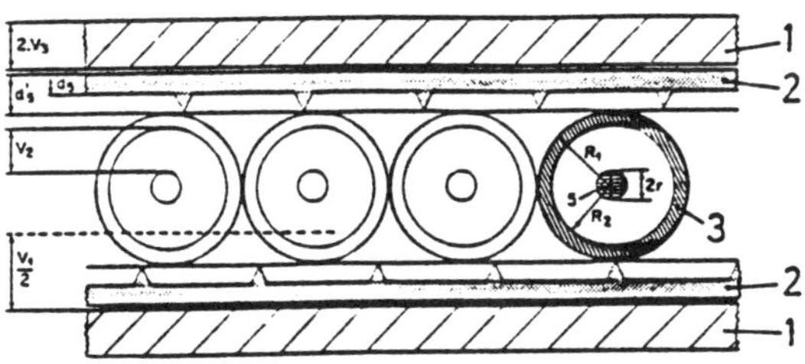

Fig. 10.12 Cross section of a positive tubular plate with two adjacent separators and negative electrodes. 1: Neg. electrodes. 2: Separator. 3: Tubes of the pos. electrode. Separators for tubular plates usually have diagonal ribs for accurate spacing, otherwise, the ribs must precisely meet the tubes.

The additional separator causes a slightly higher internal resistance of batteries with gelled electrolyte than of the same design with absorbed electrolyte (cf. Section 4.6.4.4).

10.5 Containers

The container of a lead-acid battery has first to keep the acid within the cell. But the container simultaneously provides mechanical support for the battery. In valve-regulated lead-acid batteries, the mechanical stress on the container is increased because the internal pressure can be different from the external pressure. Furthermore, proper sealing of the terminals and the connection between the container and its lid is required.

In Table 10.1, mechanical properties are listed for some plastics materials used in valve-regulated lead-acid batteries.

Abbrev.	Components	Density g/cm^3	Temp. range °C	E - modulus MPa*	Bending strength N/mm^2
ASA	Acrylic ester; Styrene;Acrylonitrile	1.07	-40 - +85	2600	70 - 100
ABS	Acrylonitrile; Butadiene, Styrene	1.02 - 1.05	-50 - +85	2000 - 2800	45 - 80
PC	Polycarbonate	1.17 - 1.22	-100 - +135	2300	100
PP	Polypropylene (Polyethylene)	0.9	(0) - +100 (pure PP)	800 - 1300	25 - 40
PVC	Polyvinyl chloride	1.2 - 1.4	-30 - +65	3000	70 - 110
SAN	Styrene; Acrylonitrile	1.08	-40 - +85	3600	110 - 140

Table 10.1 Abbreviated name, composition and mechanical properties of plastics for lead-acid battery containers. (from manufacturer catalogues and ([54])). Standard values for comparison only; data of technical products often deviate on account of various production parameters. * 1 MPa = 10^6 N/m^2; = 1 N/mm^2.

The temperatures in the column 'Temp. range' describe the interval recommended for the different polymers. At higher temperatures, mechanical strength is reduced by softening. At lower temperatures, the material becomes brittle. The lower limit of the temperature range for PP (Polypropylene) of 0 °C is inserted in brackets. This limitation is true only for pure polypropylene, because of embrittlement at temperatures below this. Technical products usually contain some polyethylene and can be applied down to minus 40 °C.

The E-modulus (elasticity modulus) describes the extension under a definite pulling force (DIN 53 457). The bending strength is specified according to DIN 53 452 and represents the force required to bend a specified sample bar to the 1.5th multiple of its diameter (cf., e.g., ([54]), p.582). Both figures characterise the stiffness of the material.

ABS is applied predominantly for valve-regulated lead-acid batteries on account of its favourable mechanical properties. ABS containers are rather stiff, as indicated by the figures 'E-modulus' and 'bending strength' in Table 10.1, which are more than double the figures for polypropylene. Consequently, batteries with ABS containers are less prone to bulging from internal under- or overpressure. Furthermore, ABS can be heat-welded excellently, which gives a leak-proof connection between the container and the lid. The disadvantage of the fairly high permeability to water vapour was shown in Table 2.13, p.61. In unfavourable environmental conditions (high temperature, low humidity), attention has to be paid to possibly increased water loss through the container walls.

SAN is another container material for valve-regulated lead-acid batteries. It is in widespread use for conventional stationary lead-acid batteries, e.g. in the batteries in

Fig. 10.28, p.347, and those in Fig. 10.29, p.348. SAN is the mechanically strongest of all the materials listed in Table 10.1 . It has the highest values for E-modulus and bending strength. Thus, even tall battery containers do not need mechanical support. Another advantage of styrene-based materials is that they can be glued, which does not apply for polypropylene. This gives a more economical connection between container and lid, especially for batteries manufactured in a limited quantity. However, for valve-regulated lead-acid batteries, welding is, in general, preferred to ensure proper sealing.

Polypropylene is the predominant container material in the field of starter batteries on account of its excellent impact resistance, weldability and moderate price, but is also used for valve-regulated lead-acid batteries. Small block batteries are often equipped with polypropylene containers, e.g. the batteries in Fig. 10.34, p.353. Polypropylene is less stiff, and the internal pressure must not exceed rather limited values (100 to 200 mbar, ≈ 1 to 3 p.s.i.); otherwise, bulging of the battery might be a danger.

Polycarbonate (PC) and **Polyvinylchloride (PVC)** are container materials specified for conventional stationary lead-acid batteries when extraordinary fire resistance is desired. But polycarbonate is expensive, and polyvinylchloride is inferior at elevated temperatures, as indicated in Table 10.1. Furthermore, in case of fire, although PVC does not keep the fire burning, it is decomposed and forms hydrochloric acid (HCl), which can cause serious damage by corrosion. In battery practice, it is preferred to use ABS, SAN or PP container materials in fire-protected versions offered by different manufacturers. This 'fire protection' is achieved by additives that reduce the flammability, but these additives also influence the mechanical characteristics of the material.

10.6 Valve Design

The valve is an essential element of the valve-regulated lead-acid battery, and for this reason a special Section is devoted to this subject.

In conventional lead-acid batteries, the vent plug permits the exchange of gas between the inside of the cell and the surrounding atmosphere. If p_{int} represents the internal pressure and p_{ext} the pressure of the surrounding atmosphere, the situation for the vented battery is:

$$\boxed{p_{int} = p_{ext}} \tag{10.1}$$

Normal vent plugs have only a few narrow holes to keep the evaporation rate of water low, which is proportional to the open surface area (cf. Section 2.6.5). Most vent plugs are equipped with mechanical traps to return droplets of acid and reduce the acid fumes escaping from the cell. Furthermore, vent plugs are frequently fitted with porous discs or cylinders to prevent ignition sparks or flames from entering the cell from the outside ([55]) (cf. also Fig. 10.15, p.332).

The valve allows only gas to escape from the cell. It has to prevent overpressure inside the cell. A too-high overpressure leads to deformation of the container (bulging of the cell) and may cause problems in the contact pressure in the plate block. The valve must, however, totally prevent any flux or diffusion of air into the cell. If air penetrates, the oxygen content of 21% by volume is reduced at the negative electrode and causes chemical self-discharge (cf. Section 4.4.1).

The valve opens when the overpressure in the cell exceeds a certain value:

$$\boxed{p_{int} \geq p_{ext} + \Delta p} \tag{10.2}$$

with Δp: opening pressure of the valve.

p_{int}: internal pressure of the battery.

p_{ext}: surrounding atmospheric pressure (external pressure).

The valve remains closed as long as the following relation holds:

$$\boxed{p_{int} < p_{ext} + \Delta p} \tag{10.3}$$

As long as Eq.(10.2) applies, gas escapes from the battery until the internal pressure p_{int} is reduced far enough for the valve to close again. Actually, the opening and closing of the valve do not happen exactly at a definite overpressure Δp. On account of hysteresis the closing pressure is always somewhat lower than the opening pressure.

Δp represents the maximum overpressure that can occur within the battery and be exerted on the container walls. Δp is independent of the absolute value of p_{ext} (cf. Section 10.8.6).

A corresponding limitation for the minimum pressure within the cell does not exist. When the battery is cycled, a slight underpressure can develop inside during the charging period, because the oxygen content of the internal atmosphere is consumed by chemical self-discharge at the negative electrode. But gradually the hydrogen evolution causes overpressure again.

The use of a tightly sealing valve, of course, only makes sense if the cell itself has no leakage. Tightness of the container wall, faultless welding between cell container and lid, and perfect pole sealing are therefore essentials for proper operation, especially in view of the high flow rate of gas through a leakage (cf. Table 2.8, p.51).

One design of valve, the BUNSEN valve, is based on a rubber ring or cap that closes the opening of the cell in normal conditions, but is blown open when the internal pressure exceeds a specific limit. This design is widely used in small valve-regulated lead-acid batteries. Such a ring can be seen in Fig. 10.22, p.342; a cap is applied in the battery shown in Figs. 10.23 and 10.24 (more clearly seen in Fig. 10.13). Opening and closing pressures are determined by the toughness of the rubber. Selection of suitable material for the cap is important for the performance of the valve after prolonged service life. Usual opening pressures range between 20 and 200 mbar (2000 and 20 000 Pa). They are determined by the stiffness of the elastic cap.

Fig. 10.13 shows the cross section of such a valve in a cylindrical cell.

In the valve design, shown in Fig. 10.14, a rubber body is pressed against a sealing surface by a plastics wedge. The rubber disk has an edge that seals. To improve the sealing, this surface is covered with a thin silicon-oil film. The void space at the bottom protects the sealing area against acid and gel particles that might otherwise enter the valve when the battery is seriously overcharged.

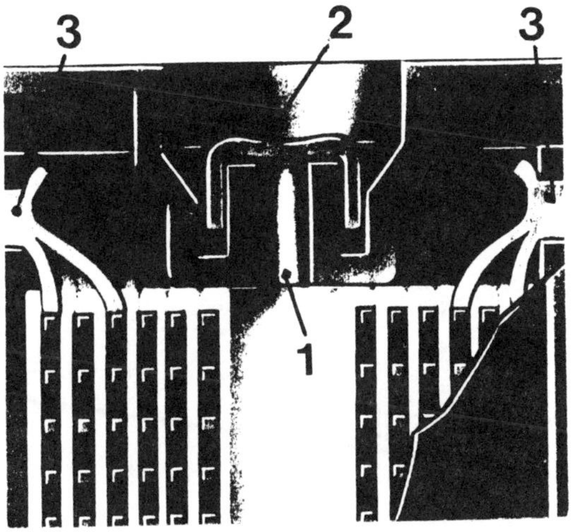

Fig. 10.13 Bunsen valve with synthetic rubber cap. Section of the block battery in Fig. 10.24, p.343. Opening pressure 6 - 8 p.s.i. (= $4.1 \cdot 10^4$ - $5.5 \cdot 10^4$ Pa= 410 - 550 mbar).
1: Opening tube. 2: Synthetic-rubber cap. 3: Intercell connection.
(GATES Energy Products, now HAWKER Energy Products Inc.)

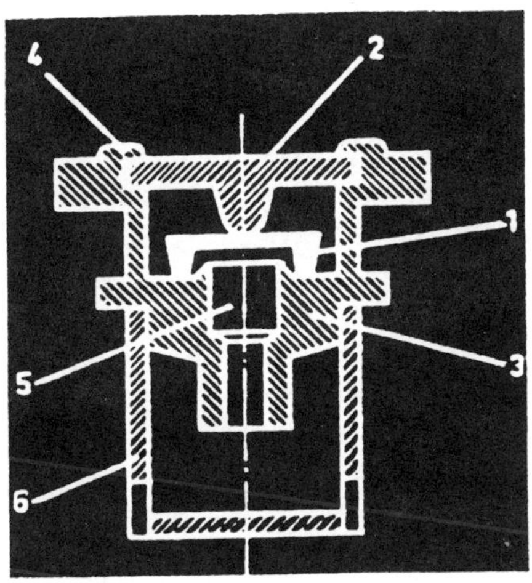

Fig. 10.14 Section of a valve with a rubber disc as sealing element.
1: Rubber disc. 2: Cover with pressing plug. 3: Valve body with sealing area.
4: Welding bulge. 5: Cell opening. 6: Protecting cage.
(SONNENSCHEIN dryfit ([56]).)

The ultrasonically welded cover bears some small holes for gas escape.

Fig. 10.15 shows a valve design where an acid-resistant steel spring causes the pressure required for sealing. The sealing element is a synthetic-rubber plate pressed onto a circular edge. This valve works precisely within narrow limits between opening and sealing, and these limits can be varied by spring tension. Usually, internal pressure ranges from 100 to 200 mbar are used for batteries equipped with this valve.

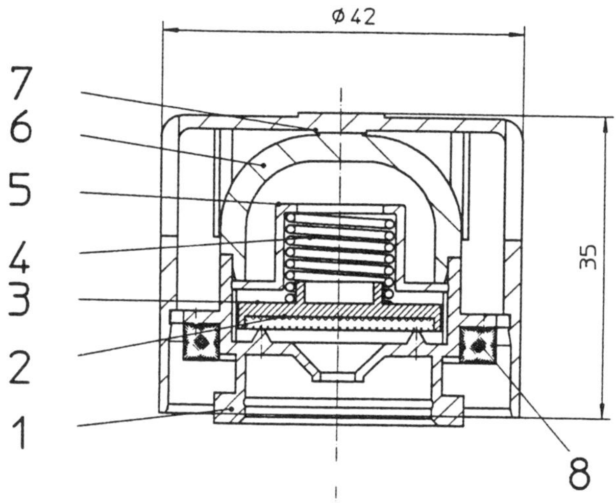

Fig. 10.15 Spring controlled valve.
1: Basis with sealing edge.
2: Sealing disc (EPDM = ethylene-propylene-diene elastomere).
3: Support for the sealing disk (polycarbonate).
4: Spring (Hastelloy).
5: Mount of the spring.
6: Filter that only allows the escape of gas and vapour (no droplets) and simultaneously acts as flameproof protection.
7: Outer casing.
8: Gasket (O-ring).
(VARTA OPzV, now VHB in the HAWKER Group.)

The valve in Fig. 10.15 has the advantage that it can be adjusted precisely to a certain opening pressure by choice of the spring, and this opening pressure is not changed during service life. A disadvantage is the size that is required for a disc with a sufficiently large diameter. Such a diameter is needed because of the small pressure differences that have to be controlled in a lead-acid battery. For this reason, such valves cannot be used in small batteries. A further disadvantage is the comparatively high price.

The valve shown in Fig. 10.16 is based on the compression of a rubber block. It opens at about 70 kPa (= 700 mbar) and reseals between 15 and 20 kPa (= 150 and 200 mbar).

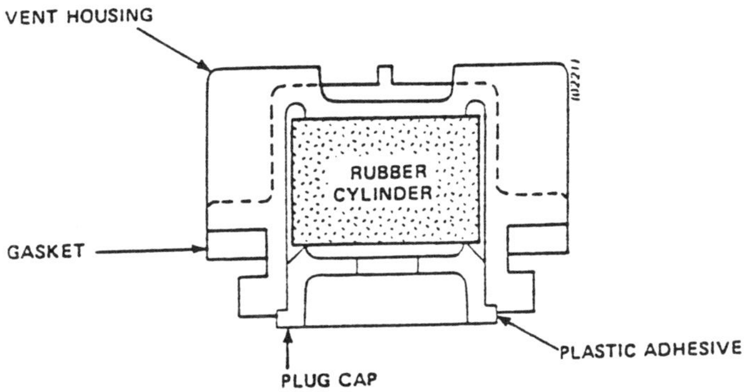

Fig. 10.16 Section of a self-resealing vent, based on the compression of rubber. (GNB Absolyte ([57]).)

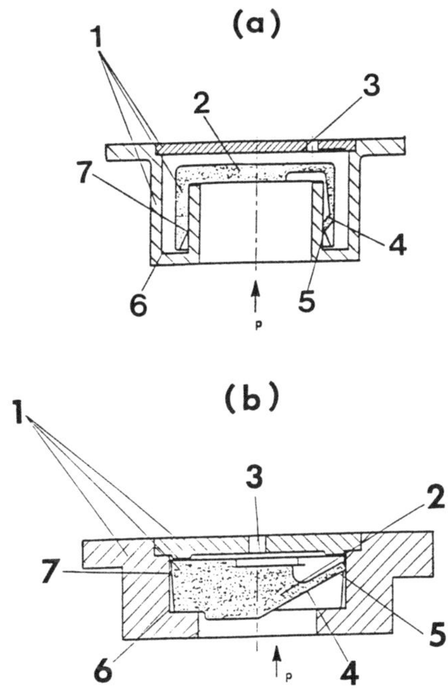

Fig. 10.17 Valves for batteries, based on the deformation of elastomeres.
1: Both valves are made up of only three parts
2: Elastic cap (a) or base (b) of EPDM or FPM
3: Outlet 4: Enlarged area for pressure response
5: Dynamic seal 6: Casing
7: Permanent seal (large area (a) of sealing bead (b)
Busak + Shamban ([58]).

Two valves that have been developed for mass production under the aspect of high reliability are shown in Fig. 10.17. Both are made up of only three parts, but these parts have to be made with high precision. So, for the plastic parts a certain surface roughness must not be exceeded.

The valve in Fig. 10.17a follows the same principle as that in Fig. 10.13. But here, in the cap three grooves are formed that enlarge the surface area and so enhance the pressure response (the cut through in the Figure shows one of them). Furthermore, the seal is reduced in this section to a small edge. This is beneficial for a definite opening pressure, since it reduces the danger that the cap sticks to the plastic tube.

The elastic base in Fig. 10.17b forms a butterfly valve. A bead seals the rest of this base so that air intake is absolutely prevented.

Oil can be used to improve the situation in both valves. The reliability during service life depends on the selection of the elastomers. EPDM (ethylene-propylen-diene elastomer) is used in the temperature range -50 to 150 °C; FPM (Fluororubber) for applications at higher temperatures (-20 to 200 °C). Both valves can be placed into a vent cup or into the lid of the container.

10.7 The Manufacturing Process

Fig. 10.18 is a generalised picture of the steps in the manufacture process of lead-acid batteries with pasted (nontubular) plates. (A detailed survey of the processes and the required equipment is given in the book by Reasbeck and Smith cited under ([38]).)

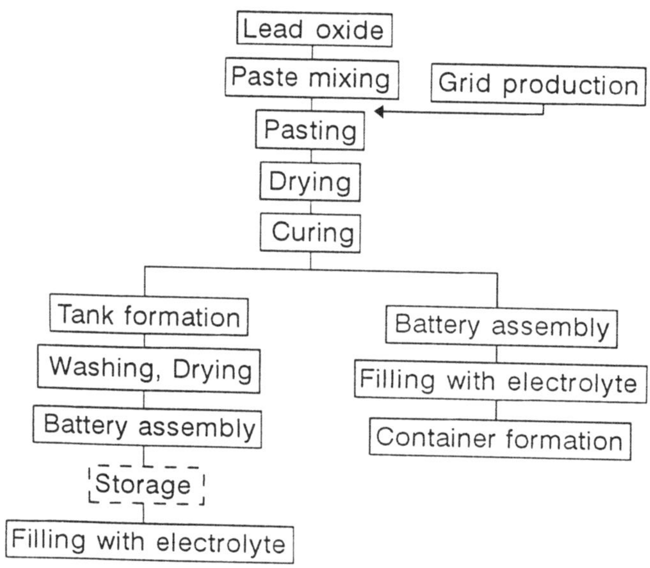

Fig. 10.18 Manufacturing process of pasted-plate batteries.

The production of active material for both electrodes starts with the same substance, a mixture of lead oxide (PbO) and metallic lead called grey oxide or lead dust. The lead (Pb) content is 20 to 30 wt.%. The grey oxide can be produced by a **milling process,** which, strictly speaking, does not mill the material. Flakes are sheared, subsequently crushed and at the same time partly oxidised until the desired powder, the grey oxide, results.

Another process, the **Barton process**, is based on molten lead. Fine droplets of lead are produced by heavy stirring of the melt. The surface of each droplet is transformed by oxidation into a shell of PbO. These oxidised particles are, if small enough, carried away by an air stream; otherwise, they fall back into the melt and the process is repeated.

A short description of both processes is given in ([59]). The physical and chemical features of the grey oxide and the substances involved in the subsequent production steps are described in detail in ([2]), pp. 162, and environmental problems are discussed in ([60]). Nowadays, the Barton process is preferred, since it can be better controlled ([61]).

Paste mixing means the addition of sulphuric acid and water. The result is a fairly stiff paste of density between 1.1 and 1.4 g/cm^3 containing 8 to 12 wt.% of lead sulphate ([62]). The water content of this mix determines the achievable porosity of the active material (cf. 'curing'). In the paste, a mixture of lead sulphate and lead oxide is formed, having two variants:

- tribasic lead sulphate ($3 \cdot PbO \cdot PbSO_4$) and
- tetrabasic lead sulphate ($4 \cdot PbO \cdot PbSO_4$).

In the usual mixing process between room temperature and 50 °C, tribasic lead sulphate is formed. The generation of the tetrabasic modification ($4 \cdot PbO \cdot PbSO_4$) is favoured at temperatures above 70 °C ([63]).

To a certain extent, the formation of the tetrabasic variant is desired, because $4 \cdot PbO \cdot PbSO_4$ forms fairly large crystals when transformed into lead dioxide (PbO_2). This results in a mechanically stable active material. But there are disadvantages, because it is more difficult to transform this material into lead dioxide; i.e. the formation process (cf. below) is more expensive (and takes longer) and the initial capacity is slightly reduced. (cf., e.g., ([64])). For 'long-life batteries' (Bell Syst. cell), a special process has been developed to produce pure tetrabasic material ([65]).

Sometimes red lead or minium (Pb_3O_4) is added to the paste destined for positive active material. The addition is usually 5 to 10 wt.%, and is mainly made for easier formation of the final compound lead dioxide (PbO_2) (cf. 'curing' and 'formation', below). Pb_3O_4 is a more oxidised state than PbO that contains Pb^{2+} and Pb^{4+} ions (cf. ([2]), p. 17).

A general problem in open mixing machines is that the exothermic process causes temperature increase of the mix. Evaporation of water can be used for cooling, but then water is lost and the composition of the mix changes. For this reason a certain surplus of water is scheduled in the paste recipe. But evaporation rate depends on environmental parameters like temperature and humidity, and this limits the possibility to maintain the specified water content of the paste, which is important in regard to the porosity of the active material.

For accurate mixing, independent of environmental conditions, vacuum mixers are used as shown in Fig. 10.19.

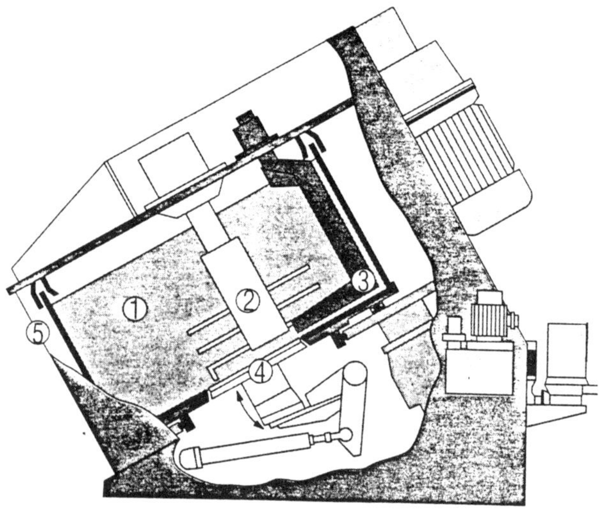

Fig. 10.19 Vacuum mixer for lead paste.
1: Mixing compartment. 2: Fast rotating mixing tools.
3: Material deflecting plate. 4: Discharge opening.
5: Static, vacuum sealed enclosure. (Eirich, (66).)

In the vacuum mixer, water evaporation is also used for temperature control, since the evaporation rate can be influenced by the grade of the vacuum. The water vapour, however, does not escape from the mixer, but is condensed and returned into the mix. So, the composition of the mix is not changed.

At the end of the mixing process, the paste contains about 10 wt.% of metallic lead and about 50 vol.% of water. The water is evaporated during the subsequent production steps, and the resulting void space represents the pore volume of the dried active material.

Pasting means that the paste and the grids are supplied to a machine that smears the mix into the grid. Single plates are superficially dried after pasting, to prevent sticking when they are stacked afterwards. Continuously cast grids leave the pasting machine as an endless ribbon, usually enveloped by paper. Pasting of flat elements or foils (cf. Section 10.8.1.1) is achieved with a slurry instead of a stiff paste (cf., e.g., (67)).

The subsequent production step, the '**curing**', is especially important for the positive plate, because the structure of the active material can be influenced by the environmental conditions like temperature and humidity (68). During the curing step, the lead content in the active material is reduced by gradual oxidation from about 10 to less than 3 wt.%. Furthermore, the water content (about 50 vol.%) is evaporated. This evaporation must be done quite carefully, to ensure that the volume occupied by the water actually forms porosity and is not lost by shrinkage, which again might lead to the formation of cracks. As in the paste-mixing process, the transformation into tetrabasic lead oxide ($4 \cdot PbO \cdot PbSO_4$) is favoured at curing temperatures above 70 °C ([69]) which is of prime interest for the later mechanical stability of the positive active material.

Single plates are usually cured in special devices (curing ovens or curing chambers, cf., e.g., ([70])) that control humidity as well as temperature. In continuous plate production, the drying of the pasted ribbon is correspondingly controlled. Furthermore, in continuous manufacture final curing can occur after the plates are separated and inserted into the containers.

10.7.1 Tank Formation, Dry Charged Negative Plates

The production path in Fig. 10.18 is split at this stage. The left branch concerns **tank formation,** which means that the cured positive and negative 'raw plates' are inserted alternately in special tanks filled with fairly dilute sulphuric acid (generally in the range 1.1 to 1.15 g/cm^3) and positive and negative plates connected in parallel with a rectifier. The **formation process** means that the active material of the plates is electrochemically transformed into the final stage, namely:

- lead dioxide (PbO_2) in the positive electrode (Section 10.1.1) and
- spongy metallic lead (Pb) in the negative electrode (Section 10.1.2).

(A survey on formation techniques is given in ([71]).)

Because of the porous material in the raw plate, both substances are produced in a spongy state with a porosity of about 50 vol.%. Tank formation takes between 8 and 48 hours, depending on plate thickness and formation schedule.

A brief comparison of the advantages and disadvantages between tank formation and its alternative, the container formation, is given at the end of Section 10.7.3.

When the formation process is finished, the plates are washed and dried again. The drying of negative plates is not possible without precautions, because of the tendency to spontaneous oxidation described in Section 4.4.1. This oxidation reaction is much accelerated by water. When the plate is dry, only a thin layer of oxide covers the surface of the active material, and prevents further oxidation. So, a common feature of the various methods to achieve **dry charged negative plates** is prevention of access of oxygen as long as the plates are wet. As a result of the superficial oxidation, a loss of about 10% of capacity is always connected with the dry charge process, regardless of the method applied. But oxidation does not proceed further when the dry plate is stored properly (cf. Section 15.1.2). Additives like oil support the stability of the charged material, as mentioned in Section 10.1.2.

Ways to get charged negative electrodes dry are:

- **Vacuum drying,** which is expensive and lengthy, because of the slow heat exchange in a vacuum.

- **Inert-gas drying** uses oxygen-free gases to dry the plates. The inert gas is generally produced by burning natural gas, and contains mainly nitrogen (N_2) and carbon dioxide (CO_2). This method is in widespread use.
- **Superheated-steam drying** is also much applied. It uses water vapour as the drying agent. A special version of this method is called **contact drying** or **hot plate drying**. The wet plate is lightly pressed between heated metal surfaces, e.g. two steel belts, when the process is run continuously. The water vapour evaporated from the pores of the active material forms the protecting cover around the plate as long as it is wet.
- **Petroleum drying** uses heated petroleum or corresponding agents that form azeotropic mixtures with water vapour, which carries away the water.

The dried plates can be stored for a practically unlimited time without losing capacity or ageing. This is true also for complete batteries that are assembled but not yet filled with electrolyte (cf. Section 15.1.2).

10.7.2 Battery Assembly

Battery assembly includes all processes that complete the battery: the plate stack is formed with the separators aligned between the plates unless the separators form pockets. The conducting elements are welded to the plate lugs. Finally the plate stack is inserted into the container and the lid heat sealed or glued. With increasing numbers of manufactured batteries, these steps are more and more automated (cf., e.g., ([72])).

10.7.3 Container Formation

The production path shown on the right in Fig. 10.18 is another sequence in general use for vented and for valve-regulated lead-acid batteries. The fundamental difference from the first alternative is that the battery is assembled first, then filled with electrolyte, and finally the **formation process** is carried out with the complete battery.

During the formation process, the battery is considerably overcharged and generates both hydrogen and oxygen with resulting water loss. The concentration of the filling acid is adjusted in such a way that the desired final acid concentration is approximated at the end of the formation, and only minor corrections are required. Another method includes dumping of the acid and refilling of the battery during the formation step ('one-shot formation', 'two-shot formation', cf., e.g., ([71]), p. 17).

The advantages of the container formation are obvious:

- The production step 'washing, drying' is eliminated.
- This elimination means lower production costs, especially in view of effluent problems with washing processes on account of the acid and lead traces in the washing water.

Advantages of tank formation are:

- Formation conditions like acid concentration and temperature can be optimised.

- Dry charged plates can be produced by tank formation, and these plates, as well as batteries with these plates, can be stored for practically unlimited time. Service life of the battery starts with electrolyte filling. For this reason, manufacturers using tank-formed and dry-charged negative plates sometimes print the filling date on the battery ([73]).

As pointed out in Chapter 9, which of the two methods is to be preferred depends not only on the manufactured battery type but also on the particular situation of the manufacturer.

10.7.4 Batteries with Tubular Plates

The production of tubular positive plates is in principle similar to that of pasted plates. A number of manufacturers use the same grey oxide as basic filling substance. Sometimes the share of red lead or minium (Pb_3O_4) is increased above 25 or even to 100 wt.%. The latter is more economic when the manufacturer himself runs a minium plant. Then the expense of the chemical oxidation of lead oxide (PbO) to minium (Pb_3O_4) may be compensated by reduced formation cost. Furthermore, curing is not required, because of the high oxidation state, and the battery starts with full capacity when formed.

A general picture of manufacture is given in Fig. 10.20.

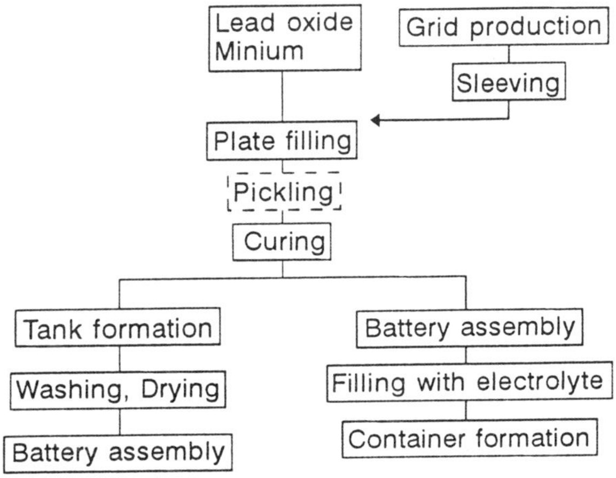

Fig. 10.20 Manufacture of tubular-plate batteries.

Different methods are in use for plate filling. The material can be filled as a powder with the aid of vibrators. Other techniques use a slurry of lead oxide or even a paste, as described above ([74]).

When dry material or a slurry has been filled, 'pickling' is required, which means that the plate is stored in sulphuric acid for a short time. The material is soaked by the acid and transformed, at least partly, into lead sulphate ($PbSO_4$), as in the paste mixing process (Section 10.7). When minium is used, also lead dioxide is formed according to

$$\boxed{Pb_3O_4 + 2{\cdot}H_2SO_4 \Rightarrow 2{\cdot}PbSO_4 + PbO_2 + 2{\cdot}H_2O} \tag{10.4}$$

The subsequent procedures, formation, washing, drying, and battery assembly are similar to those described above.

10.7.5 Electrolyte Filling

The **electrolyte filling** method for valve-regulated lead-acid batteries with absorbed electrolyte is similar to that for conventional lead-acid batteries. Either a definite amount of acid is filled per cell, or the battery is slightly overfilled and the excess acid removed by suction, while the battery is overcharged at a specified rate. In both cases, the aim is to provide enough void space in the separator to establish the internal oxygen cycle.

A sufficiently high rate of the internal oxygen cycle is approached automatically, because of the increased water loss, if the battery initially contains too much electrolyte. This is a fundamental difference from sealed nickel/cadmium batteries, discussed in Section 8.3. The valve-regulated lead-acid battery does not maintain its initial layout parameters like charge reserve and discharge reserve.

The filling process with gelled electrolyte is slightly complicated because of the thixotropy of the mixture of sulphuric acid and silicious acid (SiO_2). Low viscosity can be maintained only while the mixture is agitated intensively. As soon as the mixture is at rest, it is transformed within less than half an hour into a fairly stiff gel. Special stirring devices are required for this reason. Some manufacturers apply organic additives, like polyacrylamide emulsion, a polymer, to reduce the content of silica (SiO_2). Solidification period and initial viscosity of the electrolyte mixture can be influenced by such additives for easier control of the filling process; furthermore, it is claimed that the stiffness of the gel finally reached is increased, and that the gelled substance ceases to be thixotropic ([75]).

When the gel electrolyte is solidified, cracks are formed by shrinkage. This effect provides void space for the transport of gaseous oxygen immediately after filling. For this reason, it is not required to decompose a corresponding amount of water until the internal oxygen cycle is established.

The initial efficiency of the internal oxygen cycle is slightly inferior for gelled electrolytes compared to batteries with absorbed electrolyte that have a properly adjusted amount of electrolyte. However, a maximum rate of about 0.5 A per 100 Ah, initially reached in gelled electrolyte, is sufficient in all normal operational conditions (cf. Note in Section 8.3.6).

10.8 Lead-Acid Battery Design

Chapter 9 pointed out that there are many ways to design a battery. The following pages give a few examples only to demonstrate some of the basic design principles.

10.8.1 Portable Batteries

At first, the valve-regulated lead-acid battery was developed to compete with the sealed nickel/cadmium battery, mainly for economic reasons (76), and in the field of small batteries this competition still exists to some extent. Initially, these batteries were mainly used in portable devices like tape recorders or tools, but also stationary applications, like small types of emergency lighting equipment, were soon supplied by such batteries. Later, larger and larger types were developed.

Fig. 10.21 shows examples of valve-regulated lead-acid block batteries. The larger types are not portable in the literal sense. They are used in stationary and traction applications. Corresponding batteries are also offered as single cells, but the block design is often preferred, because most of the applications require a voltage of more than 2 V. Fig. 10.21 demonstrates the great number of different types offered by a single manufacturer.

Fig. 10.21 Valve-regulated lead-acid batteries with gelled electrolyte as block batteries. Blocks with 3 and 6 cells (6 and 12 V resp.). Capacity range: 1 to 130 Ah. SONNENSCHEIN (77). (cf. also (78).)

The constructional details of such block batteries are demonstrated in Figs. 10.22 and 10.23. Both batteries are equipped with microporous glass-mat separators (absorbent type). A corresponding battery with gelled electrolyte has the same basic constructional features, but instead of the glass mat a conventional microporous separator is inserted to space the electrodes (cf., e.g., Fig. 10.31, p.350).

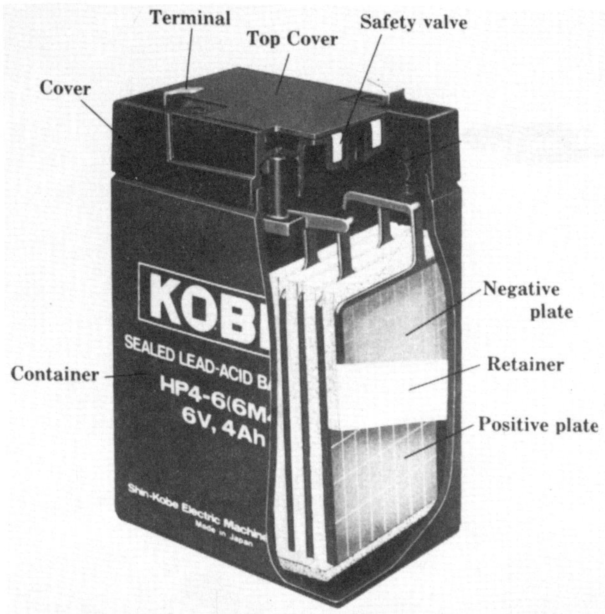

Fig. 10.22 6 volt block battery (absorbent glass-mat type). SHIN KOBE ([79]).

The plate groups in the cutaway cells of both Figures have the same essential components as the plate group in Fig. 10.10. They are arranged in different ways in both batteries: three cells side by side in Fig. 10.22; two lines of three cells each in the 12 V version shown in Fig. 10.23.

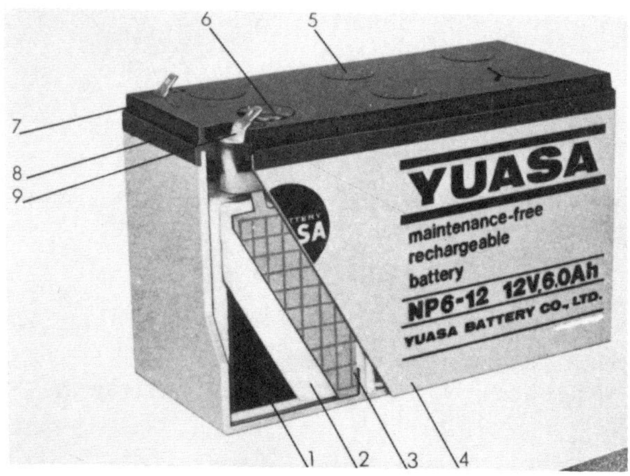

Fig. 10.23 12 volt block battery (absorbent glass-mat type).
1: Positive plate. 2: Microporous glass mat. 3: Negative plate.
4: ABS container. 5: Cover of the valve. 6: Valve.
7: Terminal seal. 8: ABS battery cover . 9: Terminal.
YUASA Battery Co., Ltd. ([80]).

In the cutaway of Fig. 10.22, above the plate group, the pole bridges can be seen welded to the (fairly long) plate lugs. The connection between the individual cells within the bloc is welded as an 'up and over design' shown in Fig. 10.11, p.326. The seal of the valve is a rubber ring that closes the opening.

The batteries in Figs. 10.22 and 10.23 are equipped with individual valves for each cell. Opening pressures of the valves are between 10,000 and 80,000 Pa (100 and 800 mbar or 1.5 and 12 p.s.i. resp.).

A different design principle that is used for batteries in the capacity range between 2.5 and 25 Ah, is shown in Fig. 10.24: on the left, a cross section of a monocell battery; on the right a 12 V 'block' composed of six of these cells. Each cell contains one positive and one negative electrode, which are separated by an absorbent glass-mat separator and spirally wound to form the cell element. Details of this design are to be seen in the foreground on the right of Fig. 10.24. Similar monocells were developed earlier using expanded lead as grid material for both electrodes ([81]).

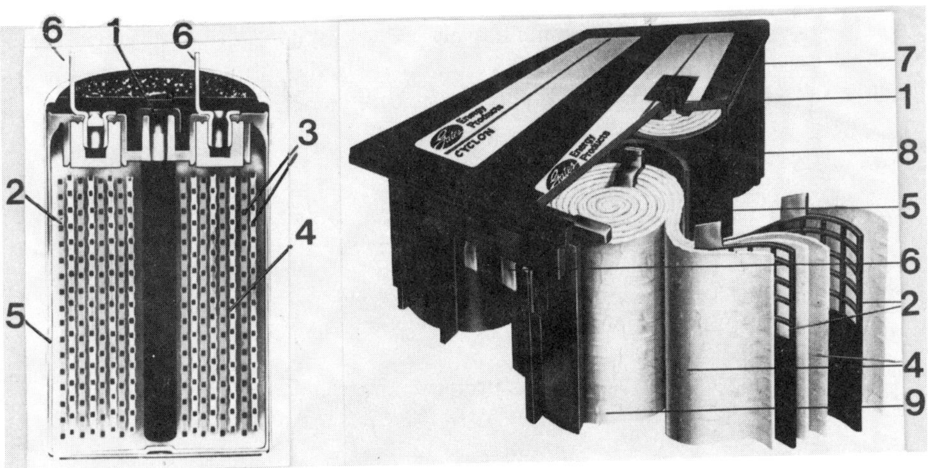

Fig. 10.24 Batteries with spirally wound electrodes.
Left: Monocell; right: 12 volt block.
1: Valve (cf. Fig. 10.13, p.331). **2**: Grid, punched from lead.
3: Positive and negative electrodes. **4**: Glass-mat separator.
5: Polypropylene container (as a monocell encased by a metallic housing).
6: Terminal (bobbin contacts or thread (for 25 Ah only)).
7: Monoblock cover. **8**: Intercell connector.
9: Spirally wound element of the two electrodes and the separator (4 layers).
(Gates Energy Products ([82]), now HAWKER Energy Products Inc.)

The cross section on the left of Fig. 10.24 shows the version as a cylindrical monocell. The grid is punched from a lead sheet (no alloying additives, except a little amount of tin), soft enough to be wound as in the cutaway cell of the block battery. The cross section shows that the grid is surrounded by the active material as in Fig. 10.6.

The sealing of the cell is achieved by a thin-walled polypropylene container mechanically supported by a metallic casing. The polypropylene container is welded to a cover of the same material. A number of lugs on the top edge of the electrodes are welded to the terminal (cf. Fig. 10.13, p.331).

The monocell, as well as the block battery in Fig. 10.24 are equipped with 'bobbin contacts'. In single cells of 25 Ah size, the terminals form threads to screw the connector eye on account of the fairly high current density that may be reached.

The valve consists of a synthetic rubber cap (Fig. 10.13) which in the monocell opens at 3 to 4 atm of internal overpressure (40 - 60 p.s.i. or $2.76 \cdot 10^5$ - $4.14 \cdot 10^5$ Pa). This high internal pressure is permitted in view of the stable housing of the battery.

The block battery or monoblock shown on the right in Fig. 10.24 consists of six cylindrical cells in a polypropylene container (without metallic support). Different combinations of number of cells and configurations of the battery are possible. The cutaway cell on the right shows also the spirally wound electrodes and the four layers of the separator. Within the battery, the cells are connected directly through the container wall (cf. Fig. 10.11, p.326). One common valve serves all the cells in this battery. This valve opens at between $4.1 \cdot 10^4$ and $5.5 \cdot 10^4$ Pa (410 and 550 mbar or 6 and 8 p.s.i.) because of the reduced mechanical support of the cells compared to the monocell.

A battery also based on wound electrodes, but designed for extremely high loads is shown in Fig. 10.25.

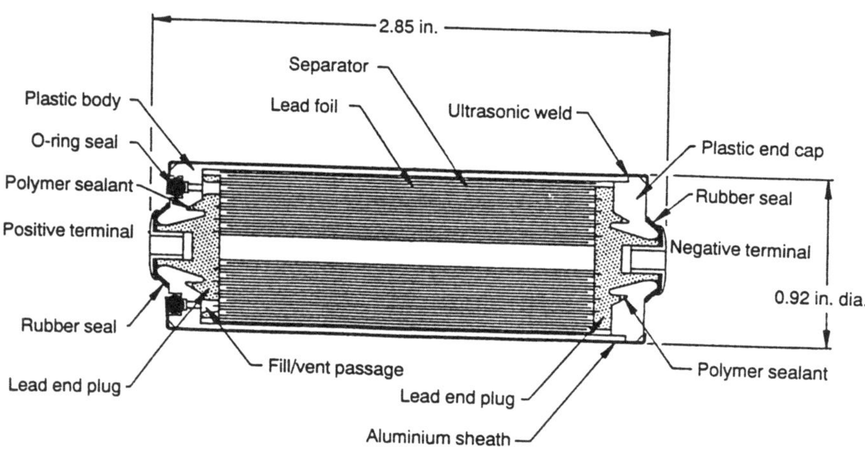

Fig. 10.25 Cylindrical valve-regulated lead-acid battery with foil electrodes manufactured by Bolder Battery, Inc. 2 V, 1.45 Ah at the 1.5-hour rate (≈ 37 Wh/kg; ≈ 95 Wh/dm^3). From ([83]).

The electrodes of the battery in Fig. 10.25 are coated thin foils (0.05 mm) of a lead/tin alloy. The active material is coated on both sides in 0.075 mm layers. The separator (absorbent glass-mat) is 0.18 mm thick. Cast-on end caps ('lead end plugs' in the Figure) make contact with the whole length of the positive and negative electrodes.

This battery has been developed for the application in tools. 400 cycles to 100% depth of discharge are claimed at the 2-hour rate.

10.8.1.1 Miniature Batteries

The trend to minimise the weight and dimensions of portable electronic devices resulted in the development of batteries suited for such applications. Thus, small and flat valve-regulated lead-acid batteries have been developed in the range 0.1 to 2 Ah ([84]). Some of these batteries are flexible; so they can be mounted on a curved surface.

Fig. 10.26 shows the construction of such a flat cell. Basically, it corresponds to the usual design of flat-plate cell (Fig. 10.10, p.325), but each cell consists of only one pair of electrodes. Positive and negative active materials are pasted as a slurry into a small grid plate. The chemical composition of the active material corresponds to that of conventional types. The thin grid plates are pressed to a metallic plate (lead/calcium alloy) that provides mechanical support and acts as the current conductor. Either these plates are injection moulded into plastics parts that form the outer cover of the battery, or the plate group including the metallic plates is encapsulated by a plastic film.

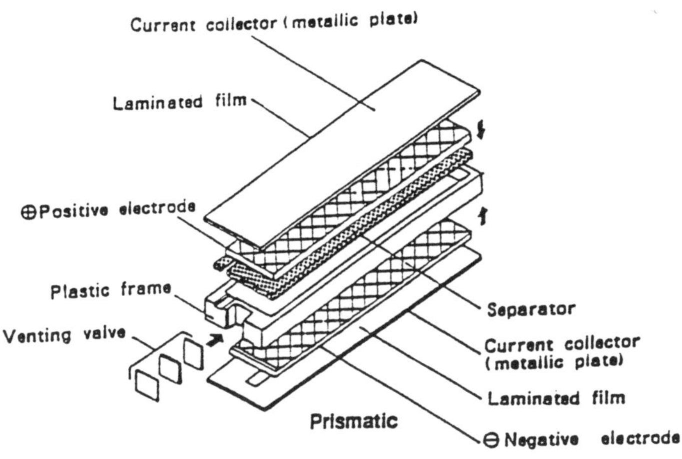

Fig. 10.26 Miniature valve-regulated lead-acid battery. (From ([85]).)

Flat valve-regulated lead-acid batteries (thickness 0.8 cm) in the range between 1 and 11 Ah are manufactured by the same design principles ([86]).

10.8.2 Stationary Applications

The stationary application of valve-regulated lead-acid batteries was strongly stimulated when British Telecom decided to decentralise the d.c. power supply in telephone exchanges by distribution to a number of power-equipment racks that can be installed in situ with the telecommunication equipment (87). In this concept, the batteries had to be arranged underneath the electronic equipment within the racks.

This could only be achieved with valve-regulated lead-acid batteries, since acid fumes must not be produced by the batteries when they are installed together with the electronic components. Otherwise, serious corrosion problems are to be expected.

Furthermore, the low hydrogen evolution of the valve-regulated lead-acid battery (cf. Section 8.3) is beneficial to its ventilation. However, this question is of minor importance, because the gassing rate of modern flooded lead-acid batteries for stationary application is also low. If the hydrogen-evolution rate is considered, vented lead-acid batteries can also be installed close to electronic equipment.

Nowadays valve-regulated lead-acid batteries are in widespread use for stationary applications (88). Uninterrupted-power-supply systems are mostly equipped with such batteries. Increasingly, also, large installations like load-levelling applications are expected to be supplied with valve-regulated lead-acid batteries (89).

Typical examples for valve-regulated lead-acid batteries for stationary applications are shown in Figs. 10.27 and 10.28.

Fig. 10.27 Block battery for stationary applications. This battery is offered as valve-regulated and as vented design (VBV and VB/UPS, respectively), as 4 V, 6 V and 12 V blocks, and in the capacity range from 18 to 256 Ah (10-hour rate). The valve-regulated design has gelled electrolyte. VARTA, now VHB in the HAWKER Group.

The batteries in Fig. 10.27 are made as blocks with different voltages to limit the weight that is to be handled. 12 V blocks are offered in capacities up to 108 Ah; 6 V blocks from 128 to 192 Ah; while the largest batteries are made as 4 V blocks.

Larger capacities, usually up to 3000 Ah per cell, are offered as single cells. Fig. 10.28 shows an example.

Fig. 10.28 Valve-regulated lead-acid batteries for stationary applications. Single cells with gelled electrolyte; tubular plates; 2 V; 200 to 3000 Ah; (10-hour rate); VARTA OPzV. From ([90]). (Now VHB in the HAWKER Group.)

When batteries are equipped with microporous glass-mat separators, a limitation of the cell height is recommended on account of the filling profile caused by gravity (cf. Section 4.6.4.1). For tall cells, occasionally the manufacturer prescribes operation of the battery only on its side, which automatically reduces cell height ([91]) (cf. Fig. 10.29).

Another possibility is to use large block-battery containers with cells switched in parallel. Fairly low-profile batteries can be made by this method for capacities up to 3000 Ah ([92]).

One advantage of valve-regulated lead-acid batteries is the possibility to install them together with electronic equipment. Another advantage is that the blocks or cells of the battery can be operated in the horizontal position which is space saving and also eases the checking of block or cell voltages. Both effects result from the possibility of operating these batteries in any position. Fig. 10.29 shows an example that speaks for itself.

Fig. 10.29 Stationary valve-regulated lead-acid battery. Gelled electrolyte. 48V, 1000 Ah (2 strings in parallel). SONNENSCHEIN dryfit A 600.

In stacking the batteries corresponding to Fig. 10.29, however, heat problems may arise if the current flow is too high for the cooling effects. According to Section 2.5.2.3, spacing of at least 1 cm has to be provided around each block or cell if heat dissipation by air convection is to be effective. The cells in Fig. 10.29 are spaced correspondingly.

Fig. 10.29 clearly demonstrates that, despite of the compact installation, easy access is given to each terminal for voltage measurements. But attention has to be paid to the high weight of lead-acid batteries. It may exceed the load capacity of the floor.

10.8.3 Traction Applications

Valve-regulated lead-acid batteries are also used in traction applications. The design corresponds basically to that of conventional traction batteries, except for the immobilised electrolyte and grid alloys with extremely low or zero antimony content.

The larger types are assembled from single cells within a steel tray in the same way as conventional traction batteries. Two examples are shown at the rear of Fig. 10.30. The reduced utilisation on account of the immobilised electrolyte can be compensated by enlarged plate dimensions, because neither 'mud space' is required at the bottom of the cell nor volume for surplus acid at the top of the plates. Thus, the capacity per volume compares to that of conventional types of corresponding size. But the weight of the valve-regulated lead-acid battery design is slightly higher.

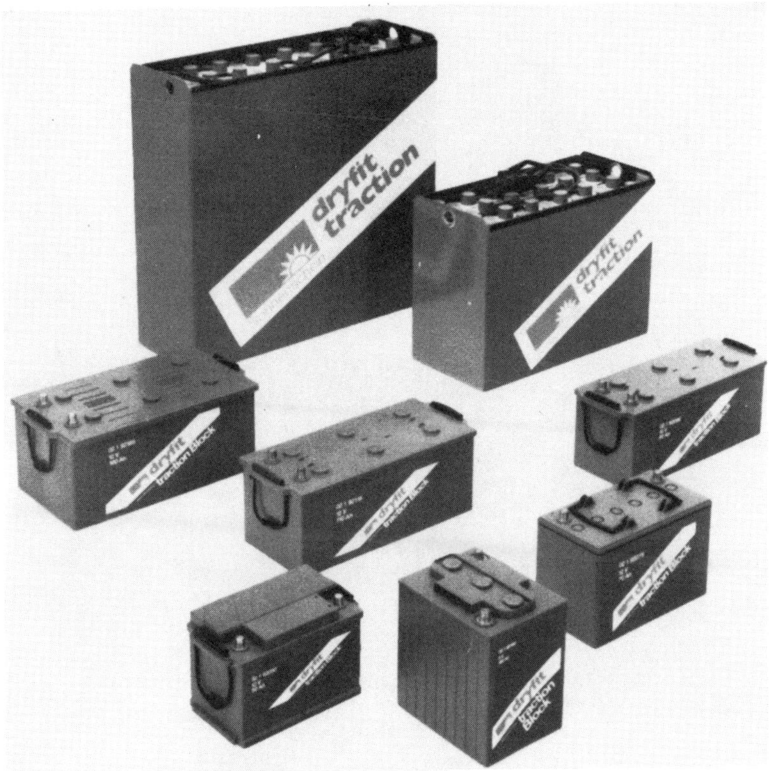

Fig. 10.30 Valve-regulated lead-acid batteries for traction applications with gelled electrolyte. Block batteries with pasted-grid plates. Capacity range 50 to 160 Ah (5-hour rate). Trays with single cells with tubular positive electrodes of capacity range 110 to 1200 Ah (5-hour rate). SONNENSCHEIN dryfit (cf. ([93])).

Fig. 10.30 also shows some block batteries in the foreground. For economic reasons containers of starter batteries are used for some of these blocks; but a specially designed block battery is also to be seen in the Figure that is shown in more detail in Fig. 10.31.

Block batteries, as in Fig. 10.31, are used in all kinds of electric vehicles e.g. passenger cars, but also battery-powered handcarts. A special field of application is wheel-chairs, where the fact that acid cannot be spilled is of particular importance, because they are used in the normal home with carpeted floor.

The propulsion of electric road vehicles is a successful application, because of superior cycle performance to corresponding batteries of conventional design ([94]) *(> 700 cycles (80% depth of discharge) compared to >400 for similar batteries of vented design*).

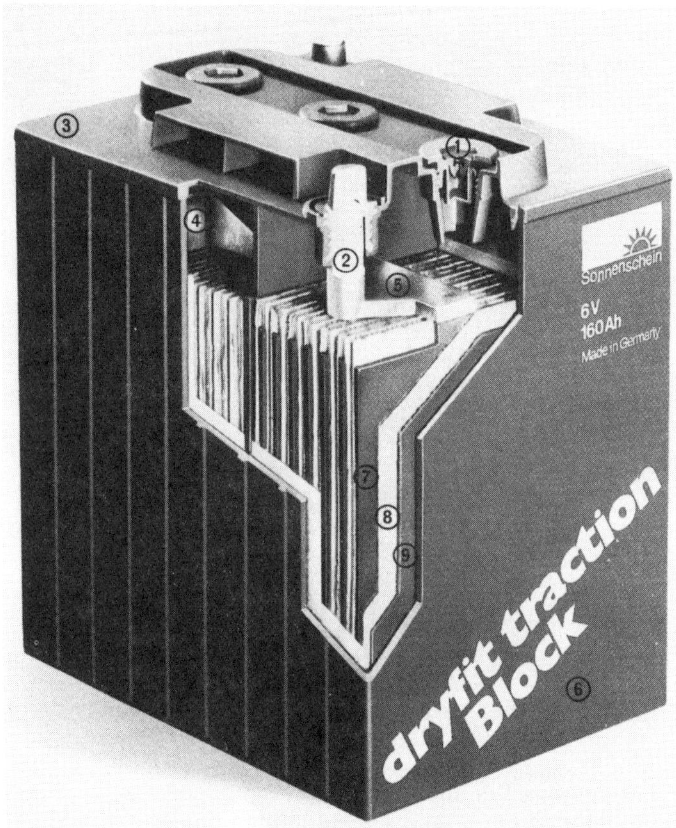

Fig. 10.31 Valve-regulated lead-acid battery with gelled electrolyte for traction. 3 cells (6 volt), 160 Ah.
1: Valve (opening pressure ≈ 10000 Pa = 100 mbar).
2: Positive terminal.
3: Cell cover (polypropylene).
4: Intercell connection.
5: Pole bridge.
6: Container (polypropylene).
7: Positive plate.
8: Microporous separator with glass felt adjacent to the positive electrode (not a microporous glass mat).
9: Negative Plate.
(SONNENSCHEIN.)

An example for such an application is shown in Fig. 10.32, with the battery standing besides the car. This battery is equipped with a water cooling system. Some of the tubes are to be seen on the top of the battery. This cooling system consists of pockets of plastic material that are inserted between the monoblocks. Water flows through the pockets and the side walls of the blocks are used as cooling areas. As mentioned in Section 2.5.3, the purpose of such forced cooling is not only increased heat dissipation; it is even more important for keeping all blocks of the battery at uniform temperature.

Fig. 10.32 Electric vehicle (Golf-Citystromer) and its lead-acid battery (other systems can also be used, cf. Fig. 12.4, p.402). 16 monoblocks 6 V/160 Ah (Fig. 10.31) form the battery of 96 V/160 Ah.

10.8.4 Starter (SLI) Application

Valve-regulated lead-acid starter batteries for automobiles were introduced in the 1970s [95]. But up to now they have not been successful in this field, mainly for two reasons:

- The modern conventional starter battery does not need a water refill during its normal service life (cf. Section 7.5.1.1). For this reason the customer would not notice any advantage.
- The valve-regulated starter battery is more expensive.

Further arguments, such as the absence of gas evolution or the safety argument that acid cannot be spilled in a crash, do not outweigh the higher price, although the same performance can be achieved as with flooded batteries [96].

However, this situation may change in the future, as the valve-regulated battery is absolutely maintenance-free and can be used in any position. So, the future car designer may prefer it, because it can be installed anywhere in the car and in any position.

Starter batteries of valve-regulated design are applied to some extent for cars running in special conditions, meaning that the battery is used not only for cranking, lighting and ignition, but also to supply additional electrically powered equipment that causes cycling of the battery. Buses for travelling are such applications, but also military vehicles. The low self-discharge of valve-regulated lead-acid batteries and their ability to withstand certain deep discharge periods, is advantageous for vehicles that are subject to longer standing periods.

10.8.4.1 Starter Batteries for Motorcycles

Fig. 10.33 shows a valve-regulated lead-acid battery for motorcycles.

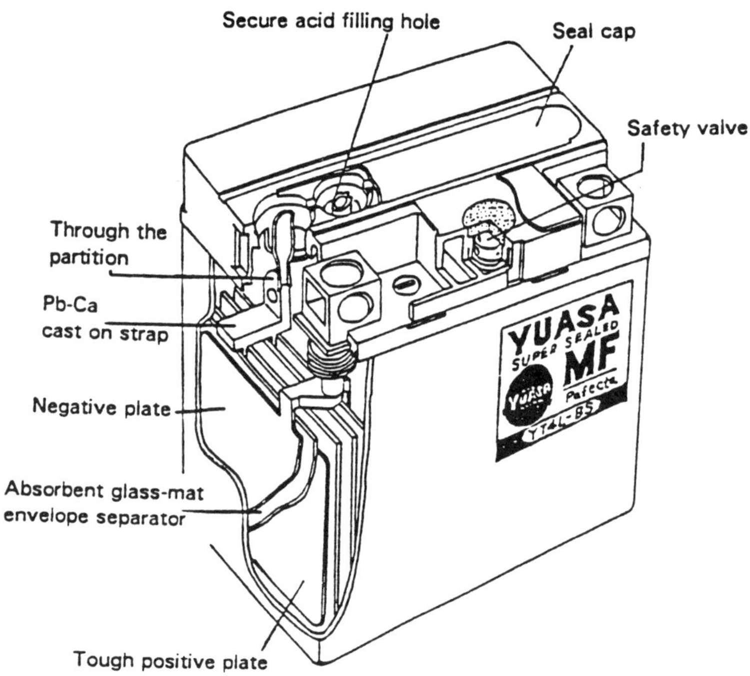

Fig. 10.33 Valve-regulated lead-acid battery for motorcycles. From (97).

The battery in Fig. 10.33 is delivered in the dry charged state (cf. Section 10.7.1) and filled by the motorcycle dealer when put into service. As with the block battery shown in Fig. 10.24, one common valve is used for the six cells. The pole bar is lead/calcium alloy to withstand high vibration.

Motorcycle batteries of valve-regulated design offer several advantages in the special application:

- No leakage of electrolyte can occur, even if the motorcycle overturns.
- The design is superior when the motorcycle is stored for extended periods and

the battery is not charged (cf. Section 15.1.4).

- The design always has minimum self-discharge, giving better cranking when the motorcycle is not used continually.

10.8.5 Hobby and Leisure Applications

An important market for valve-regulated lead-acid batteries is in hobby and leisure. A series of batteries, specially designed for the power supply of boats or caravans, is shown in Fig. 10.34. Here, for economical reasons, the container corresponds to that of a starter battery. On account of its weight, the larger types are equipped with carrying aids.

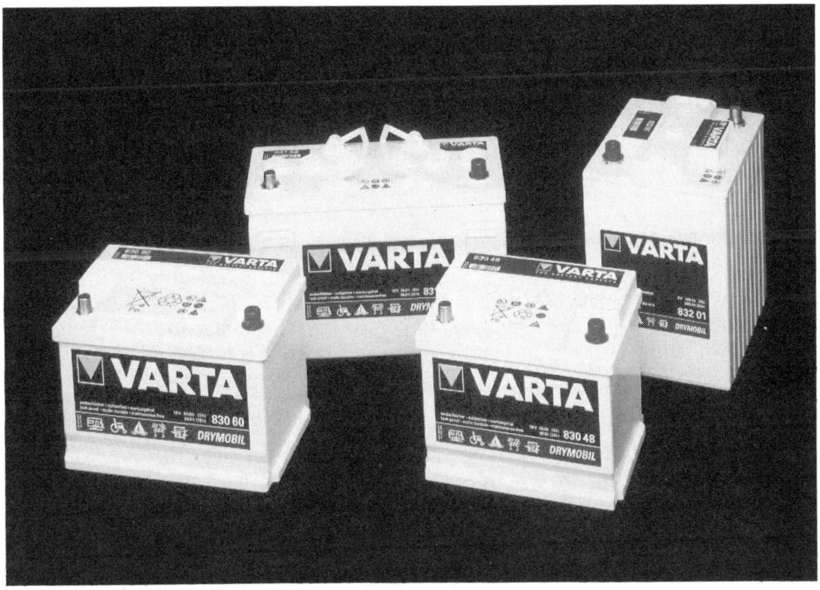

Fig. 10.34 Valve-regulated lead-acid batteries for power supply of boats, caravans, cleaning machines, golf carts, etc. Microporous glass-mat separator, 6 V/200 Ah; 12 V/48 - 85 Ah (20-hour rate). 6 V/160 Ah; 12 V/40 - 75 Ah (5-hour rate). (VARTA *DRYMOBIL.*)

A special designed starter battery for leisure applications, like boots and motorcycles, is shown in Fig. 10.35. This battery is equipped with lead/tin grid alloys and suffers very low self-discharge. After 6 month of storage more than 85% of the rated capacity are available and after 2 ½ years of storage at 25 °C without charging, still 50% of the rated capacity is claimed to be available.

10.8.6 Aircraft Batteries

Aircraft batteries fulfil the same functions as the starter battery in a motor car. Larger aircraft usually employ specially designed vented nickel/cadmium batteries ([98]) for high currents which are required to start the engines or auxiliary power units, and also for the extremely low temperatures to which they may be exposed. But in small aircraft, lead-acid batteries have always been used for economic reasons. They have also been used in most UK-manufactured civil aircraft.

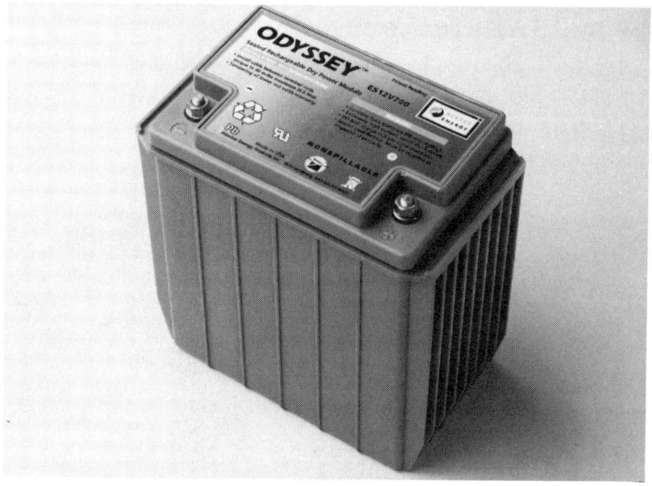

Fig. 10.35 Valve-regulated starter battery with absorbent glass-mat separation. 12 V monobloc, 17 Ah (10-hour rate). Hawker Energy Products Inc, Warrensburg, MO, USA ([99]).

Special valves have been developed for conventional lead-acid batteries so they may be used in any orientation during flight. These valves are closed automatically as soon as the battery is turned over, to prevent any escape of acid ([100]). The increase of internal pressure while the valve is closed can be tolerated, because these orientations are only temporarily.

The valve-regulated lead-acid battery seems destined for this application, because it can be used in any position without precautions. They are currently installed in quite a number of military aircraft, and meet all specifications ([101]).

The low external pressure at which aircraft batteries are operated for extended periods, in practice does not affect the function of the valve, which always maintains the pressure difference Δp, the opening pressure, between the interior of the battery and its surrounding, because the opening pressures of the valve designs described in Section 10.6 are not affected by the external pressure; they are determined by elastic forces of the 'rubber' or the spring.

The internal pressure of the battery is (cf. Eq.(10.2), p.331).

$$\boxed{p_{int} \leq p_{ext} + \Delta p} \qquad (10.5)$$

with p_{int} : internal pressure

p_{ext} : external pressure

Δp : pressure difference that opens the valve

This means that the internal pressure may be lower than the external pressure, but it can never exceed the sum of external pressure p_{ext} and valve opening pressure Δp.

The reduction of the external pressure with increasing flight altitude can be approximated to the barometric-pressure gradient

$$\boxed{p = p_0 \cdot \exp\left(-\frac{h\ [km]}{7.99}\right)} \tag{10.6}$$

The curve drawn in Fig. 10.36 shows the exponential decay of the pressure with altitude. p_o describes the pressure on the ground (h = 0). At the altitude of 5.54 km the pressure is reduced to one half.

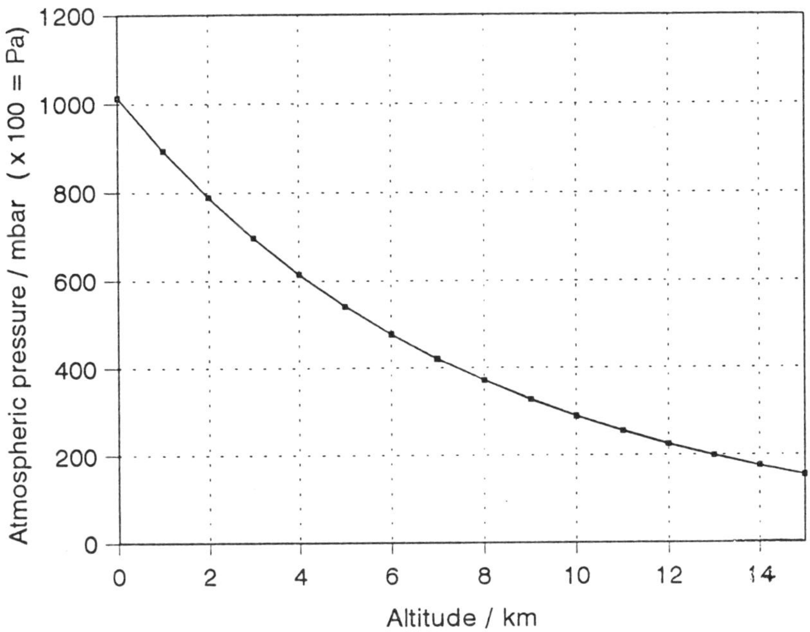

Fig. 10.36 Atmospheric-pressure gradient with altitude. Calculated from Eq. (10.6), (0 °C).

When the external pressure is reduced as the aircraft climbs, p_{int} will exceed the sum $p_{ext} + \Delta p$. The valve opens and gas escapes until balance is re-established. Then the valve closes and the battery operates in the usual way, but at a reduced internal pressure. This is true even if the external pressure were reduced to zero and only Δp was left as internal pressure.

This pressure reduction does not affect the operation of the battery, except for a trivial reduction in the rate of the internal oxygen cycle (cf. Section 8.1, Fig. 8.4) and negligible influence on gas evolution (cf. Section 2.6.1.3). The charge/discharge reactions of the lead-acid battery do not depend on pressure, because gaseous substances are not included (cf. Eq. (4.5), p.99). The low internal pressure would only cause a critical situation, if it fell below the vapour pressure and caused increased water loss due to boiling. However, according to Eq. (10.5) at a height of about 13 km the external pressure is about 200 mbar. If Δp is 100 mbar, the internal pressure p_{int} cannot fall below 300 mbar, even at that altitude. According to Table 2.8, p.51, pure water needs more than 60 °C to reach the boiling point at this pressure, and the water vapour of the acid is even lower (cf. Fig. 2.14, p.58). So, normally the boiling point is not reached, even at a minimum opening pressure of the valve.

The maximum pressure load on the container walls is not altered, because the difference Δp between the internal and external pressures remains constant.

The above also applies to valve-regulated lead-acid batteries on high mountains. There is no difference from operation at or near sea level.

Additional water loss occurs on account of the gas which escapes from the battery while the aeroplane is climbing. Most of it is hydrogen according to the composition of the gaseous phase in the battery. But some oxygen will escape also and this counts as water loss, because it is not available for reduction at the negative electrode and this reduces the effectiveness of the internal oxygen cycle.

The gas that leaves the battery during climbing depends on the altitude reached and the void volume in the battery. Assuming that the battery contains 300 cm^3 of void space per cell per 100 Ah of nominal capacity and the aircraft climbs to 5.54 km (half pressure), then 150 cm^3 of gas escapes from each cell. If it is also assumed that 20 vol.% is oxygen (a high share) then 30 cm^3 or 0.001 mole of oxygen escapes which corresponds to a water loss of 0.018 g per cell and per 100 Ah.

The electrolyte of valve-regulated lead-acid batteries contains about 800 g of water per 100 Ah. Thus, in climbing to an altitude of 5.4 km, about 0.002% of the water is lost. So, if the aeroplane completes 1000 such climbs, only about 2% of the water content of the electrolyte is lost. As this value is based on the high oxygen content of 20 vol.% in the gaseous phase, the actual figure should be significantly lower. Accordingly, the additional water loss of a valve-regulated lead-acid battery due to the special conditions of aircraft service should be negligible. This is confirmed by the fact that even after years of service, weight loss could not be detected in aircraft batteries of valve-regulated design ([102]).

Note: *The above estimation only applies when the battery is exposed to the atmospheric pressure. Often the battery is installed in the pressurised part of an aircraft where the pressure is only reduced to some extent. Then the water balance is affected to a correspondingly lesser degree.*

References

(1) D.Pavlov, E.Bashatavelova and V.Iliev, Proc. Symp. 'Advances in Lead Acid Batteries', The Electrochemical Society, Pennington, N.J., Proc.Vol. 84-14 (1984), p.16
J.Yamashita, H.Yufu and Y.Matsumaru J. Power Sources, **30** (1990), p.13

(2) H.Bode, 'Lead-Acid Batteries', John Wiley & Sons, New York ,1977, pp.8 - 84

(3) VARTA Battery AG (Editor), 'Bleiakkumulatoren', 11th Ed., VDI-Verlag, Düsseldorf, 1986, p.79 (in German)

(4) J.P.Carr and N.A.Hampson, The Lead Dioxide Electrode, Chem.Rev., **72** (1972), p. 679

(5) V.H.Dodson, J.Electrochem.Soc., **108** (1961), p. 401

(6) J.Burbank, A.C.Simon and E.Willihnganz, 'The Lead Acid Cell' *in* Adv. Electrochemistry and Electrochemical Engineering, Vol.8, John Wiley & Sons, New York, 1971, p. 170

(7) D.Pavlov, E.Bashatavelova and V.Iliev, loc.cit., (1), p.19

(8) K.Kordesch and K.Tomantschger, Proc. Symp. 'Advances in Lead Acid Batteries', The Electrochemical Society, Pennington, N.J., Proc.Vol. 84-14 (1984), p.323

(9) S.Atlung and B.Zachau-Christiansen, J. Power Sources, **30** (1990), p.131

(10) A.F.Hollenkamp, K.K.Constati, A.M.Huey, M.J.Koop and L.Aputeanu, J. Power Sources, **40**, (1992), p.125

(11) E.Meissner and H.Rabenstein, J. Power Sources, **40** (1992), p.157

(12) A.Winsel, E.Voss and U.Hullmeine, J. Power Sources, **30** (1990), p. 209, and esp. Fig.12, p.220

(13) U.Hullmeine, E.Voss and A.Winsel, J.Power Sources, **25** (1989),p. 27

(14) D.Pavlov, J. Electrochem. Soc., **139** (1992), p. 3075

(15) S.Tudor, A. Weisstuch and S.H.Davang, Electrochem. Techn., **4** (1965), p. 406, and esp. Fig.2, p.408

(16) E.Nann, J. Power Sources, **33** (1991), p. 93
D.Calasanzio and D.H.McClelland, J. Power Sources, **33** (1991), p. 59

(17) T.G.Chang Proc. Symp. 'Advances in Lead Acid Batteries', The Electrochemical Society, Pennington, N.J., Proc.Vol. 84-14 (1984), p. 86

(18) J.Burbank, 'Power Sources 3', Oriel Press, Newcastle upon Tyne, 1971, p. 13

(19) D.Pavlov, A.Dakhouche, and T.Rogachev, Journal Power Sources, **42** (1993), p. 71

(20) B.Monahov and D.Pavlov, J. Electrochem. Soc., **141** (1994), p. 2316
D.Pavlov, Journal Power Sources, **48** (1994), p. 179

(21) C.Drotschmann, 'Bleiakkumulatoren', Verlag Chemie GmbH, Weinheim, 1951, p.161

(22) S.Tudor, A. Weisstuch and S.H.Davang, Electrochem. Techn., **5** (1967, p. 21, and esp. Fig.3, p. 23

(23) German Patent 16 71 693 (Accumulatorenfabrik Sonnenschein GmbH, Büdingen, 1.12.67)

(24) K.R.Bullock and D.H.McClelland, J.Electrochem.Soc.,**124** (1977), p. 1478
K.R. Bullock, ibid.,**126** (1979), pp. 360 and 1848

(25) J.Garche, H.Döring and K.Wiesener, J. Power Sources, **33** (1991),p. 213

(26) E.Voss, J. Power Sources, **24** (1988), p. 171

(27) A.C.Simon, S.M.Caulder, P.J.Gurlusky and J.R.Pierson, Electrochimica Acta, **19** (1974), p. 739

(28) S.Gust, E.Hämeenoja, J.Ahl, T.Laitinen, A.Savonen and G.Sundholm, J. Power Sources **30** (1990), p. 185

(29) A.Winsel and Coworkers *in* BMFT-report BMFT-FB-T 83-225, pp.57 (in German)

(30) S.B.Hall, G.A.Wright and I.G.Mawston, J.Power Sources, **30** (1990, p.193

(31) W.Borger and U.Hullmeine, Proc. Symp. 'Advances in Lead Acid Batteries', The Electrochemical Society, Pennington, N.J., Proc.Vol. 84-14 (1984, p.302

(32) VARTA (Editor), 'Blei-Akkumulatoren', VDI Verlag, Düsseldorf, 1986, p.87

(33) D.Pavlov, Journal Power Sources, **53** (1995), p. 9

(34) J.O.Wirtz, Batteries International, Issue 3, April 1990, p. 16

(35) D.Berndt and H.Franke, Proc. 9th Intelec Meeting, Stockholm, 1987, p. 115

(36) H.Tuphorn 'Progress in Batteries & Solar Cells', Vol.6, (1987), p.145

(37) H.Tuphorn, Proc. Third Intern. Lead-Acid Battery Seminar, Orlando, 1989, ILZRO Res. Organisation Inc., 1989, p.191

(38) D.W.H.Lambert, Batteries International, Issue 13, October 1992, p.32
Detailed description in P.Reasbeck and J.G.Smith 'Batteries for Automotive Use', © Research Studies Press Ltd., Taunton, 1997, Section 3.2

(39) B.D.McNicol and D.A.J.Rand (Edit.), 'Studies in Electrical and Electronic Engeneering, Vol.11, Power Sources for Electric Vehicles', Elsevier Science Publishing Co., Amsterdam, 1984, pp.249

E.M.Valeriote, 'Expanded Grids for Lead-Acid Batteries', Proc. 3rd Intern. Lead-Acid Battery Seminar, Orlando, 1989, ILZRO Research Organization Inc., 1989, pp. 17 - 64

Cominco Ltd., Batteries International, Issue 15, April 1993, p. 49

(40) M.D. Achtermann and M.E.Greenle, J. Power Sources, **33** (1991), p.87

(41) C.Knowles, Batteries International, Issue 3, April 1990, p.12

(42) J.R.Pierson and C.E.Weinlein, *in* 'Power Sources 9', Proc. 13th Intern. Power Sources Symp., Brighton, 1982, Edited by J. Thomson, Academic Press, London, 1983, pp. 52, 53

(43) H.Laig Hörstebrock *in* BMFT report T 80-151 (VARTA, 1980), p.63 (in German)

(44) US Patent 4509-581-A, WIRTZ Manufacturing Co. Inc., Port Huron, 27.7.83

(45) J.O.Wirtz, Batteries International, Issue 26, January 1996, p. 56

(46) L.E.Gardiner, Batteries International, Issue 13, October 1992, p. 40

Mac Engineering, and Equipment Co., Batteries International, Issue 22, Jan. 1995, p. 44

(47) D.Prengamen, Proc. 3rd Intern. Lead-Acid Battery Seminar, Orlando, 1989, ILZRO Res. Organization Inc., p.1

(48) E.Y.Weissman *in* 'Batteries', edited by K.Kordesch, Marcel Dekker, New York, 1977, p.143, Fig.71

(49) B.Szakacs, Batteries International, Issue 24, July 1995, p. 96

(50) M. Shimpo and H.Nakashima, Proc. 4th Intern. Lead-Acid Battery Seminar, San Francisco, Ca, 1990, p.61

(51) G.C.Zguris, Proc. 18th Intelec Meeting, Boston 1996, © IEEE Communications Soc., p.344

(52) M.J.Weighall, 4th Intern. Lead-Acid Battery Seminar, San Francisco, 1990, ILZRO Res. Organization Inc., p.81

(53) B.D.McNicol and D.A.J.Rand (Edit.), 'Studies in Electrical and Electronic Engeneering, Vol.11, Power Sources for Electric Vehicles', Elsevier Science Publishing Co., Amsterdam, 1984, p. 366, Table 25

(54) H.J.Saechtling, 'Kunststoff Taschenbuch', 7th Edition, Hanser Verlag, München, 1989, p 553 and 554 (in German)

(55) H.Krakowski *in* 'Battery Technology Handbook', edited by H.A.Kiehne, Marcel Dekker, Inc., New York (© 1989 by Expert-Verlag, Eningen/Germany), p.171, Fig.7.3

(56) K.Eberts, ibid, p.347

(57) J.Szymborski and M.L.Eggers, Proc. Intelec Meeting, Washington, DC, 1982, © IEEE Publishing Inc., p.410

(58) 'Ventseal' published by Busak + Shamban GmbH & Co, Handwerkstraße 5 - 7, D-70565 Stuttgart

(59) Batteries International, Issue 21, October 1994, p. 84

(60) R.Marx and D.Hardy, Batteries International, Issue 13, October 1992, p.68

(61) E.A.Busdieker and K.W.Maurer, Batteries International, Issue 16, July 1993, p. 56

(62) A.I.Rusin, Batteries International, Issue 13, October 1992, p.60

(63) D.Pavlov and G.Papazov, J. Appl. Electrochem., *6* (1976), p. 339

(64) D.Pavlov and N.Kapkov, J. Electrochem. Soc., **137** (1990), p.16

(65) R.V. Biagetti and M.C.Weeks, Bell System Technical Journal, **49** (1970), p.1305

(66) 'Processing Technology for Lead Acid Battery Paste', Maschinenfabrik Gustav Eirich, Postfach 1160, D-74732 Hardheim
H.J.Vogel, Journal Power Sources, **48** (1994), p. 71

(67) H.Tamura, 'Progress in Batteries and Solar Cell', Vol.7, (1988), p.205

(68) L.T.Lam and D.A.J.Rand, Batteries International, Issue 13, October 1992, p.52

(69) D.Pavlov and N.Kapkov, J. Electrochem. Soc.,**137** (1990), p.21

(70) G.Clerici, J. Power Sources, **33** (1991), p.67, Fig.11

(71) R.Kiessling, 'Lead Acid Battery Formation Techniques', © 1992, Digatron/Firing circuits, Digatron, Industrie Elektronik GmbH, Tempelhofer Str. 12, D-5100 Aachen, Germany Firing Circuits Inc, Box 2007, Norwalk, CT 06852, USA

(72) D.Johnson, Batteries International, Issue 23, April 1995, p. 44

(73) D.Berndt, H.Franke and E Meissner, VARTA special report 3/1990e, VARTA, P.O. Box 210 540, D-3000 Hannover

(74) W.Lüdecke, Batteries International, October 1990, p.60
F.X.Mittermaier, Batteries International, Issue 28, July 1996, p. 43

(75) G.Pösch, J. Power Sources, **33** (1991), p.127

(76) K.Eberts, Der Elektriker **15** (1976), pp.293-296 and 319-327 (in German)

(77) AKKUMULATORENFABRIK SONNENSCHEIN GmbH, D-W-6470 Büdingen, Germany

(78) K.Eberts *in* 'Battery Technology Handbook', edited by H.A.Kiehne, Marcel Dekker, Inc., New York 1989, p.340
(identical with K.Eberts *in* 'Portable Batteries', edited by H.A.Kiehne,© Expert-Verlag, Ehningen bei Böblingen 1989, p.39)

(79) Shin Kobe, 'Technical Handbook for Sealed Lead-Acid Batteries', p.3, Shin Kobe Electric Machinery Co.Ltd., P.O.Box 218, Mitsui Bldg., Shinjuku-Ku Tokyo, 163 Japan

(80) YUASA (Editor), 'NP series maintenance-free rechargeable battery Application Manual', p.2, YUASA BATTERY Co. Ltd., 12-11, Higashi Shimbashi 2-Chome, Minato -Ku, Tokyo, 105, Japan

(81) J.A.Carcone, Proc. Intelec Meeting, Washington, D.C., 1982, © IEEE Publishing Inc., p. 416

See also: General Electric Company (Editor), 'The Sealed Lead Acid Battery Handbook', General Electric Company, Gainesville, Florida, U.S.A., 1979, p. 5

(82) 'Prospects of Gates Energy Products', Gates Battery Application Manual, Gates Energy Products Inc., 1050 S.Broadway, Denver, Colorado 80217, U.S.A.

Description of the monocell: M.R.Harrison, Proc. Intelec Meeting, Washington, D.C., 1982, © IEEE Publishing Inc., p.424

(83) Batteries International, Issue 19, April 1994, p. 36

(84) Y.Kobayashi, Y.Sakata and S.Fukuda, 'Progress in Batteries and Solar Cells', Vol.6 (1987), p.133

H.Tamara, 'Progress in Batteries and Solar Cells, Vol.7' (1988), p. 205

(85) C.W.Seitz, 'Advances in Portable Rechargeable Batteries', Battery Council International, 102nd Convention, May 1990

(86) Portable Energy Products Inc., Batteries International, January 1991, p.28

A.Komaki and T.Hayakawa, Proc. 13th Intelec Meeting, Kyoto, 1991, © IEEE Communications Soc., p.37

T.Horie, M.Takahashi, H.Hirayama and T.Ogata, ibid., p.45

(87) R.New and B.A.Wittey, Proc 4th Intelec Meeting, Washington, D.C., 1982, © IEEE Publishing Inc., p. 270

A.I.Harrison, N.E.Bagshaw and E.Thompson, Proc 4th Intelec Meeting, Washington, D.C., 1982, © IEEE Publishing Inc., p. 291

(88) K.Eberts, Proc. Intelec Meeting, New Orleans 1984, © IEEE Publishing Inc., p.45

(89) G.M.Cook and W.C.Spindler, J. Power Sources, **33** (1991) p.145

(90) D.Berndt, H.Franke and E.Meissner, VARTA special report 3/1990e. Published by VARTA, P.O.Box 210 540, D-3000 Hannover 21

(91) CHLORIDE Powersafe VH Range

(92) T.Matsushima, M.Ichimura and T.Take, Proc. 9th Intelec Meeting, Stockholm, 1984, © IEEE Publishing Inc., p. 132

S.Sasabe, M.Sasaki, M.Kishomoto, Y. Kasai and K.Fuchida *in* 'Lead Battery Power for the 90s', Paris, 1988, Paper Nr. 13

(93) H.Tuphorn, J.Power Sources, **31** (1990), p.57, and esp. p. 66

(94) J. Strebe, B.Reichman, B.Mahato and K.R.Bullock, J. Power Sources, **31** (1990), p.43

(95) B.Culpin, K.Peters and N.R.Young *in* J.Thompson (Editor), 'Power Sources 9', Proc. 13th Intern. Power Sources Sypmposium, Brighton, 1982, Academic Press, London, 1983, p.129

(96) M.Tsubota, S.Osumi and.M.Kosai, J. Power Sources, **33** (1991) p.105

(97) T.Isoi, Y.Nakayama, S.Nakao and H.Furukawa, J. Power Sources, **33** (1991), p.117

(98) S.U.Falk and A.J.Salkind,'Alkaline Storage Batteries', John Wiley & Sons, New York, 1969, pp.466

(99) Hawker Energy Products Inc., 617 N Ridgeview Drive, Warrensburg, MO 64093-9301, USA

(100) G.W.Vinal, 'Storage Batteries', John Wiley & Sons, New York, 1951, p.430

(101) D.G.Vutetakis, Proc. 34th Intern. Power Sources Symp., 1990, © IEEE Inc., East 47th Street, New York, p.61

D.G.Vutetakis, Batteries International, Issue 28, July 1996, p. 54

(102) N.Bagshaw, J. Power Sources, **31** (1990) p.23, and esp. p. 28

11. Nickel/Cadmium Battery Technology

The technology for sealed nickel/cadmium batteries is basically the same as for the vented design. Most of the components are used in both types, and the basic features of the nickel/cadmium system described in Chapter 5 apply to sealed and to vented nickel/cadmium batteries. An advantage of the nickel/cadmium system is that the electrolyte, the KOH, is only insignifcantly involved in the electrode reaction, as has been mentioned in Section 5.1.1 (the same applies to nickel/metal hydride batteries). Thus it is not necessary to provide a certain space between the electrodes to place a sufficient amount of electrolyte. Instead, the spacing between the plates can be reduced as far as possible and one has only to consider insulation between the electrodes. A small distance between the plates means low resistance of the electrolyte. This is one of the reasons that make nickel/cadmium batteries suited for extremely high loads. Another advantage of nickel/cadmium batteries is the fact that they can be sealed and no gas escapes from the battery, since the equilibrium potential of the cadmium electrode is positive compared to that of hydrogen evolution.

On account of the various features, the nickel/cadmium system allows the design of small sealed batteries with excellent performance data that are in widespread use for the power supply of countless cordless devices, like telephones. In contrast to lead-acid batteries, here the sealed version is mostly used, while vented nickel/cadmium batteries are produced on a comparatively small scale. In the field of vented batteries, competition with lead-acid batteries is more difficult on account of the higher price. Nevertheless, vented nickel/cadmium batteries are an important battery system. Besides their capability at extremely high loads, they are superior in special applications, e.g. at low temperatures, since their electrolyte does not freeze, or at continuously high temperatures when the lead-acid battery suffers serious corrosion problems.

11.1 The Active Material

The active material in the positive electrode is nickel hydroxide,

in the charged state as NiOOH (*nickel (III) hydroxide* with three-valent nickel ions (Ni^{3+})),

and in the discharged state as $Ni(OH)_2$ (*nickel (II) hydroxide* with bivalent nickel ions (Ni^{2+})).

In the negative electrode, cadmium (Cd) constitutes the charged state of the active material and cadmium hydroxide ($Cd(OH)_2$) the discharged state.

Various technologies are applied to produce the active material, largely depending on the electrode type. For a great number of electrode types, the active material is processed separately and subsequently shaped to an electrode or filled into a substrate.

- For button cells (cf. Section 11.5), electrode tablets are pressed from powdered active material.
- For pocket plates, that are used for vented and 'semi-sealed' batteries (cf. Section 11.5.4), powder or pre-pressed small blocks are filled into perforated pockets of nickel-plated steel.
- For 'plastic-bonded electrodes', the powdered active material is mixed with binders and subsequently rolled or pressed to form the electrode.
- For electrodes based on fibre or foam substrates (Section 11.2.5), the active material is filled as a paste or a slurry into the substrate.

Another way to form the active material is by precipitation from a solution within the substrate for sintered electrodes and to some extent also for electrodes based on fibre or foam substrates, either by chemical precipitation or by electrochemical precipitation (cf. Section 11.2.4.1).

As a further possibility, the active material of the negative electrode, the cadmium, can be deposited electrochemically as a porous body on the surface of a nickel-plated steel ribbon. In the resulting electro-deposited cadmium electrode, the active material itself forms the porous structure (cf. Section 11.2.8).

11.1.1 The Positive Active Material ($NiOOH/Ni(OH)_2$)

Chemical precipitation is used, in general, when the active material is processed separately. (*With sintered electrode this precipitation occurs within the pores of the substrate, cf. Section 11.2.4.1.*) Solutions of soluble nickel salts, like nickel sulphate ($NiSO_4$), are the parent materials. Addition of sodium hydroxide (NaOH) causes the precipitation of finely dispersed nickel hydroxide $Ni(OH)_2$. In usual conditions, particle sizes of 10 to 100 μm are the result. The material is fairly amorphous and the crystalline appearance largely depends on the processing parameters.

For high-capacity electrodes, special precipitation processes produce nickel hydroxide as balls with diameters of about 10 μm (Fig. 11.1). The tapping density of this material is quite high and leads to a correspondingly high specific capacity for the volume of the electrode concerned.

Fig. 11.1 SEM photographs of nickel hydroxides.
(a) Conventional $Ni(OH)_2$ powder. (b) High-density $Ni(OH)_2$ powder. (From ([1]).)

Cobalt is added to the active material in the manufacture of most nickel electrodes ([2]), improving the performance, especially at elevated temperatures. At a fairly high cobalt content (20 wt.%), stabilisation of the α/γ-couple of nickel hydroxide (Eq. (5.26), p.189) was observed ([3]). In the usual practice, cobalt and cadmium are added to the positive active material in the range of 1 to 5 wt.%.

Cadmium and zinc addition reduces the swelling of the active material when charged. Measures against swelling are especially important for the high density material as shown in Fig. 11.1. Presumably this swelling is caused by the formation of γ-NiOOH together with a drastic change in crystal density (cf. ([1])). According to these experiments, cadmium addition suppresses the formation of γ-NiOOH, but the mechanism is not yet clear.

According to different authors ([4]), cadmium also increases the oxygen overvoltage which simultaneously favours better charge acceptance.

Iron has fairly detrimental effects and leads to capacity decay. The unfavourable effect of iron can to some extent be compensated by cobalt addition ([5]). Special structures of nickel/iron hydroxides which contain layers with nitrate (NO_3^-), sulphate (SO_4^{2-}) or phosphate (PO_4^{3-}) are said to withstand 'iron poisoning' ([6]).

Conducting aids have to be added when the material is used in pressed or rolled form on account of its poor conductivity, especially in the discharged state ($Ni(OH)_2$). Formerly, graphite was mainly used in pressed active material and is still used in pocket plates (cf. the following Section). For tubular plates in vented batteries specially processed nickel flakes were used (cf. ([7]), p.96). Today, in button cells metallic nickel as powder is added for this purpose.

11.1.1.1 Graphite Oxidation

Graphite as a conducting aid has the disadvantage that it is gradually oxidised to carbon dioxide (CO_2), which is absorbed by the electrolyte and forms carbonate ions CO_3^{2-} according to

$$\boxed{CO_2 + 2 \cdot OH^- \Rightarrow CO_3^{2-} + H_2O} \qquad (11.1)$$

By this reaction, OH^- ions are replaced by CO_3^{2-} ions, which reduces the conductivity of the electrolyte, because the conductivity of CO_3^{2-} solutions is reduced to about 30% of that of OH^- solutions of comparable concentration, as shown by Table 11.1. (*This effect is important for vented nickel/cadmium batteries with graphite in the positive electrodes, since it makes it necessary to exchange the electrolyte after a certain service life. Otherwise, the increased resistance of the electrolyte reduces the available capacity on account of a too-high voltage drop. For detailed discussion see ([7]), Fig. 8.19, p.597 and p.623 f.*)

Temp. / °C	0	18	25
OH^-	105	174	197
½ CO_3^{2-}	36	60.5	74

Table 11.1 Equivalent conductivities of OH^- and CO_3^{2-} ions in $\Omega^{-1}\cdot cm^2\cdot equiv.^{-1}$ ([8]). Maximum values at infinite dilution.

11.1.2 The Negative Active Material ($Cd/Cd(OH)_2$)

Cadmium oxide (CdO) or cadmium hydroxide ($Cd(OH)_2$) usually forms the basis for the negative active material. The formerly widespread production of mixed Cd/Fe material (([7]), p.57) by co-precipitation of cadmium and iron is hardly applied any longer.

Addition of some wt.% of nickel is in general use to improve conductivity, as far as pressed or rolled active material is concerned. Graphite is also applied to improve the conductivity. The above-mentioned problem of carbonate formation does not apply for the negative electrode, because graphite is not oxidised at the potential of this electrode.

Fig. 11.2 shows the fine crystals that form the surface of the cadmium electrode. On account of the solution mechanism (cf. Fig. 5.2, p.177), this electrode is prone to a gradual growth of crystals when discharged and charged repeatedly. This growth of crystals means a reduction of active surface and causes a decay of capacity. As in the lead electrode (cf. Section 10.1.2), additives, called expanders, are used to reduce crystal growth in the active material. Such expanders can be organic chemical compounds like polyvinyl alcohol (PVA), methyl cellulose (MC) or carboxyl methyl cellulose (CMC). But inorganic chemical substances like iron oxide (Fe_2O_3), nickel hydroxide ($Ni(OH)_2$) and sodium silicate (compounds of Na_2O and SiO_2, like Na_2SiO_3 or Na_4SiO_4) are also used ([9]). The addition of nickel has already been mentioned as a conducting aid; it also acts as an expander. Most manufacturers no longer use iron in view of its detrimental effects on the positive electrode, mentioned above.

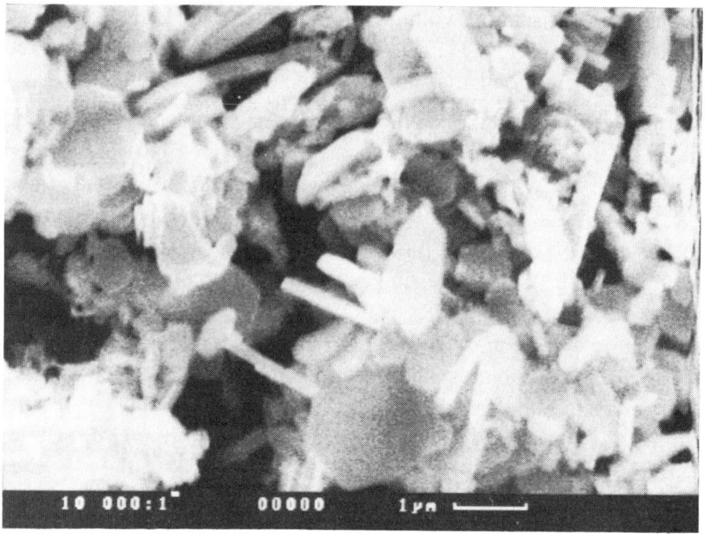

Fig. 11.2 SEM photograph of cadmium hydroxide $Cd(OH)_2$ in a discharged electrode.

11.1.3 The Memory Effect

The 'Memory effect' is a vague description of a temporary loss of capacity after:

1. Subjecting the battery to a larger number of unvarying partial discharge-charge cycles.
2. Extended storage periods without recharge or periods with insufficient charge.
3. Prolonged constant-current charging (float charging), e.g. in emergency-lighting applications.

The 'genuine' memory effect, as far as it can be defined, concerns the first case, namely the experience that a nickel/cadmium battery loses the ability to deliver full capacity if it is utilised only partially for a prolonged period of time. But extended storage periods as well as prolonged float-charging periods according to the Items 2 and 3 may have similar effects. In practice, each kind of temporary capacity loss is often called 'memory effect'.

The temporary capacity loss can be caused by two factors connected with the cadmium electrode:

- Cadmium crystals of growing size are observed when the battery is recharged at a low current rate ([10]). This crystal growth can be enhanced when the battery is subjected to repeated partial discharge and recharge, and the recharge occurs at a moderate current rate. But increased crystal size means reduced surface area of the active material which causes an additional voltage drop during discharge. So, the voltage level of the whole discharge curve is reduced and the end-of-discharge voltage is reached earlier, which results in a reduced capacity (cf. Fig. 3.1, p.71). Complete discharge and subsequent recharge with not too-low current densities eliminate the effect.

- A further reason for the memory effect is connected with a certain nickel content of the cadmium in the negative electrode. This is the memory effect mostly observed with sintered electrodes. The nickel may have been added to the active material deliberately as a conducting aid and also as an expander. But the presence of nickel in the active material can also be due to corrosion reactions of the substrate during precipitation of the cadmium hydroxide into the sintered substrate (cf. Section 11.2.4.1). Nickel and cadmium form intermetallic compounds (Hume-Rothery phases) like Ni_2Cd_5 or Ni_5Cd_{21}. The share of these compounds can grow during the repeated partial discharges and recharges in the unaffected part of the active material ([11]). If subsequently a complete discharge is carried out, the cadmium bound in the intermetallic phases or at least a part of it has to be used as well. But the intermetallic phases are discharged at a more positive potential than the cadmium electrode and this causes a reduction of about 150 mV of the battery voltage during discharge. Fig. 11.3 shows examples of such a voltage drop, which is also called the 'second discharge step'.

This 'memory effect' diminishes when the electrode is deeply discharged and recharged again, because the intermetallic compounds are destroyed. But they are formed again in similar operating conditions of the battery. Modern manufacturing methods of sintered electrodes aim to prevent solution of nickel from the substrate to avoid nickel contamination of the active cadmium material.

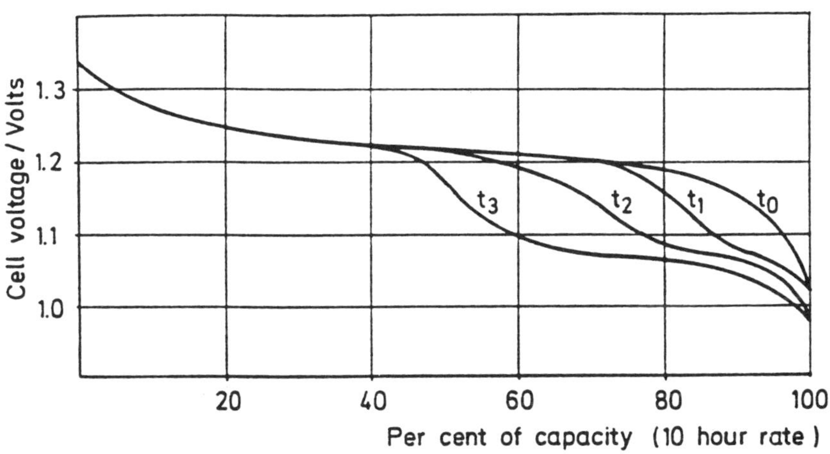

Fig. 11.3 Discharge curves showing the reduction of the discharge voltage after prolonged charging periods with low current density ('trickle charging). (From ([12]).)

The memory effect mainly is a problem of the cadmium electrode, but the positive electrode can also contribute a certain capacity reduction under the conditions described by the Items 1 to 3:

- 'Deactivation' of the active material of the positive electrode can occur, which has been observed during artificial ageing ([13]). This means that the charged material converts gradually to a lower-energy form that is discharged at a lower voltage (cf. Section 5.2.2.2).
- The reduced capacity can also be due to gradual discharge of the positive electrode, especially when the batteries are operated at high temperatures.

In the latter case, it is not an additional voltage drop that reduces the discharge voltage so that the end-of-discharge voltage is reached earlier. Rather, the discharge period is actually reduced on account of the partially discharged positive electrode. This effect can be avoided by suitable charging methods (cf. Section 13.2).

11.2 Types of Electrodes

Compared to the design of lead-acid batteries, much more variety is possible in the nickel/cadmium battery, since the corrosion of nickel is usually negligible, as described in Section 5.6. Thus, thin layers or foamlike structures of nickel can be used as stable substrates, since they are not destroyed by corrosion. The large surface area of such substrates has no relevance to hydrogen-evolution rates, despite the comparatively low hydrogen overvoltage of nickel, since the equilibrium potential of the cadmium electrode is 20 mV above the hydrogen equilibrium potential. Thus, hydrogen evolution is only possible when the negative electrode is, correspondingly, polarised.

11.2.1 Tubular Electrodes

Tubular electrodes (([7]), p.64 f.) are a design that goes back to Edison. The active material, mixed with nickel flakes as a conducting aid, is encapsulated by a perforated tube of nickel-plated steel. Today, tubular electrodes are hardly used anymore since their manufacture is expensive and the utilisation of the active material is poor.

11.2.2 Pocket Electrodes

Pocket electrodes are in widespread use in vented nickel/cadmium batteries (cf. Fig. 11.29, p.394). They consist of strips of perforated nickel-plated iron that form 'pockets', filled with the active material (([7]), p. 59f.). A number of such strips are assembled to form a flat plate. Both electrodes usually contain graphite as a conducting aid. In 'genuine' sealed nickel/cadmium batteries, pocket plates are not used, but they are in so called 'ultra-low maintenance batteries', described in Section 11.5.4.

11.2.3 Pressed Electrodes

Pressed electrodes are mainly used in button cells (Fig. 11.13, p.381). This electrode construction is also called 'mass electrode'. Thin tablets of the active material are pressed and enclosed in a nickel-wire gauze that simultaneously provides mechanical support and acts as a current collector (in Fig. 11.13 sections of the wire gauze are removed from both electrodes to show the active material). The conductivity of the active material itself is improved by the addition of nickel powder (carbonyl nickel).

11.2.4 Sintered Electrodes

The widespread use of the sintered plate arises from the high utilisation of the active material and the suitability of this electrode type for high-rate discharges. Both features are caused by the close contact between the active material and the large surface of the current-conducting substrate and also by the fact that the surface of the electrode is not partly covered up by the conducting material, as in tubular and pocket plates, where the active material is enclosed by the perforated conducting metal.

Sintered plates can be manufactured with a thickness of up to several millimetres, but nowadays fairly thin plates are generally used with a thickness in the range of 0.5 to 1 mm. The core of the plates is a perforated, nickel or nickel-plated iron strip (0.05 to 0.1 mm thick) coated on both sides with the sintered nickel.

To produce single sintered plates, a mould is half-filled with nickel powder and a perforated nickel screen added so that it later forms the central current collector and simultaneously provides mechanical support. Then the mould is filled completely with the powder and heated to about 1000 °C in a reducing atmosphere to prevent oxidation of the nickel powder, and the powder is sintered.

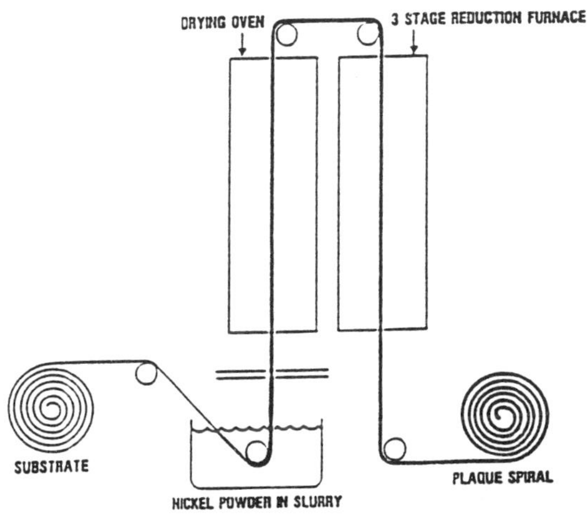

Fig. 11.4 Continuous sintering process. (From ([14]).)

Fig. 11.4 illustrates the continuous sintering process, which has the same processing steps but in a continuous sequence. By far the majority of sintered plates are manufactured today by this process.

A coil of perforated metal strip is seen on the left of Fig. 11.4. This strip runs through a vessel filled with a slurry of carbonyl-nickel powder and methyl cellulose. Carbonyl nickel is an extremely fine powder produced by thermal decomposition of gaseous nickel tetracarbonyl $Ni(CO)_4$. The coated strip passes a wiper that controls the thickness. After drying in the oven on the left, the coated strip passes through a reduction furnace (about 1000 °C) where the methyl cellulose is destroyed and the nickel powder becomes sintered. The result is a spongy sintered body with fine pores and a porosity of about 80 per cent by volume.

11.2.4.1 Filling the Sintered Substrate with Active Material

The process of filling the sintered plates with active material is called 'impregnation'. Different methods are used; in all of them, the active material is deposited from a dissolved state within the pore system of the substrate. Otherwise, it would not be possible to fill the fine pores. Furthermore, a close contact between the substrate and the active material is achieved by such a precipitation.

Chemical impregnation is used by most manufacturers. Fig. 11.5 illustrates the basic elements of this method, which has many variations.

The ribbons of sintered material are coiled up, and the coils seated on horizontal grids. A pile of seven such grids loaded with the sintered material is shown on the left of Fig. 11.5.

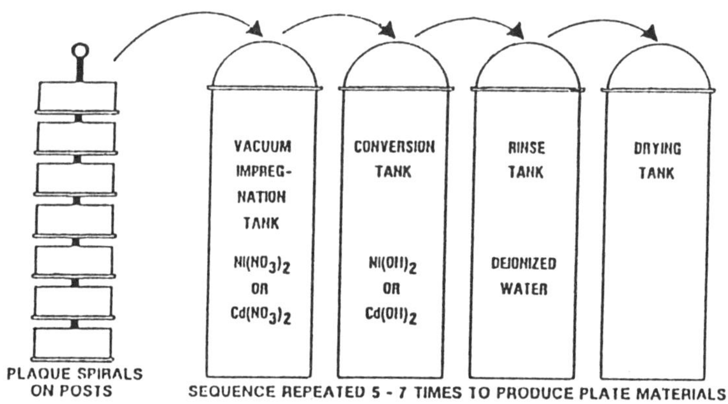

Fig. 11.5 The impregnation of sintered substrates by precipitation of the active material. (From ([14]).)

The pile with these coils is then inserted into a tank, which is evacuated and then filled with:

- nickel nitrate solution ($Ni(NO_3)_2$) when positive active material is impregnated or

- cadmium nitrate solution ($Cd(NO_3)_2$) when negative electrodes are processed.

Temperatures between 50 and 80 °C are applied, depending on the kind of solution used. On account of the evacuation, the sintered material is soaked completely with the nitrate solution, i.e. $Ni(NO_3)_2$ or $Cd(NO_3)_2$.

The tank with the pile of coils is then emptied, and the surplus solution, not retained within the capillary system, drips off.

After drying with steam, the coils are transferred to the next tank which is filled with sodium hydroxide (NaOH). The precipitation of nickel hydroxide

$$\boxed{Ni(NO_3)_2 + 2{\cdot}NaOH \Rightarrow Ni(OH)_2 + 2{\cdot}NaNO_3} \qquad (11.2)$$

occurs in positive-electrode impregnation.

When cadmium hydroxide had been impregnated, the reaction is:

$$\boxed{Cd(NO_3)_2 + 2{\cdot}NaOH \Rightarrow Cd(OH)_2 + 2{\cdot}NaNO_3} \qquad (11.3)$$

When this reaction is completed after about a quarter of an hour (depending on the thickness of the substrate), the coils are transferred to the next tank, where they are rinsed thoroughly with water. Then they are dried.

The described sequence is called the 'impregnation cycle', and this cycle is repeated four to eight times (five to seven times in Fig. 11.5), depending on the type of electrode and the amount of active material required in the plate. Additives like cobalt or cadmium are added to one or several of the impregnation steps. Sintered plaques produced as single plates can be impregnated similarly.

As mentioned above, many variations of this process are applied; e.g. the material is not transferred to different tanks but the solutions are exchanged while the coils remain in the same vessel.

Continuous impregnation processes have been tested repeatedly. As in Fig. 11.4, the ribbon of sintered substrate runs through the impregnation bath. For the positive electrode, a continuous process has been proposed with repeated short impregnation and drying periods, using molten nickel hydroxide ($Ni(OH)_2{\cdot}6{\cdot}H_2O$) for the impregnation, and subsequent precipitation with KOH ([15]).

Electrochemical impregnation is another possibility for filling the sintered substrate with active material. It also is based on nitrate solutions, nickel nitrate $Ni(NO_3)_2$ or cadmium nitrate $Cd(NO_3)_2$, but the precipitation is achieved by a cathodic current that changes the pH value of the solution ([16]).

When the sintered substrate, filled with the nickel or cadmium nitrate solution, is polarised by a cathodic current, the electrochemical reactions listed in Table 11.2 may occur. Comparison of the potentials shows that a cathodic current that polarises the electrode to more negative values, primarily only causes reduction of the nitrate ions (NO_3^-), because the equilibrium potential of this reaction (top line in Table 11.2) is much more positive compared to the other competing reactions. A correspondingly high polarisation of the electrode that is sufficient to evolve hydrogen can be expected with a high current. The precipitation of nickel or cadmium affords even higher polarisation values.

Reaction	Equilibr. Potential E_o in Volts
$2 \cdot NO_3^- + 8 \cdot H^+ + 6 \cdot e^- \Rightarrow 2 \cdot NO + 4 \cdot H_2O$	+ 0.96
$2 \cdot H^+ + 2 \cdot e^- \Rightarrow H_2$	0
$Ni^{2+} + 2 \cdot e^- \Rightarrow Ni$	- 0.250
$Co^{2+} + 2 \cdot e^- \Rightarrow Co$	- 0.277
$Cd^{2+} + 2 \cdot e^- \Rightarrow Cd$	- 0.403

Table 11.2 Reduction reactions in cadmium or nickel nitrate solutions.

The electrochemical reduction of the nitrate ion causes a shift of the pH value on account of the H^+ ion consumption, which leaves a surplus of OH^- ions near to the electrode surface. This shift to a more alkaline electrolyte within the pores of the electrode is intensified because the reduction of the nitrogen compounds is continued until the reduction state NH_4^+ is reached. The complete reduction reaction can be written as:

$$NO_3^- + 10 \cdot H^+ + 8 \cdot e^- \Rightarrow NH_4^+ + 3 \cdot H_2O \qquad (11.4)$$

That this complete reduction is actually reached can be confirmed by the amount of NH_4^+ detected ([17]). This reduction reaction consumes a great number of H^+ ions and increases the change of the pH value of the electrolyte in the pore system from an acidic to an alkaline solution. So the solubility limits of cadmium hydroxide ($Cd(OH)_2$) or nickel hydroxide ($Ni(OH)_2$) are exceeded and the hydroxide is precipitated. *Both are not precipitated according to the reactions in Table 11.2, but by the change of their solubility due to the alkaline pH values.*

The bulk of the electrolyte is not affected and supplies the vicinity of the electrode surface with protons (H^+ ions) as well as cadmium (Cd^{2+}) or nickel ions (Ni^{2+}) by diffusion. Thus, the electrochemical impregnation process, which is often called the '**Kandler process**', is influenced by a number of diffusion processes, and precise conditions must be fulfilled ([18]).

Co-impregnation of nickel hydroxide with the hydroxides of cobalt and cadmium is possible when mixed solutions are applied, on account of the fairly similar solubilities of these substances. Such a co-impregnation of the additives cobalt and cadmium is, e.g., used for fibre plates to improve performance at high temperatures ([19]).

Electrochemical impregnation has been revised and improved repeatedly ([20]). It has not gained the importance of chemical impregnation, but occasionally it is used for electrodes based on fibre supports (cf. Section 11.2.5).

A third way to fill the positive active material into the sintered substrate is **thermal decomposition**, mentioned in ((7), p.129). The substrate is soaked with nickel nitrate ($(Ni(NO_3)_2 \cdot 6{\cdot}H_2O)$, which then is subjected to thermal decomposition at 180 to 250 °C. Quite high capacity values are obtained with a single impregnation cycle, which indicates that parts of the sintered substrate may have been converted into nickel hydroxide. More recent variations of this method are described in (21). The negative electrode, e.g., is impregnated with cadmium acetate, which is decomposed at 400 °C.

11.2.4.2 Formation of the Electrodes - Cell Balance

The impregnation process leads to the discharged state for the positive and negative electrodes. A charging treatment is required for several reasons:

- The active material contains impurities, e.g. carbonate or nitrate ions, that should be removed before the cell is assembled. Otherwise, they may cause electrochemical problems, like increased internal resistance as a consequence of carbonate (K_2CO_3) formation or increased self-discharge on account of the nitrate shuttle (cf. Section 5.3).
- Charging after impregnation is also required to remove loosely adhering particles of active material, which increase the danger of short-circuits when the cell is assembled.
- The fact that both electrodes should have reached full capacity when the cell is assembled is a third reason for the preceding charge. If further development of the electrode capacities occurred after cell assembly, it would not be possible to establish the cell balance precisely (cf. Section 8.1, esp. Figs. 8.6 and 8.8).

Different charging processes have been developed by manufacturers. Usually, the electrodes are charged in a surplus of electrolyte. Often such formation schedules include discharge periods.

Electrochemical and chemical discharge processes are used to achieve the state of charge of both electrodes required for the proper cell balance. The state of charge of the positive electrode can also be reduced by the addition of organic substances, which themselves are oxidised, e.g. methyl cellulose. An example of the adjustment of the state of charge of the cadmium electrode by the reduction of H_2O_2 is given at the end of Section 11.2.8.

Precise implementation of the cell balance is important for nickel/cadmium batteries since this balance is maintained during the full service life, at least to a close approximation, as mentioned in Section 8.1.4.

- Hydrogen evolution does not occur in normal operational conditions. Thus, the safety valve never opens during normal operation; oxygen and water vapour do not escape from the cell.
- Corrosion of the current-conducting elements in the positive electrode can practically be neglected.
- Furthermore, gas or water vapour is not lost by permeation on account of the metallic housings of sealed nickel/cadmium batteries.

The unavoidable self-discharge of the negative electrode and the gradual water loss that are characteristic for valve-regulated lead-acid batteries are not relevant for the sealed nickel/cadmium battery. Only small amounts of water can be lost by separator oxidation or oxidation of the substrate. Thus, in the sealed nickel/cadmium battery, the initially established cell balance and electrolyte content remains nearly unchanged during the whole service life.

11.2.5 Fibre Electrodes

Repeated efforts have been made to replace the fairly expensive sintered substrate. The **composite nickel electrode** is one result. Felts of graphitic-fibre material are coated with a layer of nickel of thickness 0.6 to 0.8 μm. These coated fibres are sintered together and then welded to a supporting thin nickel grid (usually of expanded metal). The completed plate is impregnated by the electrochemical process described above ([22]). The advantage of this electrode is that it is lighter than the sintered type.

Fibre or **felted** **support** structures are manufactured by precipitation of a nickel layer onto specially prepared fibres. The organic or carbon fibres (cf. Fig. 11.6) are left underneath the coating or are removed by thermal decomposition ([23]). Fig. 11.6 shows a fibre substrate not yet filled with active material. The porosity of fibre supports is 85 - 90 vol.%.

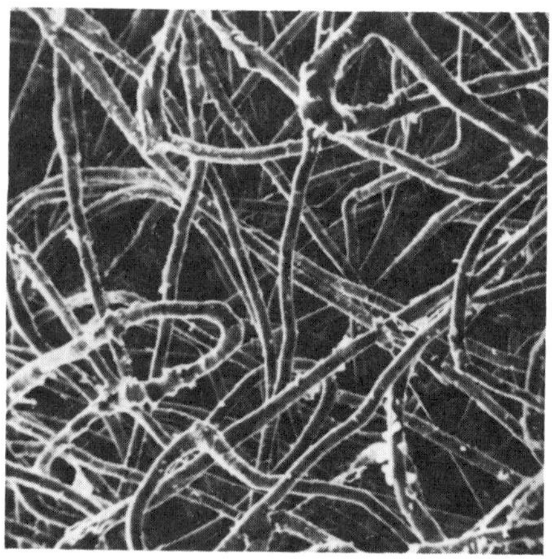

Fig. 11.6 Detail of a fibre electrode support. Nickel fibres at about x100 magnification. (From ([24]).)

Impregnation of the fibre substrate can be by:

- electrochemical impregnation (KANDLER process),

or by

- a pasting process where the fibre is filled with a slurry of separately processed active material.

The advantage of electrodes based on fibre substrates is that they can be manufactured with thicknesses between 0.6 and 10 mm, with approximately constant performance features, because the conducting substrate increases proportionally to electrode thickness. The substrate also retains some flexibility even when the fibres are coated with nickel, and this flexibility helps when the active material swells.

The fibre substrate itself can be used as an auxiliary electrode to increase the oxygen-reduction rate (cf. Section 11.5.3.2).

11.2.6 Foam Electrodes

Foam substrate (spongy nickel) is used for positive electrodes of high specific capacity. The basis is foamed urethane resin, which is nickel plated. After plating the urethane is removed by burning and a nickel skeleton is left with about 95% of void space (porosity), as shown in Fig. 11.7.

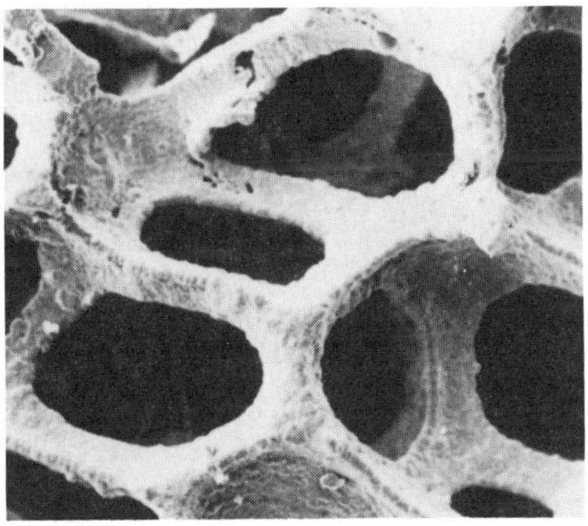

Fig. 11.7 Foamed nickel matrix. (From [25].)

The active material is usually filled as a paste into this framework of nickel. The final thickness of the plate is achieved by pressing after pasting and drying.

11.2.7 Plastic-Bonded Negative Plates

Plastic-bonded electrodes were developed to replace the more expensive sintered electrodes. The active material is mixed with 5 to 10 wt.% of PTFE (polytetrafluorethylene) and pressed or rolled on a substrate, usually a nickel-wire net [26]. They are mainly used for larger batteries in traction applications.

To achieve sufficient porosity of these active materials, graphite has to be added, which forms a fine pore system and simultaneously increases the conductivity [27]. But graphite is oxidised in the positive electrode and so, after a rather short service time (6 months [28]), the carbonate in the electrolyte exceeds tolerable values (cf. Eq. (11.1)). So, plastic-bonded electrodes are applied only as negatives.

11.2.8 Electro-deposited (Negative) Electrodes

The cadmium electrode can also be made by galvanic precipitation on a nickel strip or a nickel-plated steel strip, similar to that used as substrate for the sintering process ([29]). The metal strip runs through a bath containing Cd^{2+} ions and is connected to the negative pole of a rectifier. The positive electrode consists of metallic cadmium (Cd), which is dissolved when the current flows.

The current density has to be adjusted so that the precipitation of cadmium and evolution of hydrogen, which occur simultaneously, form a spongy cadmium layer, which is compressed by rolling to the desired thickness.

Fig. 11.8 shows the dendritic structure at the end of the production process. On account of the large surface area, these electrodes provide high energy densities ([33]).

Fig. 11.8 SEM photograph of an electrodeposited cadmium electrode (charged state). Tadiran Battery Div. ([30]).

The manufacturing process for electro-deposited negative electrodes leads to completely charged negative electrodes that are combined with charged positives when the cell is assembled. To establish the required charge reserve (cf. Fig. 8.6, p.276), the negative electrodes have to be discharged to a definite state of charge. This can be achieved by addition of a certain quantity of hydrogen peroxide (H_2O_2), which oxidises cadmium:

$$\boxed{Cd + H_2O_2 \Rightarrow CdO + H_2O \Rightarrow Cd(OH)_2} \qquad (11.5)$$

and creates the desired amount of discharged active material ($Cd(OH)_2$) ([31]).

11.3 Containers

Containers of nickel-plated or stainless steel are mostly used for sealed nickel/cadmium batteries. As for the nickel-plated substrates in the electrodes, corrosion can be neglected and the low hydrogen overvoltage of nickel has no relevance due to the potential of the cadmium electrode.

Advantages of the metallic container are:

- Mechanical stability to withstand fairly high internal pressures, especially in small cells.
- Weldability providing proper sealing.
- Impermeability to gases or water vapour.

Disadvantages of the metallic container are its price, mainly due to manufacturing costs, and its weight for larger cells. Some further disadvantages are caused by the electrical conductivity of the metallic container. Precautions have to be taken for the safety of the operating crew, and against short-circuits from contact between containers of different cells.

Containers of plastics material, polyamide or polystyrene, are often used for vented nickel/cadmium batteries. Occasionally they are also used for sealed prismatic cells (cf., e.g., Fig. 11.25, p.391).

Some mechanical properties are listed in Table 11.3. Polyamide is superior in mechanical stability at high temperatures. However, its high permeability to water vapour (Table 2.13, p.61) would cause considerable water loss from sealed batteries.

Abbrev. name	Components	Density g/cm^3	Temp. range °C	E - modulus MPa*	Bending strength N/mm^2
SB	Styrene; Butadiene	1.05	<-50 - +70	1800 - 2500	40 - 80
PA	Polyamide	1.14	-30 - +100	1400	50

Table 11.3 Abbreviated name, composition and mechanical properties of plastics used for nickel/cadmium batteries (from manufacturer catalogues and ([32])). Standard values for comparison; technical products mostly represent variations. * 1 MPa = 10^6 N/m^2 = 1 N/mm^2. The definition of E-modulus and bending strength is given in Section 10.5.

11.4 Valve Design

Nickel/cadmium batteries can be sealed completely, as pointed out in Section 8.1. In theory, they should need no valve, but can virtually be sealed. Nevertheless, for safety reasons, sealed nickel/cadmium batteries usually are equipped with valves to prevent explosion from unexpected increase of internal pressure, usually caused by abuse of the battery, e.g. during overcharge with too high a current.

Two kinds of valve are used:

- genuine 'safety vents' destroyed in an emergency (Fig. 11.9),
- resealing valves, which open at a certain overpressure but close again when the internal pressure drops below the critical value (Figs. 11.10 and 11.11).

Fig. 11.9 is a cross section of a safety vent. The sealing element is a metallic membrane sealed within the cell cover, which bulges upwards with increasing internal pressure. A pointed piece of metal is bent down from the outer cell cover, and its point pierces the membrane when its bulge is excessive.

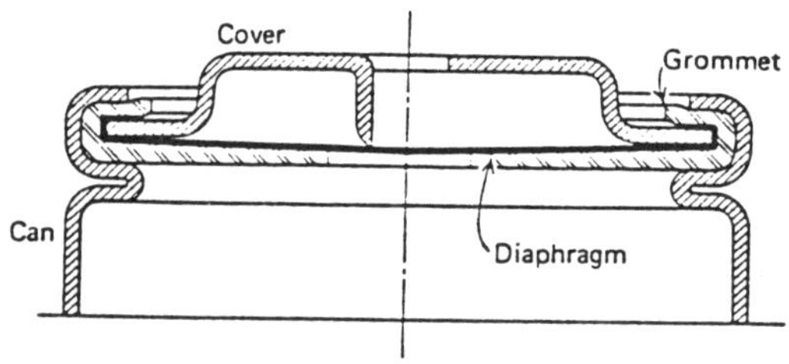

Fig. 11.9 Safety vent in a cylindrical sealed nickel/cadmium battery (([7]) Fig. 2.59, p.212, *reprinted by permission of The Electrochemical Society, Inc .*).

Safety vents as in Fig. 11.9 were used largely during the first decades of the manufacture of sealed nickel/cadmium batteries. Absolutely tight sealing can be achieved with this vent. But as soon as the membrane is pierced, the sealing is completely destroyed. Then the cell operates like a vented one, which means that water loss limits the service life in a short time. Thus, an occasional overcharge might spoil the whole battery.

For this reason safety vents as in Fig. 11.9 have been largely replaced by resealing valves. Today, the safety vent is used only with batteries that have to be sealed absolutely, e.g. for space application or with button cells where a point of fracture in the cup acts as a safety vent (cf. Section 11.5.1).

Two examples of resealing valves are shown in Fig. 11.10. In both cases, a steel spring provides the sealing forces, while the sealing itself is achieved by a pad of rubber, usually a synthetic elastomer.

In the top example, a coil spring presses an elastic disc against a seat. In the bottom example, a flat spring presses a small rubber block against a seat with a hole.

Improvements of elastomers, 'synthetic rubber', enabled the development of simpler but also reliable valves that are used to a large extent today. One example of various similar designs is shown in Fig. 11.11.

The shaped rubber part is pressed on a seat with a hole by a metallic cap, welded on top of the cell cover. This cap is welded only at certain points, and slots are left for the gas to escape between the cell cover and the metallic cap. When the internal pressure exceeds the critical value, the rubber part is compressed so much that a slot is formed between the steel cover and the rubber, and the gas can then escape.

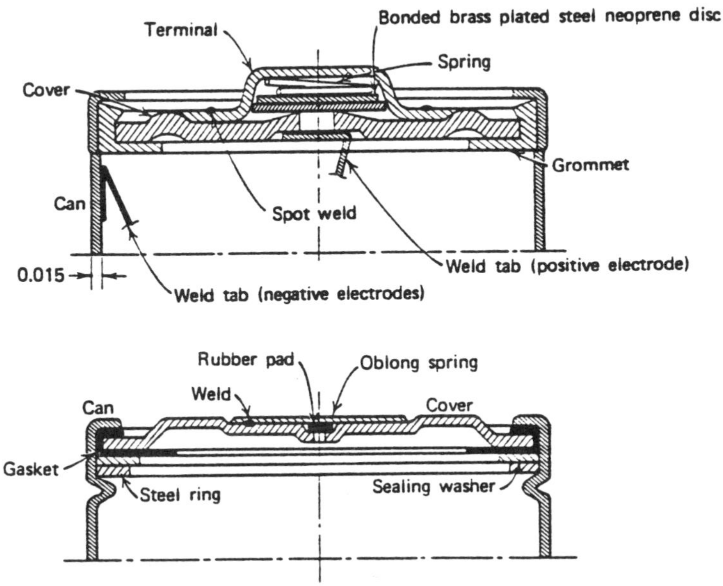

Fig. 11.10 Two resealing valves for sealed nickel/cadmium batteries. In both examples, the sealing force is provided by a steel spring (([7]), Fig. 2.60, p.214, *reprinted by permission of The Electrochemical Society, Inc.*).

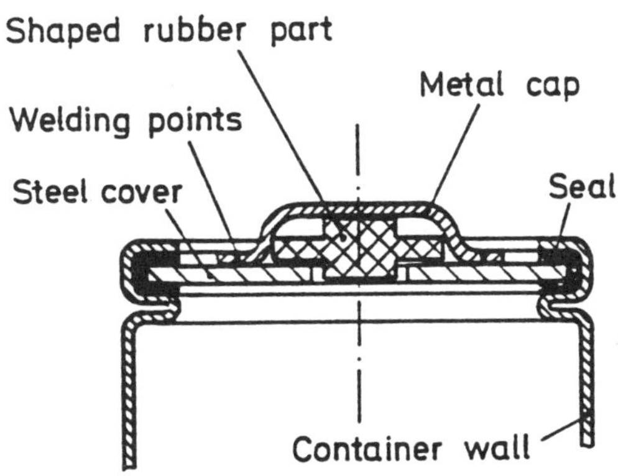

Fig. 11.11 Resealing valve for nickel/cadmium batteries based on a compressed rubber part.

The advantage of the resealing valve is obvious: occasional abuse of the battery causes some water loss; but, if this loss is limited, it will not influence the performance significantly. The battery will operate as before and can reach the usual service time. Hydrogen that has been evolved will even be gradually removed, as mentioned in Section 7.5.3.1.

A disadvantage of the resealing valve is that it cannot be sealed as perfectly as can the safety vent based on a metallic membrane as in Fig. 11.9. However, for usual battery applications the advantages outweigh the disadvantages. Especially in batteries composed of a number of cells, occasional overcharging or reversal of single cells often cannot be excluded. In such cases, the battery as a whole is better protected by resealing valves.

11.5 Battery Design

A wide spectrum of sealed nickel/cadmium batteries is available with capacities between 10 mAh and 100 Ah. As for the valve-regulated lead-acid battery (Chapter 10), just a few examples can be shown to illustrate the basic design principles. Fig. 11.12 demonstrates the variety.

Fig. 11.12 Different sealed nickel/cadmium batteries. Foreground: Button cells. Middle: Stacked button cells and cylindrical cells. Background on the left: monoblock, 6 V, 3 Ah. Background on the right: prismatic cell, 1.2 V, 15 Ah. (From (33).)

Fig. 11.12 shows the three basic cell designs which are used for sealed nickel/cadmium batteries: button, cylindrical and prismatic. The Figure also shows some batteries composed of several cells. Because any number of cells can be used, many variations are offered by different manufacturers. Some general features are outlined below. Detailed information, especially dimensions and performance data, is beyond the scope of this book. The reader is referred to the catalogues and handbooks of battery manufacturers ([34]).

11.5.1 Button Cells

Button cells are the oldest sealed nickel/cadmium batteries. They were introduced onto the market in 1950 and are in general use nowadays. For a time, they were manufactured in sizes upto 3 Ah of nominal capacity. Nowadays, usual sizes are between 4 mAh and 1000 mAh.

Fig. 11.13 shows the construction of a button cell. In this example, one pair of pressed electrodes forms the plate group. Stacks of several parallel electrodes, as in Fig. 11.14, are occasionally used to increase the electrode surface.

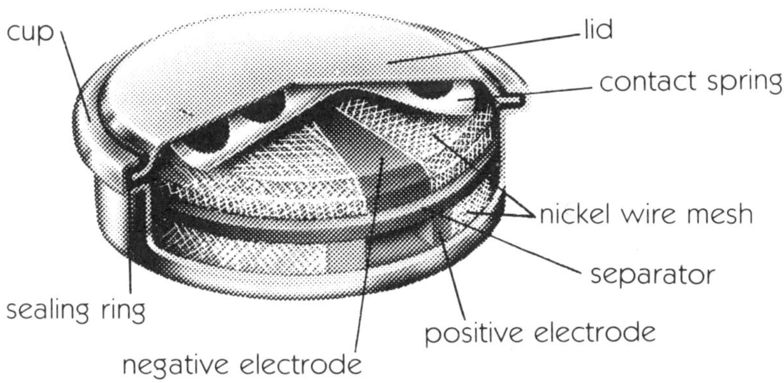

Fig. 11.13 Construction of a button cell (VARTA).

The upper electrode in Fig. 11.13 is the negative one; the positive electrode can be seen at the bottom. The separator is a felt disc between the electrodes. It consists of nonwoven polyamide (trade name Nylon) or polypropylene as described in Section 5.5.3.

The cup of the nickel-plated steel casing acts as the positive terminal. The positive sign embossed on the can (to be seen in Fig. 11.12) represents a safety vent, because the wall thickness is reduced by this sign to a point of weakness that breaks when the internal pressure exceeds tolerable values. In case of abuse (forbidden overcharge rates), explosion of the cell is thus prevented.

The steel spring between the negative electrode and the cup provides the desired pressure for the separator and compensates for swelling, especially of the positive electrode (cf. Section 11.1.1). Furthermore, the spring forms void space for fast oxygen transport to the negative-electrode surface.

The positive electrode is pressed in the cup of the metallic container, while electronic contact is established between the negative electrode and the lid by the spring. The two parts of the housing, insulated by the gasket ring between them, form the terminals of the cell.

Sintered electrodes (Section 11.2.4) are occasionally used in button cells, as shown in Fig. 11.14. On account of the large surface area of these thin electrodes and their high conductivity due to the sintered body, performance is better than that of cells with pressed electrodes, especially at high-rate discharges. However, the production of sintered plates is more expensive which reduces the market for such cells to special applications.

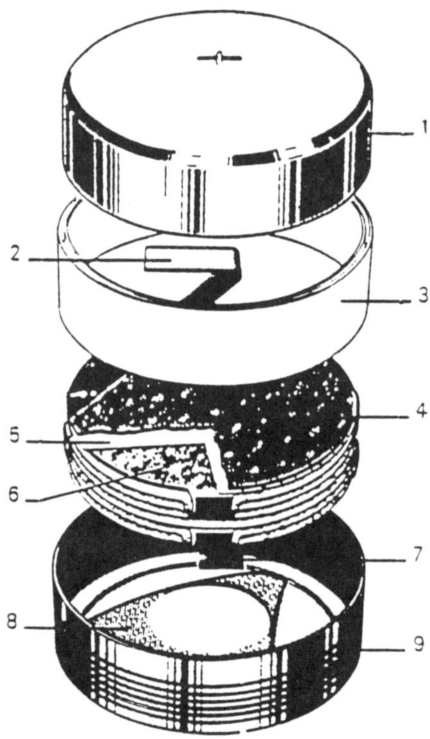

Fig. 11.14 Exploded view of a button cell with a stack of sintered electrodes.
1: Positive cup. 2: Positive connection (welded to cup 1).
3: Insulating ring. 4: Positive plate.
5: Separator. 6: Negative plate.
7: Negative connection. 8: Spring.
9: Negative cup. (SAFT from ([35]).)

11.5.1.1 Auxiliary Electrodes in Button Cells

To improve the internal oxygen cycle, auxiliary electrodes (Section 7.5.3.4) are widely used in button cells. Actual auxiliary electrodes are not employed, but only layers of substances that act in the same way. To achieve the effect of an oxygen-consuming auxiliary electrode in button cells, the nickel-wire net at the top of the negative electrode in Fig. 11.13 is covered by a mixture of active carbon and lampblack. The overcharge current, which usually has to be limited to the 10-hour rate, can be doubled when the cells are equipped with these auxiliary electrodes on account of the increased oxygen-reduction rate.

11.5.1.2 Button Cells for Float-Charging Applications

Button cells for memory backup, specially designed for continuous use at high temperature, have sodium hydroxide electrolyte (NaOH) instead of the usual KOH ([36]) (cf.Section 5.5.2).

11.5.1.3 Stacked Button Cells

Most battery applications require voltages which can only be provided by a number of cells connected in series. To meet these demands, batteries are offered consisting of a number of single cells. Fig. 11.15 shows two batteries assembled by stacking button cells.

Fig. 11.15 Button-cell batteries. On the left: 4 cells, 4.8 V; 250 mAh in shrink sleeve. Terminals with solder tags (VARTA 4/600 DKZ). On the right: 2 cells, 2.4 V; 170 mA in plastic case with pin solder tags (VARTA 2/170 DK).

Four single cells forming the stack can be seen in the battery on the left in Fig. 11.15. The cells within the stack are welded together by tags similar to those in the terminals, to exclude contact problems. Battery stacks of up to 10 cells are in use, as well as batteries with several cell stacks beside one another.

The ISO-recycling sign (cf. Section 15.3.1) seen on the left example indicates that this battery should not be disposed of with usual waste, but should be collected and recycled.

On the right in Fig. 11.15, a similar battery is shown, but with a plastics housing instead of the shrink sleeve. The number of cells within one battery of this type is usually limited to five. Otherwise swelling of the cells during charge may cause problems.

11.5.1.4 Connector Design

The terminals of both batteries in Fig. 11.15 are equipped with solder tags. Various other connector types may be used instead, e.g. a press button at the positive and a contact plate at the negative terminal, or a crown connector, usually also combined with a contact plate at the opposite terminal. Fig. 11.16 shows some examples of connector design.

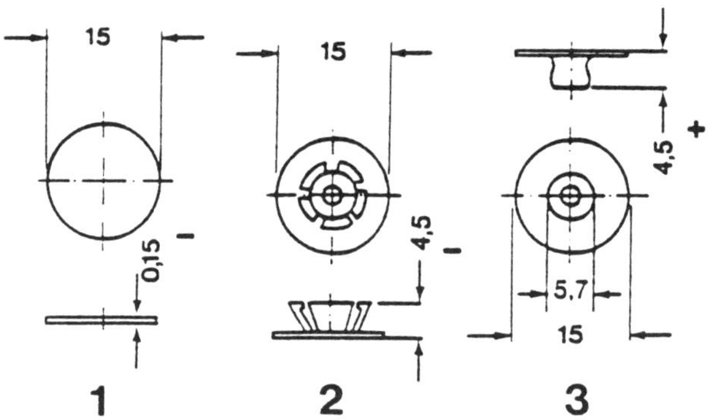

Fig. 11.16 Connector designs, other than solder tags, used with sealed Ni/Cd batteries. 1: Contact plate. 2: Crown connector. 3: Ball or stud connector. Dimensions in mm. (VARTA.)

11.5.1.5 Batteries in Plastics Cassettes

Many button cells are assembled in batteries within special plastics cassettes or casings. Fig. 11.17 shows examples designed for direct soldering into printed-circuit boards.

Fig. 11.17 Batteries assembled from button cells for direct soldering in printed-circuit boards. 1: 3 cells, 3.6 V, 60 mAh. 2: 3 cells, 3.6 V, 100 mAh. 3: 2 cells, 2.4 V, 100 mAh.

Batteries of this design are mainly used for the protection of electronic memories memory backup). The extremely flat battery in Fig. 11.18 is designed for the same application, when a particular low battery is required.

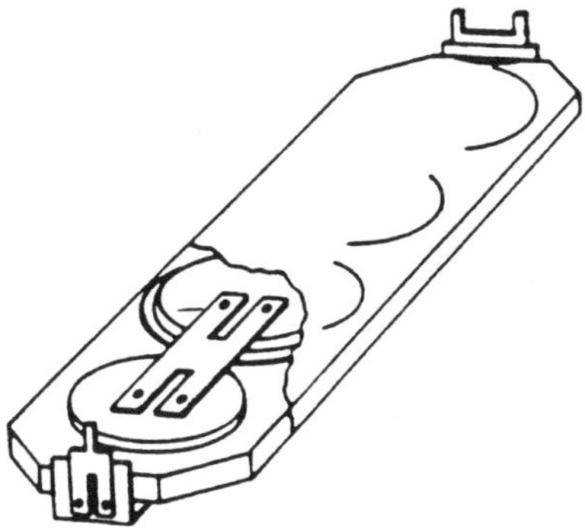

Fig. 11.18 Flat battery assembled from 4 button cells e.g. 4 cells, 4.8 V, 170 mAh (VARTA).

A battery, also assembled from button cells, used in small devices like transistor radios or dictating machines, is shown in Fig. 11.19. It is completely encapsulated into the plastics housing. The connectors at the terminals are a crown contact at the negative terminal (left hand side) and a ball contact at the positive terminal, as shown in Fig. 11.16 . Both connectors fit a corresponding double plug in the device.

Fig. 11.19 Small block battery assembled from button cells. 9 V, 100 mAh. (VARTA.)

The battery in Fig. 11.19 is interchangeable with a corresponding primary battery, as applies for many smaller sealed nickel/cadmium batteries. A special design is used for the positive terminal (right hand side in Fig. 11.19) to make sure that the primary battery cannot be charged inadvertently when it is inserted into the charging device instead of the rechargeable battery. In a rechargeable nickel/cadmium battery, the ball connector is mounted on a small metallic plate that is used as the contact for charging if this option is integrated in the electronic device. The primary battery does not have this metallic plate. So, charging is not possible, because the charging contact does not meet the contact plate, but the insulating wall of the cell housing.

11.5.1.6 Coin Cells

Very thin button cells are sometimes called 'coin cells'. There is no significant difference between button cells and coin cells; the terminology depends on the manufacturer. The term 'coin cell' is not standardised, but is fairly common for primary button cells and describes a cell that is thin compared to its diameter.

They are designed for compact equipment like pocket calculators or small dictating machines. Often the coin cell is equipped with sintered electrodes and designed for high-rate discharge ([37]).

11.5.2 Cylindrical Cells

Cylindrical cells are widely used for the power supply of devices when capacities are required between 150 mAh and about 5 Ah. Nearly all types of electrodes, namely sintered, fibres, foam and plastics electrodes are used in this cell design.

The characteristic constructional features are shown in Fig. 11.20.

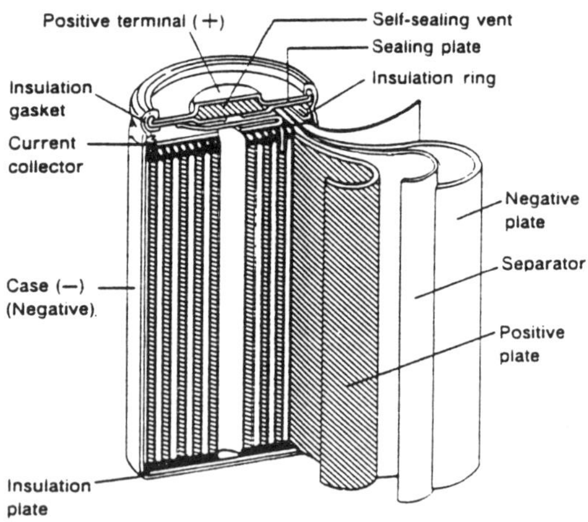

Fig. 11.20 Cross section of a sealed cylindrical nickel/cadmium battery. Panasonic, from ([38]).

The electrodes, one positive and one negative, are spirally wound together with the separator. A nickel-plated steel case encloses the electrode group and withstands the fairly high internal pressures that may occur during overcharge, as shown in Fig. 8.4, p.274. The cup acts as the negative terminal; a flat button in the cover is the positive terminal in Fig. 11.20. Different types of connectors are used, as described above for batteries assembled from button cells (Fig. 11.16, p.384). The lateral face of the cells is usually covered with a shrink sleeve of plastics to prevent short-circuits.

Inside the cell, both electrodes are equipped with a number of metal strips that act as current conductors comparable with the lugs on top of the grids shown in Fig. 10.24, p.343 for the lead-acid battery. The strips of the positive electrode are welded to the cover and establish a connection of low resistance to the positive terminal. The negative electrode is connected in a similar way with the bottom of the cell case.

The cell shown in Fig. 11.20 is equipped with a resealing valve of the sort shown in Fig. 11.11, p.379, based on the compression of a rubber pad. The performance data demanded from varying applications of this type of battery are very different and cannot be achieved by one 'universal' battery design. So, most manufacturers offer several series of cylindrical sealed nickel/cadmium batteries. Such basic types are:

- **Standard types**, which represent the most economic solution and can be applied 'universally' for applications that do not require special features.
- **High-current types,** characterised by low internal resistance, mainly achieved by special design of the conducting elements within the cell. They are used for high-load applications like motor drives.
- **High-capacity batteries,** specially designed for high loads. They combine a maximum amount of active material with low internal resistance.
- **High-temperature batteries**, characterised by stabilised active material in the positive electrode, usually by increased cobalt content and by heat resistant separators. They may have an electrolyte that contains sodium hydroxide (cf. Section 5.5.2). Emergency lighting is one field of application.
- **Fast-charge types**, suitable for recharging within one hour or less. This feature is mainly demanded for cordless tools. It is achieved with layers of active carbon, as described in Section 7.4.3.4, which are rolled onto the negative-electrode surface and increase the rate of oxygen consumption. But the 'auxiliary electrode' requires volume that cannot be used for active material. So, maximum capacity per volume cannot be achieved with this design.
- **Memory-backup batteries** for standby operation and for use at elevated ambient temperatures. Such batteries are equipped with temperature-resistant separators. Special seals prevent leakage. Often these batteries have a sodium hydroxide containing electrolyte. The main field of application is the protection of electronic memories (memory backup). As a consequence of the reduced conductivity of the electrolyte (cf. Fig. 5.17, p.208), the internal resistance is somewhat higher than that of the above mentioned types.

This listing illustrates that special cylindrical sealed nickel/cadmium batteries are available for the various applications. The reason is that a 'standard battery' would never be able to fulfil all the extremely high demands for special performance data.

11.5.2.1 Cylindrical Cell Assemblies, Power Packs

In general, sealed nickel/cadmium batteries are not used as single cells, because of the higher voltage usually required. For this reason, several cells are usually assembled for a battery. There are many ways of arranging the cells in a battery. Some examples of possible combinations are shown in Fig. 11.21.

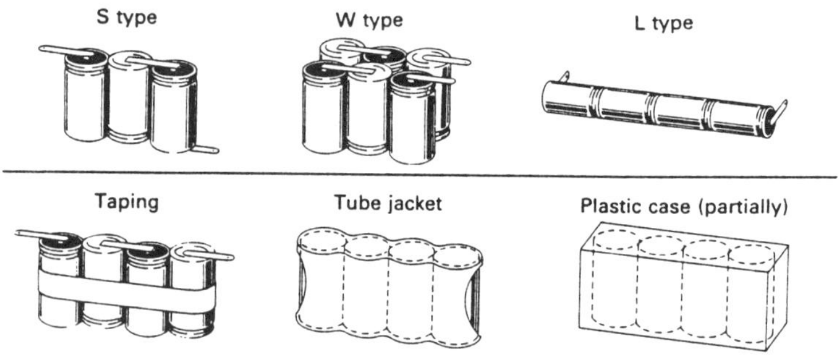

Fig. 11.21 Examples for the connection and fixing of cylindrical cells that are assembled to multicell batteries, also called 'power packs' (Sanyo Cadnica ([39])).

The upper line in Fig. 11.21 shows various ways to connect cells. Contact between the individual cells is achieved by spot welding with the aid of corresponding tags. The battery on the right hand side that is formed by stacking the cells, corresponds with that for button cells, shown in Fig. 11.15. Such stacking is mainly used for smaller cells. Sometimes two or more strings form a slim battery that can be placed in the handle of tools. To avoid short-circuits, the battery usually is covered completely by insulating plastic material, except for the soldering tags.

Another method to combine cylindrical cells that corresponds to the W-type connection in Fig. 11.21, is shown in Fig. 11.22. The individual cells are insulated by shrink sleeves and joined by spot-welded strip connectors. The whole battery is fitted into plastic end-caps that insulate the intercell connectors. Strip solder tags or flat contacts of various designs provide electrical output connection.

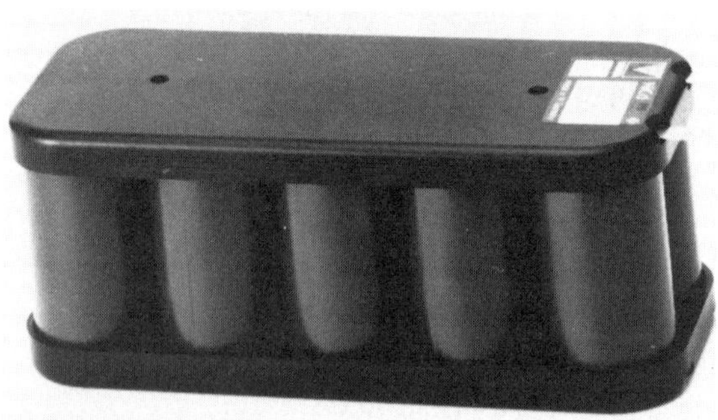

Fig. 11.22 Sealed nickel/cadmium battery in plastics end-caps with soldering tags. 10 cells, 12 V, 4 Ah. The single cells are insulated by shrink sleeves (VARTA 10/RSH 4).

The shrink sleeve serves as a cell insulation and simultaneously gives stable fitting to the caps. It is made of impact-resistant PVC and provides electrical insulation up to 20 kV/mm.

Mostly power packs are completely encapsulated by plastic material, as shown on the right hand side in the lower line of Fig. 11.21. An example of such a battery is shown in Fig. 12.2, p.399. Such packs are offered in a wide variety since each battery often is tailored to its application. The fast growing market for cordless telephones and tools, and for camcorders and toys, caused such a development (cf. also Section 14.3).

Cells combined in a battery pack usually are matched for capacity. Otherwise, individual cells may suffer deep discharge or inadmissible overcharge. This is true especially for batteries operating in rigorous conditions. For this reason, the customer should not combine single cells in batteries himself, especially for heavy-duty batteries.

11.5.3 Prismatic Cells

The design of prismatic sealed nickel/cadmium batteries is illustrated by Figs. 11.23 and 11.24. As in the prismatic lead-acid battery (cf. Fig. 10.10, p.326), a number of positive and negative electrodes are paralleled to form the plate group. Nickel or nickel-plated steel is used for the connecting elements and the terminals. An advantage of the prismatic design is the better utilisation of a given volume, since mostly a rectangular shaped volume is offered for the battery.

Fig. 11.23 shows a design that is used for smaller batteries with nominal capacities between 0.4 and 1.1 Ah.

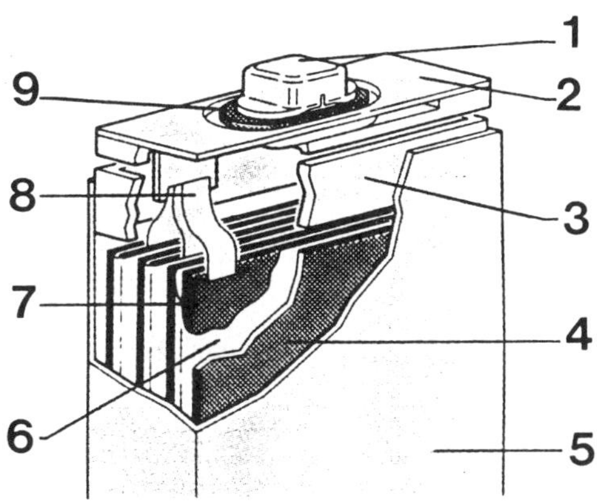

Fig. 11.23 Prismatic nickel/cadmium battery.
1: Positive cap with safety valve. 2: Cover plate. 3: Spacer.
4: Negative electrode. 5: Casing (negative terminal).
6: Separator (forms pocket around the positive electrode).
7: Positive electrode. 8: Positive current collector.
9: Gasket. (Sanyo CADNICA SLIM, ([39]).)

Unlike the design in Fig. 11.23, and unlike the usual custom for cylindrical cells, the metallic container of the prismatic cell shown in Fig. 11.24 is not connected with the negative pole, rather the housing of the cell is insulated. (*This applies also to the cell shown in Fig. 11.27.*) The insulating bottom and side layers are seen in the Figure.

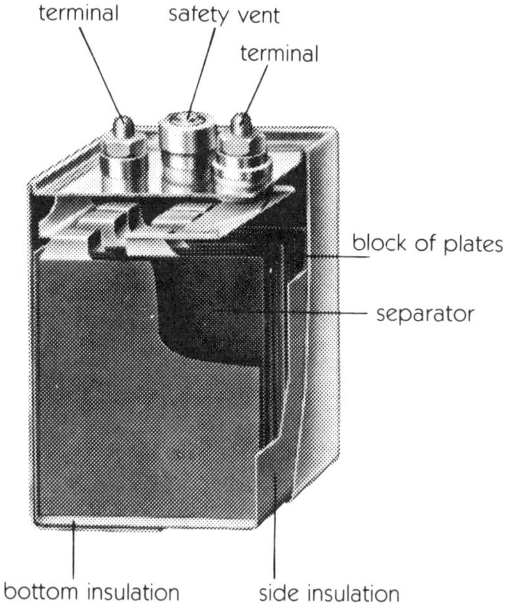

Fig. 11.24 Prismatic sealed nickel/cadmium battery. 1.2 V; 15 Ah. (VARTA, SD 15.)

The battery in Fig. 11.24 is equipped with sintered electrodes. Thus, its internal resistance is low (cf. Table 3.7, p.90), and it is suitable for high-rate discharge. Stable screw terminals support the high-rate application. This design is manufactured in a range up to 15 Ah of nominal capacity. This limitation has two reasons:

- In larger cells, the internal pressure during overcharge might cause bulging of the container walls.
- Heat management of large cells is difficult when the battery is discharged at high rates on account of the reversible heat effect, which amounts to 10.5% of the energy drawn from the cell (cf. Section 5.1.3), combined with the fairly low heat capacity of sealed cells (cf. Table 5.6, p.186).

An exception are the batteries shown in Section 11.5.3.3 where these problems are largely avoided by auxiliary electrodes and 'controlled charging'.

11.5.3.1 Monoblock Batteries

A block battery consisting of sealed nickel/cadmium cells is shown in Fig. 11.25. It is designed for portable devices like tape recorders or television devices. It has sintered electrodes like the prismatic cell shown in Fig. 11.24. The moulded case of plastic holds five plate assemblies, which are joined by strip connectors. As a whole, the construction is similar to that of lead-acid monoblocks shown in Figs. 10.22 and 10.23, p.342.

Each cell has a safety valve to prevent bulging of the plastics case by excessive internal pressure. The terminal connectors for the example in Fig. 11.25 are suitable for soldering as well as for AMP snap connectors.

Fig. 11.25 Monoblock sealed nickel/cadmium battery. 6 V (5 cells), 6 Ah. (VARTA.)

11.5.3.2 Prismatic Cells with Auxiliary Electrodes

Prismatic sealed nickel/cadmium batteries based on fibre electrodes with auxiliary electrodes are offered as 'low pressure batteries' ((40). The layout of the cell stack is shown in Fig. 11.26.

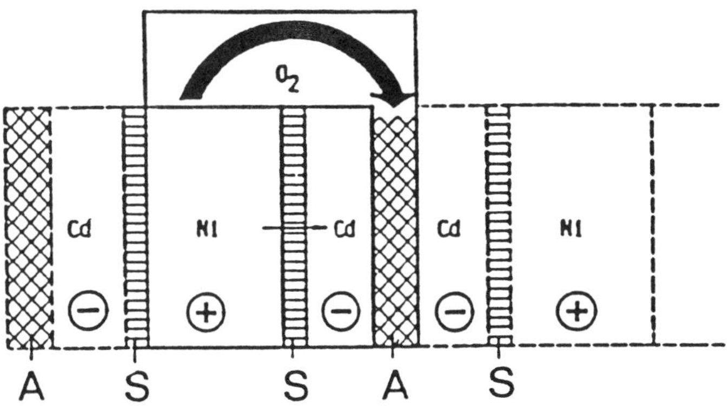

Fig. 11.26 Sealed nickel/cadmium batteries with auxiliary electrodes: array of the electrode stack . A: Auxiliary electrode (fibre electrode without impregnation). S: Separator. DAUG/HOPPECKE; from (41).

The negative electrode (Cd) is split into two plates, with an unfilled nickel-fibre electrode (A) interposed between the adjacent cadmium electrodes. The unfilled plaque acts as a catalytic site for rapid oxygen reduction. The arrow indicates that the main pathway for the oxygen to reach the unfilled nickel substrate is the gaseous phase. Oxygen transport is enhanced by the fairly large pores of the unfilled material, which are not soaked with electrolyte.

Fig. 11.27 shows such a cell. The arrangement of the electrodes can be seen in the opened part: a positive electrode in the foreground and, in the cut part of the next separator, the double negative (Cd) electrode with the auxiliary electrode lying between. The terminal on the right is connected with the positive electrode. The plate lugs are bolted, as usual, in larger nickel/cadmium batteries of the vented type.

Fig. 11.27 Sealed nickel/cadmium battery with fibre structure electrodes and auxiliary fibre mats for oxygen reduction, 1.2 V; 19 Ah. Array of the electrode stack as in Fig. 11.26 (DAUG/HOPPECKE).

The increased oxygen-reduction rate allows fast recharge of the battery. But for thermal reasons this can only be done under controlled charging conditions. Fig. 11.28 shows 1-hour rated charge/discharge cycles with the corresponding internal pressure and temperature curves. The cell was charged at the 1-hour rate (charge factor 1.18) and discharged at the same current. There was a pause of 20 minutes after charging.

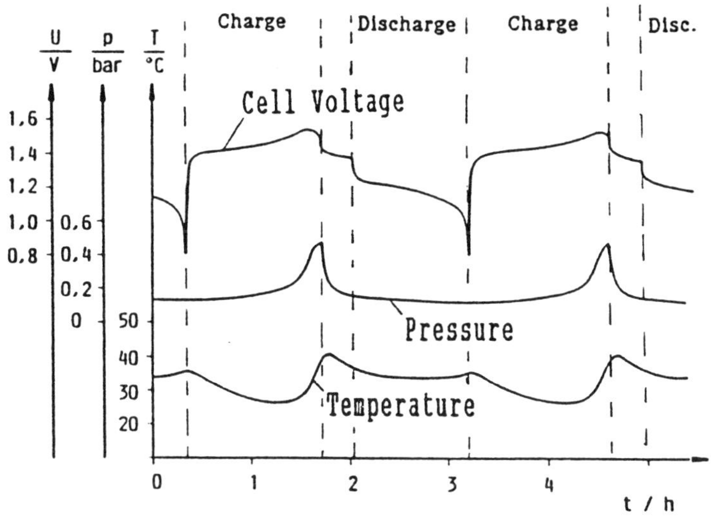

Fig. 11.28 Sealed nickel/cadmium battery with auxiliary electrodes. Constant-current charging and discharging (1-hour rate). Cell voltage, internal pressure and temperature during cycle service. Single cell, HOPPECKE FNC, nom. capacity 19 Ah. (From ([42]).)

The voltage shows the characteristic increase during charge. Its decrease at the end of the charging period is mainly due to the increased share of the current used for the internal oxygen cycle. The simultaneous temperature increase also contributes to the decrease of the voltage.

The temperature curve shows the typical behaviour of nickel/cadmium batteries. During the charge, a slight temperature decrease is observed on account of the fairly high (and negative) reversible heat effect (cf. Section 5.1.3), before the (heat intensive) internal oxygen cycle consumes most of the current at the end of the charging process. During discharge, only a slight increase of the temperature is observed. This temperature effect was fairly small since heat dissipation from the free-standing single cell was quite high in this experiment.

The pressure curve reflects the efficiency of the auxiliary electrodes. In spite of the high charging current, the pressure remains approximately constant at a low level. Only during a short period at the end of the charging process is an increase to about 500 mbar (50 000 Pa) observed. At this stage of the charging regime, nearly all the current is used for oxygen evolution.

Batteries like this are manufactured in sizes upto 100 Ah nominal capacity. For such large batteries controlled charging (cf. Section 13.2.2) is required to avoid thermal problems.

On account of the low internal pressure light metallic containers can be used. (*Regarding mechanical stress, even plastics containers could be used, but the fairly high permeation rates for water vapour (Table 2.13, p.61) and smaller heat dissipation rates through the container wall would be disadvantageous, especially when these batteries are used in the strong cycle regime of electric vehicles*).

11.5.4 Semi-sealed Batteries

The battery design described in this Section in principle is a vented nickel/cadmium battery. But it takes advantage of the internal oxygen cycle to reduce water loss. Fig. 11.29 shows the construction diagram that in principle corresponds to that of a vented cell, except that the vent cap that allows gas exchange in both directions, is replaced by a valve through which gas only can escape.

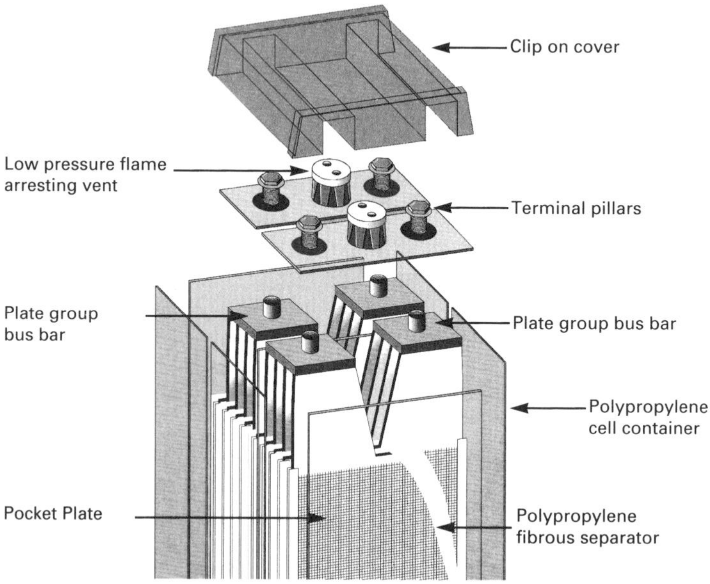

Fig. 11.29 Construction diagram of a semi-sealed nickel/cadmium battery. 1.2 V single cells with nominal capacities between 8 and 476 Ah. Blockbatteries up to 12 V per block. ALCAD 'Vantage' ([43]).

The battery in Fig. 11.29 uses pocket plates as positive and negative electrodes. The 'pockets' form horizontal stripes that are clamped by U-formed vertical metal strips; (*these details of the pocket plate are not to be seen in the Figure*). The 'low pressure flame arresting vent' is a valve with an opening pressure of about 0.2 bar. The battery is filled with liquid electrolyte. The separator is a polypropylene fibrous material that is claimed to support the oxygen diffusion from the positive to the negative electrodes. (Gas bubbles reduce the distance to be passed by the dissolved gas.) An efficiency of the internal oxygen cycle between 85 and 95% is claimed (at low float voltages). *Presumably a considerable share of the oxygen reduction occurs at the conducting parts that are not immersed into the electrolyte, but are in contact with the fibrous separator which is wet with electrolyte*

In Section 2.6.3.1, the equivalent current for the internal oxygen cycle in liquid electrolyte was estimated in the order of 2 mA/100 Ah under the usual conditions of a vented lead-acid battery. The diffusion rate through an alkaline electrolyte is similar, as shown in Table 2.10, p.53. But, an advantage of nickel/cadmium batteries is the smaller spacing between the electrodes that reduces the diffusion length. A further reduction of the diffusion length is achieved when gas bubbles are retained in the separator, as mentioned above. Nevertheless, the maximum rate of the internal oxygen cycle is limited in the liquid electrolyte, which is stated by the published values of the effect of charging voltage on the refilling periods, shown in Fig. 11.30.

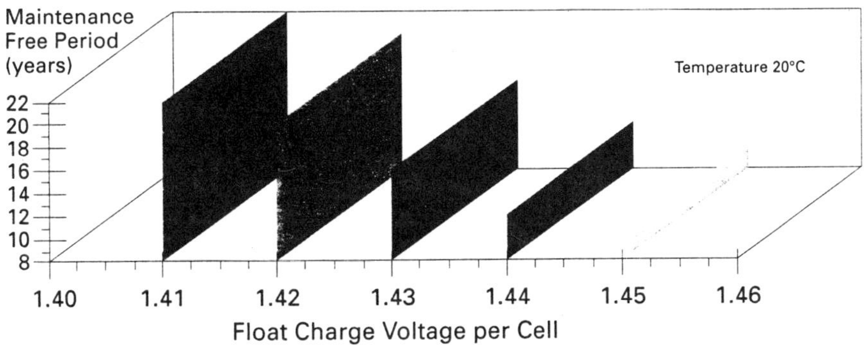

Fig. 11.30 Semi-sealed nickel/cadmium battery. Effect of charging voltage on maintenance free period. ALCAD Vantage (44).

An increase of the float voltage by 40 mV (1.41 to 1.45 V/cell) reduces the refilling interval from about 22 to 10 years. This means that the water loss is more than doubled, although a current increase by only the factor 1.5 is to be expected.

11.5.5 Aerospace Applications

Sealed nickel/cadmium batteries are used in aerospace applications, since they can be sealed hermetically. Special demands concern the leakage rate on account of the surrounding vacuum: they have to be sealed virtually hermetically with a leak rate of$<10^{-10}$ bar·dm^3·s^{-1}. Thus, a safety vent as in Fig. 11.9 has to be used instead of a resealing valve. Furthermore, absolute sealing of the terminals is achieved by ceramic or glass-to-metal seals (cf., e.g., (12), Fig. 6/8, p. 136).

References

(1) M.Oshitani, K.Takashima and Y.Matsumara *in* D.A.Corrigan, 'Nickel Hydroxide Electrodes', Proc. Vol. 90 - 4, The Electrochem. Soc.Inc., Pennington, U.S.A., 1990, p. 197

(2) J.L.Weininger *in* R.G.Gunther and S.Gross, 'The Nickel Electrode', Proc. Vol. 82 - 2, The Electrochem. Soc.Inc., Pennington, U.S.A., 1982, p. 1

(3) C.Delmas, Y.Borthomieu and C.Faure *in* D.A.Corrigan, 'Nickel Hydroxide Electrodes', Proc. Vol. 90 - 4, The Electrochem. Soc.Inc., Pennington, U.S.A., 1990, p. 119

(4) A.H.Zimmermann *in* T.Keily and B.W.Baxter (Editors), 'Power Sources 12', Intern. Power Sources Symp. Committee, 1988, p.235

M.E.Unates, M.E.Folquer, J.R.Vilche and A.J.Arvia *in* D.A.Corrigan, 'Nickel Hydroxide Electrodes', Proc. Vol. 90 - 4, The Electrochem. Soc.Inc., Pennington, U.S.A., 1990, p. 134

(5) A.K Sood *in* D.A.Corrigan, 'Nickel Hydroxide Electrodes', Proc. Vol. 90 - 4, The Electrochem. Soc.Inc., Pennington, U.S.A., 1990, p. 163

(6) O.Glemser, J.Bauer, D.Buss, H.-J.Harms and H.Löw *in* T.Keily and B.W.Baxter (Editors), 'Power Sources 12', Intern. Power Sources Symp. Committee, 1988, p. 165

(7) S.U.Falk and A.J.Salkind, 'Alkaline Storage Batteries', John Wiley & Sons, New York, 1969

(8) H.Staude, 'Physikalisch chemisches Taschenbuch', Akademische Verlagsgesellschaft, Leipzig, 1949, p.1552

(9) M.Z.A.Munshi and A.C.C.Tseung, J. Power Sources, **18** (1986, p.33

(10) J.P.Hari vel, B.Morignat and J.Migeon *in* D.H.Collins (Editor) 'Batteries 2', Proc. 4th Intern. Symp., Brighton, 1964, Pergamon Press, London 1965, p.107, esp. Fig. 10

(11) R.Barnard, J. Appl. Electrochem. **11** (1981), p.217, esp. p.231

(12) VARTA Batterie AG (Editor), 'Sealed Nickel Cadmium Batteries', © VDI Verlag GmbH, Düsseldorf, 1982, p.110

(13) R.D.Armstrong and H.Wang, Electrochimica Acta, **36** (1991), p.759

(14) G.Halpert *in* D.A.Corrigan, 'Nickel Hydroxide Electrodes', Proc. Vol. 90 - 4, The Electrochem. Soc.Inc., Pennington, U.S.A., 1990, p. 3

(15) M.Paszkiewicz and G.Mlynarek *in* D.A.Corrigan,'Nickel Hydroxide Electrodes', Proc. Vol. 90 - 4, The Electrochem. Soc.Inc.,Pennington, U.S.A., 1990, p. 222, esp. Fig. 7, p.232

(16) German Patent 1 133 442, L.Kandler (Sept. 3rd, 1958)

(17) E.Häusler in D.H. Collins (Editor) Power Sources 1966, Proc. 5th Intern. Symp., Brighton, 1966, Pergamon Press Ltd., 1967, p.287, esp. p.295

(18) E.Häusler, ibid pp.289

(19) B.Bugnet, R.Rouget and D.Doniat *in* T.Keily and B.W.Baxter (Editors), 'Power Sources 12', Intern. Power Sources Symp. Committee, 1988, p.157

(20) H.N.Seiger *in* H.C.Maru, 'Porous Electrodes', Proc. Vol. 84 - 8, The Electrochem. Soc.Inc., Pennington, U.S.A, 1984, p.50

(21) M.Paszkiewicz and G.Mlynarek *in* D.A.Corrigan, Nickel Hydroxide Electrodes, Proc. Vol. 90 - 4, The Electrochem. Soc.Inc, Pennington, U.S.A, 1990, p. 222

(22) W.A.Ferrando and R.A.Sutula *in* R.G.Gunther and S.Gross, 'The Nickel Electrode', Proc. Vol. 82 - 2, The Electrochem. Soc.Inc, Pennington, U.S.A., 1982, p. 271

(23) B.Bugnet and D.Doniat in L.J.Pearce (Editor), 'Power Sources 11', Intern. Power Sources Symp. Committee, 1986, p.179

(24) G.Benczúr-Ürmössy, G.Berger and F.Haschka, etz **104** (1983) 1098 (in German)

(25) I.Matsumoto, H.Ogawa, T.Iwaki and M.Ikeyama *in* T.Keily and B.W.Baxter (Editors), 'Power Sources 12', Intern. Power Sources Symp. Committee, 1988, p.203

(26) J.Jindra, J.Mrha, K.Micka, Z.Zábransky, V.Koudelka and J.Malík, J.Power Sources, **4** (1979), p.227

(27) J.Mrha, I.Krejcí, Z.Zábransky, V.Koudelka and J.Malík, J.Power Sources, **4** (1979), p.239
(28) S. Kulcsár and Géza Csath *in* T.Keily and B.W.Baxter (Editors), 'Power Sources 12', Intern. Power Sources Symp. Committee, 1988, p.197
(29) German Patent DE 2822-821-C, Société Piles Wonder, 24.5.1977
(30) M.Babai and C.Bezalel, Proc. 34th Intern. Power Sources Symp., 1990, © IEEE Service Center, 445 Hoes Lane, Piscataway, N.J., p. 295
(31) European Patent EP 0 160 845, VARTA Batterie AG, 4.4.85
(32) H.J.Saechtling, 'Kunststoff Taschenbuch', 7th Edition, Hanser Verlag, München, 1989, pp.582
(33) VARTA, 'Sealed NC Cells, Sales Programme and Technical Handbook', VARTA Batterie AG, Daimlerstr.1, D-7090, Ellwangen/Jagst
(34) Examples of such publications are
Gates Energy Products (Editor), 'Sealed Rechargeable Batteries, Application Manual', Gates Energy Products Inc., Gainesville, Florida, 1989 (now Hawker Batteries)
'Panasonic Nickel Cadmium Batteries Technical Handbook 89/1', Edited by Matsushita Battery Industrial Co. Ltd., 11-66 Honsyuku-cho, Chigasaki, Kanagawa 253, Japan
'SAFT Engineers' Handbook, Nickel-Cadmium Batteries',
SAFT, 156 avenue de Metz, 93230 Romainville, France
SAFT(UK) Ltd., Station Road, Hampton, Middlesex TW12 2BY, Great Britain
SAFT America Inc., 711 Industrial Boulevard, Valdosta, Georgia 31601, USA.
VARTA Batterie AG (Editor) 'Sealed Nickel Cadmium Batteries', VDI-Verlag GmbH, Düsseldorf, 1982
VARTA Sealed Ni-Cd-Batteries,
'Sales Programme and Technical Handbook',
'Battery Assemblies'
both edited by VARTA Batterie AG, Am Leineufer 51, D-3000 Hannover, Germany
(35) 'SAFT Engineers' Handbook, Nickel-Cadmium Batteries', (SAFT (UK) Ltd., Station Road, Hampton, Middlesex, p.10
(36) S.Tateishi, Y Takahashi and K.Fuchida, Proc. 5th Intelec Meeting, Tokyo 1983, © IEEE
(37) S.Miyagishi, Journal of the Electronics Industrie **37** (1990) p.38
(38) 'Panasonic Nickel Cadmium Batteries Technical Handbook 89/', Edited by Matsushita Battery Industrial Co, Ltd., 11-66 Honsyuku-cho, Chigasaki, Kanagawa 253, Japan
(39) 'Sanyo Rechargeable CADNICA', Sanyo Electric Co., Ltd., Sumoto-city, Hyogo, Japan
(40) 'HOPPECKE SPECIAL III/June 1986', published by Akkumulatorenwerke HOPPECKE Carl Zoellner & Sohn GmbH, P.O.Box 1140, D-W-5790, Germany
(41) F.Haschka and G.Benczur-Ürmössy, Proc. 34th Power Sources Symp., 1990, © IEEE Service Center, 445 Hoes Lane, Piscataway, N.J., p. 286
(42) G.Benczúr-Ürmössy, 'Dechema Monographs, Vol.109', 1987, p.379
(43) Vantage Technical Manual, Alcad Limited, Castle Works, Station Road, Hampton, Middleessex, TW12 2BY England
(44) J.N.Taylor, Proc. 17th Intelec Meeting, The Hague 1995, © IEEE Communications Soc., p.778

12. Nickel/Metal Hydride Battery Technology

The technology of nickel/metal hydride batteries can be described very briefly, since:

- The positive electrode is identical with that of the nickel/cadmium battery. Usually foam electrodes (cf. Section 11.2.6) are applied, to balance the comparatively large capacity of the hydrogen storage electrode.
- The separator is the same as in nickel/cadmium batteries.
- The electrolyte is identical in both systems.
- Due to corresponding cell voltage and discharge/charge behaviour (cf. Section 13.2.3), nickel/metal hydride batteries can replace nickel/cadmium batteries. For this reason, they are manufactured in the same dimensions. Button cells ([1]) and cylindrical and prismatic cells ([2]) are offered, but also battery or power packs and 'smart batteries' (cf. Section 14.3)

Figs. 12.1 and 12.2 show some examples (cf. also Fig. 14.1, p.447).

12.1 Negative-Electrode Materials

Most batteries, offered at the moment, are based on modified AB_5 alloys (cf. Section 6.3.1). The manufacture of these materials is briefly described in Section 6.5.2.

The pulverised material is mixed with binders like PTFE, hot pressed to a substrate, rolled and pressed together with a binder to a wire mesh ([3,4]), or it is spread on the substrate as a slurry.

12.2 Nickel/Hydride Battery Design

Cylindrical and prismatic single-cell nickel/hydride batteries were the first on the market. In the meantime button cells have been developed, as shown in Fig. 12.1, and a great number of battery packs or 'power packs' is offered for the fast growing market of portable devices, like telephones (Fig. 12.2).

Fig. 12.1 Nickel/metal hydride button cells and batteries assembled of button cells. VARTA ([1]).

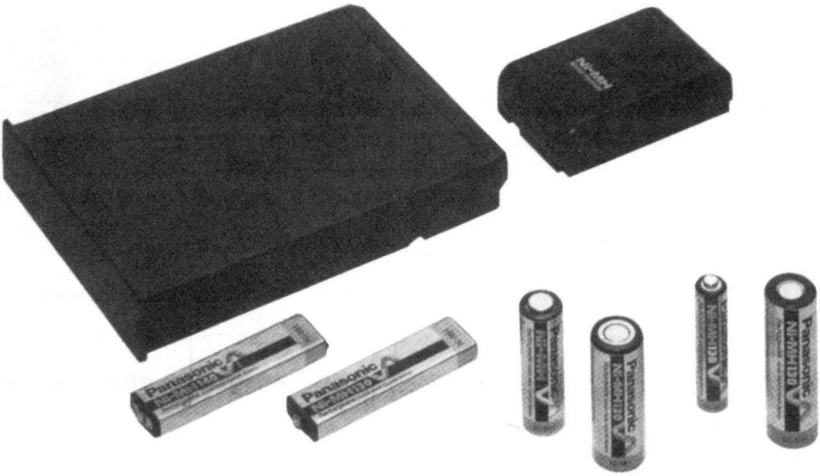

Fig. 12.2 Nickel/metal hydride cylindrical and prismatic cells and power packs. Panasonic ([5]).

Constructional details of cylindrical and the prismatic nickel/hydride batteries can be seen in Fig. 12.3. The arrangement of the electrodes in cylindrical cells is similar to that of cylindrical lead-acid and nickel/cadmium batteries (Figs. 10.24, 10.25 and 11.20, p.344 and p.386, respectively): a positive and a negative electrode are spirally wound, together with the separator of nonwoven fibre material. In the prismatic cells, a number of positive and negative electrodes are paralleled and form the plate group as in Fig. 10.10, p.326.

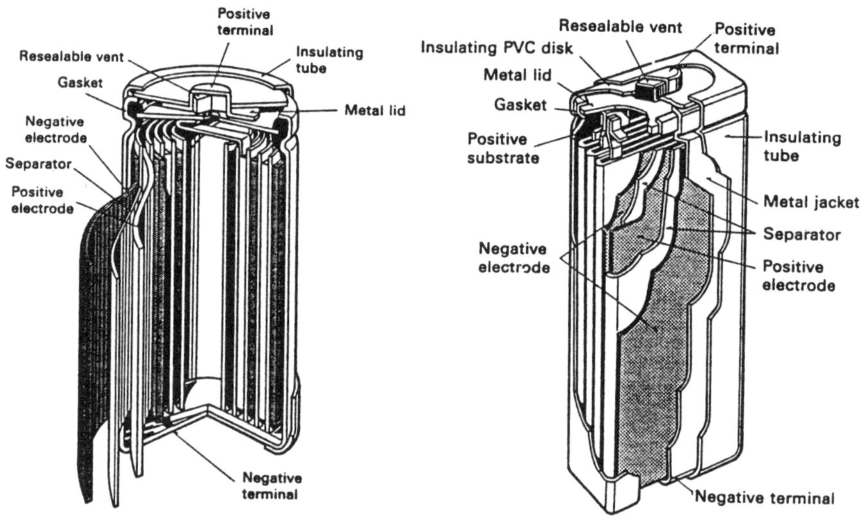

Fig. 12.3 Cross sections of cylindrical and prismatic nickel/hydride batteries. VARTA (6).

The electrode assembles are inserted into a nickel-plated steel can. The cell is sealed by crimping a lid to the can with an intermediate plastic gasket. Perfect sealing is of great importance because of the fugacity of hydrogen.

The can acts as the negative pole, while the lid is the positive terminal; both are insulated by a plastic gasket. The lid contains the re-sealable valve similar to that described in Fig. 11.11, p.379. As in the case of nickel/cadmium batteries, this valve opens only in an emergency to prevent inadmissible pressure increase. In normal operational conditions the nickel/hydride battery remains completely sealed, and hydrogen does not escape.

12.3 Electric Vehicle Applications

Due to their high specific energy content, nickel/metal hydride batteries are promising candidates as storage systems for electric vehicles. The capability of high discharge rates and rapid charging allows the design of batteries especially suited for hybrid vehicles that combine an ignition engine with electric drive. Longer distances are covered by the gasoline engine; within the cities the vehicle is driven by battery power. This requires a comparatively small battery that can be heavily loaded. Fig. 12.4 shows such a module. The fan belongs to the air cooling system.

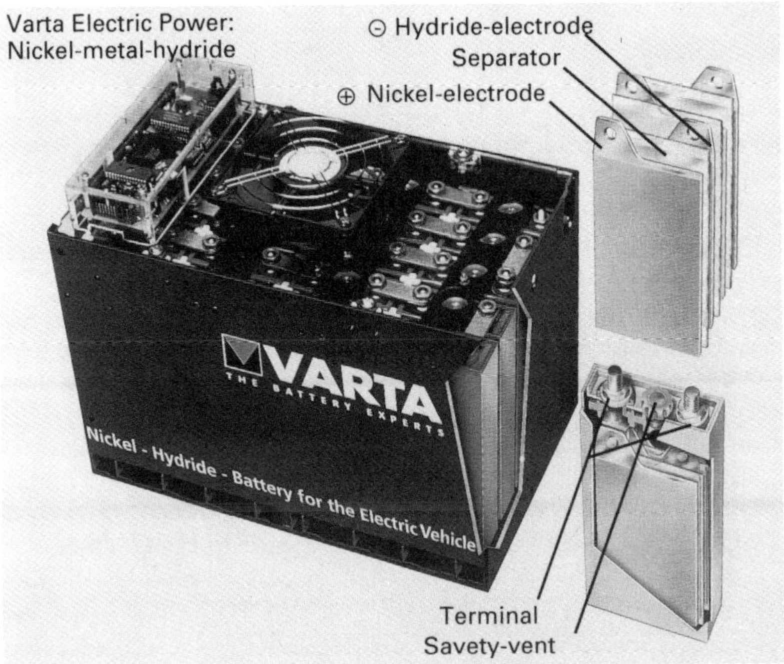

Fig. 12.4 Nickel/metal hydride battery. Module for electric vehicle applications with battery management system and air cooling (7).

Examples of prismatic nickel/metal hydride battery types developed for electric vehicles are given in Table 12.1. They correspond to the cells in the module of Fig. 12.4.

	High power cells		High energy cells	
Cell type	NP 40	NP 60	NP 80	NP 150
Capacity (2 hours)	40 Ah	60 Ah	80 Ah	150 Ah
Specific energy	50 Wh/kg	60 Wh/kg	70 Wh/kg	80 Wh/kg
Energy density	130 Wh/dm^3	155 Wh/dm^3	190 Wh/dm^3	200 Wh/dm^3
Specific power	180 W/kg	200 W/kg	160 W/kg	150 W/kg
Power density	550 W/dm^3	550 W/dm^3	450 W/dm^3	420 W/dm^3

Table 12.1 Prismatic nickel/metal hydride batteries for electric vehicles (8).

The high power cells are the above mentioned as especially suited for hybrid vehicles. Expected cycle life of these batteries is ≥2000 cycles (100% DOD).

An appreciable number of such batteries are being tested at the moment (1996) in vehicles.

References

(1) VARTA, 'Ni-MH Button Cells', Sales Program and Technical Handbook', © by VARTA Batterie AG, 1995, mailed by VARTA Batterie AG, Daimlerstr.1, P.O. Box 1363, D-7090 Ellwangen/Jagst, Germany

(2) VARTA, Ni-MH Cylindrical and Prismatic cells, Sales Program and Technical Handbook', © by VARTA Batterie AG, 1995

(3) German Patent DE 2941774 C2, Oct. 16, 1979 (VARTA Batterie AG)

(4) Z.Ye. Sakai, D.Noréus, E.Rosen, and N.E.Bärring, in Proc. Symp. 'Hydrogen and Metal Hydride Batteries, Editors D.Bennett and T.Sakai, © The Electrochemical Society, Inc, Proceedings Vol. 94-27, 1994, p.184

(5) Panasonic Nickel Metal Hydride Batteries, Technical Handbook, ©Matsushita Battery Industrial Co., Ltd. 11-66, Honsyukucho, Chigasaki, Kanagawa 253, Japan

(6) VARTA, ' Sealed Ni/MH Cells, Sales Program and Technical Handbook', © by VARTA Batterie AG, Toshiba Battery Co., Ltd and DURACEll International Inc. 1992, mailed by VARTA Batterie AG, Daimlerstr.1, P.O. Box 1363, D-7090 Ellwangen/Jagst, Germany

(7) VARTA (Editor), Electric vehicles- current situation and perspectives, © VARTA Batterie AG, Am Leineufer 51, D-30419 Hannover

(8) K.Brandt, U.Köhler, E.Niggemann, F.Böttcher, F.J.Kruger, Proc. 13th International Seminar on Promary and Secondary Battery Technology and Application, Boca Raton /Florida, 1996

13. Charging Techniques

The three battery systems described in this book are secondary batteries, and the energy drawn from them can be restored by reversal of the electrochemical reactions. This reversal is the charging process, and the battery and its charger have to be regarded as a system to be matched with an application. A brief description of basic charging parameters was given in Section 3.5, which shows that the efficiency of the charging process depends on a number of parameters, like charging time and available charging current, which again depend on the specific circumstances of the application. For this reason, a general charging schedule does not exist, rather the charging process has to be adapted to the specific conditions. Often, the charging method represents a compromise, since optimum charging conditions may require too much time for the recharge or the desired schedule may demand too high investment costs. Independent of the applied method, the charger and the charging schedule are important components of the system, since the possible number of recharges of a battery, as well as its service life, depend on the charging method.

Although the charging methods for all three battery systems are based on the same parameters, those that are applied for lead-acid batteries mostly are different from those for sealed nickel/cadmium and nickel/metal hydride batteries on account of their specific charging behaviour and their different fields of application. For this reason, the charging methods for these battery systems are treated in different Sections, although this causes some overlap.

13.1 Charging of Lead-Acid Batteries

One advantage of lead-acid batteries is that they can be recharged and kept fully charged at a voltage slightly above the equilibrium value, because at a definite cell voltage, the active material can attain either the 'charged' or the 'discharged' state. There is no range of varying oxidation states, as for the nickel electrode (Section 5.2.2.1).

On account of this simple charging behaviour, it would seem obvious to charge lead-acid batteries always with constant voltage. This, however, has the disadvantage that extended charging periods would be required, since the charge acceptance and so the charging current are reduced more and more when the battery approaches full charge (cf. Section 3.5.1 and Fig. 13.2). Thus, a complete recharge at a low constant voltage takes quite a long time, which is not acceptable for applications where the battery has to be recharged within a certain period. On the other hand, charging at increased voltage accelerates the unwanted side reactions of water loss and grid corrosion. As a compromise, different voltages or constant currents often are applied during different stages of the charging process. This is one reason for the existing great number of charging methods.

13.1.1 Constant-Voltage Charging, Float Charging

Constant-voltage charging is called **float charging** when the constant voltage is applied continuously. It is the most common method for stationary vented and valve-regulated lead-acid batteries, at least for those applications where the battery is used as a standby battery and is only discharged for emergency energy supply if the network fails. Usually the permanent overcharge, the 'float charge', is accomplished by connecting the battery to the d.c. power supply bus as shown in Fig. 13.1.

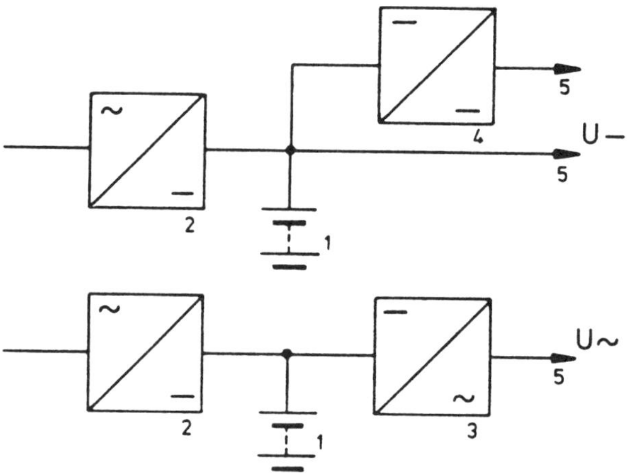

Fig. 13.1 Circuit diagrams for d.c. and a.c. backup power supply with the battery under float charge conditions.
1: Battery.
2: Rectifier.
3: Inverter.
4: D.C. converter.
5: Load.
From ([1]).

The number of cells in the battery has so be matched to the voltage of the bus:

- so that the desired float voltage is achieved under mains supply,
- and the battery voltage under load is sufficiently high to supply the load.

Under these conditions, the battery normally is kept fully charged, and, if the main network fails, it provides the required energy for the load without interruption. The extended time period until 100% recharge is no problem for this application because the battery is discharged only occasionally. Furthermore, the autonomous periods mostly are short so that the battery is only partly discharged. (*To achieve voltage balance between bus-bar and battery during charging and discharging, occasionally additional cells are used that are automatically connected with the battery when the main supply is interrupted. Such 'additional cells' are sometimes simulated by electronic devices.*)

The principal current and voltage characteristics during constant-voltage charging are shown in Fig. 13.2.

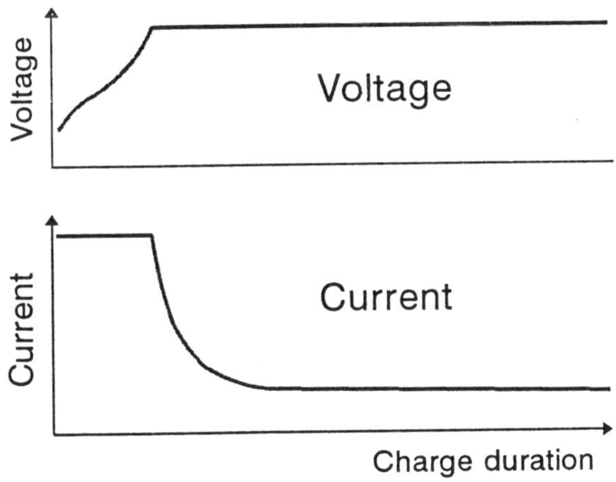

Fig. 13.2 Charging at constant voltage (float charging). Voltage and current characteristics versus charging time.

When charging at a constant voltage starts, usually the maximum current that can be supplied by the charger limits the charging rate because of the high charge acceptance of the discharged battery (cf. Section 3.5.1). Later, the current acceptance of the battery limits the charging current that gradually decreases with the progressing recharge. Finally, a constant value of the current is attained that remains unchanged. Then the charging reaction is completed, and the float current is caused solely by the side reactions. In vented batteries hydrogen evolution and (negligible) oxygen reduction occur at the negative electrode, and oxygen evolution and grid corrosion at the positive electrode (cf. Fig. 7.6, p.244). In valve-regulated lead-acid batteries, the internal oxygen cycle mainly determines the float current (cf. Section 8.3).

Depending on the initial charging current, the float voltage is reached sooner or later (as shown in Fig. 13.10, p.418).

13.1.1.1 Float Voltage

From the values in Table 4.2, p.103, a lead-acid battery filled with acid of density 1.24 g/cm^3 should be kept charged at a float voltage just above 2.08 V, while an acid density of 1.32 g/cm^3 would require at least 2.17 V per cell. Simultaneously, with such a minimum float voltage the rate of the side reactions would be kept as low as possible.

In practice, however, float voltages between 2.23 and 2.38 V per cell are prescribed, considerably higher than the above 2.08 and 2.17 V per cell. Reasons for the increased float voltage are:

- Polarisation of positive and negative electrodes may not be symmetrical, and one of them may not be polarised sufficiently at very low float voltages, although the cell voltage seems to be sufficient. This applies especially to valve-regulated batteries as has been pointed out in Section 8.3.
- Float charge usually does not concern a single cell, but a battery of a number of cells switched in series.

In the second case the charging voltage is determined as an average voltage for the whole battery, but the current acceptance deviates between the individual cells, owing to the scatter of hydrogen- and oxygen-evolution rates. As a consequence of this varying charge acceptance, the individual cell voltages deviate, as shown in Fig. 13.3.

The top part of Fig. 13.3 shows this deviation from the average value of 2.38 V. The distribution of the voltages, plotted in the bottom part, results when a classification is made at 15 mV interval. This distribution corresponds quite well with the normal distribution represented by the curve. This indicates that the cell voltages actually obey a statistical distribution and the deviations are not caused by systematic errors, e.g. by the fact that some of the cells were not fully charged. With such errors, the distribution would be split into different groups of voltages and would not match a normal curve.

Voltage interval ref. to the average value		Included values %
± 1 x standard deviation	$\pm 1\cdot\sigma$	68
± 2 x standard deviation	$\pm 2\cdot\sigma$	95
± 3 x standard deviation	$\pm 3\cdot\sigma$	99.7

Table 13.1 Relationship between multiples of the standard deviation and the percentage of values included in a corresponding range (for a normal distribution of values).

Note: The relationships listed in Table 13.1 apply only to a so called normal distribution according to the Gaussian error integral. If the distribution of the measured data is different from such a normal distribution, the standard deviation (calculated by the usual formula) will also describe a deviation, but the percentages included in certain intervals may be different from those in Table 13.1. So it is always necessary to check whether the measured data correspond with a normal distribution before conclusions are reached. A simple (approximate) method for this check is to plot the data as in the lower part of Fig. 13.3.

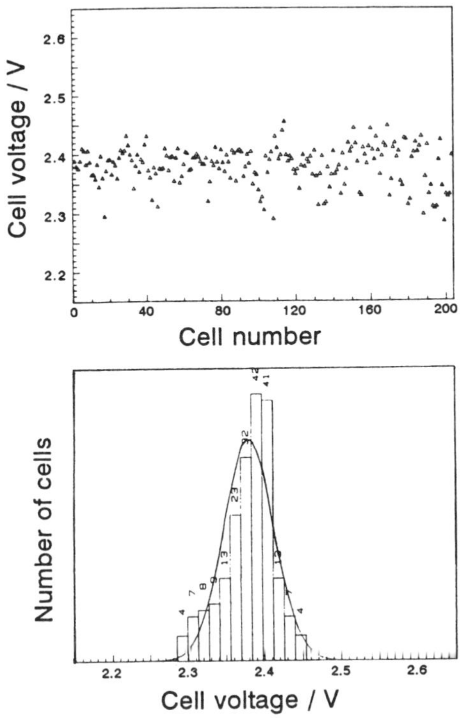

Fig. 13.3 Cell-voltage deviation in a random sample of 203 cells of a battery consisting of 2 x 180 cells of the valve-regulated design . Average voltage: 2.380 V/cell. Minimum: 2.285 V; Maximum: 2.455 V; Standard deviation: 33 mV. Top part: Individual cell voltages. Bottom part: Distribution of the individual cell voltages. The curve represents the normal distribution calculated for this sample. From (2).

The standard deviation (σ) is an important statistical parameter. It is a measure for the scattering of values within a statistical distribution, and can be used to estimate the minimum and maximum values that are to be expected. An often used quantity is $3\cdot\sigma$ that includes 99.7% of all values, which means that almost all values will be within the interval $\pm 3\cdot\sigma$ referred to the average. The evaluation of the standard deviation for a number of measured data is usually programmed in pocket calculators. The relations between σ and the lower and upper bounds of a standard distribution are listed in Table 13.1 (cf. the Note below that Table).

For the example in Fig. 13.3, the standard deviation σ is 33 mV, and the interval of average voltage $\pm 3\cdot\sigma$ (which contains practically all the values) is 2.380 V ±99 mV. This means that, for an individual cell, a minimum float voltage of 0.1 V below the average voltage is to be expected. Similarly, the maximum is expected to be 100 mV above the average voltage. To ensure that even the cell with the lowest voltage remains properly charged, the average float voltage must exceed the minimum float voltage, mentioned at the beginning of this Section, by at least 100 mV.

For block batteries, the relation between the standard deviation of the monoblock and that of the individual cells is:

$$\text{Stand. dev. per block} = \sqrt{n} \cdot (\text{stand. dev. per cell}) \qquad (13.1)$$

with n: number of cells per block.

This means that the standard deviation for a 6 V block (3 cells) is $\sqrt{3} = 1.73$ times the standard deviation of a single cell, while for a 12 V block the factor is $\sqrt{6} = 2.45$.

If the cells in Fig. 13.3 formed block batteries, 99.7% (±3 x stand. deviation) of the individual block voltages would fall within:

7.14 ± 0.173 V for 3-cell blocks;

14.28 ± 0.245 V for 6-cell blocks.

Relative to the block voltages, the deviation is reduced with an increasing number of cells per block by the factor $\sqrt{n}$, owing to the equalisation within the blocks. So the battery formed by 3-cell blocks would require an average float voltage of 173 mV above the average block voltage, corresponding to 58 mV per cell. For the battery formed by 30 monoblocs of 6 cells, the corresponding relation would be 245 mV per bloc or 41 mV per cell.

13.1.1.2 Temperature Compensation of Float Voltage

Temperature influence on float charge has been discussed in detail in Milners basic paper in 1970 ([3]). He did not recommend temperature compensation, but he already observed that at the normal float voltage, the polarisation of the positive electrode is reduced with increasing temperature, while that of the negative electrode is increased (cf. Section 8.3, especially Fig. 8.19, p.295, in this book). Temperature compensation is rarely used for vented lead-acid batteries. When float voltage is adjusted to temperature it is due to the acid concentration that occasionally is more concentrated for permanent operation at low temperatures (*'arctic electrolyte'*) and more diluted at increased temperature (*'tropic electrolyte'*). The reason for using different electrolytes is that increased acid concentration improves the capacity at low temperatures while decreased acid concentration allows lower float voltage and so reduces water loss and grid corrosion at elevated temperatures.

The discussion on temperature compensation of the float voltage became stimulated anew when valve-regulated lead-acid batteries were introduced. Mainly this has three reasons:

1. The valve-regulated battery is more sensitive to temperature increase since its water content is limited and it loses too much water at elevated temperature.
2. The internal oxygen cycle is a strong heat source that contributes to heating of the battery. Under unfavourable conditions valve-regulated batteries may even suffer thermal runaway (cf. Section 4.5.3.6 and the end of this Section).
3. The valve-regulated battery is often not installed in special rooms or cabinets but has to operate within the electronic equipment where it is exposed to elevated temperatures.

Adjustment of the float voltage to the less favourable conditions seemed to be necessary, and nearly all manufacturers published such compensation factors. These factors, however, show a wide variation from 2 mV/°C to 5 mV/°C (always per cell), and it cannot be said that the size of this factor is directly linked to battery design or type (4). Thus one suspects that these factors are not the result of experience but are based on plausibility.

Table 13.2 and Fig. 13.4 are examples for such float-voltage compensation factors. The data in Table 13.2 correspond to a temperature compensation of about -4 mV per °C.

Mean Operating Temperature Range°C	0-9	2.33/2.35	**Recommended Applied Voltage per cell minimum/ maximum**
	10-14	2.30/2.33	
	(15-24)	(2.27/2.30)	
	25-29	2.25/2.27	
	30-34	2.23/2.25	
	35-40	2.21/2.23	

Table 13.2 Recommended voltage settings for float charging at various temperatures. (CHLORIDE (5).)

The recommendation in Fig. 13.4 means that temperature compensation is required only when the battery is operating above 35 °C or below 5°C. Between these temperatures, compensation is not afforded. This example allows a fairly wide range of temperatures without compensation. Quite a number of battery manufacturers recommend temperature compensation only outside 10 and 30 °C.

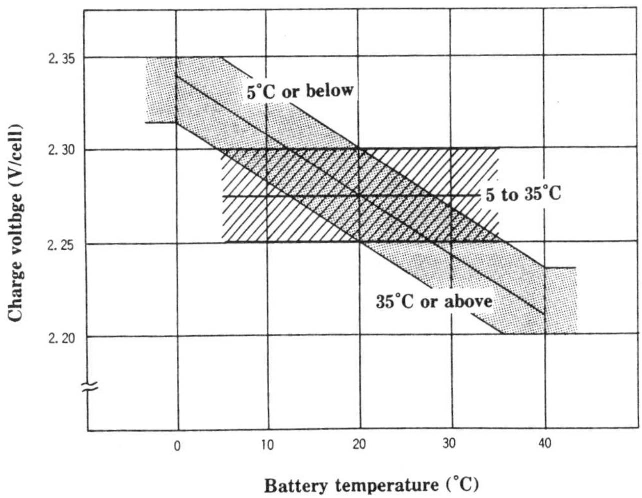

Fig. 13.4 Recommendations for temperature compensation of float voltage. Wide hatching: temperature range 5 to 35 °C without compensation. Shaded area: extended range with -3.25 mV/°C. (Shin-Kobe (6).)

For lead-acid aircraft batteries of valve-regulated design a compensation of the voltage setting is recommended from 2.58 V/cell at or below -10 °C to 2.25 V/cell at 60 °C or above (7). This is about -5 mV/°C. But the aircraft battery is charged, although with constant voltage, only for short periods which is quite different to the usual stationary application with continuous charging.

Reasons for float-voltage compensation

Changes of the equilibrium voltage with temperature can usually be neglected in practice, since the temperature coefficient of the lead-acid system is only 0.23 mV/K (Eq. (4.17)).

But the rates of the reactions depend on temperature. This affects the charging reaction and also the side reactions such as gas evolution and corrosion. As long as the different reactions are subject to a similar temperature dependence, the balance between them is not changed. This can be assumed as a first approximation, because most reactions more or less obey Eq. (2.68), p.33, which is derived from the ARRHENIUS equation and means that the reaction rate is approximately doubled for each temperature increase of 10 °C (cf. Section 2.3.6). Thus all reactions in the battery are similarly accelerated, the current balance is not significantly changed with temperature, and the battery remains fully charged with the normal float voltage, independent of temperature (cf., e.g., Fig. 8.18, p.294).

But the unwanted side reactions of grid corrosion, hydrogen and oxygen evolution (and as a consequence also the internal oxygen cycle) increase with temperature, and the goal of float compensation is the limitation of these reactions to their unavoidable minimum. But temperature effects can never be virtually compensated by voltage changes, because the dependence on voltage of the various reactions is different.

If the factor 2 for a temperature change of 10 °C is to be compensated by a corresponding potential shift (to achieve the same reaction rate or current), this means:

- A temperature compensation of -2.7 mV per °C is required to keep the oxygen-evolution rate constant, when a slope of 90 mV/decade is assumed for the TAFEL line ($E_1 - E_2 = 0.09 \cdot \log(2) = -27$ mV per 10 °C).
- Hydrogen evolution with the usual b-factor of 120 mV would require -3.6 mV/°C.
- The corresponding evaluation for the corrosion reaction with the b-factor between 0.140 and 0.300 V, as in Fig. 7.5, means that a potential reduction between -4 and -9 mV/°C would reduce the corrosion rate by the required factor 2 to compensate for a temperature increase of 10 °C. When the electrode potential is below the minimum of corrosion, even a positive compensation factor would be necessary.

So, each side reaction would require a different factor for voltage compensation. Especially for the corrosion reaction, the required factor would depend on electrode polarisation and either would be far too large, and reduce the float voltage far too much, or it would have to be reversed when the polarisation of the positive electrode is below the corrosion minimum.

Thus, the temperature compensation of the float voltage can only be partial and a compromise between the demands of the various reactions. A further problem arises from the fact that a sufficient polarisation of positive and negative electrodes has to be observed. That float-voltage compensation can be critical in this respect and is illustrated for a vented lead-acid battery in the Figures 13.5 and 13.6.

Calculated current vs. float voltage curves are shown in Fig. 13.5. This Figure corresponds to Fig. 7.4, p.241, but is based on slightly different kinetic data (different cell type). The overcharging current increases by one order of magnitude when the cell voltage is increased by 210 mV. So, the curves in Fig. 13.5 correspond fairly well to the general rule that float current of vented lead-acid batteries is raised to its tenfold value when the cell voltage is increased by about 200 mV, and that current is approximately doubled at unaltered cell voltage per each 10 °C of temperature increase.

Fig. 13.5 shows that the compensation factor of -3 mV/°C does not really compensate for the temperature increase, but only reduces the increase of hydrogen and oxygen evolution. The current increase is reduced to 8.2 and 12.4 mA/100 Ah at 30 and 40 °C, respectively, compared to the non-compensated values of 11.3 and 23.2 mA/100 Ah. And even with the compensation factor of -5 mV/°C complete compensation is not achieved as shown by the slight increase of the current to 6.6 and 8.2 mA/100 Ah at 30 and 40 °C, respectively.

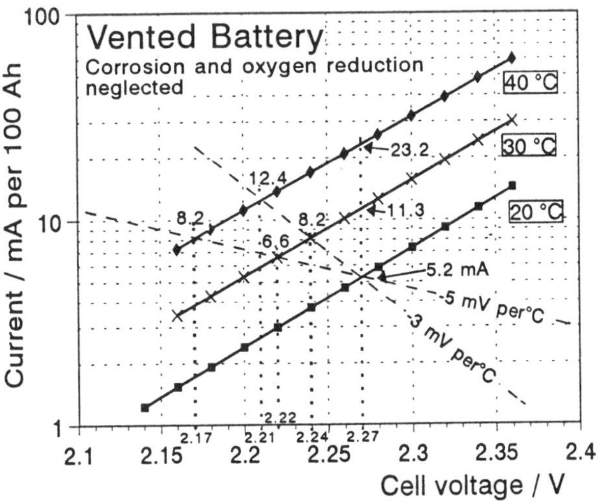

Fig. 13.5 Overcharging current versus cell voltage calculated for a vented battery, and based on the same data as the TAFEL lines in Fig. 7.2, p.239. (These TAFEL lines are shown in Fig. 13.6.) The broken lines and inserted numbers indicate the effect of cell-voltage reduction with temperature by -3 and -5 mV per °C. (Fig. 3 in ([8]).)

But even such a partial compensation of temperature causes the general problem that is illustrated by Fig. 13.6. This Figure shows the TAFEL lines for hydrogen and oxygen evolution that form the basis of Fig. 13.5 and the result that the reduced float voltage would have: at 30 °C and the float voltage of 2.22 V/cell, the polarisation of the positive electrode is reduced to $\eta_+ = 30$ mV, and so presumably below the minimum of corrosion. The situation is aggravated at 40 °C, when the float voltage of 2.17 V/cell reduces the polarisation of the positive electrode so far that significantly increased grid corrosion is to be expected and the electrode is in danger of being no longer kept fully charged.

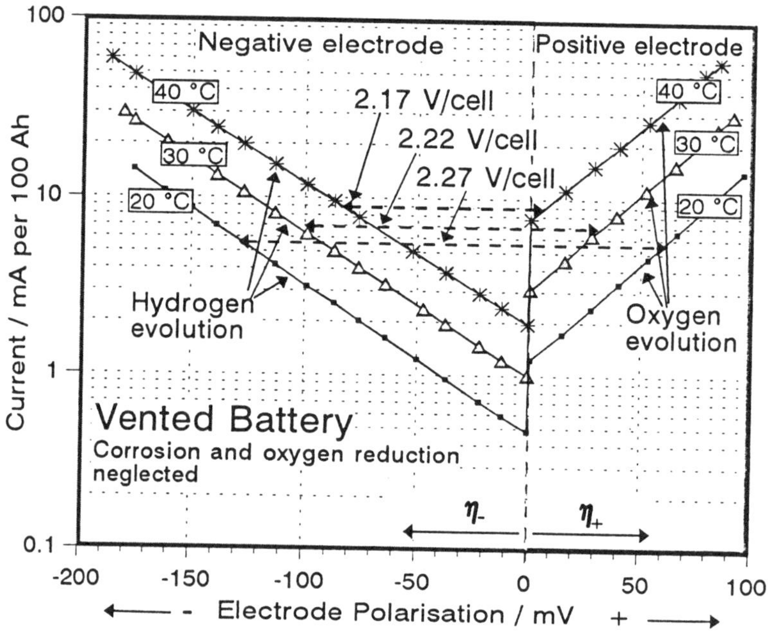

Fig. 13.6 Hydrogen- and oxygen-evolution rates versus electrode potential. Kinetics according to Table 2 in ([3]). The arrows indicate temperature compensation of cell voltage by -5 mV/°C. Horizontal axis includes open-circuit voltage of the battery. (*2.08 V/cell at the assumed acid density of 1.24 g/cm*^{3}.)

Valve-regulated batteries

The above described effects apply also to valve-regulated batteries. A fundamental difference compared to the vented design is that the polarisation of the negative and the positive electrodes are no longer determined by hydrogen- and oxygen-evolution rates. With increasing efficiency of the internal oxygen cycle, electrode polarisation is more and more a result of the balance between hydrogen evolution and grid corrosion, as described in Section 8.3. As a consequence, at low float voltages, polarisation of both electrodes is no longer proportionally increasing with float voltage as shown in Fig. 8.18, p.295. Very small or even zero polarisations have to be expected if the polarisation of the positive electrode is below the corrosion minimum (E_m in Fig. 8.19). On the other hand, the corrosion minimum corresponds to an 'ideal float voltage' since not only is corrosion at its minimum, but also hydrogen evolution is minimised as a comparison between Figs. 8.18 and 8.19 shows. This is a further objective, to keep the polarisation of the positive electrode always a little above the corrosion minimum, which means: $\eta_+ > 40$ to 80 mV.

Since the polarisation of the positive electrode is reduced with increasing temperature, float voltage should be increased with temperature. In practice, however, such an increase is not advisable since it accelerates the internal oxygen cycle and contributes additional heat that again raises battery temperature (*cf. the following Section 'Thermal runaway'*). As a compromise, it seems appropriate to choose, at room temperature, a float voltage corresponding to a positive electrode polarisation slightly above the corrosion minimum and keep this voltage unaltered at elevated temperatures.

The polarisation of the negative electrode, on the other hand, is always increased with increasing temperature. Thus, in regard to the negative electrode, float voltage could be reduced with increasing temperature. But with decreasing temperature, the polarisation of the negative electrode is reduced (cf., e.g., Fig. 18.19, p.296). Since the negative electrode in the valve-regulated battery is always prone to be depolarised too much by the internal oxygen cycle, increased float voltage at low temperatures is advisable.

Fig. 13.7 shows a recommendation based on the preceding deliberations. For the sake of the positive electrode, the float voltage is kept constant at elevated temperatures. At temperatures below 10 °C the float voltage is increased by -3 mV/°C to ensure that the negative electrode remains sufficiently polarised.

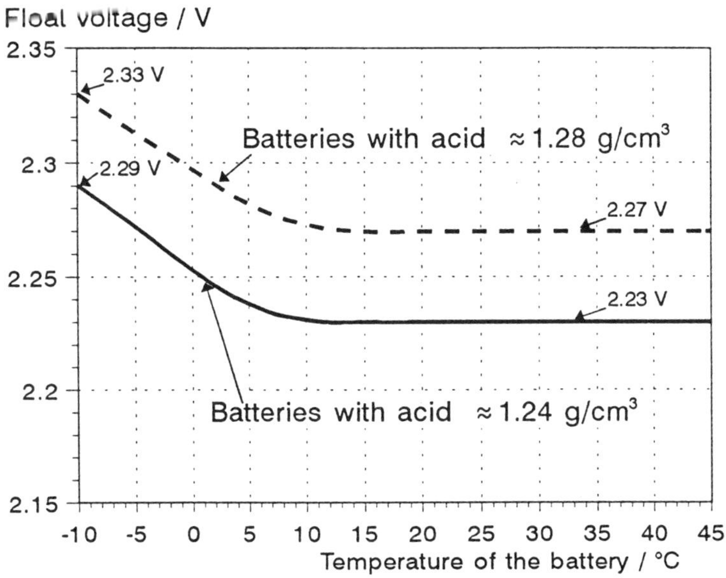

Fig. 13.7 Temperature compensation of float voltage recommended for batteries based on acid densities d ≅ 1.,28 g/cm³ and 1.24 g/cm³. Only required for longer lasting abnormal operational temperatures. (VARTA, now VHB in Hawker Batteries, cf. Fig. 21 in ([8]).)

An increase of float voltage with decreasing low temperatures has the further advantage that it ensures rechargeability of the battery within a reasonable period of time. On the other hand, increased float voltage at low temperatures can be accepted, since water loss and grid corrosion are slow even at increased cell voltage. Furthermore, no significant heat effects are to be expected, since the rate of the internal oxygen cycle is also small.

Note: *In any case, float charging at elevated temperatures should be avoided because dramatic reduction of service life is the inevitable consequence that cannot be compensated by any means, except by cooling the battery.*

Thermal runaway

Temperature compensation is often claimed as a remedy against thermal runaway. Actually, it can only contribute a marginal effect. Temperature compensation of the float voltage can neither prevent a thermal runaway nor can the lack thereof cause thermal runaway. A simple calculation shows this:

Fig. 13.8 shows the effect of a compensation factor of -3 mV/°C in a valve-regulated battery (the Figure corresponds to Fig. 8.20, p.296). A 'basis' float voltage of 2.27 V/cell is assumed at 20 °C. The compensated float voltages are 2.24 and 2.21 V/cell at 30 and 40 °C, respectively. The resulting current values are inserted at the corresponding intersections with the current curves. As far as the current is concerned, the compensation for 40°C is very effective, since the current curve is met within the section of steep slope. However, according to the above discussion, this means that the potential of the positive electrode is below the corrosion minimum (cf. Figs. 8.19 and 8.20, p.295 and p.296, respectively).

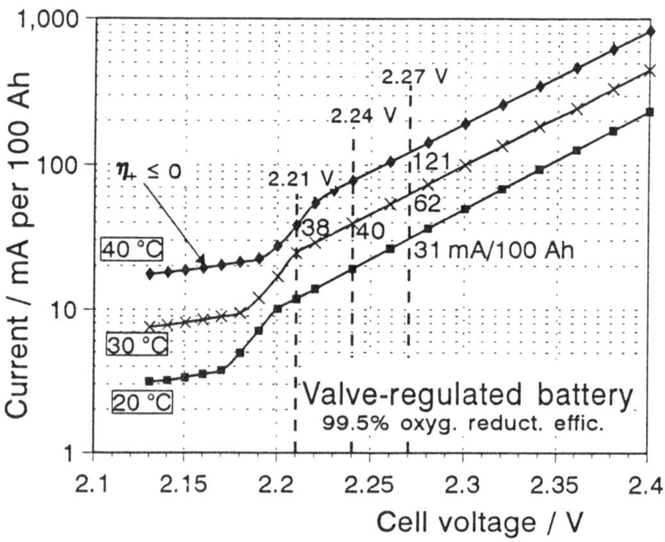

Fig. 13.8 Current curves at various temperatures vs. float voltage as in Fig. 8.20, p.296. Inserted are voltages and currents that result from temperature compensation of the float voltage by -3 mV/°C.

The current values for the various float voltages and derived figures for heat generation are compiled in Table 13.3.

Temp.	Non-compensated				Compensated by -3 mV/°C			
	V/cell	mA*	mW*	Ratio	V/cell	mA*	mW*	Ratio
20 °C	2.27	31	70	1	2.27	31	70	1
30 °C	2.27	62	141	2	2.24	40	90	1.3
40 °C	2.27	121	275	3.9	2.21	38	84	1.2
* always referred to 100 Ah of nominal capacity								

Table 13.3 Float voltages and currents marked in Fig. 13.8 and heat generation derived under the assumption that 100% of the current is consumed by the internal oxygen cycle.

Table 13.3 shows that with non-compensated float voltage at 40 °C, about four times as much heat is generated than at 20 °C. When the float voltage is compensated by -3 mV/°C only a slight increase of heat generation is observed.

When, however, these figures are compared with heat dissipation figures, it becomes significant that this effect can only have a negligible influence on thermal runaway. According to Fig. 2.12, p.57, normally about 5 W/m^2 of heat dissipates by combined radiation and air convection per °C of temperature difference between the battery and its surrounding. The surface of a cell within a block battery is in the order of 0.1 m^2 per 100 Ah (Table 4.12, p.131). Consequently, 500 mW/100 Ah would cause an increase of the temperature difference between the battery and its surrounding by about one degree. According to the values in Table 13.3, at 40 °C the increase of battery temperature between the 'compensated' and the 'non-compensated' float voltage would be less than 1 °C.

This comparison emphasises the statement that thermal runaway cannot be influenced significantly by temperature compensation of the float voltage. It may only contribute a negligible effect. The main reasons for thermal runaway are lack of heat dissipation or markedly increased float voltage on account of bridged cells in a string. Measures against thermal runaway have been described at the end of Section 4.5.3.6.

13.1.1.3 Number of Cells per Battery

When a single cell is float-charged with constant voltage and the charging voltage is determined precisely by the charging device, the current is automatically adjusted to its optimum. With two cells, the average voltage determines the current. With more cells, increasingly, the bulk of the cells determines the current, and individual cells that deviate much from the average may be over- or undercharged. Then the above statistical rules must be used to determine the required float voltage.

Recommended variation of the float voltage with length of cell string ([9]) is thus relevant only for batteries of a small number of cells.

The number of cells connected in series is not limited and does not affect the float situation, provided that the correct float voltage is supplied. No special circuit arrangements are required, but safety measures have to be observed for the operating staff in view of the high voltage of the battery.

13.1.1.4 Individual Cell Voltage Regulation

Individual voltage and charge regulators are occasionally used to reduce deviations in cell voltages or state of charge within a battery string.

One possibility is a shunt regulator that is paralleled to each cell or block and in principle consists of a resistor and transistor in series. The current that flows through this combination must be a multiple (≈ 15 x) the normal float current.
The basic circuit diagram is shown in Fig. 13.9.

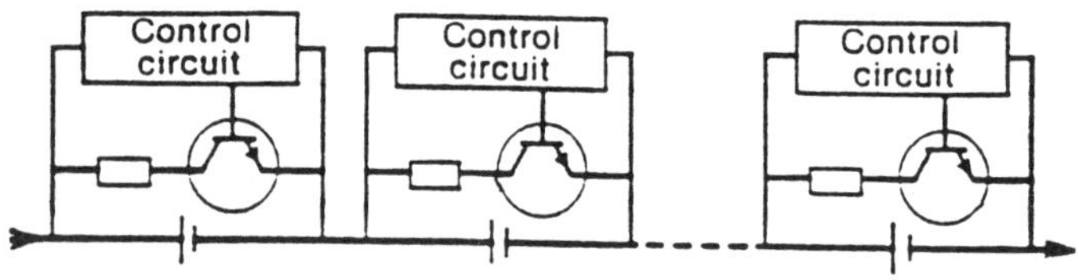

Fig. 13.9 Individual voltage regulator for each cell of a stationary battery. (ERICSSON POWER SYST. ([10]).)

The control circuit ensures that the voltage drop along the resistor/transistor combination always remains identical with the desired float voltage per cell. For a cell that requires more current than the average, the current through the shunt decreases until the voltage drop reaches the rated value. Without the regulator, the voltage of this cell would decrease.

The advantages of such a regulator system are obvious:

- Each cell or block battery is floated exactly at the desired voltage, like a single cell with an individual charger. The voltage deviation within the battery is almost zero.
- Thus, the float voltage can be reduced to values appropriate for single cells or monoblocs. Overcharging is minimised, and optimum lifetime is asserted for the battery.
- Even cells with fairly high current input remain charged properly, unless the drawn current exceeds that of the shunt. The safety factor of the battery is improved.
- The battery can be used for applications that do not allow an increased float voltage. The gap between charge and discharge voltage can be minimised.

Disadvantages of the regulator system are:

- The shunt current must be quite high, compared to the current that flows through the cells (≈ 15:1), or the range of regulation would be too small. So the system is limited to small batteries (it has been developed for 25 Ah batteries).

- The voltage regulator masks cells that draw too much current e.g. on account of an internal short-circuit, unless the current exceeds the range of regulation.
- The system is expensive, because each cell or block battery needs its individual shunt circuit.

Such regulators are of interest for special applications that require such a low float voltage that the voltage reserve of 3·σ, described in Section 13.1.1.1, cannot be observed and single cells or blocks in the battery string would be in danger of not being kept fully charged.

More ambitious solutions are 'charge equalisers' that equalise the individual cell or block voltage during charging and discharging by balancing the current ([11]). Such equalisers have mainly been developed for battery banks in photovoltaic applications where desired charging conditions often cannot be achieved and the batteries are subject to discharge cycles that include deep discharges. These kind of 'equalisers' correspond to 'battery management systems' as they are used for vehicle applications (cf. Section 14.2).

13.1.1.5 Rate of Charging, Initial Charging Current

The valve-regulated lead-acid battery can be recharged completely at the usual float voltage. The time required to restore most of the withdrawn capacity (> 90%) depends mainly on the initial charging current, as shown in Fig. 13.10. In this example, charging currents between 5 A (16-hour rate) and 30 A (2.7-hour rate) were applied, referred to a nominal capacity of 100 Ah (3-hour rate).

The decreasing curves in Fig. 13.10 represent the reduction of the charge acceptance from the high initial value to the rather small float current that finally is reached at such low charging voltages. As the Figure shows, charge duration can be reduced markedly at a high initial current without increased side reactions compared to the float charge situation, since the final voltage is not exceeded. Because of the reduced charge acceptance when full charge is approached, it takes a longer time (≈ 3 days) to recharge the last few %.

Especially for larger batteries the method is expensive due to the high investment cost of the rectifier.

The recharge period can also be reduced to a certain degree by increased charging voltage as shown in Fig. 3.11, p.84. For valve-regulated batteries, the increased float charging voltage should not exceed 2.40 V/per cell and should be applied only occasionally; otherwise, the life of the battery is reduced due to increased water loss. Furthermore, the charging current should be limited to the 5-hour rate while increased charging voltage is applied, to avoid overcharging of single cells at too high a rate. Otherwise, the maximum rate of the internal oxygen cycle can be exceeded and water loss increased. Recharge of 100% can be achieved by this method within less than 24 hours.

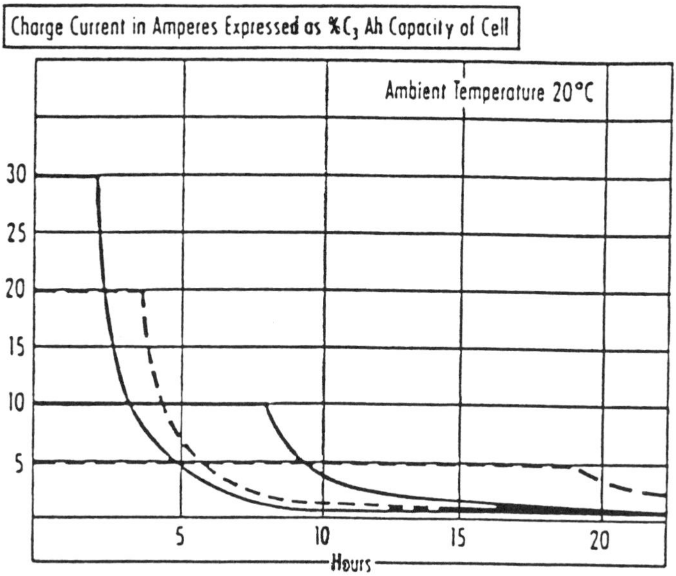

Fig. 13.10 Constant-current /Constant-voltage recharge of a maintenance-free lead-acid battery after (complete) discharge. Charging voltage 2.27 V/cell. Discharge current referred to the nominal capacity (3-hour rate ≈ 30 A if C_3 = 100 Ah). (CHLORIDE, POWERSAFE Battery([12]).)

13.1.1.6 Float Charge of Paralleled Batteries

Often, batteries are not available in the desired capacity. Parallel connection of a number of cell strings offers a simple solution. The essential rules for such a parallel connection of battery strings are:

- Only entire strings (of the full length) should be connected in parallel. There should be no cross-connections between the different strings, even if the strings are very long. (*Cross connections mask faulty cells and may cause dangerous overload of parts in the parallel strings.*)
- Only batteries of similar design should be connected in parallel (same battery type, similar plate size). *This is of importance when the battery is discharged repeatedly, since then a string with a lower internal resistance is discharged more deeply.*
- The environmental situation should as far as possible be identical for the different strings (temperature, heat dissipation). So the paralleled strings should not be installed at different levels within the same cabinet to avoid a gradient of temperature between the strings.
- The circuit arrangements for the paralleled strings (bus-bar, connecting-cable length) must provide identical resistance between the rectifier and the battery for each string, to ensure uniform current distribution.

Battery manufacturers usually recommend to limit the number of paralleled strings to 4. There is no theoretical reason for this limitation, but experience shows that it becomes increasingly difficult to secure identical conditions for the paralleled strings above a certain number.

13.1.1.7 Hydrogen Evolution During Float Charging

The float current of modern lead-acid batteries with liquid electrolyte (vented types) is (at room temperature):

10 to 30 mA/100 Ah of nom. capacity at a float voltage of 2.25 volt/cell.

According to Table 2.7, p.49, at 25 °C this is equivalent (per cell) to:

5 to 14 cm^3 of hydrogen per hour (w.r.t. 100 Ah nom. cap.).

In valve-regulated lead-acid batteries and at normal float voltage, the polarisation of the negative electrode remains small and hydrogen evolution corresponds more or less to the self-discharge rate of 1 to 3 mA/100 Ah. This is equivalent to:

0.5 to 2 cm^3 of hydrogen per hour (w.r.t. 100 Ah nom. cap.).

This means that hydrogen evolution is reduced by a factor between 5 and 10. The actual difference increases with temperature increase, because both values have to be doubled for a temperature rise of 10 °C.

Extended measurements with batteries in float conditions gave values that correspond to quite low self-discharge values ([13]). This could mean that a portion of the hydrogen may be oxidised at the positive electrode, if no other factors cause hydrogen loss (cf. Note at the end of Section 7.5.3.1, p.263), or the hydrogen evolution is reduced by a partially discharged negative electrode. Based on these measurements, limits for hydrogen evolution, e.g. < 0.3 cm^3/100 Ah per hour, are specified (2 cm^3/(cell/Ah and 30 days) ([14]), corresponding to an equivalent current of 0.6 mA/100 Ah.

Gassing measurements in lead-acid batteries, however, encounter two general problems. The first is complete sealing of the container inclusive wall openings for tubes and reference electrode, but if this is achieved, the second problem is the permeation of hydrogen gas through container walls, sealings, and tubings when the latter are made of plastic (cf. Table 2.13, p.61). In experiments, where all precautions had been observed (copper tubings), hydrogen loss occurred from the measuring set-up in the order of an equivalent of about 0.5 mA/100 Ah ([15]).

Such a hydrogen loss causes the general problem that reliable measurements are only possible when hydrogen evolution exceeds a certain value (at least the threefold of the above H_2 loss).

Note 1: *The described considerable hydrogen loss, in spite of all precautions, indicates how erratic gas measurements can be without corresponding care. For this reason, it seems questionable to include gas measurements in battery standards, unless it can be ensured that adequate measuring equipment is available.*

Note 2: *Extremely low hydrogen-evolution rates can be expected only as long as the battery is operating properly. In abnormal charging conditions (overcharge at too high a voltage, too high a temperature) the situation can be changed dramatically (cf., e.g., Fig. 4.19, p.140).*

13.1.2 Constant-Current Charging

Constant-current charging is also applied for valve-regulated lead-acid batteries, mainly for smaller types used for occasional discharges, like flashlights, garden tools or similar portable applications. Simple constant-current chargers are inexpensive, because they do not require any controlling circuit, except time-limiting equipment. The method is, however, not advisable for general use in valve-regulated batteries, since the voltage can increase more or less uncontrolled. But voltage increase always means increased hydrogen evolution and grid corrosion. Furthermore, water loss is additionally increased when the current exceeds the maximum rate of the internal oxygen cycle. Compared to nickel/cadmium batteries, this maximum rate is rather low in lead-acid batteries that can even reach critical cell voltages when they are overcharged with constant current at the 10-hour rate (I_{10} or $0.1 \cdot C(A)$) (cf., e.g., ([16])).

An improvement can be achieved when the increased constant current is applied for only a limited period and then switched to a lower value (split-rate charging) or to constant-voltage charging.

13.1.2.1 Trickle Charging

Trickle charge means continuous charge at a low rate, approximately equivalent to the internal losses of the battery and suitable to maintain the battery in a fully charged state. The current required for trickle charge amounts to < 100 mA/100 Ah, which is less than the 1000-hour rate.

Trickle charging may be used as the second step in a split-rate charging. Trickle charging is sometimes used for batteries in storage (cf. Section 13.1.6).

As has been mentioned in Section 13.1.1.2, a battery that is in danger of thermal runaway may be switched to trickle charge that secures full charge but determines energy input at a low level.

13.1.3 Taper Charging

Taper chargers (in German: W-chargers) are common for vented lead-acid batteries in traction applications, mainly because the charging device is inexpensive, especially when it is unregulated.

The charging current of a taper charger decreases with increasing battery voltage, and the curves form the 'taper', shown in Fig. 13.11, that gives this charging method its name. The reduction of the current is determined by the internal resistance of the transformer-output winding (the internal resistance of the charging device). But the ratio between the initial and final currents is not sufficient to allow the battery to be connected with a taper charger for extended periods. Furthermore, in unregulated rectifiers, fluctuations of the mains voltage cause corresponding or even stronger fluctuations of the charging current. So the taper charger is not suitable for general use with valve-regulated lead-acid batteries. It is an inexpensive charging device for batteries that are charged only occasionally, and is preferred for smaller batteries used, e.g., for hobby applications.

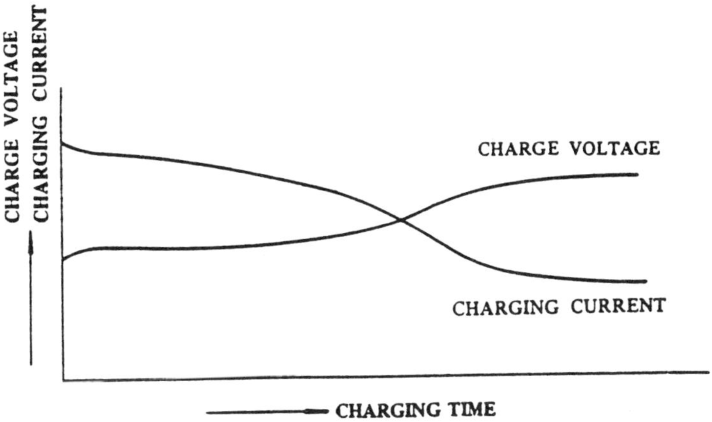

Fig. 13.11 Charging voltage and charging current of a taper charger. (From ([17]).)

13.1.4 Charging Methods for Cycle Applications

Charging time is important when the battery is applied for a cycle regime, e.g. for traction, and its full capacity has to be restored within a limited period.

Increased initial charging currents are generally used to take advantage of the fairly high charge acceptance of a discharged battery (cf. Fig. 13.10). However, a too-high initial charging current may cause increased oxygen loss, as shown in Fig. 4.16, p.128. This can be more critical when a partly discharged battery is connected to the charging device. (To avoid this oxygen loss, initial steps at reduced voltage are sometimes used.)

A general problem is the differences between the charge acceptances of the individual cells within a battery, owing to the scatter of hydrogen- and oxygen-evolution rates, as discussed for float charging in Section 13.1.1.1, Fig. 13.3, p.407. Even if these differences are marginal, they add up with increasing number of charge/discharge cycles. So slight overcharging has to be accepted, to make sure that each cell of the battery becomes fully charged within the limited time. Otherwise, individual cells do not reach full capacity, and lose more and more of their actual capacity during subsequent cycles. Thus charging methods for cycle applications always represent a compromise between several contradictory parameters, like charging time, charging efficiency and water loss. Two such charging methods are shown in the Figs. 13.12 and 13.14.

The charging schedule in Fig. 13.12 is a combination of two constant-current/constant-voltage steps. The two voltage levels are determined by the battery temperature, a probe being installed in the centre of the battery. The temperature coefficient, which is not linear, is about -5 mV/°C for this schedule.

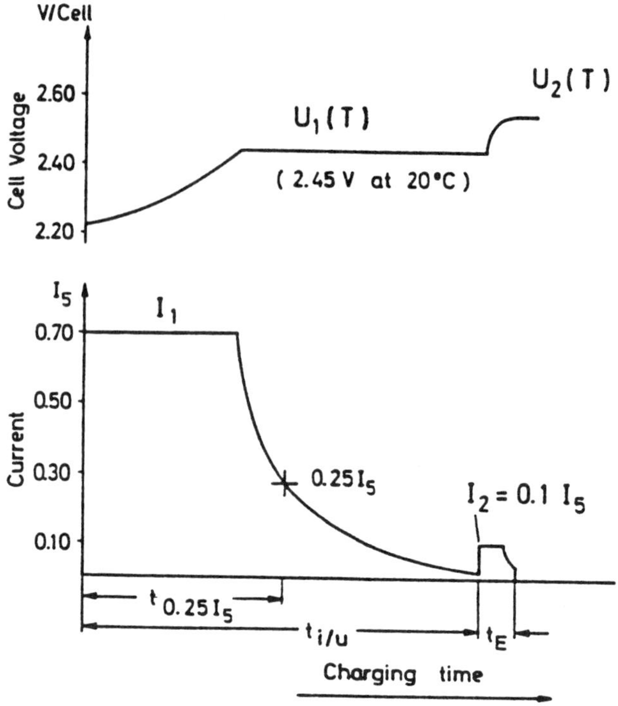

Fig. 13.12 Charging schedule for valve-regulated traction batteries, gelled electrolyte. (VARTA carat, characteristic of the VARTA multi charger.) $I_1 = 0.70\,I_5$; $I_2 = 0.1\,I_5$; The voltage levels U_1 and U_2 are determined by the battery temperature; $U_2 = U_1 + 0.12$ V. The charging time is determined by decay of the current to $0.25\,I_5$: $t_{I/U} = 1.35\,t_{0.25\,I_5} + 3$ (h). Duration of the equalising charge $t_E = 1$ hour.

The charging time $t_{I/U}$ is determined by the time for the current decay to reach $0.25{\cdot}I_5$ ($t_{0.25{\cdot}I_5}$). This current decay is a suitable measure to determine charging time to achieve an approximately constant charging factor, independent of the state of charge at the beginning. When a completely discharged battery is connected to the charging device, more time is taken before the critical current level is reached, and an extended charging time results. If a charged battery is connected to the charger, a sudden current decay reduces the charging time.

The 'equalising charge' t_E automatically follows the main charging period. It is a one-hour charge at increased cell voltage, while the charging current I_2 is limited to the fairly low level of $0.1{\cdot}I_5$ to avoid undesirable overcharge. The necessity of this equalising step is illustrated by Fig. 13.13.

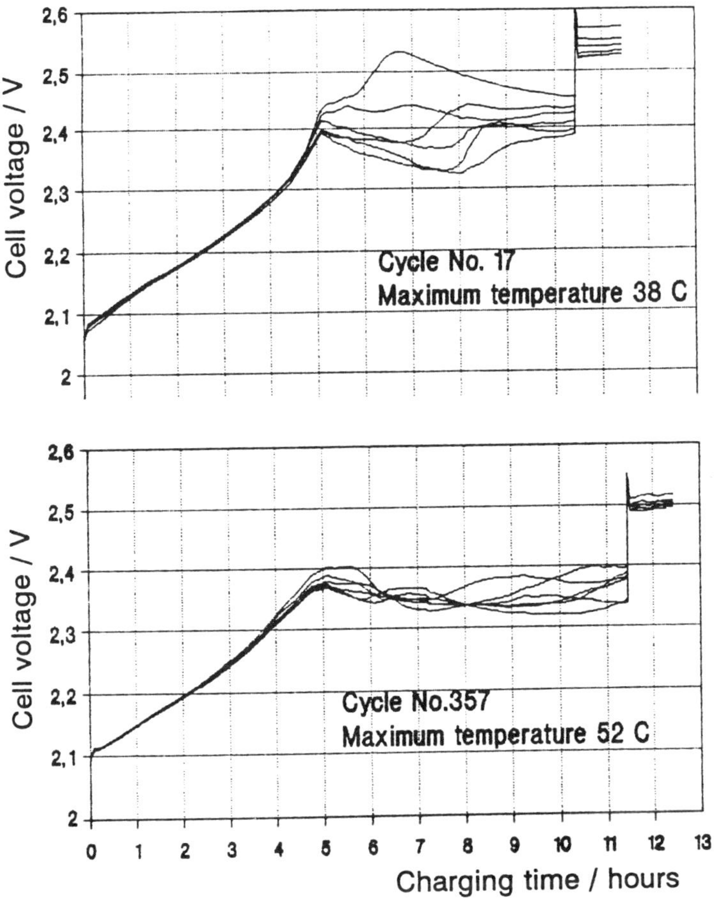

Fig. 13.13 Charging curves of 6 individual cells in a 24 V/440 Ah valve-regulated lead-acid battery (VARTA carat) while it is charged to a schedule as in Fig. 13.12.

The upper part of Fig. 13.13 shows charging curves of six out of twelve cells of a fairly new battery. The rate of the internal oxygen cycle varies markedly between the different cells, which becomes obvious during the charging period at constant voltage. Cells with an already well established internal oxygen cycle would accept more current than is provided on account of the average cell voltage. So these cells attain a fairly low voltage. Cells that still contain more liquid electrolyte, partly operate as vented types since the rate of the internal oxygen cycle is low. In these cells the negative electrodes must be sufficiently polarised for an adequate hydrogen-evolution rate. The consequence is a fairly high cell voltage, which can be observed for one cell in Fig. 13.13. Gradually, the increased water loss connected with hydrogen evolution leads to an increased maximum rate of the internal oxygen cycle and equalises the situation, as can be observed in the lower part of Fig. 13.13, where the same curves are shown after the completion of 357 cycles.

The generally increased rate of the internal oxygen cycle becomes apparent in Fig. 13.13 by the fact that a higher temperature is reached with the aged battery in comparable circumstances. This is caused by the more depolarised negative electrodes which means lower cell voltage and a correspondingly increased charging current that again increases heat. Furthermore, the charging time is extended for the aged battery, because the parasitic oxygen cycle consumes a larger portion of the charging current. Consequently, the voltage decay to $I_{0.25 \cdot I_5}$ requires a prolonged period and the charging time is extended correspondingly to the schedule described in Fig. 13.12. The charging factor gradually increases from 1.07 for the new battery to about 1.2 for the battery after about 1000 cycles.

Fig. 13.13 indicates why the equalising charge is required at the end of each of the repeated charging cycles. If the charging process were discontinued after the U_1 stage (cf. Fig. 13.12), cells with a fairly high rate of internal oxygen cycle would not be charged completely and would lose a little charge during each cycle. This situation would be aggravated dramatically if these 'weak cells' (the cells with the highest rate of internal oxygen cycle) were extremely deeply discharged or even reversed. So the equalising step is required within the charging schedule, even though gas evolution and corresponding water loss are increased by the step, as shown in Fig. 4.16, p.128.

Another charging schedule for valve-regulated traction batteries is shown in Fig. 13.14. The basic elements of this schedule are similar to those in Fig. 13.13. A current-controlled stage and a subsequent voltage-controlled period are the main parts, followed by an additional and an (optional) equalising charge with rising cell voltage. Charging time is also largely determined by current decay.

The detailed approach, however, is different to that in Fig. 13.12. At the beginning, constant-current charge is applied until the average cell voltage of 2.35 V is reached. The current is similar to that in Fig. 13.12 (about $0.7 \cdot I_5$); the subsequent constant-voltage charging occurs at a slightly lower level (2.35 instead of 2.45 V/cell). When the current falls to $0.06 \cdot I_5$, the 'additional charging' begins, characterised by constant-current pulses of decreasing duration. The voltage is not limited during this step and increases to about 2.6 V/cell. This period corresponds to some extent to the equalising charge in Fig. 13.13 and also is aimed to reach full charge. The current is about $0.1 \cdot I_5$ and is low enough not to exceed the internal oxygen cycle, to keep water loss as small as possible.

The final 'equalising' period shown in Fig. 13.14 is optional and occurs normally when the battery is not used after recharging, usually during the weekend; it is limited to 48 hours.

Figs. 13.12 and 13.14 illustrate how complex the charging schedule must be to achieve both:

- complete recharge of all cells within a limited time,
- minimum water loss during recharge.

In spite of sophisticated charging schedules, the valve-regulated lead-acid battery is inferior to conventional lead-acid batteries as far as rapid charging is concerned. The main reason is that water loss, inseparably linked with rapid charging, cannot be compensated for in valve-regulated batteries.

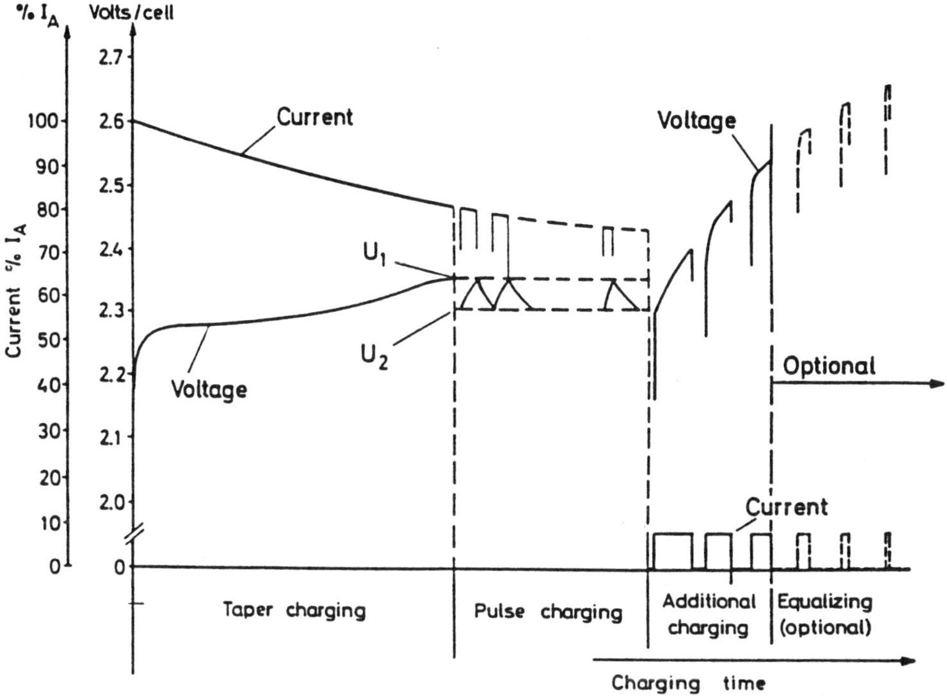

Fig. 13.14 Charging schedule for valve-regulated traction batteries (DETA ([18])). I_{nom}: nominal (maximum) current of the charger (≈ 6-hour rate).

Pulsed currents are often applied, as shown in Fig. 13.14, but there is no basic difference from continuous charging. One advantage of the pulse method is that during the current interruptions, the open-circuit voltage can be measured. This is helpful in the recharge of deeply-discharged batteries ([19]), because, owing to increased internal resistance, the voltage of such batteries exceeds the cut-off voltage immediately when the charging current is switched on. Only when the cell voltage is measured at open-circuit periods is constant-voltage recharging possible.

13.1.5 Boost Charging

Boost charging has different meanings in the field of batteries.

When stationary batteries were used with about 9% of antimony in the positive grid, '**boost charging**' or '**equalising**' described overcharging at increased voltage to restore the polarisation of the negative electrodes. 'Antimony poisoning' of the negative electrode occurred when antimony, released from the positive grid by corrosion, had been precipitated at the negative electrode. The result was heavily increased hydrogen evolution due to the reduced hydrogen overvoltage at the antimony surface (cf. Section 7.2.3.2). This 'antimony poisoning' could be overcome to some extent by regularly repeated 'boost charging' or 'equalising', since antimony and hydrogen form together stibine (SbH_3) according to the reaction:

$$Sb + 3 \cdot H^+ + 3 \cdot e^- \Rightarrow SbH_3 \qquad (13.2)$$

which escapes as a gas. This gas is very unstable and to a great extent already decomposed in the vent cap where it forms a brown antimony film. But the electrode is 'cleaned'. Since the equilibrium potential of reaction (13.2) is E^o = -0.51 V, stibine is only formed at a corresponding polarisation of the negative electrode. For this reason this kind of 'boost charging' required cell voltages above 2.6 V/cell. The benefit of such an overcharge is limited. To a certain extent it robs Peter to pay Paul, since the strong overcharging causes increased grid corrosion and accelerated ageing of the battery. But with high-antimonial batteries it is the smaller evil.

'Boost charging' of valve-regulated lead-acid batteries is also a terminated overcharge at increased voltage. It is applied to regain full recharge of partly discharged negative electrodes ([20]). It is done with such an increased voltage that the current exceeds the maximum rate of the internal oxygen cycle, so that the battery partly operates like a vented one (cf. Fig. 8.14, p.288). Then, besides hydrogen, oxygen also escapes at an appreciable rate. So the efficiency of the internal oxygen cycle is reduced and the negative electrode polarised to values $\eta_- < 0$ and recharged if it has been partly discharged before. Such boost charging is inconsistent with the principle of the valve-regulated battery since it causes increased water loss. It is not necessary, as long as the capacity of the battery is stable, even if the negative electrodes are limiting. Then balance between hydrogen evolution and grid corrosion is established, as described in Example 2 of Section 8.3.3. Forced charging would only cause renewed self-discharge of the negative electrodes. But such boost charging is the only way to recharge negative electrodes that suffer continual discharge, e.g. by oxygen intake through a leakage (cf. the last paragraph in Section 14.1.3).

In the field of traction applications, especially in connection with automatic guided vehicles, '**boost charging**' or '**opportunity charging**' ([21]) means partially charging, usually at a high rate for a short period. It is employed when the capacity of the battery is not sufficient for a full day's work or the full programme of the vehicle. Especially for automatic guided vehicles, this charging method is quite common, because there are always rest periods at the different stations where the vehicle plugs itself automatically into a charger ([22]).

This kind of boost charging is usually terminated at a fairly low voltage to avoid overcharging. The battery is partially recharged repeatedly; a complete recharge is carried out only occasionally, at least once a day. The range of the vehicle between two complete recharges can be much extended by this method.

Such boost charging is rather critical with the conventional lead-acid battery from the point of view of acid stratification (cf. Section 4.6.3), which is caused by a sequence of intermediate charges without gassing. In valve-regulated lead-acid batteries acid stratification does not occur in gelled electrolyte, and is hardly observed with absorbent glass-mat separation. Thus boost charging can be applied.

13.1.6 Refresher Charges During Storage

Refresher charges are required, if a valve-regulated lead-acid battery is disconnected from the float voltage for prolonged periods, or a new battery is stored for a long time before it is commissioned. For flooded batteries, a refresher charge is recommended every 6 months, but the length of this interval largely depends on temperature, as shown in Fig. 4.8, p.119. Due to the lower self-discharge rates of valve-regulated batteries (cf. Fig. 4.9) the interval between refresher charges may be doubled in comparison to flooded batteries.

Constant-voltage charging can be used for refreshing at the usual float-voltage setting. The charging period should be at least 48 hours. Adequate constant-current charges can be applied as well.

For the storage of valve-regulated lead-acid batteries, see Section 15.1.4.

13.2 Charging of Sealed Nickel/Cadmium Batteries

The charging methods are determined by the fundamental differences between the nickel/cadmium and lead-acid batteries systems:

With sealed nickel/cadmium batteries,

- the secondary reactions:
  - hydrogen evolution at the negative electrode and
  - corrosion of the conducting elements of the positive electrode

  can be neglected in normal operational conditions.

  So the portion of the current not consumed for charging is exclusively used for the internal oxygen cycle. The capacity balance of the electrodes, and the amount and concentration of the electrolyte, do (nearly) not change with service time.
- The oxygen-reduction rate at the negative electrode is much higher in the sealed nickel/cadmium battery than in lead-acid batteries. So overcharging is possible at much higher rates with sealed nickel/cadmium batteries without losing water.
- An increased charging potential is required for the nickel electrode to achieve complete recharge. Charging with float voltage is not sufficient in normal temperature conditions.

- At increased voltage or elevated temperatures, charging with constant voltage without surveillance is dangerous on account of the high rate of the internal oxygen cycle, which may cause thermal runaway.

The first two statements express that prolonged overcharging is less critical for sealed nickel/cadmium batteries, since the gradual water loss that is connected to overcharging of valve-regulated lead-acid batteries has no relevance. The next two are arguments against the simple float charging, generally applied for valve-regulated lead-acid batteries in stationary applications.

For these reasons, sealed nickel/cadmium batteries are not used in float charging applications, but are simply charged by constant currents unless methods of 'controlled charging' are applied.

13.2.1 Charging with Constant Current

The classical method of charging sealed nickel/cadmium batteries is constant-current charging with the ten-hour rate (I_{10} or $0.1 \cdot C(A)$), usually with a charge factor of 1.4.

The first 50% of capacity is recharged at a fairly low cell voltage, regardless of the charging rate, on account of the high initial charge acceptance. With proceeding recharge, the charge acceptance of the battery decreases and an increasing share of the current has to be carried by the internal oxygen cycle. Thus, the voltage/current relation is more and more determined by the oxygen-evolution curve (Fig. 7.9, p.253), provided that the maximum rate of the internal oxygen cycle is not exceeded. Otherwise, the negative electrode is also polarised and hydrogen evolution takes place as in vented cells (cf. Fig. 8.5, p.275).

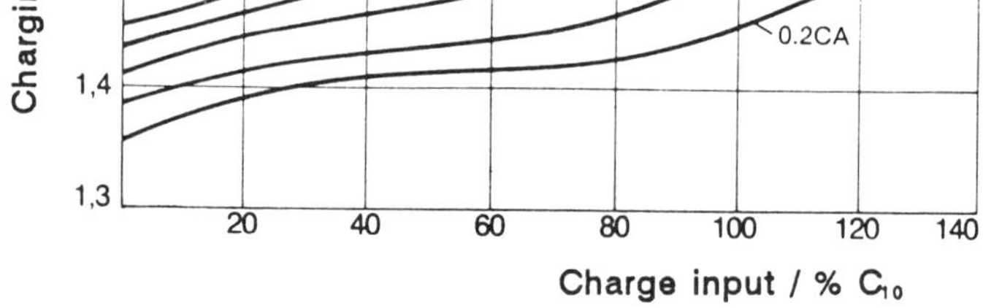

Fig. 13.15 Constant-current charging of cylindrical sealed nickel/cadmium batteries. Charging voltage versus charge input. (VARTA RS 500 (23), p.72.)

Fig. 13.15 shows charging curves of sealed cylindrical nickel/cadmium batteries, supplied with constant currents between 0.2 C(A) (5-hour rate) and 2 C(A) (less than the half-hour rate). Polarisation of the positive electrode causes the voltage increase that can be observed at charging rates between 0.2 and 1 C(A). At the higher charging rates, the internal oxygen cycle is obviously no longer sufficient to carry the overcharging current, as indicated by the steep increase in charging voltage of the curves for 1.5 and 2 C(A). This increase is caused by a corresponding polarisation of the negative electrode that now evolves hydrogen. (Definition of C(A) cf. Eq. (3.4), p.73.)

The horizontal axis of Fig. 13.15 shows the recharged energy as a percentage of the nominal capacity. 140% recharge is usually applied with constant-current recharge of sealed nickel/cadmium batteries, corresponding to a charge factor of 1.4 (cf. Table 3.4, p.76).

Curves as in Fig. 13.15 depend on temperature. At reduced cell temperatures, both the charge acceptance and the rate of the internal oxygen cycle are reduced. Charging at the 1-C(A) rate may then already cause hydrogen evolution.

As soon as the rate of the internal oxygen cycle is exceeded, hydrogen evolution occurs at the negative electrode and an equivalent amount of oxygen can no longer be reduced. This leads to an increase in internal pressure until the safety vent of the cell opens. Fig. 13.16 illustrates the pressure increase when inadmissible high overcharge rates are applied. The increase starts at between 80 and 55% of recharge, depending on the rate of charge. Up to that time, the charge acceptance of both electrodes is sufficient to carry the current.

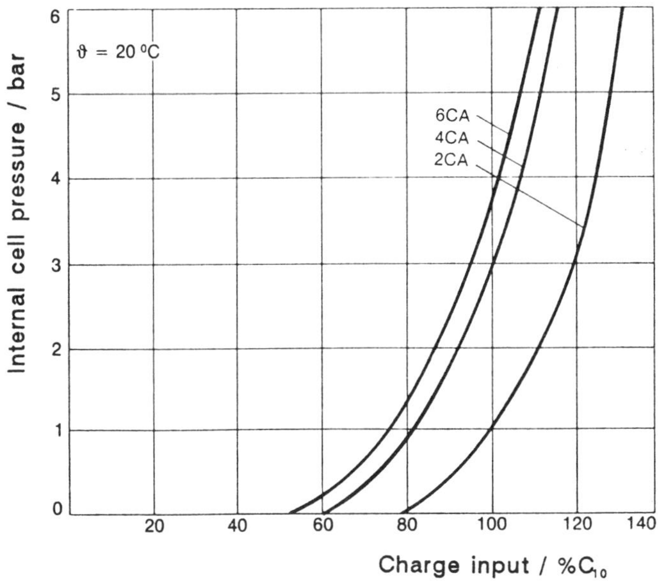

Fig. 13.16 Pressure increase during inadmissible rapid constant-current overcharging of cylindrical sealed nickel/cadmium batteries. (VARTA, RS 500 (23), p.74.)

The situation in Fig. 13.16 leads to severe water loss of the battery and reduces its service life considerably. It even destroys a button cell not equipped with a safety valve but only with a breaking point. But the Figure shows also that, during the initial stage of the charging process, such high currents are accepted. This is used in the rapid-charging methods described below.

Pressure increase during normal constant-current charging is shown in Fig. 13.17. The charging current does not exceed the rate of the internal oxygen cycle, and consequently the internal pressure remains below the critical value.

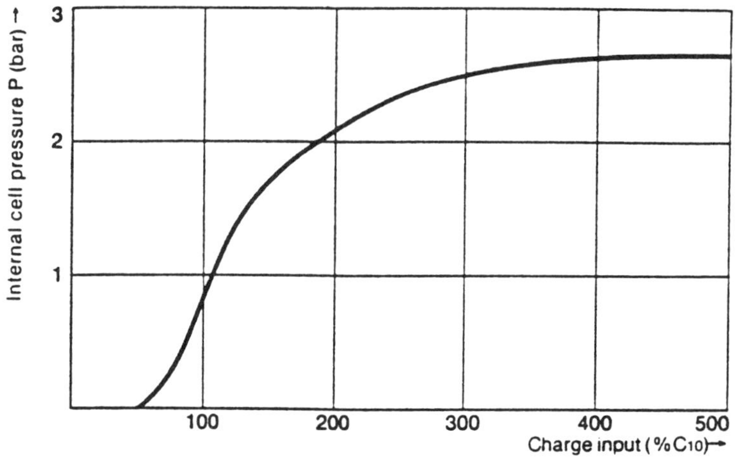

Fig. 13.17 Pressure increase during charging and overcharging of a cylindrical sealed nickel/cadmium battery with constant current 0.3 C(A). (VARTA, RS 500, ([23]), p.74.)

The increase of internal pressure starts when charge acceptance of one of the electrodes is no longer sufficient to carry the current. On account of the cell balance, shown in Fig. 8.6, p.276, in normal operation the positive electrode starts gassing (oxygen evolution). The oxygen is reduced at the negative electrode, and the internal oxygen cycle is established. At the beginning of the oxygen evolution, charging of both electrodes continues and the internal oxygen cycle carries only a small portion of the charging current. But, gradually, with increasing internal oxygen pressure, more and more of the current is used for the internal oxygen cycle, until this cycle is the only reaction occurring (cf. Fig. 5.13, p.201). A balance between gas pressure and oxygen-reduction rate is then established and the gas pressure is stabilised at a certain value, as in Fig. 13.17.

13.2.2 Accelerated Charging Methods

Acceleration of the charging process requires increase of the charging current, but inadmissible overcharging has to be avoided, which means that the charging current has to be reduced as soon as the cell reaches a certain stage of recharge. So additional parameters are required to determine this critical stage of the charging process.

Such parameters are:

- **Amount of charge**, measured directly or calculated as the product of charging current and charging time.
- **Cell temperature** that increases significantly when the internal oxygen cycle has started, on account or the connected heat generation.
- **Cell voltage** is an often used parameter. Cell-voltage decay indicates the starting internal oxygen cycle when the positive electrode is fully charged. If the charging current is not constant, the varying voltage drop has to be considered, e.g. by measuring the cell voltage during periods of current interruption.
- **Internal pressure** reflects also oxygen evolution at the positive electrodes. Its measurement is, however, rather expensive and seldom applied in practice.

The behaviour of the three parameters, cell voltage, cell temperature, and internal pressure, during constant-current charge is shown in Fig. 13.18. All three parameters are plotted versus charging time during an accelerated recharge at 1 C(A) (≈ the one hour rate). Current and temperature curves correspond fairly well when the same charging method is applied for nickel/metal hydride batteries, as shown in Fig. 13.22, p.436.

In Fig. 13.18, the cell voltage increases with charging time and reaches a maximum shortly after about 100% of the nominal capacity has been restored. But the battery is not recharged completely at this point, because a portion of the current has been used for the internal oxygen cycle. The subsequent decline of the cell voltage is caused by the rising cell temperature, mainly on account of the internal oxygen cycle. *The increasing temperature accelerates the oxygen-evolution reaction at the positive electrode and thereby reduces its polarisation.*

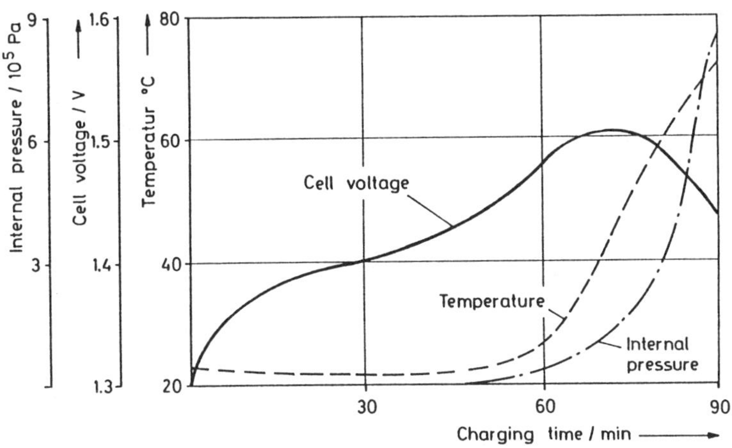

Fig. 13.18 Cell voltage, cell temperature and internal pressure of a sealed nickel/cadmium battery versus charging time during recharge at 1·C(A) (aprrox. one-hour rate) ([24]).

The temperature curve in Fig. 13.18 shows a slight decrease during the first stage of the charging process, caused by the negative reversible heat effect (cf. Eq.(5.17), p.182). With the onset of the internal oxygen cycle, heat production is increased and the cell temperature rises.

The internal pressure is related to the balance between the oxygen-evolution rate at the positive electrode and the oxygen-reduction rate at the negative electrode (cf. Section 8.1).

13.2.2.1 Voltage- and Temperature-Controlled Charging

A voltage between 1.40 and 1.45 V/cell is used for float charging of vented nickel/cadmium batteries. In the sealed nickel/cadmium battery, voltage-controlled charging without supporting measures, like temperature control or measuring the voltage decay, is dangerous. The reason is the negative slope of the temperature/time characteristic caused by heat generation on account of the internal oxygen cycle (cf. Section 5.4.2). Fig. 13.18 confirms this situation. After a charging period of about 70 minutes, the cell voltage reaches its maximum. If the cell voltage were used to terminate the charging time and the critical voltage were higher than this maximum, charging would continue with no limit, because of the falling voltage with time, on account of the further increasing temperature. If no other parameter terminated the charging process, this situation would finally lead to thermal runaway.

One decisive reason for thermal runaway is a high rate of the internal oxygen cycle, combined with the heat generated by this cycle (cf. Section 5.4.2). This heat evolution accelerates the temperature increase within the cell, which again leads to reduced cell voltage and increased charging current. In view of this danger, sealed nickel/cadmium batteries should always be equipped with a temperature sensor to cut the charging current when charged at a rate that exceeds the usual constant-current rate.

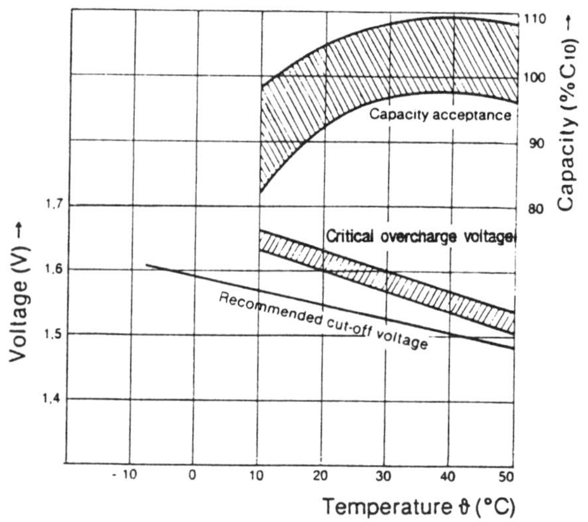

Fig. 13.19 Parameters that have to be observed during voltage-controlled charging at 1 C(A). Recommended maximum voltage and cut-off voltage are reduced with increasing temperature. VARTA SD type ((23), p.82.)

An example of the parameters that have to be observed during voltage-controlled accelerated charging with constant current is shown in Fig. 13.19. The maximum overcharge voltage should not be exceeded in any cell of the battery. The difference between the maximum voltage and the cut-off voltage describes the voltage decay after the maximum has been passed. This difference is also reduced with increasing temperature of the battery. The temperature compensation of the cut-off voltage is -2.2 mV/°C. 90 to 100% of the rated capacity can be restored within about one hour, as shown by the hatched area.

13.2.2.2 Rapid-Charging Methods

Rapid charging is wanted for many applications, e.g. battery-powered tools, which are often used with two or more battery packs that can be exchanged for recharge. Complete recharge of such a battery pack should take less than an hour, to keep the tool running without interruption.

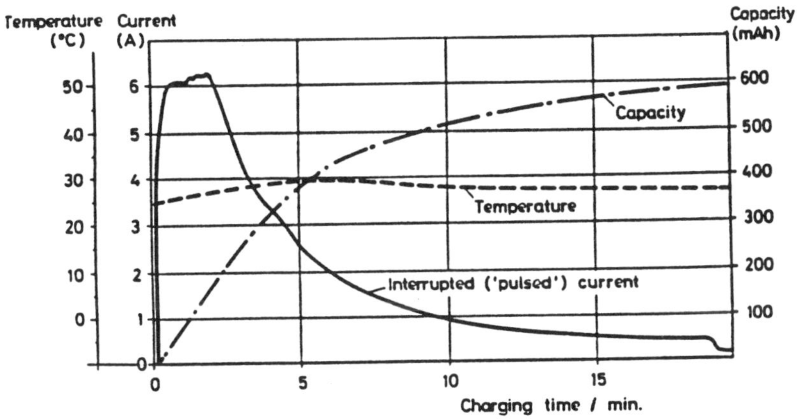

Fig. 13.20 Voltage-controlled rapid charging of a sealed nickel/cadmium battery with pulsed current. (VARTA RSE, 600 mAh.)

Several special schedules have been developed to achieve such rapid charging of sealed nickel/cadmium batteries. In all of them, the charging voltage is measured while the charging load is switched off (resistance-free voltage). This presupposes that the battery is not charged continuously, but with interruptions, easily achieved when a pulse current is applied (cf., e.g., ([25])). Fig. 13.20 shows characteristic curves for such a schedule.

The cell voltage during charging is controlled by the cell (or battery) voltage reading 10 to 30 ms after current interruption. Basically, this is a constant-current / constant-voltage regime, but it is not the voltage under load that is used as a controlling parameter, but the open-circuit voltage including concentration factors not levelled out within the interruption period.

After an initial increase, the current is approximately constant except for a slight increase due to the increasing temperature. When the pre-set voltage is reached, the current is reduced, and the voltage setting is adjusted with temperature so that the cell temperature remains nearly constant, even when the starting temperature is high.

Fig. 13.20 demonstrates that complete recharge is possible within twenty minutes by this method. However, it is recommended that this charging method be applied only to batteries made up of selected cells that have been matched in their capacity.

13.2.2.3 Time-Controlled Charging

Time-controlled charging allows rapid charging of sealed nickel/cadmium batteries, and the first attempts to introduce fast charging methods were mainly based on this principle.

A timer terminates the charge when the required amount is restored. The main problem is that the depth of the previous discharge must be known in order to recharge properly. If the previous discharge was shallow, the battery is overcharged if the full length of the charging period is applied.

To ensure a definite state of charge, the cell usually is discharged to a definite level (e.g. 0.9 V/cell) prior to recharge. On account of the definite depth of charge, a fairly high current can be used for the recharge (up to 5 C), so that it can be done within about a quarter of an hour. However, this method, the **dump-timed charge**, is recommended only for single cells ([26]). In a multicell battery, the depth of discharge may differ too much between individual cells when charging starts.

13.2.2.4 Coulometric-Controlled Charging

Coulometric-control of the charge/discharge balance is one way to avoid overcharging. Coulometric devices based on electrochemical processes as well as pure electronic devices are available, which can measure the amount of charge removed from the battery as well as the recharged amount. Both have to be balanced to evaluate the state of charge.

In practice, however, the unknown efficiency of the internal oxygen cycle complicates this balance, and a correct reset of the coulometer is only possible if sufficient time is available to complete the recharge at a tolerably low rate. But this condition is inconsistent with accelerated charging.

13.2.3 Charging of Nickel/Metal Hydride Batteries

The charging behaviour of nickel/metal hydride batteries corresponds broadly with that of nickel/cadmium batteries. Fig. 13.21 shows charging curves when constant currents between the ten- and five-hour rates were applied at room temperature. Corresponding curves are shown for nickel/cadmium batteries in Fig. 13.15.

So, the most common method of charging nickel/metal hydride batteries is the constant-current charge at 0.1 C(A) (10-hour rate) with time-limited termination (cf. Section 15.2.1). The recommended charge factor between 1.5 and 1.6 is slightly higher than that for nickel/cadmium batteries, and nickel/metal hydride batteries should not be overcharged at this rate for more than 100 hours. At high temperature, charge efficiency is reduced on account of the positive electrode (the nickel hydroxide electrode), as in nickel/cadmium batteries (cf. Section 5.2.2.1).

Charging time can be reduced when a charging current of 0.3 C(A) is used, but overcharge should then be limited to 150% and temperature of the battery controlled, to cut off the charging current when the temperature exceeds 60 °C.

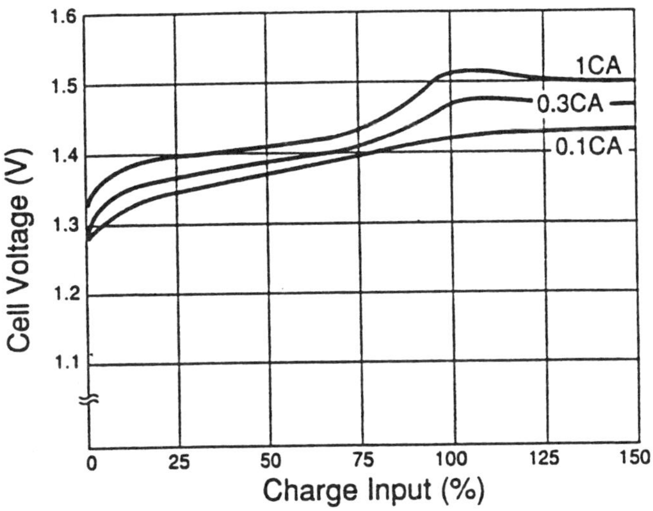

Fig. 13.21 Charging-voltage curves of cylindrical nickel/metal hydride batteries versus charge rate in percentage of the capacity (C_{10}). Constant currents: 0.1 C(A) (≈10-hour rate) 0.3 C(A) (≈ 3.3-hour rate) 1 C(A) (≈ 1-hour rate) (VARTA ([27]).)

The similar charge behaviour allows the replacement of nickel/cadmium batteries in many applications by nickel/metal hydride batteries without changing the charging device.

Rapid charging is also possible. Fig. 13.22 shows voltage and differential temperature output at a charging current > 0.5 C(A) (shorter than the two-hour rate). Charge termination by differential temperature output together with maximum temperature control are recommended to achieve the best cycle life. But temperature rise as well as voltage decrease (ΔV) can also be used for charge termination (cf. Section 15.2.2).

The shapes of cell-voltage as well as temperature curves are similar to those in Fig. 13.18, p.431, for nickel/cadmium batteries. The slight decrease of temperature during the initial stage with nickel/cadmium batteries does not occur here, because the negative reversible heat effect is masked by the heat generated by hydrogen absorption (and the i·R drop).

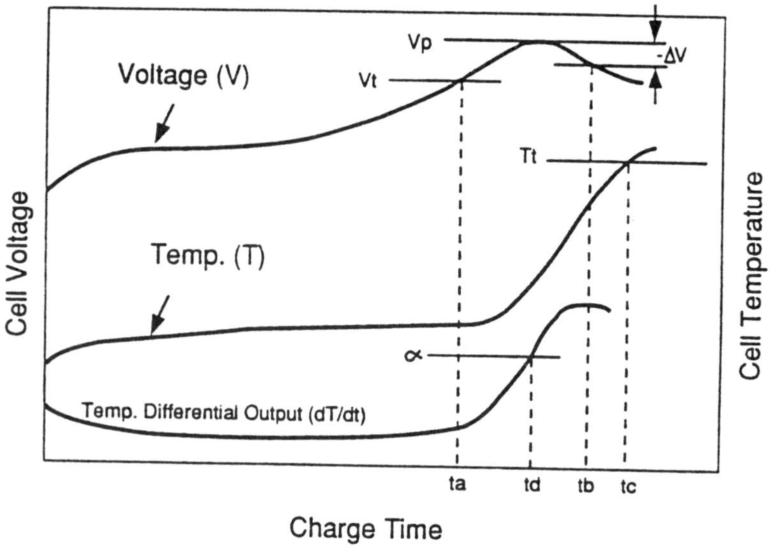

Fig. 13.22 Charging voltage, cell temperature and differential temperature output versus charging time during rapid charging of a nickel/metal hydride battery at constant current. (VARTA (27).)

References

(1) K.Boettger in H.A.Kiehne (Editor), 'Batteries', © Expert Verlag, Ehningen, 1989, p.110

(2) D.Berndt, Proc. 7th Intelec Meeting, Munich,1985, © IEEE Comm. Soc.,1985, p. 125, esp. Fig.10, p.132

(3) P.C.Milner, 'Float Behavior of the Lead-Acid Battery System', in The Bell System Techn. Journ. 49, 1970, p. 1321

(4) S.S.Misra and A.J.Williamson, Proc. 18th Intelec Meeting, Boston 1996, © IEEE Communications Soc., p. 25

(5) CHLORIDE POWERSAFE, 'Specifiers' Manual', Chloride Industrial Batteries Group, Clifton Junction, Swinton, Manchester M27 2LR, England,1990, p.18

(6) KOBE (Editor), 'Technical Handbook for Sealed Lead Acid Batteries', Shin-Kobe Electric Machinery Co. Ltd., P.O. Box 218, Mitsui Bldg., Shinjuku-Ku, Tokyo, 163, Japan

(7) D.G.Vutetakis, Proc. 34th Intern. Power Sources Symp. 1990, © IEEE Inc., East 47th Street, New York, p.61, esp. p.62

(8) D.Berndt, R.Bräutigam and U.Teutsch, Proc. 17th Intelec Meeting, The Hague, 1995, © IEEE Communications Soc., p.1

(9) Gates Energy Products (Editor), 'Sealed Rechargeable Batteries, Application Manual', Gates Energy Products Inc., Gainesville, Florida, 1989, p.160, Fig.4-41

(10) S.Bergvik and L.Bjorkstrom, Proc. Intelec Meeting, New Orleans, 1984, © IEEE Catalog No. 84,Ch. 2073-5, p. 63
B.Lindemark, Proc. 13th Intelec Meeting, Kyoto, 1991, © IEEE Communications Soc., p.196
(11) H.Schmidt and C.Siedle, Proc. Power Conversion Meeting, 1993, p.461
H.Schmidt and C.Siedle Proc. 12th Intelec Meeting, Paris 1993, © IEEE Communications Soc., p.146
N.H.Kutkut and D.M.Divan, Proc. 18th Intelec Meeting, Boston 1996, © IEEE Communications Soc., p.514
(12) CHLORIDE POWERSAFE; 'Specifiers' Manual', CHLORIDE Industrial Batteries Group, Clifton Junction, Swinton, Manchester M27 2LR, England, 1990, p.19
(13) A.I.Harrison and B.A.Wittey, 'Power Sources 10', Intern. Power Sources Symp. Committee, 1985, p. 447, and esp. Table III, p. 453
(14) 'Eurobat Guide to the Specification of Valve Regulated Lead-Acid Stationary Cells and Batteries', Edited by the Association of European Accumulator Manufacturers, Secretarial Office PO Box 5032, CH-3001 Bern, Switzerland, 1991, Table 2
(15) D.Berndt, R.Bräutigam and U.Teutsch, Proc. 17th Intelec Meeting, The Hague 1995 © IEEE Communications Soc., p.1, esp Section 4.2
(16) Gates Energy Products (Editor), 'Sealed Rechargeable Batteries, Application Manual', Gates Energy Products Inc., Gainesville, Florida, 1989, Fig.4-38, p. 156
(17) 'YUASA Application Manual', YUASA Battery Co. Ltd., 12-11 Higashi-Shimbashi, 2-Chome Minato-ku, Tokyo, 105, Japan, p. 15
(18) DETA Akkumulatorenwerke GmbH, P.O.Box 180, D-W-3422 Bad Lauterberg, Germany
(19) K.Eberts and O.Jache, 'Lead 65', Proc. 2nd International Conference on Lead, Arnhem, 1965, Pergamon Press, p 199, esp. p.201
(20) J.M.McAndrews and R.H.Jones, Proc. 18th Intelec Meeting, Boston 1996, © IEEE Communications Soc., p. 507
(21) Batteries International, Issue 3, April 1990, p. 60 (Excerpt from Electric Vehicle Association's Manual, UK)
(22) P.Preuß *in* H.A.Kiehne (Editor) , 'Batteries', Expert Verlag, 1989, p. 46 f.
(23) VARTA Batterie AG (Editor), 'Sealed Nickel Cadmium Batteries', © VDI-Verlag GmbH, Düsseldorf, 1982
(24) W.Raudszus *in* 'VARTA spezial report 6/1989', Edited by VARTA Batterie AG, P.O. Box 210540, D-3000 Hannover, p.1892 (in German)
(25) European Patent Application 0 311 460 (A2), Norvik Inc., 10.10.88
(26) Gates Energy Products (Editor), 'Sealed Rechargeable Batteries, Application Manual', Gates Energy Products Inc., Gainesville, Florida, 1989, p.63
(27) VARTA, ' Sealed Ni/MH Cells, Sales Program and Technical Handbook', © by VARTA Batterie AG, Toshiba Battery Co., Ltd and DURACEll International Inc. 1992, mailed by VARTA Batterie AG, Daimlerstr.1, P.O. Box 1363, D-7090 Ellwangen/Jagst, Germany

14. Battery Monitoring and Battery Management

Battery monitoring is of special interest for standby batteries, to get information of the condition of the battery under float and its ability to provide the reserve needed when a power outage occurs. 'Monitoring' covers a wide range of possibilities, depending on the grade of supervision. Quite simple voltage measurements and comparison of the data represent a kind of 'battery monitoring' as well as expensive installations that continuously measure many parameters and analyse the data.

Battery management systems are employed to achieve maximum service life of batteries under a cycling regime. Their main fields of application are electric vehicles and photovoltaics. A battery management system provides proper charge and discharge and prevents abuse of the battery, like deep discharges or detrimental overcharges. This may include special treatment of single cells or blocks within the battery.

Compared to a monitoring system, a battery management system interferes more actively with the charging and discharging process. However, the borderline between both terms is not precisely defined, and monitoring systems do not only collect and compare data, but their routine often also includes capacity tests and special charging procedures.

14.1 Battery Monitoring

The amount of energy that can be stored in a battery is its capacity, as described in Section 3.2. It depends on the discharge conditions and is gradually changed by ageing effects. These gradual changes are included in the 'state of health' of the battery that describes the general condition of the battery concerned. The actually available share of this capacity is influenced by a number of parameters given by the discharge conditions and by the 'state of charge'.

The actual or available capacity can exactly be determined only by capacity tests. The aim of monitoring is to deduce corresponding information without discharge from other parameters of the battery including its previous 'history'.

The state of a battery includes two aspects:

- The '**state of charge**' describes the grade of recharge in the present situation of the battery. It corresponds with the share of the capacity that actually can be drawn from the battery.
- The '**state of health**' compares the actual situation of the various components with the original design. This includes the remaining capacity, compared to that of the new battery, when fully charged. Moreover, the 'state of health' describes the general state of the battery, especially its 'ageing'. Thus, the state of health allows judgements on future performance.

The 'state of charge' and the 'state of health' cannot be determined by direct readings, because there is no parameter directly connected to the stored energy and accessible for measurement, as with a petrol gauge in a car, and the influence of the ageing processes is complex and depends not only on battery design but predominantly on operational conditions

The state of charge depends largely on the balance of the whole system; i.e. whether the current supply is always sufficient to keep the battery fully charged. Unnoticed periods of battery load are often the reason for partial discharge.

The 'state of health' concerns parameters indicating ageing effects. So, the actual capacity of the battery is subjected to changes during service life on account of factors like the loss of active material or recrystallisation effects. In the lead-acid battery, corrosion of the conducting elements (cf. Section 4.7) is an important ageing process that increases the internal resistance and influences mainly the high-rate performance. With valve-regulated lead-acid batteries, the gradual water loss described in Section 7.2 constitutes another ageing process. Water loss changes the cell balance and increases the internal resistance (cf. Figs. 4.24 and 4.25, p.150/151). Its extent cannot be determined in the valve-regulated design, because neither the amount of the residual electrolyte nor its concentration can be measured, as is usual with vented lead-acid batteries.

14.1.1 The Urgency of Battery Monitoring

The flat slope of discharge curves (cf. Fig. 3.1, p.71) does not enable simple voltage measurement during partial discharge to determine the state of charge (or discharge). Only when a considerable portion of the capacity is already drawn and the battery is approaching the end of discharge, does the increasing slope of the discharge curve allow the the discharge voltage to be used as a reliable signal. The high rate discharge curve in Fig. 3.1 has a steep slope, but the available capacity under such a load is reduced so far that, in this case again, the capacity available under moderate load cannot be deduced from partial discharges.

Devices that predict the remaining capacity while the battery is discharged, are either based on a processor that stores the discharge data of the battery, compares the

actual discharge situation with these data, and calculates the remaining discharge time under the given load (an example was the 'LOGISTRONIC ([1])); or the actual discharge situation is compared to a model of the discharge behaviour based on general kinetic data that has to be adjusted to the specific data of the battery concerned by some characteristic discharge data ([2]). But such systems already represent ambitious monitoring systems.

Capacity measurements are the only way to determine exactly the available capacity exactly, especially for valve-regulated batteries. *For vented lead-acid batteries, the state of charge can be evaluated from density measurements of the participation of the acid in the cell reaction. But density readings are not possible in valve-regulated lead-acid batteries.*

Repeated capacity measurements are not expedient for several reasons:

- Repeated discharges cause accelerated ageing of the battery.
- During the capacity measurement and the subsequent recharge period the battery is not available for an emergency.
- Capacity tests are tiresome, and can be dangerous with high-voltage batteries.
- Capacity tests are often not possible, because the required testing equipment is not available.

So capacity measurements are to be avoided whenever possible and should be replaced by battery monitoring. But, as already mentioned, even quite expensive monitoring systems can only derive data indirectly. Occasional capacity measurements are required even if partial discharges are included in the monitoring routine.

To ascertain the true condition of the battery, information concerning 'the state of health' is always required to support the information provided by capacity measurements. Only on the basis of such additional information it can be judged whether the actual capacity can be maintained into the future.

14.1.1.1 The Urgency for Valve-Regulated Lead-Acid Batteries

Monitoring of valve-regulated lead-acid batteries is especially important for several reasons:

- Valve-regulated lead-acid batteries are in widespread use as stationary batteries, usually not as large central batteries but as small units and often in outdoor or remote installations. Under these conditions, they are much more exposed to widely varying temperatures.
- Valve-regulated lead-acid batteries are often closely packed in battery cabinets, because they need no water refill and therefore open space is not required above the cells. Sometimes it is troublesome, and with high-voltage batteries also dangerous, to measure the individual voltages of cells or blocks.
- Moreover, valve-regulated lead-acid batteries are predominantly designed as block-batteries so that individual cell voltages cannot be determined.
- The scattering of the individual cell voltages within the battery during float

(cf. Fig. 13.3, p.407) exceeds that of conventional batteries markedly, especially when the battery is new, since the current/voltage relation is influenced by the internal oxygen cycle. So float-voltage measurements provide less accurate information with valve-regulated lead-acid batteries ([3]).

Monitoring of its temperature is of special importance for the valve-regulated lead-acid battery for two reasons:

- The water loss of the battery increases with temperature, and water loss means loss of performance. The user should be warned about the temperature exceeding specified limits.
- The valve-regulated lead-acid battery is more sensitive to heat effects. The internal oxygen cycle acts as an additional heat source because the current consumed by this cycle is totally converted into heat within the battery. In abnormal operational conditions, this may even lead to thermal runaway (cf. Sections 4.5.3.6 and 13.1.1.2).

14.1.1.2 Demands on Monitoring Systems

The monitoring system is expected to provide confidence that the battery will supply power for the required period of time, not only at that moment but for a reasonable period in the future. This presupposes that the 'state of charge' is known, but also requires information on the 'state of health' of the battery.

As long as the battery is in a proper state, it will achieve the required performance when it is fully charged. So:

- **the monitoring system should tell whether the battery is fully charged.**

Usually a battery consists of a number of cells or blocks connected in series; and weak cells can terminate the discharge period prematurely. So:

- **it has to be checked that each cell of the battery is properly charged (no abnormal voltage deviations).**

Problems in respect to the 'state of health' are often indicated at an early stage by gradual changes with time. So:

- **the monitoring system should enable the operator to register time-dependent changes of battery parameters.**

To preserve the 'state of health' of the battery, the monitoring system should prevent operational conditions (high temperature, too high a float current) that might be detrimental for the battery and jeopardise its proper operation in the future. This means:

- **abnormal operating conditions should release an alarm and (or) cut off the current supply.**

Deep discharges (cf. Section 3.4.1), with the danger that individual cells are reversed, may seriously reduce the service life of the battery. So:

- **abnormal operational conditions that might be detrimental to the battery should be registered to warn the operator.**

14.1.2 Off-Line Monitoring

Off-line monitoring means that the monitoring device is not continuously connected to the battery; rather, contact with the battery or its elements is established only when the parameters are measured. This kind of monitoring is usually included within the maintenance schedule of conventional stationary lead-acid batteries and is as old as these batteries themselves, although the term 'monitoring' has been introduced more recently for automatic surveillance of the battery.

14.1.2.1 Conventional (Vented) Lead-Acid Batteries

For vented lead-acid batteries, off-line monitoring is well known and standard precepts are established. Such batteries, for example in power stations or telephone exchanges, are subject to regularly repeated checking patrols, looking for irregularities like leakage, excessive gassing and low electrolyte level. Readings of the individual cell voltage, sometimes complemented by acid-density readings, are repeated at periods of some months to find out whether all the cells are properly floated.

Short-circuits within a single cell can be detected by a gradual decrease of the cell voltage below the usual float range (cf. Section 13.1.1.1, especially Fig. 13.3, p.407). The short-circuit consumes a portion of the float current and the cell requires an increased current to maintain the polarisation of both electrodes. But the float current is determined by the average of all the cell voltages within the battery, and thus the short-circuited cell cannot get enough current to supply the short-circuit and simultaneously maintain the polarisation of positive and negative electrodes for 'normal' hydrogen and oxygen evolution. As a consequence, electrode polarisation is reduced and the cell voltage is decreased. As long as positive and negative electrodes remain sufficiently polarised, only gas generation is reduced. But if the polarisation of one or both electrodes falls below the open-circuit value, the faulty cell is gradually discharged by the short-circuit.

The voltage check can provide much information. If, for example, a number of cells show voltages below or above the tolerated limit, which usually covers the average cell voltage ± 50 mV, this indicates that

- the battery may not be charged properly, or
- a number of cells suffer internal short-circuits.

This battery should be properly floated for about two more weeks and then the voltage reading repeated. Three different results can be expected:

1. All cell voltages are within the desired range. This indicates that the battery is now in a proper state and was not fully charged when the first voltage reading took place.
2. The situation is in principle not changed after the two weeks. But now other cells may show too low or too high a voltage, and there may even be more cells that deviate too much from the desired average.

   This indicates that the charge/discharge balance of the battery is insufficient, e.g. that the battery is being used unnoticed for peak shaving.

3. The situation has not changed; the same cells show too-low voltages. This indicates that the 'low-voltage cells' are deficient. For these weak cells, acid-density readings should be taken to check whether the state of charge is yet affected. As soon as the weak cells show reduced acid-density, they should be replaced by new ones.

For vented lead-acid batteries that are checked regularly according to the above recommendations, capacity tests are required only after prolonged service periods. The Deutsche Bundespost, for example, starts capacity tests with a normally operating and continually checked battery only when the battery is approaching the end of its life. This means that capacity measurements are started after about 13 years of service life ([4]).

Note: *Acid-density readings are not required for cells that are within the voltage tolerance, since it cannot be that such a cell has a too-low acid density, unless the battery had not been commissioned properly and started with different acid densities in various cells. Such superfluous density readings are not only time consuming, but often the battery is spoiled by spilled acid. For this reason, acid-density readings should only be made when the cell voltage indicates a critical state of the concerned cell.*

14.1.2.2 Valve-Regulated Lead-Acid Batteries

The above checks can in principle be transferred to valve-regulated lead-acid batteries. However, as mentioned in Section 14.1.1, a number of effects reduce the evidence of the tests:

- The scattering (cf. Fig. 13.3, p.407) of the individual cell voltages is larger than in the vented design on account of the internal oxygen cycle. The reason is that the depolarisation of the negative electrode is mainly influenced by the efficiency of the internal oxygen cycle. This efficiency changes with time on account of water loss ([5]).
- Valve-regulated lead-acid batteries are mostly used as block batteries of three or six cells (6 and 12 V blocks, respectively). The standard deviation is increased by factors of 1.73 and 2.45, respectively. ($\sqrt{n}$, as shown in Section 13.1.1.1.)

Both factors (the increase of the voltage deviation inherent in valve-regulated lead-acid batteries, and the fact that these batteries are predominantly used as block batteries) reduce the worth of voltage readings. Furthermore, acid-density readings which provide an important complementary parameter, are not possible. So capacity measurements are required more often with valve-regulated lead-acid batteries when the state of the battery has to be known exactly.

14.1.2.3 Automated Off-Line Measuring Devices

The above method of checking a battery is time-consuming and has the disadvantages that the results have to be transferred into files and that it is cumbersome to compare the data of the different checks to derive information on the dependence of the data on service time, required to determine the 'state of health'. Many attempts have been made to automate the monitoring process.

One solution was presented at the 1990 Intelec Meeting (6). This battery maintenance-system incorporates a portable measuring and data-logging unit that stores the voltage readings in an electronic memory. Additional data, like acid density or amount of added water, can be fed into the logging unit manually. Later, these data can easily be transferred to a personal computer and treated by a specially developed program for data analysis. The software enables the operator to compare data of subsequent measurements and analyse the development with time of the different parameters and also of the individual cells (or blocks).

Off-line monitoring systems are of special interest where a great number of multi-cell batteries have to be supervised and permanent installation of a monitoring system would be too expensive, or where the additional wiring for a monitoring system would not be allowed for safety reasons.

14.1.3 On-Line Monitoring

On-line monitoring means that the supervising installation is continuously connected to the battery. On-line monitoring gives improved reliability, achieved by continuous measurement, and meets the difficulties in measuring the required data on account of the usual compact installation of valve-regulated lead-acid batteries, mentioned above. Furthermore, the last three of the demands in Section 14.1.1.2, concerning the avoidance or at least registration of battery abuse, can only be fulfilled by on-line monitoring.

Different methods can be used for this kind of monitoring:

A fairly simple method of on-line monitoring is used in sealed nickel/cadmium batteries for the 'state of charge' display in electric shavers. The battery voltage is measured under load and used to evaluate the state of charge. The accuracy of this method is rather limited, but the result is sufficient to inform the user, especially to warn him that the battery will soon be exhausted.

The dilemma of on-line monitoring systems is that simple systems do not provide enough information, while rather sophisticated systems are too expensive for the usual application. So, on-line monitoring systems are mostly looking for compromises between both extremes.

A monitoring system for all kinds of batteries that can be used for off-line as well as on-line monitoring has been described in (7). Up to 300 cell voltages can be monitored, and the overall voltage of the battery and the current and temperature are continuously monitored. Automatic discharge tests are also possible when corresponding devices are installed.

On-line monitoring of stationary valve-regulated lead-acid batteries is used on a large scale for decentralised telephone exchanges applied by British Telecom (8). Three valve-regulated lead-acid batteries (48 V) are installed in a power-supply rack with the electronic equipment, including the charger. The system records a number of measurements, automatically including partial discharges. For a variety of discharge currents, the measured data are compared with the curves published by the battery manufacturer. Furthermore, comparison of the individual discharge currents of three paralleled batteries detects divergences.

Monitoring individual cell voltages is advantageous because cells working at an increased voltage cannot mask others with too low a voltage. But, as mentioned above, the investment costs for such a system are high ([9]). To reduce the cabling that connects each cell or block to the monitoring system, groups of cells or batteries are often used for monitoring instead of individual units. Such a reduction is desirable not only for economic but also for safety reasons, because additional wiring is always a possible source of short-circuiting. The simplest way of forming groups is to divide the battery into two groups (mid-point voltage method). To detect discrepancies, the voltages of the two halves of the battery are compared, but only quite significant deviations can be detected by this method. In connection with partial discharges, cells with too-large capacity deviations can be detected by comparison of the mid-point voltage of the battery halves ([10], [11]).

A monitoring system is described in ([12]) that divides the battery into a number of groups (at maximum 18) for comparison and complies with the above listed demands (Section 14.2.2). The measured data: group voltages, current, and the temperature of two cells, are regularly collected and stored by devices that are fixed to the battery. Changes of the data with time are interpreted for error detection and to produce error signals. Alarm signals are also stored and can be identified later. The monitoring device can be connected to an interfacial unit, like a personal computer. To determine the remaining capacity during discharge, the measured voltage of the battery is compared with characteristics for different discharge rates, supplied by the manufacturer. A forerunner of this method has been used in systems for special emergency power supplies ([1]).

The monitoring systems described or mentioned in this Chapter are but a few examples of a great number of comparable devices on the market. More and more often, monitoring systems are included in the power-supply system.

Occasionally, boost charging is included in monitoring routines for valve-regulated lead-acid batteries ([11]). This is more or less a revival of the boost charging for stationary batteries that was usual with high-antimony batteries as mentioned in Section 13.1.5. But now it is not the antimony poisoning that is to be removed, but depolarisation of the negative electrode above its equilibrium potential by oxygen that causes partial discharge (cf. Section 4.4.1). Frequently such a depolarisation is caused by oxygen intake through a leakage. Normally, valve-regulated lead-acid batteries do not require such boost chargings, that reduce service life by increased water loss. But it is the only way to recharge the negative electrodes. It should, however, be considered in the particular case, whether it would be more economic to replace the faulty cells or blocks than to regularly subject the whole battery to such ageing processes.

14.1.4 Impedance Measurements

Continual monitoring of the float voltage indicates at an early stage impending failure of the battery when short-circuits are formed by dendrites between the plates or direct contact between the grown positive plate and the negative bus-bar. Both represent the usual reasons for premature failure of cells in stationary, flooded lead-acid batteries.

Voltage measurements, however, cannot indicate two additional failure modes that occur in valve-regulated batteries. One is the loss of electrolyte, the other failure mode is caused by corrosion either of the conducting elements in the negative plate group or the intercell connectors in block batteries. Corrosion of intercell connectors occurs, of course, also in flooded monoblocs, but they are not used so often in stationary applications, compared to the valve-regulated version.

Loss of electrolyte and corrosion of the conducting elements influence the internal resistance of the cell. But this cannot be detected by float-voltage measurements, since the voltage drop caused by the internal resistance is small and its contribution to the float voltage negligible. For example, a float current of 30 mA/100 Ah causes in a cell with an internal resistance of 1 mΩ (cf. Table 3.7, p.90) a voltage drop of 30 μV.

Conductance or impedance measurements that reflect the internal resistance, as described in Section 3.6.4, can be used to determine changes of this resistance. In recent years, a huge number of conductance measurements have been carried out and these have shown that faulty cells can be detected with great accuracy ([13]). Mostly the battery strings or groups are compared. Corresponding to the mid-point voltage method, mid-point conductance monitoring compares the two halves of a battery string ([14]). Usually, single frequency impedance measurements are applied. The whole impedance spectrum of a battery provides more information, but it is difficult to distinguish between capacity reductions caused by lack of charge and those due to ageing of the battery ([15]).

14.2 Battery Management

The main task of a battery management system is to secure maximum cycle performance of the battery. This means that charge as well as discharge have to be controlled to prevent operations that would reduce service life. This includes control of the uniformity of all cells and their performance, since cells that are deeply discharged or overheated compared to the rest of the battery would limit the service life of the whole battery. On account of their participation in the interplay between battery, charger, and load, battery management systems usually are tailored to the battery and its application.

Objectives of a battery management system are:

- Monitoring and protecting the battery. This means, in detail:
  - Voltage measurements of cells, blocks or groups of both. Comparison of these data not only between the cells or groups, but also comparison with previously measured data.
  - Charge equalisation to prevent weak cells or blocks being increasingly over-discharged, which may cause preliminary failure of the battery.
  - Monitoring the temperature of the battery and its uniformity by corresponding regulation of heating or cooling systems. Adjusting the charging process to battery temperature and, if necessary, interrupting the charging process.
  - Overcharges must be prevented by interrupting the charging process or

reducing the charging voltage.
  * Deep discharges must be prevented by disconnection of the battery. This, however, requires warnings before the battery is switched off, since that may provoke critical situations, especially in the case of electric vehicles.

- Information for the user:
  * Indication of the state of charge.
  * During discharge, indication of the still available capacity.
  * Indication of critical cells or blocks in the battery.
  * Fault diagnosis.

Battery management systems often include charge equalisation between cells or blocks during discharging and charging ([16]). This means that equalisation currents flow between 'strong' and 'weak' cells so that the cells with lower capacities are neither deeply discharged nor overcharged.

14.3 Smart Batteries

'**Smart batteries**' is a new term and means batteries or battery packs with a certain 'intelligence' ([17]). The idea is to have connected to the battery and included in its casing a processor that stores data of the battery, like performance as a function of temperature, or charging data. It is more or less a monitoring system that is included in the battery casing. In combination with a host system, the battery itself can provide information of its remaining capacity, and the battery can control the charging regime. The idea is that the battery itself maintains its performance data and also data that concern its 'state of health'. Its 'intelligence' always achieves optimum discharging and charging conditions since it carries the required data with it. With such batteries, adjustment of the users devices is not required when he changes the battery system.

Fig. 14.1 shows such a battery pack that includes a processor.

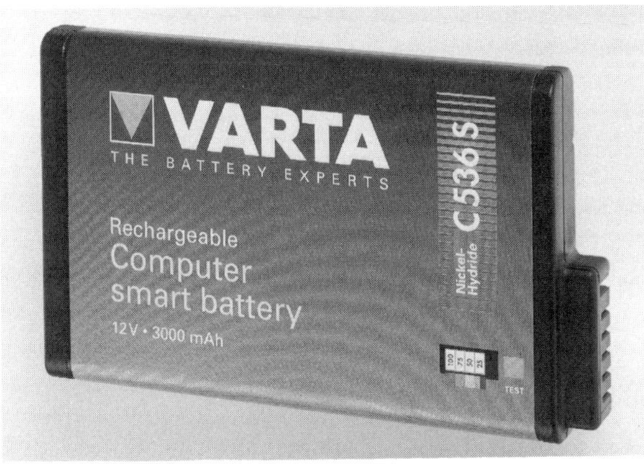

Fig. 14.1 Nickel/metal hydride battery pack as a 'smart battery'. Cylindrical cells, 12 V; 3000 mAh (VARTA).

The display on the screen of a notebook for a corresponding battery (VARTA 515S) during charging is shown in Fig. 14.2. This information is provided by the battery itself. It concerns the 'state of charge' (in the bottom, on the left hand side), but also the history of the battery (in the top, on the right hand side). Furthermore the actual 'state of charge' is displayed.

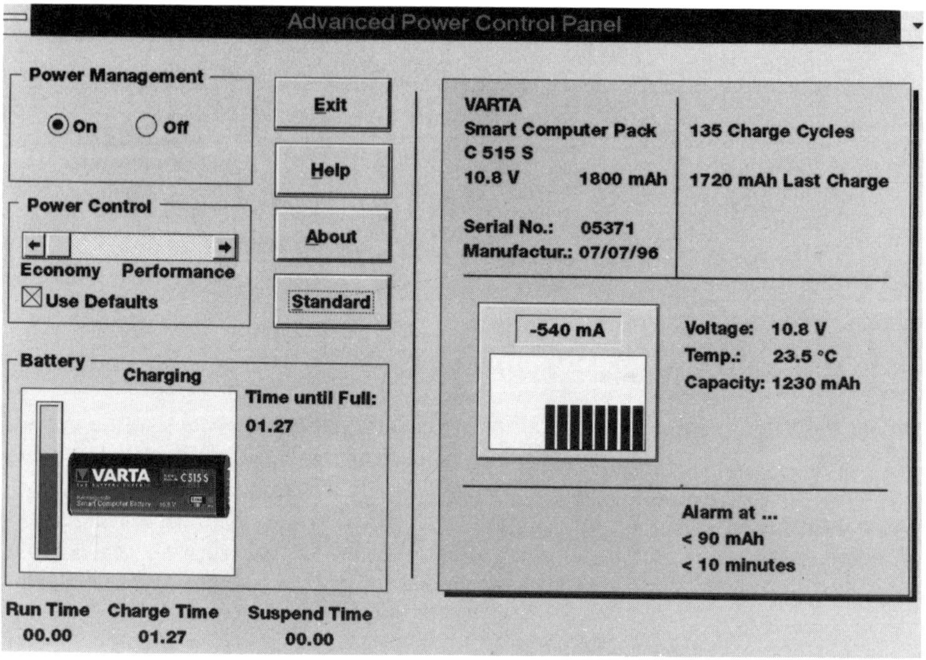

Fig. 14.2 Display of battery data on the screen of a notebook, based on the information by the 'smart battery'.

References

(1) D.Berndt and H. Tappeiner, Proc. 8th Intelec Meeting, Toronto, 1986, © IEEE Communications Soc., p.585

(2) P.K.Ng, M.Hirsch and T.V.Nguyen, Proc. 18th Intelec Meeting, Boston 1996, © IEEE Communications Soc., p.616

(3) R.F.Nelson, Proc.12th Intelec Meeting, Orlando, 1990, p. 165
D.Berndt, Proc. 7th Intelec Meeting, Munich, 1985, p.125
both IEEE Publishing Inc. publications

(4) H.Krakowski, *in* H.A.Kiehne (Editor), 'Batteries' Expert Verlag, 1989, p. 172

(5) F.J.Vaccaro and R.E.Landwehrle, Proc. 13th Intelec Meeting, Kyoto, 1991, IEEE Communications Soc., p.20, esp. p.24

(6) V.Maunula and T.Keränen, 'Computerized Battery Log', Proc. 12th Intelec Meeting, Orlando, 1990, p.242

Short review in Batteries International, Issue 6, January 1991, p. 77

(7) T.Williams and M.Tebbutt, Batteries International, Issue 8, July 1991, p. 44

(8) A.I.Harrison, N.E.Bagshaw and E.Thompson, Proc. Intelec Meeting, Washington, D.C., 1982, IEEE Publishing Inc., p. 291

J.A.O'Connor, 'Remote Monitoring of Sealed Lead Acid Batteries in Telephone Exchanges', Proceedings of the 5th ERA Battery Seminar, ERA Technology Ltd., Cleeve Road, Leatherhead, Surrey KT22 7SA, England, 1989, Paper 2.2.1

(9) J.S.Parsons and P.Sanctuary, Proc. 13th Intelec Meeting, Kyoto, 1991, ©IEEE Communication Soc., p.171

(10) A.Glad, P.Waltari and T.Suntio, Proc. 13th Intelec Meeting, Kyoto 1991, ©IEEE Communication Soc., p.189

(11) K.E.White, Proc. 18th Intelec Meeting, Boston 1996, © IEEE Communications Soc., p.624

(12) D.Berndt, E.Meissner, F.J.Kruger and R Luetkehoelter, Proc. 13th Intelec Meeting, Kyoto, 1991, © IEEE Communications Soc., p.181

(13) D.O.Feder, Proc. 17th Intelec Meeting, The Hague 1995, © IEEE Communications Soc., p. 22

(14) M.J.Hlavac and D.O.Feder, Proc. 18th Intelec Meeting, Boston 1996, © IEEE Communications Soc., p.632

(15) J.M.Hawkins and R.G.Hand, Proc. 18th Intelec Meeting, Boston 1996, © IEEE Communications Soc.

(16) H.Schmidt and C.Siedle, Proc. Power Conversion Meeting, 1993, p.461

H.Schmidt and C.Siedle Proc. 12th Intelec Meeting, Paris 1993, © IEEE Communications Soc., p.146

N.H.Kutkut and D.M.Divan, Proc. 18th Intelec Meeting, Boston 1996, © IEEE Communications Soc., p.514

(17) Batteries International, Issue 27, April 1996, p. 42

15. Storage, Ageing, and Disposal of Batteries

15.1 Storage of Batteries

Batteries often have to be stored for a number of reasons. The storage can concern:

1. New batteries that may have to be stored before they are delivered. This happens in the warehouse of the manufacturer or the dealer. Small batteries (mainly primary batteries) are also stored in the home of the customer. With larger stationary batteries it may also happen that they have to be stored on account of delays at the site where the battery is to be installed. Occasionally new batteries are stored for prolonged periods as spare cells or blocks.
2. Batteries that already have been in service may have to be stored for interim periods, for example, with farm equipment, military vehicles, sailing boats or many hobby applications. Occasionally, even large stationary batteries are stored after use for a certain period when they are moved to another site.

15.1.1 Storage of 'Wet' Lead-Acid Batteries

As soon as the lead-acid battery is filled with electrolyte, corrosion of the positive grid and self-discharge begin (described in Sections 4.7 and 4.4, respectively). Prolonged storage without refresher charges not only results in gradual self-discharge that may end with a deeply-discharged battery (Section 3.4.1), but also is detrimental for two further reasons:

- The positive grid of a deeply-discharged battery is prone to increased corrosion attack, owing to degradation of the passivation layer, as described in Section 4.7.1.

- The self-discharge causes dilution of the sulphuric acid, which finally may reach such a low concentration that the solubility of lead sulphate is markedly increased (cf., e.g., ([1])). When the battery is recharged, the high Pb^{2+} concentration favours the growth of lead dendrites from the negative electrodes that can bypass the separation and may cause short-circuits.

For both reasons, lead-acid batteries, filled with electrolyte, require refresher charges about every 6 months (cf. as indicated in Fig. 4.8, p.119 and also in Section 13.1.6), first of all to preserve the protecting layer of lead dioxide on the grid surface.

The above problems always arise when lead-acid batteries are stored after service, because batteries, once filled with electrolyte, can never be converted again into the 'dry' state, described in Section 10.7.1, but have to be stored in the 'wet' state.

15.1.2 Storage of Lead-Acid Batteries 'Charged and Dry'

To avoid the above described problems during storage, conventional lead-acid batteries are often delivered 'charged and dry', which means that the battery is delivered with charged electrodes but without electrolyte. It requires that the negative electrodes have been specially treated to preserve nearly full capacity. This treatment is briefly described in Section 10.7.1.

Such 'dry charged' batteries keep their capacity for a practically unlimited time in proper storage conditions. When filled with electrolyte, they can deliver about 80% of their normal capacity without any recharging, and their service life starts with the filling date.

Proper storage conditions mean mainly that the battery is protected against moisture, because water catalyses the oxidation of the negative electrode and annuls the measures for charge preservation. For this reason, large variations in temperature should be avoided, because of the possible condensation of water within the battery when it is cooled down. Furthermore, the storage temperature should not exceed 45 °C, because all degradation processes are accelerated by temperature.

15.1.3 Storage of 'Emptied' Lead-Acid Batteries

'Charged and dry' batteries can be manufactured only by tank formation (cf. Section 10.7.1). When 'container formation' according to Section 10.7.3 is applied, the required special treatment of the negative electrodes is not possible. For this reason, container-formed batteries are always 'wet', i.e. filled with electrolyte. When certain storage periods are to be expected, e.g. prolonged shipping periods, these batteries often are 'emptied', to reduce the problems during storage. It means that they are discharged to about 50%. Then the battery is emptied of electrolyte and sealed by a valve that prevents the access of oxygen. The benefits of this treatment are:

- Self-discharge is reduced by the half-discharged state and stops when the residual sulphate is consumed.

- The increased dissolution of lead cannot occur, because of the minimal residual volume of electrolyte.

When these batteries are taken into service, they have to be refilled with acid of appropriate concentration.

This method, however, is limited to storage periods of a few months. Otherwise, the battery suffers as mentioned in Section 15.1.1.

15.1.4 Storage of Valve-Regulated Lead-Acid Batteries

Valve-regulated lead-acid batteries are always delivered filled with electrolyte, and the self-discharge and corrosion processes, mentioned in 15.1.1, are unavoidable from the beginning. The situation is relieved by the fact that the design of valve-regulated lead-acid batteries automatically reduces self-discharge and corrosion rates to minimum values. Nevertheless, during long-term storage, valve-regulated lead-acid batteries require periodic refresher charges as described in Section 13.1.6.

Some manufacturers store valve-regulated lead-acid batteries in the 'charged and dry' state in the factory before they are filled with electrolyte, recharged and shipped. Then, the filling date, representing the beginning of service life, is often marked on the label of the battery ([2]).

15.1.5 Determination of the State of Charge

In lead-acid batteries, the acid dilution connected with self-discharge reduces the open-circuit voltage with ongoing discharge according to the relation between acid density and equilibrium voltage shown in Fig. 4.2, p.107. Thus it suggests determining the residual state of charge by measuring the open-circuit voltage ([3]).

Basically, this is possible when the amount of electrolyte per Ah and the cell balance are known, but the accuracy of this method is limited for several reasons:

- The contribution of the different self-discharge mechanisms, shown in Table 4.8, p.119, is not exactly known. When only the negative electrode suffers self-discharge, half the amount of acid is consumed by capacity loss, compared to uniform self-discharge of both electrodes.
- The open-circuit potentials of the electrodes, as 'mixed potentials', (cf. Section 6.3) cannot be measured as accurately as true equilibrium potentials.

Thus the accuracy of this evaluation is limited.

15.1.6 Storage of Ni/Cd and Ni/MeH Batteries

Nickel/cadmium batteries, filled with electrolyte, can be stored for a practically unlimited time. They are gradually discharged by self-discharge reactions, described in Section 5.3. But this does not cause either permanent capacity degradation, or reduction of life expectancy. Corrosion problems, as in lead-acid batteries, do not exist.

Concentration changes of the electrolyte are small, and solubility problems, as in lead-acid batteries, are not observed. So vented nickel/cadmium batteries are often completely discharged before being stored or shipped (which reduces the risk of short-circuits by mechanical contact between cells). Subsequently, when the batteries subsequently are recharged, nearly full capacity is reached. Only after long-term storage does the electrode material have to be reactivated by a few charge/discharge cycles.

The same situation applies to the storage of sealed nickel/cadmium batteries and nickel/metal hydride batteries. They lose capacity much faster than lead-acid batteries (cf. Sections 5.3.1 and 6.6.1, respectively). But neither capacity degradation nor reduction in life expectancy is observed, when they are stored for long periods in the completely discharged state.

Occasionally, larger vented nickel/cadmium batteries are delivered in the dry state (without electrolyte), because they can be shipped more easily. But this state cannot be compared to that of the 'dry charged' lead-acid battery. Instead, after installation and filling with electrolyte, they have to be commissioned according to the recommendations of the manufacturer. This means that the active material has to be activated by a rather laborious charge/discharge treatment.

15.2 Ageing of Batteries, Life Expectancy

Some basic aspects and definitions that concern the ageing of batteries have been shown in Section 3.7. Many parameters influence the service life of a battery. Some of these ageing factors, like grid corrosion and separator degradation, depend largely on calendar time, which means that even batteries that are only occasionally discharged are subject to ageing.

Battery design is important for service life. Each battery has to be adapted to the special requirements of its application, and this results in a great number of battery types, as indicated in Chapters 10 to 12. These battery designs are always compromises. Life expectancy, for example, has to be sacrificed when extremely high-load performance and low weight are indispensable attributes, as in starter batteries for motor cars.

For this reason, it is quite impossible, to quote general figures for live expectancy. Life expectancy of batteries largely depends on the special type of battery and the operational conditions. Some reference points are:

* For vented lead-acid and nickel/cadmium batteries under normal float conditions (20 to 25 °C) a service life in the order of 15 years is expected at an actual capacity ≥ 80% of the original value.
* Valve-regulated lead-acid batteries are on the market with expected service lives under float charge in the wide range between 3 and more than 10 years, as indicated by Tables 16.1 and 16.2, p.471/472.
* Vented lead-acid batteries for usual traction applications reach more than 1,500 test cycles with the depth of discharge of 75% per cycle.

* Valve-regulated block batteries, as used in electric vehicles (e.g. Fig. 10.31, p.351), reach about 700 of such test cycles; corresponding batteries with absorbent glass-mat separation about 400 cycles.
* As a standard cycle number for cylindrical nickel/cadmium batteries 600 cycles are quoted for a capacity ≥ 80% of the nominal value, but occasionally, 1000 cycles are claimed.
* The cycle figures of nickel/metal hydride batteries are somewhat higher compared to nickel/cadmium. As the standard value 1000 cycles are quoted.

All these cycle figures are derived by continual charge/discharge regimes. They only indicate the ability of such batteries to undergo repeated discharges and charges. In practice, quite different 'cycle numbers' may be reached, since these figures are markedly influenced by the specific kind of the discharges, their frequency, and calendar life of the battery.

In the following, the influence of some parameters that are important for ageing or life expectancy of a battery will be described in more detail.

15.2.1 Maintenance

A Section on 'Maintenance' seems a contradiction in a book on maintenance-free atteries. But maintenance is a parameter that significantly influences the service life of a battery. Often it is the failure of a single cell that jeopardises the function of the whole battery when not detected.

The concept 'maintenance-free', as used in this book, means a battery design based on the internal oxygen cycle, with the consequence that the regular replacement of water, characteristic of conventional lead-acid and nickel/cadmium batteries, is not required. Thus, most of the maintenance expenditures are eliminated. But this concept 'maintenance-free' must not be equated with the expression 'install and forget!'; otherwise, especially with standby batteries, unpleasant surprises may result.

The urgency of battery surveillance is described in Section 14, including methods to check the state of the battery. Even with quite small batteries, a minimum of maintenance effort should be employed to exclude unexpected failure. Voltage controls of the battery and charger are important, as are checks of battery temperature. Mechanical problems, like improperly bolted connectors or terminals, may also cause battery failure. Dirt deposited on the battery may reduce the insulation between the cells or blocks and to ground, and lead to partly discharged batteries.

For all these reasons, operational conditions should be checked periodically, even with maintenance-free batteries. Otherwise life expectancy and reliability of the battery can dramatically be reduced.

15.2.2 The Influence of Temperature on Battery Life

Temperature is a decisive parameter, because the rate of most electrochemical processes is approximately doubled when the temperature is increased by 10 °C (cf. Section 2.3.6).

Such an influence of temperature applies also to the ageing factors, since ageing of the battery is caused by unwanted electrochemical reactions. It means that the life expectancy of a battery is approximately halved when the average operational temperature is increased by 10 °C. If the battery is operated at 40 °C instead of the normal 20 °C, about one-quarter of the normal service life can only be expected (cf., e.g., ([4])).

Note: *The strong influence of the temperature, even at values only slightly above room temperature, is often underestimated, since the service life of the usual components of electronic devices is little affected at temperatures below 60 °C. The reason is that chemical reactions normally do not occur in such components, as mentioned at the beginning of Section 3.7.*

When the battery is operated for only limited periods at elevated temperatures, these periods have to be related to the whole operation time to estimate their influence on service life.

Increased temperature may not always be caused by environmental conditions. Heat can also be generated by the battery itself. In stationary batteries under float charging for example, heating may be a consequence of some defective cells. Short-circuits in such cells reduce the cell voltage and may even cause bridging. So the voltage of the remaining battery is too low, but the rectifier supplies the normal voltage. As a consequence, the float current is increased, and this again increases heat generation. Such a situation is especially dangerous with valve-regulated or sealed batteries, since the increased float current is consumed by the internal oxygen cycle that totally converts the energy input into heat, and may even cause thermal runaway (cf. Section 4.5.3.6).

15.2.3 Ageing Effects in Lead-Acid Batteries

The main ageing processes in lead-acid batteries are shown in the Tables 15.1 and 15.2. Besides the ageing effects, some countermeasures are listed.

Table 15.1 comprises general ageing processes in lead-acid batteries that, of course, concern VRLA batteries as well.

The first two lines concern ageing processes of the active material (cf. Section 10.1): **disintegration of the active material** is a typical ageing process in the positive electrode, while the active material in the negative electrode is prone to lose its capacity on account of **re-crystallisation**. Such re-crystallisation effects are mainly observed at elevated operational temperatures. The expander is exhausted in such conditions, and the growth of larger lead sulphate crystals is no longer prevented.

For standby batteries, these factors are of minor interest, since they are caused by stress of the active material, which mainly happens by repeated charge/discharge cycles. For this reason, disintegration of lead dioxide and re-crystallisation of lead are life-limiting in cycled batteries. When such batteries are operated at elevated temperatures, it is mainly the negative electrode that suffers by formation of large lead sulphate crystals ('sulphatation').

	1	2	3
	Main ageing effect	Countermeasures	Remarks
1	Positive active material		
	Disintegration of the active material, known as 'premature capacity-loss' (partly reversible).	Improved structure of active material by curing process. Limited utilisation. Suitable charging regime. Grid alloys with antimony or tin.	General problem with PbO_2 electrodes when cycled. More marked with low-antimony or antimony-free grid alloys. Tin addition is often used as a remedy.
2	Negative active material		
	Re-crystallisation.	Expanding additives.	Aggravated at higher temperature.
3	Positive grid and conducting elements		
	Corrosion.	Oversized grid design. Appropriate alloys and grid manufacturing processes.	Life limiting in standby batteries. Release of alloying components, like antimony increases hydrogen evolution.

Table 15.1 General ageing effects in lead-acid batteries.

Grid corrosion at the positive electrode is the most critical parameter during the float charge of lead-acid batteries, and grid corrosion is most important for service life, expressed in calendar time. Therefore oversized positive grids are characteristic for long-life stationary batteries. Grid corrosion is increased with temperature, but increased grid corrosion can also be induced by overcharging at too high a cell voltage. Advanced grid corrosion means increased internal resistance of the battery and a correspondingly reduced voltage, especially when it is discharged at a high rate. Furthermore, grid corrosion may cause grid growth and finally result in short-circuits between the grown positive plate and the negative pole bridge. As a side effect, the alloy composition not only influences the corrosion rate but also hydrogen evolution (cf. Section 4.7.3).

Table 15.2 shows additional ageing factors that are observed in the valve-regulated design. From the point of view of service life, the immobilised electrolyte is a restricting factor on account of the limited amount of electrolyte and the fact that its component 'water' cannot be refilled, as in vented batteries. Thus in valve-regulated batteries, all factors that increase water loss reduce service life and act as additional ageing factors ([5]).

	1	2	3
	Ageing effect	Countermeasures	Remarks
	Water loss of electrolyte (means 'ageing of the electrolyte')		
1a	Hydrogen evolution.	Very-low-antimony or antimony-free grid alloys. Adequate float voltage.	Elevated temperature accelerates both reactions and reduces service life.
1b	Grid corrosion.	Corrosion-resistant alloys.	
1c	Evaporation.	Observance of suitable operational conditions (temperature, humidity of surrounding atmosphere); Metallic or metallised containers.	Critical only, when the battery is exposed to 'desert climate'.
	Negative active material		
2	Oxidation by oxygen intake.	Container thoroughly sealed; Suitable valve design and material.	Oxygen intake acts like a continuous discharging current of negative electrodes.
	Negative grid and conducting elements		
3	Corrosion, at bus-bar and plate lugs.	Appropriate alloys (usually lead/tin alloys); Embedding of top lead in (gelled) electrolyte.	Caused by loss of 'cathodic corrosion protection'.

Table 15.2 Additional ageing effects in VRLA batteries.

Hydrogen evolution and grid corrosion are the two unavoidable secondary reactions that cause water loss (cf. Section 7.2.2). They are kept as low as possible by the battery design, based on antimony-free and corrosion-resistant alloys. Adequate float voltage is also important, since a too-high cell voltage increases hydrogen evolution and corrosion. Under this aspect, overcharging at increased voltage ('boost charging' is critical (cf. Section 13.1.5). The rate of hydrogen evolution and grid corrosion is doubled per 10°C of temperature increase. A corresponding increase occurs to water loss. For this reason, service life of valve-regulated batteries is seriously affected when they are operated at increased temperature.

Evaporation of water vapour through the cell walls can be critical, since all plastic materials are permeable to water vapour. Under normal European climate, the difference between the vapour pressure inside the battery and the relative humidity of the surroundings is small, and the resultant water loss negligible. When, however, the battery is exposed to a very dry atmosphere (< 20% of rel. humidity), evaporation can be critical (cf. Section 2.6.5.2).

Air intake (Line 2 in Table 15.2) is a critical factor in valve-regulated batteries, because of the easy access of oxygen to the negative electrode surface and the resulting oxidation according to Eq. (4.24), p.150. Air intake often causes partly discharged negative electrodes. Such a capacity loss of negative electrodes counts the more, since recharge of the negative electrode is only possible at fairly high voltages, when the battery, at least partly, operates as a vented one (cf. Section 13.1.5). But then, water loss as well as grid corrosion are increased with the consequence of reduced service life. Proper sealing of containers and terminals is therefore a must in VRLA batteries.

Air intake can also be due to a failure of the valve or safety vent required with these batteries. Their plastics parts can suffer ageing problems, especially at elevated temperature.

A further additional ageing effect is **corrosion of the negative conducting elements**, like bus-bar and plate lugs. In flooded batteries, these elements are protected by the negative-electrode potential, because they are covered by electrolyte. In VRLA batteries this protection can be lost (cf. Section 4.7.6.2).

On the whole, valve-regulated batteries are more sensitive to unfavourable operational conditions than comparable vented types, since increased loss of water as the consequence of any abuse, like excess overcharging, cannot be compensated by water refill.

15.2.4 Ageing Effects in Ni/Cd and Ni/MeH Batteries

Typical ageing effects in nickel/cadmium batteries are listed in Table 15.3. A general ageing factor is the gradual loss of capacity of the positive electrodes. This occurs in float applications as well as during cycling. Volume changes and loss of conductivity are the reasons. The cadmium hydroxide electrode tends to recrystallisation which is one of the reasons for the 'memory effect' (cf. Section 11.1.3). A gradual increase of the internal resistance is observed if graphite is used as a conducting aid, because of its oxidation. But this nowadays only concerns vented types with pocket plates. Furthermore, degradation of the separator material (cf. Section 5.5.3) is an important ageing factor.

	Ni/Cd battery
Vented types with pocket plates	Capacity decay of the positive electrode. Graphite oxidation (mainly vented types). Recrystallisation of the active material. Separator degradation.
Sealed types	Decay of capacity of the positive electrode. Re-crystallisation of the negative active material. Separator degradation.

Table 15.3 The main ageing processes of nickel/cadmium batteries.

Sealed nickel/cadmium batteries are mainly used in cycling applications. The number of cycles that can be reached usually is limited by the cadmium electrode.

Furthermore, with sealed nickel/cadmium batteries a creeping film of electrolyte may cause leakage at the safety vent and at the seal between container and cover.

Ageing effects in nickel/metal hydride batteries are identical to those in nickel/ cadmium batteries as far as the positive electrode and the separator are concerned. The negative electrode is not subjected to 'memory effects' as described in Section 11.1.3. Ageing effects of the negative electrode are mainly caused by oxidation of components that form the hydrogen-absorbing alloy. Thus the gradually progressing disintegration of the hydrogen-storing alloy reduces the storage capability.

15.2.5 Testing Procedures

The performance data of a battery are determined by test procedures carried out in precisely defined conditions. Many such procedures have been specified and are included in Standards (Chapter 16). But test procedures can only deliver comparative data indicating the suitability of a battery design in those conditions. The number of cycles reached in such a test, for example, may differ considerably from the actual service life because of divergent operational conditions.

Usually, the development of a new battery type is accompanied by extensive testing to guarantee the expected performance data. The results of these 'type tests' form the basis of performance data published by the manufacturer. Some of these tests are regularly repeated in the course of quality control. Sometimes, 100% control is carried out, as for example, in the capacity test for matching of cells assembled in special batteries (cf. Section 11.5.2.1). Customers often demand special 'acceptance tests' that ensure that the battery meets the performance data required for the special application. (A survey of different test objectives for the various lead-acid battery designs and their suitability for type or acceptance testing is given in ([6]).)

Problems of capacity testing with stationary batteries are described in Section 14.1.1.

The general desire to reduce testing time by accelerated test procedures is often frustrated by the difficulty that in accelerating operational conditions (increased voltage, elevated temperature), a failure mechanism determines service life, that is less important in normal operational conditions. Then the result of the accelerated test does not correspond to practical experience in normal conditions (cf., for example, ([7])). Nevertheless, such a test indicates that attention has to be paid to this parameter when the battery has to operate under such conditions.

There are a number of proven accelerated procedures to determine; for example, corrosion effects at elevated temperature within a fairly short period. But these accelerated tests at elevated temperature are usually restricted to evaluating the expected service life of battery components (like the positive grid in the case of corrosion). A comprehensive accelerated life test has been elaborated by Bellcore for valve-regulated batteries in stationary applications ([8]).

Difficulties generally arise when accelerated tests that are based on increased test temperature are applied to derive cycle-performance data of a battery. Cycle-performance tests can only be used for comparison under equal conditions and are only rough indications how the cycle performance of the compared batteries may be in practice. Cycle tests are very time consuming, because only two cycles can be passed per day in most of the test schedules. Acceleration to some extent is possible by increasing the load and so reducing the required time per cycle.

15.3 Disposal of Used Batteries

A question that attracts more and more attention is the fate of the battery after service life, and there are regulatory and legislative trends all over the world to reduce the risks connected with uncontrolled battery disposal and the subsequent release of heavy metals like cadmium and lead. Problems arise mainly with small batteries, because they are usually carried away with the municipal waste that is disposed of in landfills or is incinerated. Larger industrial batteries are usually exchanged by the manufacturer of the new battery, who ensures proper recycling of the batteries.

Although the contribution of worn-out batteries to municipal waste is comparatively small, the amount of heavy metals used in batteries is considerable. The world annual consumption is about 7000 metric tons of cadmium and 2,200,000 metric tons of lead ([9]). Corresponding figures for Western Europe are 1400 tons of cadmium and 700,000 tons of lead ([10]). While the recycled ratio of lead-acid batteries is fairly high (>> 95%), less than 60% of the nickel/cadmium batteries are recycled nowadays ([11]). So it is important to motivate people to collect batteries. The result is a great variety of programmes, ranging from extensive nation-wide programmes for 'battery hunting' ([12]) to the imposition of fees or deposits on the sale of batteries ([13]).

15.3.1 The ISO-recycling Sign

To support the efforts to collect batteries, a special sign has been introduced world-wide, characterised by three hunting arrows, as in Fig. 15.1. All maintenance-free valve-regulated lead-acid and sealed nickel/cadmium batteries are marked directly or indirectly (on the package) with this recycling sign. The sign means that the batteries can be reprocessed after the end of their service life.

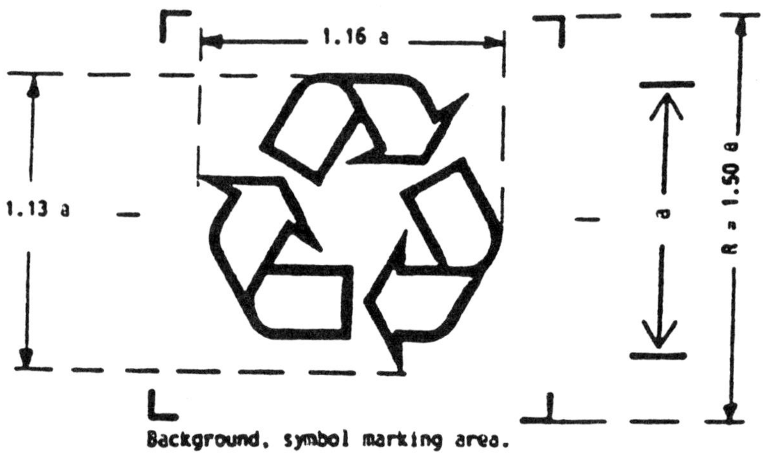

Fig. 15.1 The ISO-recycling sign (ISO Symbol 7000-1135) ([14]).

Batteries with this sign should never be disposed of with the normal household rubbish, but returned to the dealer or to special collecting points.

15.3.2 Recycling of Batteries

Collection of batteries does not solve the problem, because of the growing shortage of landfills. So the reuse of the raw materials is the goal, and this can be achieved by appropriate recycling methods.

Appropriate means in this case:

- The recycling process must result in materials that can compete with the primary materials or at least can be used instead of them. In lead-acid batteries, for example, secondary lead is still mostly used as grid material, while most manufacturers prefer primary lead for the active material. Increasingly, however, recycled lead is refined so far that it can also be used for the active material.
- The method must be economical; i.e. the price of the secondary raw material must compete, at least approximately, with that of the primary raw material. This is mainly a question of the energy required, but labour cost may also be important, for example, when the battery scrap has to be laboriously sorted.
- The method must be ecologically suitable; i.e. the recycling process itself must not cause environmental problems or generate residues that are difficult to dispose of.

The first item mainly concerns the purity of the recovered material. The achieved purity is a question of expenditure and thus is closely connected to the second item, but also to the price of the primary metal. It furthermore depends on the specification of the required material, which kind of recycling process may be suitable and how far purification has to be carried.

Overall, the economic aspect is favoured by the dramatic increase in dumping costs in most countries, and the dwindling possibilities of battery disposal in landfills greatly stimulate recycling efforts.

A fairly detailed review of the different recycling methods for batteries is given by a brochure edited jointly by the Association of European Accumulator Manufacturers (EUROBAT) and the Association of European Dry Battery Manufacturers (EUROPILE) ([15]).

15.3.2.1 Recycling of Lead-Acid Batteries

Recycling of lead-acid batteries has been well established for decades, and most lead-acid batteries are actually recycled.

Two main processes are used ([16]):

- The **blast-furnace** or **shaft-furnace process**, without previous separation of the battery components. The more or less smashed batteries, together with coke and smelting additives, are directly filled into the furnace. The organic components like separators and containers are used as additional fuel, but they are lost. Lead sulphate ($PbSO_4$) and residual sulphuric acid (H_2SO_4) are reduced and form a slag of iron sulphide (FeS) together with added iron. Chlorine, which is mainly contained in separators or spacers made of PVC, forms lead chloride ($PbCl_2$), which is volatile at temperatures above 900 °C and escapes from the furnace together with the furnace gas. It has to be collected by effective filter systems after cooling.

- The **rotary-furnace process**, requiring preceding separation of the metallic parts of the battery. This means intensive pre-treatment, including flotation processes to separate the different base materials by their different specific weights. Sulphate, which forms a part of the active material as lead sulphate ($PbSO_4$) is removed with the aid of soda (Na_2CO_3) which converts the lead sulphate into lead carbonate ($PbCO_3$) which is further reduced to the metal. Only the metal-containing components are smelted in the rotary furnace.

In both cases, the crude lead obtained has to be refined by subsequent metallurgical processes.

The advantages and disadvantages of the processes have already been indicated.

- The blast-furnace process requires practically no pre-treatment and generates no waste water that has to be neutralised and thoroughly decontaminated of residual lead. So it is economical, but all the organic components of the battery are not recycled but lost (used as fuel).
- In the rotary-furnace process, all the organic components survive and may be recycled. The main problem is that a great variety of organic substances are used in batteries which can hardly be separated, and that these materials are contaminated by lead and other metals. But the increasing necessity for recycling everything may stimulate further development of this process.

The diluted acid is a problem for economic reasons. It contains 10 to 20 wt.% of H_2SO_4; the rest remains in the partly discharged plates of the scrapped batteries. But it is contaminated, mainly with lead but also with other metals like antimony and traces of copper and occasionally nickel. So it cannot be re-concentrated, but has to be distilled to be reused as battery electrolyte. But this distillation has to be carried out in a vacuum; otherwise, sulphuric acid is decomposed before its boiling point of 338 °C is reached. Nowadays, this process is too expensive and the scrap acid is used for other chemical or metallurgical processes, like pickling, and is finally deposited as (heavy-metal-contaminated) gypsum ($CaSO_4$).

15.3.2.2 Recycling of Nickel/Cadmium Batteries

Recycling of nickel/cadmium batteries on a larger scale was started at the end of the 1970s. Some of the reasons for this rather late start were economic:

- The collection of sealed nickel/cadmium batteries usually leads to a mixture with primary batteries that requires sorting.
- The batteries have first to be dismantled.

- Iron and nickel, both of which are contained in these batteries, cannot be separated by the usual metallurgical processes. Both form an alloy that is only used for stainless-steel production. The condition that the recycling process must result in materials that can be used again for battery production cannot be fulfilled.

A further reason for the late beginning of recycling is the above mentioned problem of collecting these mostly small batteries that often are installed in devices, like tape recorders and photo flash lights, that also are disposed uncontrolled.

In the meantime the situation has changed, as is shown in a survey by H.Morrow ([17]). One reason is that an increasing share of the world wide production of cadmium is used for batteries, and it seems necessary to assure further supply by recycling.

At present, pyrometallurgical processes combined with distillation (for the cadmium) predominantly are applied ([18]). The organic components are destroyed by pyrolysis before the cadmium is distilled at temperatures above 900 °C.

References

([1]) H.Bode, 'Lead-Acid Batteries', John Wiley & Sons, New York, 1977, Table 2.6, p.27

([2]) D.Berndt, H.Franke and E.Meissner, VARTA spezial report 3/1990 e, Fig.15, VARTA spezial report, P.O. Box 210 540, D-3000 Hannover 21

([3]) Gates Energy Products (Editor), 'Sealed Rechargeable Batteries, Application Manual', Gates Energy Products Inc., Gainesville, Florida, 1989, Fig.4-42, p.164

([4]) 'Eurobat Guide to the Specification of Valve Regulated Lead-Acid Stationary Cells and Batteries' Edited by the Association of European Accumulator Manufacturers, Secretarial Offic, PO Box 5032, CH-3001 Bern, Switzerland, 1991, Table on p.5

([5]) D.Berndt, E.Meissner and W.Rusch, Proc. of the 15th Intelec Meeting, Paris, 1993, © IEEE Comm. Soc., 1993, p 139

([6]) DIN 43 539, Teil 1, Issue May 1985, Tabelle 1, Beuth Verlag GmbH, Berlin 30 (in German)

([7]) R.F.Nelson, Proc. 11th Intelec Meeting, Florence, 1989, © IEEE Communications Society, 1989, Paper 12.6

([8]) Bellcore TR-NWT-001200, 'Generic Requirements For Accelerated Life Testing Of Valve Regulated Lead Acid Batteries At High Temperatures', © Bellcore 1992, 445 Soutz Street, Room 2J125, P.O.Box 1910, Morristown, NJ 07962-1910

([9]) J.P.Wiaux and T.Nguyen, Third International Seminar on Battery Waste Management,1991, Florida Educational Seminar, 2836 Banyan Blvd. Circle N.W., Boca Raton, Fl. 33431, USA, Table 1

([10]) Eurobat - Europile, 'Realized and Projected Recycling Processes for Used Batteries', Edited by the Association of European Accumulator Manufacturers, Secretarial Office, PO Box 5032, CH-3001 Bern, Switzerland, 1991, p.3, Table 2

([11]) H.A.Kiehne, Proc. 13th Intelec Meeting, Kyoto 1991, © IEEE Communications Society, 1991, p.209, esp. Fig.2

([12]) T.Kertesz Third International Seminar on Battery Waste Management, 1991, Florida Educational Seminar, 2836 Banyan Blvd. Circle N.W., Boca Raton, Fl. 33431

([13]) D.B.Weinberg, Third International Seminar on Battery Waste Management, 1991, Florida Educational Seminar, 2836 Banyan Blvd. Circle N.W., Boca Raton, Fl. 33431, p.8

([14]) H.A.Kiehne, Proc. 11th Intelec Meeting, Florence 1989, © IEEE Communications Society 1989, Paper 17.3, Fig.8

([15]) Eurobat - Europile, 'Realized and Projected Recycling Processes for Used Batteries', Edited by the Association of European Accumulator Manufacturers, Secretarial Office, PO Box 5032, CH-3001 Bern, Switzerland, 1991

([16]) H.Niklas *in* F.Hiller (Editor) 'Batterie und Umwelt' ('Batteries and Environment'), 2nd Edition, Expert-Verlag, Ehningen bei Böblingen, 1969, Chapter 7 (in German)

Plants using both processes are presented in 'The Battery Man', Independent Battery Manufacturers Association Inc. (IBMA), Largo, Florida, Issue February 1992, pp.12 to 18

([17]) H.Morrow, Batteries International, Issue 29, October 1996, p. 51

([18]) J.C.Onuska, Jun., R.H.Manewald and M.E. Schweers, Third International Seminar on Battery Waste Management, 1991, Florida Educational Seminar, 2836 Banyan Blvd. Circle N.W., Boca Raton, Fl. 33431

16. Standards and Regulations

Standards are descriptions or test documents of a technical product. They are established by technical authorities in mutual agreement with manufacturers and users. They are furthermore available to everyone. A brief survey on battery standardisation and its history is given in ([1]).

Standards themselves are only recommendations; they are not legally binding. Often, however, they are included in rules, and then they have to be observed. So standardisation is an important principle of order in technological and economic events.

Standards are important for three main reasons:

- The world-wide production and use of technical products would hardly be possible if there were no standardisation of parts, especially for those parts that have to be replaced from time to time. Standardisation means that corresponding types of battery are interchangeable in physical dimensions as well as in electrical performance.
- Standardised dimensions, performance data and test procedures are required as the basis of sales contracts to define the subject with the aid of 'approved technical rules'. Private standards are often established for this reason by manufacturers of electronic equipment that require batteries as components. Examples for such standards and requirements are the documents published by Bellcore for stationary lead-acid batteries ([2]). Furthermore, in most countries the supply of the Armed Forces is based on special military standards.
- Standardisation includes safety rules. So standards are often used in legal regulations.

Regulations usually concern the safety aspects that have to be observed when batteries are used.

- The evolution of hydrogen and its explosive mixture with oxygen (cf. Chapter 7) represent the most important safety aspect with lead-acid and vented nickel/cadmium batteries. So there are many regulations on proper ventilation of battery rooms or cabinets, to make sure that the hydrogen concentration never reaches the critical value of 4 vol.%.

- The installation of safety valves is laid down for certain applications to avoid critical internal pressures.
- The handling of sulphuric acid and potassium hydroxide is subject to safety regulations.
- From the point of view of environmental pollution, fast-growing attention is paid to all questions of battery disposal and battery recycling, and most countries have established corresponding regulations (cf. Section 15.3).

Regulations that are legally binding always have national character.

16.1 National Standards

National standards have been developed in different countries and by different authorities for many decades. Examples are ([3]):

- ANSI American National Standards Institute
- ASTM American Society for Testing and Materials
- BS British Standards (formerly BSS, Brit. Stand. Specifications)
- DIN Deutsches Institut für Normung (German Institute for Standardisation)
- JIS Japanese Industrial Standard
- MIL Military Standards
- NEMA National Electrical Manufacturers Association, USA
- NF Norme Française (also AFNOR = Association Française de Normalisation)
- SAE Society of Automotive Engineers, USA
- UTE Union Technique de l'Électricité (France)

16.2 International Standards

Promoted by world-wide trade, many of these national standards have been harmonised during the last decade and are represented in parallel now by international standards. The most important authorities in this field are:

- ISO International Standards Organisation
- IEC International Electrotechnical Commission

ISO is mainly concerned with the standardisation of mechanical parts, while the standards for batteries are established by IEC. Both authorities have their central offices in Geneva, Switzerland ([4]).

The standardisation is carried out by technical committees, and the different countries that are members of the organisation are represented by delegates.

On rechargeable batteries, two IEC committees are decisive:

- TC 21 (Technical Committee 21) 'Rechargeable Batteries'
- SC 21 A (Subcommittee 21 A) 'Alkaline Secondary Cells and Batteries'

Standardisation is a laborious task.

- Drafts established by the committee are circulated several times to give national committees the opportunity to file objections or make suggestions.
- One major problem is that the market situation in each country has to be accepted; otherwise, the standard would create disadvantages for certain manufacturers.
- Because the standards include dimensions, it is sometimes impossible to harmonise the national standards as a whole. So international standards comprise more general requirements and test methods to establish product compatibility ([5]) (cf. Section 16.3 and Tables 16.1 and 16.2).

IEC Standards are revised from time to time and issued as a new edition to incorporate technical progress. This is indicated by a number in brackets, for example, IEC 95-1 (5) 1988 means the fifth edition. So the user has always to ensure that his edition is the latest one ([7]).

The following IEC standards have been established or prepared for rechargeable batteries by 'Technical Committee 21 - Secondary Cells and Batteries':

Lead-acid starter batteries:

- IEC 95-1 (1988) General requirements and test methods
- IEC 95-2 (1984) Dimensions of batteries and dimensions and marking of terminals
- IEC 95-4 (1989) Dimensions of batteries for heavy trucks

Lead-acid traction batteries:

- IEC 60254-1 (1997) General requirements and methods of test ([7])
- IEC 60254-2 (1997) Dimensions of cells and terminals and marking of polarity on cells ([7])
- IEC 1044 (1990) Opportunity charging of lead-acid traction batteries

Lead-acid stationary batteries:

- IEC 896-1 (1987) Vented types, General requirements and methods of test Amendments 88 .1 and 2 (1988, 1989)

- IEC 896-2 (1995) Valve-regulated types, general requirements and methods of test

  (This standard is reviewed as an example in Section 18.4.)

Portable lead-acid cells and batteries (valve-regulated types):

- IEC 1056-1 (1991) General requirements, functional characteristics, methods of test
- IEC 1056-2 (1994) Dimensions, terminals and marking
- IEC 1056-3 (1991) Safety recommendations for use in electric appliances

Vented nickel/cadmium batteries:

- IEC 623 (1990) Vented nickel/cadmium prismatic rechargeable single cells
- IEC 993 (1989) Electrolyte for vented nickel/cadmium cells

Aircraft batteries:

- IEC 952-1 (1988) General test requirements and performance levels
- IEC 952-2 (1991) Design and construction requirements
- IEC 952-3 (1993) External electrical connectors

Rechargeable sealed nickel/cadmium batteries:

- IEC 285 (1983) Cylindrical single cells
- IEC 285-1 (1989) Cylindrical single cells for permanent charge at elevated temperatures
- IEC 509 (1988) Button single cells
- IEC 622 (1988) Prismatic single cells

  Amendment No.1 (1989)

16.3 European Standards

Harmonisation of the national standards is limited, especially in dimensions and design. But political development in Europe demands common standards. So great efforts have been made during the last few years to harmonise the national standards. The European standards are created by:

- CEN European Committee for Standardisation
- CENELEC European Committee for Electrotechnical Standardisation

Both committees co-operate closely with ISO and IEC to avoid separate settlements as far as possible. The most desirable aim is to implement the international standards without any amendments in Europe. For example, IEC 986-2 will be the basis of a CENELEC Standard.

16.4 IEC 896-2 (1995) as an Example

In this Section, the IEC Publication 896-2 (Draft) is used to illustrate the substance of an international standardisation paper. The main requirements of this publication are listed in Tables 16.1 and 16.2, which are taken from a brochure published by EUROBAT ([6]). In both Tables, the second rows refer to the corresponding paragraphs of the IEC 896-2 (1995).

EUROBAT distinguishes between four groups of valve-regulated lead-acid battery:

- **10+ Year - High Integrity**. Batteries for applications where the highest security is required (Telecommunication, Power Stations).
- **10 Year - High Performance**. High-performance batteries with long service life whose requirements for performance and safety are not quite so severe.
- **5 to 8 Year - General Purpose**. Batteries with high performance and long service life whose life-related tests are not as stringent.
- **3 to 5 Year - Standard Commercial**. Batteries for the consumer market (non-stationary applications) and small emergency equipment.

This classification is **not** the object of the IEC Standard, and furthermore it is arbitrary, as the description of the four classes indicates. Many overlaps exist, and the differences between the four groups cannot be distinguished clearly. This example reveals one general problem of standardisation: a wide spectrum of valve-regulated lead-acid batteries is on the market with different features for different applications. To measure the detailed data of all these batteries by the same standard would be quite impossible. So the requirements of the standard can only be of a general character.

Note: *A side-effect of such a classification is to draw the attention of the battery user to the fact that quite different designs of valve-regulated lead-acid battery are available.*

16.4.1 Performance Influencing Parameters

In Table 16.1, the main parameters that characterise battery performance are listed.

Line 3 concerns the capacity. Clause 11 of the IEC Standard defines the Ah capacity as '**rated capacity**' (referred to a certain discharge duration) and its decisive parameters: discharge current, end-of-discharge voltage and temperature (cf. Section 3.3). Clause 18 describes how to carry out the capacity test. The recommended correction that considers the influence of the discharge temperature, the factor δ in Eq. (3.6), p.80, is 0.06, unless otherwise specified by the manufacturer. The values in this line of Table 16.1 indicate the different levels of the four battery groups in fulfilment of the performance data. Only the High Integrity Battery has to deliver full capacity from the very beginning. In the Standard Commercial Battery the gradual build-up of the capacity is left open.

Line 4 in Table 16.1 concerns **float charging** (cf. Section 13.1.1). The detailed requirements of IEC 896-2 are given in Clause 12.3. As shown in Table 16.1, all batteries have to reach full capacity after 6 months of float charging. For the voltage limits, the standard always refers to manufacturers' recommendations. Clause 19 describes the corresponding test procedure.

The **cycle test** (line 5 of Table 16.1) is of minor importance for the usual stationary applications, but may be important, for example, in solar-energy applications. Clause 13 is a general description of the cycle test; Clause 20 is a detailed schedule of the test procedure.

USER GUIDE FOR STANDBY POWER VALVE REGULATED LEAD-ACID BATTERIES							
	SPECIFICATION			TYPE TEST PERFORMANCE REQUIREMENTS BY APPLICATION GROUP			
	Specification Reference	Requirement	Description	10+ Year High Integrity	10 Year High Performance	5-8 Year General Purpose	3-5 Year Standard Commercial
PERFORMANCE	Draft IEC 896-2	Clause 11 & 18	Capacity	100% at 1st Cycle	95% at 1st Cycle 100% at 10th Cycle	95% at 1st Cycle 100% at 10th Cycle	Mnaufacturer Declares Capacity
	Draft IEC 896-2	Clause 12.3 & 19	Float (Voltage)	User States Voltage Limits	Manuf. States Voltage Limits	Manuf. States Voltage Limits	Manuf. States Voltage Limits
			(Capacity)	6 Months Capacity> Rated Capacity	6 Months Capacity> Rated Capacity	6 Months Capacity> Rated Capacity	6 Months Capacity> Rated Capacity
	Draft IEC 896-2	Clause 13 & 20	Cycles	User States Cycles and Criteria	Manuf. States No of Cycles	Manuf. States No of Cycles	Manuf. States No of Cycles
	Draft IEC 896-2	Clause 14 & 21	Retention	Manuf. States % Charge Retention	Manuf. States % Charge Retention	Manuf. States % Charge Retention	Manuf. States % Charge Retention
	Draft IEC 896-2	Clause 15 & 22	Resistance	Manuf. States Value	Manuf. States Value	Manuf. States Value	Manuf. States Value

Table 16.1 Valve-regulated lead-acid batteries. Requirements for performance in IEC 896-2, (1995) (Table 1 in EUROBAT Guide ([4])).

Line 6 of Table 16.1 concerns **retention of charge** which is the reversal of **self-discharge**. It is described in Clause 14 of IEC 896-2, while Clause 21 provides the test procedure. 90 days of storage on open-circuit are prescribed.

Clause 15 mentions that the internal resistance can be useful for safety and for equipment-protection calculations. As the bottom line in Table 16.1 shows, the standard always refers to the values provided by the manufacturer. The measuring method (Eq. (3.11), p.87) is described in Clause 22.

16.4.2 Parameters Related to Safety and Service Life

Safety aspects are also covered by the IEC 896-2. Table 16.2, again taken from the EUROBAT brochure, lists safety requirements, although the internal resistance (bottom line in Table 16.2) is also connected with safety.

Valve pressure in line 3 of Table 16.2 concerns the opening pressure of the valve (cf. Section 10.6). This parameter is of interest to the user mainly with respect to whether bulging of the container might occur with a fairly high opening pressure. Mechanical support of the container may then be required, especially at high operational temperatures (Clause 5 of IEC 896-2). Furthermore, users in special conditions may need to know how far components are internally pressurised. But internal pressures of valve-regulated lead-acid batteries are always fairly low (cf. Section 10.6).

Line 4 in Table 16.2 concerns the **flammability** of the container material (Clause 6) to be specified for certain applications.

The requirements of line 5 are to prevent explosion or fire (by sufficiently strong current-conducting elements), if the battery is **short-circuited** (Clause 8). 'Zero Volts' means that the battery is bridged for 1 minute by a resistance of 1.8 + 0.3 m (Clause 24).

User guide for standby power valve regulated lead-acid batteries							
	Specification			Type test performance requirements by application group			
	Specification Reference	Requirement Reference	Description	10+ Year High Integrity	10 Year High Performance	5-8 Year General Purpose	3-5 Year Standard Commercial
SAFETY	Draft IEC 896-2	Clause 5	Vent Pressure	Manuf. States Max Pressure	Manuf. States Max Pressure	Manuf. States Max Pressure	Manuf. States Max Pressure
	Draft IEC 896-2	Clause 6	Flammability	V0 to IEC 707	Manuf. States Rating to IEC 707	Manuf. States Rating	Manuf. States Rating
	Draft IEC 896-2	Clause 8 & 24	Short Circuits	Shall Not Explode at Zero Volts	Shall Not Explode at Zero Volts	Shall Not Explode at Zero Volts	Shall Not Explode at Zero Volts
	Draft IEC 896-2	Clause 12.4 &19.2	Gas Emission (Norm. Float)	Normal Float <10ml. (NTP)/Cell Ah C10/ 30 days	Manuf. States Value (NTP)/Cell Ah C10/ 30 days	Manuf. States Value (NTP)/Cell Ah C10/ 30 days	Manuf. States Value (NTP)/Cell Ah C10/ 30 days
			(Abuse Overcharge)	Manuf States Value (NTP)/Cell Ah C10/ 30 days	Manuf States Value (NTP)/Cell Ah C10/ 30 days	Manuf States Value (NTP)/Cell Ah C10/ 30 days	Manuf States Value (NTP)/Cell Ah C10/ 30 days
LIFE	Draft IEC 896-2	Clause 7 & 23	Seal Integrity (Thermal Cyc)	No Leaks	No Leaks	No Leaks	No Loaks
			(Mechanical)	No Leaks 5 min Discharge shall change by no more than 10%	No Leaks	No Leaks	No Leaks

Table 16.2 Valve-regulated lead-acid batteries. Requirements for safety and service life in IEC 896-2 (1995). (Table 2 in EUROBAT Guide (4).)

Line 6 in Table 16.2 deals with **gas evolution** of valve-regulated lead-acid batteries under continuous float charge. Clause 12.4 states that gas is evolved. Clause 19.2 describes a fairly tedious test procedure that takes 60 days and includes gas measurements during float charging, discharge and recharge of the battery (cf. Note 1 in Section 13.1.1.7).

The value for the '10+Year High Integrity' group of 10 cm^3 per Ah and 30 days equals 1.38 cm^3 per 100 Ah per hour, which is in the range mentioned in Section 13.1.1.7.

The standard does not specify the composition of the gas evolved during float, but it is mainly hydrogen.

The second part ('Abuse Overcharge') concerns a similar test with a voltage 8% higher ($\approx$ 0.18 V increase) than the usual float voltage. The consequences of this test are described to some detail in the following Section 16.4.2.1.

Note: *The problem of determining small amounts of hydrogen accurately was already mentioned in the Notes in Sections 7.5.3.1 and 8.3.5. On account of such difficulties, the reliability of gas measurements in these tests is limited.*

In the bottom lines of Table 16.2, requirements for design integrity are listed. This means the integrity of the seals, e.g. container/lid-, lid/pillar-seal, which must not leak gas or liquid (Clause 7 of IEC 896-2). *The problem of cell leakage is discussed in Section 4.4.1, along with its influence on chemical self-discharge.*

Clause 23 describes thermal cycling tests mechanical impact tests and vibrational tests. The thermal cycling test concern the resistance of the cell components, especially the seals to repeated temperature changes.

Tables 16.1 and 16.2 show that the standard can provide only a general framework for the required performance data. Most of the data cannot be generally specified. Nevertheless, the listing of the various features, together with the detailed test methods, represent an important tool to characterise a battery accurately.

16.4.2.1 The 'Abuse Test' in IEC 896-2

The 'abuse test' in Table 16.2 means that the battery is overcharged at a voltage 8% above the normal float voltage, as mentioned above. Fig. 16.1 illustrates such a situation when Fig. 8.16, p.292, is used as a basis, and 2.25 V/cell as the usual float voltage. The curves in Fig. 8.16 are calculated for a 'Standard Cell' with 99.5% efficiency of the internal oxygen cycle.

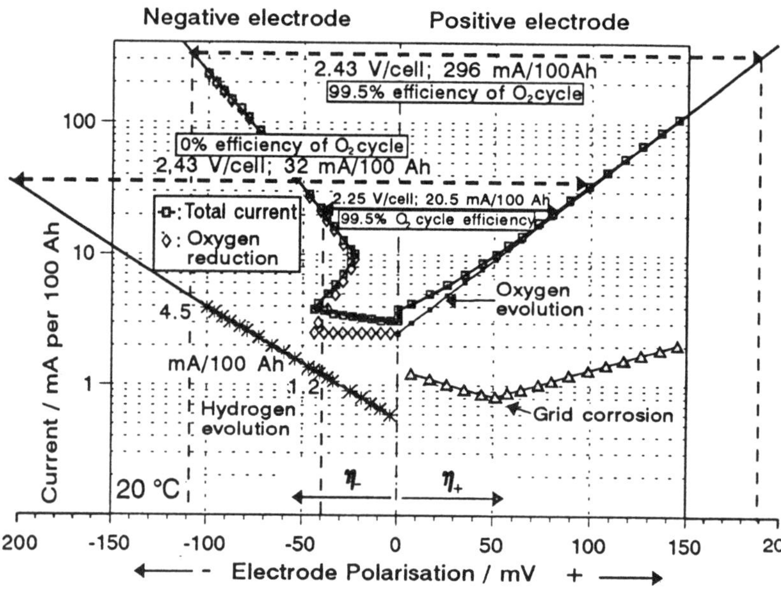

Fig. 16.1 Calculated current densities vs. electrode polarisation as in Fig. 8.16, p.291. 20 °C; acid density 1.28 g/cm^3; open-circuit voltage 2.13 V/cell. Float situation at 2.25 and 2.43 V/cell (2.25 + 8%) and 99.5% efficiency of the internal oxygen cycle and at 2.43 V/cell but 0% O_2 cycle efficiency.

When the cell voltage is increased from 2.25 to 2.43 V/cell, and the high efficiency of the internal oxygen cycle is maintained, the current is increased from 20.5 to 296 mA/100 Ah as shown by the corresponding arrows. This is a current increase by the factor of 14. Gas escape that corresponds to the hydrogen-evolution rate is increased from 1.2 to 4.5 mA/100 Ah, as indicated by the numbers at the intersections of the potential values with the hydrogen-evolution curve. It means that gas escape is increased only by a factor of 3.75. This is the result of the comparatively small polarisation of the negative electrode in the valve-regulated battery.

In an actual battery, such a high efficiency of the internal oxygen cycle would not be maintained at the high voltage, and the idea of this test is to determine how far this efficiency is reduced. In Fig. 16.1 the lower broken arrows show as the other extreme the situation when the efficiency of the internal oxygen cycle is 0% which more or less corresponds to a vented battery. Now hydrogen and oxygen evolution determine the current/polarisation relation, and the hydrogen-evolution rate corresponds approximately to that of oxygen, since the corrosion rate amounts to only about 4% of the current. In this situation the current is reduced to 32 mA/cm^3. But now hydrogen evolution corresponds to this current, and so is increased compared to the 'normal' situation by a factor of 27.

In a valve-regulated battery the efficiency of the internal oxygen cycle will not be reduced to zero at such an increased voltage, but will remain anywhere between both extremes. This test determines how far the efficiency of the internal oxygen cycle is reduced and the result depends largely on the maximum rate of that cycle.

But, as mentioned in the Note in Section 8.3.6, p.307, a high maximum rate of the internal oxygen cycle is not required in valve-regulated lead-acid batteries. Such a high rate requires correspondingly fast transport of oxygen to the negative electrode. If this transport rate is sufficient at the normal float voltage and a correspondingly smaller overcharging current, water loss does not depend on the maximum rate of the internal oxygen cycle, but is given by hydrogen evolution and grid corrosion rates. Thus, when batteries are compared, the one with the lowest gassing rate at 2.43 V/cell must not have the lowest water loss under normal operational conditions.

References

(1) F.Ciliberti, 'Battery Standardization', Chapter 4 in D.Linden (Editor) 'Handbook of Batteries', 2nd Edition, ©McGraw-Hill, Inc. 1995

(2) Bellcore, Technical Reference TR EOP-000232, 'Generic Requirements for Lead-Acid Storage Batteries'

Bellcore, Technical Reference TR-NWT-000766, 'Generic Requirements for Valve Regulated Lead Acid Cells.

Both © Bellcore, Bellcore Customer Service, 60 New Entland Avenue - ROOM 1B252, Piscataway, NJ 08854-4196, USA

(3) H.A.Kiehne, 'Battery Technology Handbook', Marcel Dekker, Inc., New York 1989, pp.287

VARTA Batterie AG (Editor),'Sealed Nickel Cadmium Batteries', © VDI-Verlag GmbH, Düsseldorf, 1982, p.250

(4) IEC: Bureau Central de la Commission Electrotechnique Internationale, 3, rue de Varembé, Geneva, Switzerland

(5) A.I.Harrison and H.Niklas 7th Intelec Meeting 1985, © IEEE Communication Soc., 1985, p.599

(6) 'Eurobat Guide to the Specification of Valve Regulated Lead-Acid Stationary Cells and Batteries', Edited by the Association of European Accumulator Manufacturers, Secretarial Office, PO Box 5032, CH-3001 Bern, Switzerland, 1991

(7) To bring the numbers of the European standards into accordance with the international systems, the numbers of IEC publications issued later than 1 January 1997 are changed by addition of 60,000. Thus, if a part 3 would be issued to the IEC 896-2, it would have the number IEC 60896-3.

Examples are the two IEC standards IEC 60254-1 (1997) and 60254-2 (1997) for traction batteries that replaced the older IEC 254-1 and 254-2, respectively.

Publications issued before 1997 keep their old numbers, and amendments made to them are also published under the old number. Thus, amendments to the IEC 896-2 will also in future be published under IEC 892-2 with the number of the amendment.

INDEX